Ship Dynamics and Hydrodynamics

Ship Dynamics and Hydrodynamics

Editors

Carlos Guedes Soares
Serge Sutulo

MDPI • Basel • Beijing • Wuhan • Barcelona • Belgrade • Manchester • Tokyo • Cluj • Tianjin

Editors
Carlos Guedes Soares
Universidade de Lisboa
Lisboa
Portugal

Serge Sutulo
Universidade de Lisboa
Lisboa
Portugal

Editorial Office
MDPI
St. Alban-Anlage 66
4052 Basel, Switzerland

This is a reprint of articles from the Special Issue published online in the open access journal *Journal of Marine Science and Engineering* (ISSN 2077-1312) (available at: https://www.mdpi.com/journal/jmse/special_issues/ship_dynamics_hydrodynamics).

For citation purposes, cite each article independently as indicated on the article page online and as indicated below:

LastName, A.A.; LastName, B.B.; LastName, C.C. Article Title. *Journal Name* **Year**, *Volume Number*, Page Range.

ISBN 978-3-0365-8106-4 (Hbk)
ISBN 978-3-0365-8107-1 (PDF)

Contents

Journal of
Marine Science and Engineering

MDPI

Editorial

Ship Dynamics and Hydrodynamics

Serge Sutulo and C. Guedes Soares *

Centre for Marine Technology and Ocean Engineering (CENTEC), Instituto Superior Técnico, Universidade de Lisboa, Av. Rovisco Pais, 1049-001 Lisboa, Portugal
* Correspondence: c.guedes.soares@centec.tecnico.ulisboa.pt

1. Introduction

Ship hydrodynamics and dynamics is a rather old, traditional branch of applied mechanics and also of naval architecture. Typically, methods and theories developed in the former are applied to the latter aiming at the development of more advanced and efficient ships, and other marine vehicles and ocean structures.

The long development of ship hydrodynamics over many decades has crystallised in the following traditional scientific disciplines: resistance, propulsion, seakeeping, and manoeuvrability. However, in the earlier stages of the development of the naval architecture, the focus was on the properties of a displacement ship, such as its floatability, static stability, and unsinkability. All these properties have been analysed in ship hydrostatics and although the mentioned properties are fundamental, ship hydrostatics has lately been provided with no attention, presuming that all hydrostatic problems are solvable with relatively small efforts and that the corresponding computations are performed routinely. However, this is not true for the remaining disciplines listed above: those which are constantly and intensely developed exploiting multiple theoretical, numerical, and experimental approaches inspired by fluid mechanics, nonlinear dynamics, theory of oscillations, data analysis, control theory, and many other areas.

The present Special Issue contains 17 articles, 1 of which deals with some specific applications of ship hydrostatics to risk analysis, 3 papers belong to the area of ship resistance, 4 revolve around ship propulsion, 5 papers cover the various aspects of seakeeping, and 4 articles are rather dedicated to ship manoeuvrability. However, as can be seen from the more detailed review that follows in the next section, some papers are interdisciplinary, and the above classification is approximate. It is also worthwhile to note that 9 papers out of 17 extensively use methods and codes belonging to computational fluid dynamics (CFD).

Citation: Sutulo, S.; Guedes Soares, C. Ship Dynamics and Hydrodynamics. *J. Mar. Sci. Eng.* **2023**, *11*, 911. https://doi.org/10.3390/jmse11050911

Received: 26 February 2023
Accepted: 25 March 2023
Published: 24 April 2023

2. Contributions' Overview

The paper written by Braidotti et al. [1] is related to the unsinkability problem; however, their scope is far beyond the classic single-case formulations. First, a scenario of the schematised accidental lateral ramming was considered and used for the generation of a damages database. For each damage case, progressive flooding was simulated, and random forest regression was used to create a flooding time predictor. Although the paper focused on passenger vessel applications, its usefulness for naval ships is also evident.

Duman et al. [2] have performed a thorough numerical study of the resistance of a fast displacement catamaran in deep and shallow water. The CFD commercial code Star CCM+ was applied to the ship advancing with a constant speed and free in sinkage and trim. The Froude number which ranges from 0.2 to 0.67 covers both the sub-critical and supercritical speed ranges in the shallow water case. For deep water, the results for the resistance coefficients and the dynamic trim and sinkage were validated against the towing-tank data, showing a good agreement. An interesting feature of this study is that not only the total resistance was considered but also the frictional, residual, form, and wave resistance components separately. The form factor for the viscous resistance was estimated with the help of additional CFD simulations with the doubled hull.

The paper by Chang et al. [3] could be attributed to the area of ship design as it considers some issues related to the optimum design of the ship hull; although, it also deals with the inverse problem of ship resistance. Only the wave-making resistance was subject to minimisation performed with the particle swarm global optimisation algorithm. Special attention was paid to the reduction in the dimensionality of the effective design factor space performed with special methods based on the analysis of the eigenvalues of the information matrix. The wave resistance for any defined hull shape was computed using the commercial code SHIPFLOW. The method was applied to optimise the hull of the well-known benchmark destroyer DTMB-5415. Two variants of the optimisation procedure were used: without and with dimensionality reduction. In both cases, a substantial and almost identical reduction in the wave resistance was achieved, but in the second case, a substantial reduction in the required computation time was observed. The optimisation was performed under the constraints imposed on variations in the hull displacement and of the longitudinal position of the centre of buoyancy. However, the optimisation was performed for a moderate Froude number rather than corresponding to the cruising speed, which is much lower than the ship's full speed when the wave-making resistance dominates.

The study presented by Li et al. [4] also belongs to the area of ship resistance and is applied for optimizing the shape of the bow of an Arctic class vessel operating both in clean water and in ice conditions. A mathematical model embracing both the hydrodynamic and ice resistance was created based on the finite volume method used in the commercial CFD code Star CCM+ for the former and on the smooth particles hydrodynamics for the latter. The model was validated using cone ice sample tests and towing tests in a refrigerating tank performed with scaled ship models. Three bow-shape parameters were considered: the waterline entrance angle, stem inclination, and the frame deadrise angle. The parameters' variations resulted in a three-level, three-factor orthogonal fractional factorial experimental design comprising nine variants. The obtained dataset was then subject to sensitivity analysis, and a minimum of three ice-resistance configurations were selected, for which the clean-water resistance was further computed. Finally, the so-called "overall resistance index" accounting for both ice and hydrodynamic resistance was estimated, which permitted the selection of a single optimal configuration.

Tadros et al. [5] studied some of the special aspects of the optimum propeller design. In particular, the effect of cupping the propeller blades was investigated. The blade cup means the deformation of the blade near the trailing edge, mimicking to some extent deflection of an aircraft wing's flap. The optimum design was based on complexing a built-in the MatLab optimiser with the specialised NavCad code. The constraints related to the cavitation inception and to the engine's limiting characteristics were taken into account. The obtained calm-water-optimal propellers were further analysed on the quasi-stationary level for a set of weather conditions for real sea routes. According to the authors' results, the heavily cupped propellers turned out to be more efficient and less prone to cavitation.

In paper [6], Tadros et al. explored a more traditional method of improving propulsive efficiency, presuming the application of contra-rotating propellers possibly also combined with cupping. The numerical study was based on the same methodology and software as used in [5], but only the calm-water condition was considered. The effectiveness of the contra-rotating propellers was confirmed, and it was demonstrated that cupping further boosts the propulsive efficiency.

Sezen and Atlar [7] studied the influence of the roughness of a propeller hub surface on the cavitation in the axial free vortex. The modelling was performed for a rather typical four-blade propeller with the CFD code Star CCM+ validated with the experimental data obtained for a smooth hub. It was found that the hub roughness only slightly reduces the open-water efficiency of the propeller while substantially mitigating the vortex cavitation noise, which is important for reducing the environmental impact and improving the stealthiness of the ship.

The study performed by Zhang et al. [8] belongs, first of all, to the area of ship machinery, but, as long as the longitudinal vibration of the propeller shaft is considered, this study also has links with the ship propulsion and marine acoustics. The paper describes a rather detailed investigation of the dynamics of a shaft line equipped with dry or Coulomb dampers in the main thrust bearing. A physical experiment and a theoretical study have confirmed the effectiveness of such damping devices, which proved to be superior to more traditional viscous dampers. The study includes the identification of the Coulomb friction coefficient based on the particle swarm global optimisation.

Pérez-Canoza et al. [9] used a highly simplified 1DOF rolling model in an attempt to analyse the capsizing of a fishing vessel.

Mohapatra et al. [10] have studied the interaction of a solitary wave with a large floating pontoon in shallow water. The elevation of the free surface and wave loads acting upon the pontoon were computed using an analytical solution developed by the authors based on the Boussinesq theory, and also using the Navier–Stokes laminar regime equations solved with the application of the OpenFoam package and the OceanWave3D fully nonlinear potential flow code. A good agreement with results obtained with different methods was found.

The paper authored by Park and Wang [11] deals with a rather different kind of floating structure consisting of a smoothed polygonal platform floating inside a similar ring with different draughts. The actual shape was formed by low-order Fourier descriptors. The application of a cylindrical coordinate system permitted to obtain an analytical solution to the radiation-and-diffraction problem using the variables separation method. The resulting solution takes the form of Fourier series with coefficients containing the Bessel, McDonalds, and Hankel functions. The numerical results obtained with this solution were compared with those provided by the commercial finite-element code ANSYS AQWA, demonstrating a good agreement. The analytical solution was further used for a systematic investigation of the influence of various parameters defining the structure's configuration.

Gao et al. [12] have performed a RANSE modelling of a seakeeping problem using the OpenFoam package. The case of oblique regular oncoming waves in shallow water was studied. All the computations were performed for the virtual benchmark KVLCC2 ship in the model scale. The surge and sway forces and the yaw moment were computed, and their time-averaging resulted in the second-order loads. The kinematic responses in heave and pitch were also computed. The CFD results were validated against the experimental data.

The study carried out by Mai et al. [13] aimed to establish the added mass and damping coefficients associated with the roll rate and acceleration, including the cross-coupling with the sway force and yaw moment. The code Star CCM+ was used to determine hydrodynamic loads for the simulated forced isolated rolling of four benchmark ships, which included a container ship, two naval combatants, and a fast displacement catamaran. A possible dependence on the roll frequency and amplitude was neglected as the results were supposed to be used not in seakeeping but in a 4DOF manoeuvring code.

Islam et al. [14] have applied a CFD code based on the OpenFoam to the investigation of the aerodynamic characteristics of a patrol ship. The study was performed with the aim of discovering further applications for manoeuvring simulations and the all-around dependencies for the surge and sway forces and the yaw moment as functions of the inflow angle were obtained. Most of the computations were performed in the model scale, but some isolated cases were also repeated for the full-scale Reynolds number. Considerable attention was paid to convergence studies and the influence of the applied turbulence model. An additional indirect validation of the results was performed through a comparison with published wind-tunnel data obtained for similar (though not identical) above-water configurations.

The article by Sutulo and Guedes Soares [15] presents a comprehensive review of publications related to the elaboration and analysis of various manoeuvring performance criteria and standards. Alongside the review, the authors performed a preliminary analysis of possible approaches to accounting for the environmental exogenous factors in standardizing manoeuvring qualities. A hypothesis was formulated that all such factors related to hydrodynamics (sea waves, hydrodynamic interaction) can be accounted for indirectly through the appropriate augmentation of the calm-water manoeuvring standards. At the same time, the wind action can only be taken into account using some special additional criteria. The paper also contains some new developments related to standardizing wind-related manoeuvring qualities.

The study performed by Lu et al. [16] is indirectly associated with ship manoeuvrability as its results may turn out to be useful for a reliable prediction of the open-water hydrodynamic characteristics of rudders. The paper represents a significant contribution to the practice of the application of CFD codes and grid generators. The commercial ANSYS codes ICEM and Fluent were used for investigating the influence of the grid's characteristics of the 2D symmetric profiles of the NACA family. The obtained results were also validated against experimental data and the data provided by several other CFD codes. It was found that the lift could be predicted with a much lower uncertainty than the drag at least in the pre-stall regime.

The study performed by Song et al. [17] de facto belongs to the area of applied control theory, aiming, however, at the application of the manoeuvring problem of heading stabilisation and path tracking. The novelty of the study consists of the developed sophisticated algorithm for a dynamic optimisation of the parameters of a PD control law. The method was validated through computer simulations performed based on a rather common 3DOF holistic dynamics model with exogenous factors simulated as artificial high-frequency perturbations of the sway force and the yaw moment. The results obtained in the simulated tracking of straight, polyline, and circular paths using the line-of-sight method combined with the heading PD control have confirmed the better quality of the tracking provided by the new tuning algorithm.

Author Contributions: Conceptualization, C.G.S. and S.S.; writing—original draft preparation, S.S.; writing—review and editing, C.G.S. All authors have read and agreed to the published version of the manuscript.

Funding: This work contributes to the Strategic Research Plan of the Centre for Marine Technology and Ocean Engineering (CENTEC), which is financed by the Portuguese Foundation for Science and Technology (Fundação para a Ciência e Tecnologia—FCT) under contract UIDB/UIDP/00134/2020.

Conflicts of Interest: The authors declare no conflict of interest.

References

1. Braidotti, L.; Prpić-Oršić, J.; Valčić, M. Effect of Database Generation on Damage Consequences' Assessment Based on Random Forests. *J. Mar. Sci. Eng.* **2021**, *9*, 1303. [CrossRef]
2. Duman, S.; Boulougouris, E.; Aung, M.Z.; Xu, X.; Nazemian, A. Numerical Evaluation of the Wave-Making Resistance of a Zero-Emission Fast Passenger Ferry Operating in Shallow Water by Using the Double-Body Approach. *J. Mar. Sci. Eng.* **2023**, *11*, 187. [CrossRef]
3. Chang, H.; Wang, C.; Liu, Z.; Feng, B.; Zhan, C.; Cheng, X. Research on the Karhunen–Loève Transform Method and Its Application to Hull Form Optimization. *J. Mar. Sci. Eng.* **2023**, *11*, 230. [CrossRef]
4. Li, H.; Feng, Y.; Ong, M.C.; Zhao, X.; Zhou, L. An Approach to Determine Optimal Bow Configuration of Polar Ships under Combined Ice and Calm-Water Conditions. *J. Mar. Sci. Eng.* **2021**, *9*, 680. [CrossRef]
5. Tadros, M.; Vettor, R.; Ventura, M.; Guedes Soares, C. Effect of Propeller Cup on the Reduction of Fuel Consumption in Realistic Weather Conditions. *J. Mar. Sci. Eng.* **2022**, *10*, 1039. [CrossRef]
6. Tadros, M.; Ventura, M.; Guedes Soares, C. Towards Fuel Consumption Reduction Based on the Optimum Contra-Rotating Propeller. *J. Mar. Sci. Eng.* **2022**, *10*, 1657. [CrossRef]
7. Sezen, S.; Atlar, M. Mitigation of Hub Vortex Cavitation with Application of Roughness. *J. Mar. Sci. Eng.* **2022**, *10*, 1426. [CrossRef]
8. Zhang, G.; Zhao, Y.; Chu, W. Longitudinal Vibration Transmission Control of Marine Propulsion Shafting with Friction Damper Integrated into the Thrust Bearing. *J. Mar. Sci. Eng.* **2022**, *10*, 1555. [CrossRef]
9. Pérez-Canosa, J.M.; Orosa, J.A.; Galdo, M.I.L.; Barros, J.J.C. A New Theoretical Dynamic Analysis of Ship Rolling Motion Considering Navigational Parameters, Loading Conditions and Sea State Conditions. *J. Mar. Sci. Eng.* **2022**, *10*, 1646. [CrossRef]
10. Mohapatra, S.C.; Islam, H.; Hallak, T.S.; Guedes Soares, C. Solitary Wave Interaction with a Floating Pontoon Based on Boussinesq Model and CFD-Based Simulations. *J. Mar. Sci. Eng.* **2022**, *10*, 1251. [CrossRef]
11. Park, J.C.; Wang, C.M. Hydrodynamic Behaviour of a Floating Polygonal Platform Centrally Placed within a Polygonal Ring Structure underWave Action. *J. Mar. Sci. Eng.* **2022**, *10*, 1430. [CrossRef]

12. Gao, Q.; Song, L.; Yao, J. RANS Prediction of Wave-Induced Ship Motions, and Steady Wave Forces and Moments in Regular Waves. *J. Mar. Sci. Eng.* **2021**, *9*, 1459. [CrossRef]
13. Mai, T.L.; Vo, A.K.; Yoon, H.K.; Park, D.K. Assessment of the Roll Derivatives of Different Surface Ships Based on Numerical Pure Roll Simulation. *J. Mar. Sci. Eng.* **2022**, *10*, 1702. [CrossRef]
14. Islam, H.; Sutulo, S.; Guedes Soares, C. Aerodynamic Load Prediction on a Patrol Vessel Using Computational Fluid Dynamics. *J. Mar. Sci. Eng.* **2022**, *10*, 935. [CrossRef]
15. Sutulo, S.; Guedes Soares, C. Review on Ship Manoeuvrability Criteria and Standards. *J. Mar. Sci. Eng.* **2021**, *9*, 904. [CrossRef]
16. Lu, S.; Liu, J.; Hekkenberg, R. Mesh Properties for RANS Simulations of Airfoil-Shaped Profiles: A Case Study of Rudder Hydrodynamics. *J. Mar. Sci. Eng.* **2021**, *9*, 1062. [CrossRef]
17. Song, L.; Xu, C.; Hao, L.; Yao, J.; Guo, R. Research on PID Parameter Tuning and Optimization Based on SAC-Auto for USV Path Following. *J. Mar. Sci. Eng.* **2022**, *10*, 1847. [CrossRef]

Journal of
Marine Science and Engineering

MDPI

Article

Effect of Database Generation on Damage Consequences' Assessment Based on Random Forests

Luca Braidotti [1,2,*], Jasna Prpić-Oršić [1] and Marko Valčić [1,3]

[1] Faculty of Engineering, University of Rijeka, Vukovarska 58, 51000 Rijeka, Croatia; jasnapo@riteh.hr (J.P.-O.); mvalcic@unizd.hr (M.V.)
[2] Department of Engineering and Architecture, University of Trieste, Via Valerio 10, 34127 Trieste, Italy
[3] Maritime Department, University of Zadar, Mihovila Pavlinovića 1, 23000 Zadar, Croatia
* Correspondence: lbraidotti@units.it

Abstract: Recently, the application of machine learning has been explored to assess the main damage consequences without employing flooding sensors. This method can be the base of a new generation of onboard decision support systems to help the master during the progressive flooding of the ship. In particular, the application of random forests has been found suitable to assess the *final fate* of the ship and the damaged compartments' set and estimate the *time-to-flood*. Random forests have to be trained using a database of precalculated progressive flooding simulations. In the present work, multiple options for database generation were tested and compared: three based on Monte Carlo (MC) sampling based on different probability distributions of the damage parameters and a parametric one. The methods were tested on a barge geometry to highlight the main effects on the damage consequences' assessment in order to ease the further development of flooding-sensor-agnostic decision support systems for flooding emergencies.

Keywords: damaged ship; progressive flooding; random forests; database generation; decision support system

Citation: Braidotti, L.; Prpić-Oršić, J.; Valčić, M. Effect of Database Generation on Damage Consequences' Assessment Based on Random Forests. *J. Mar. Sci. Eng.* **2021**, *9*, 1303. https://doi.org/10.3390/jmse9111303

Academic Editor: Decheng Wan

Received: 30 October 2021
Accepted: 18 November 2021
Published: 21 November 2021

1. Introduction

In previous decades, the lack of decision support during the progressive flooding of a damaged passenger ship has been highlighted by many accidents. Large passenger vessels have a complex non-watertight subdivision, limited stability reserve, and limited freeboard at the bulkhead deck, leading to a difficult prediction of the flooding consequences without the aid of a computer system. Therefore, after a collision or grounding, the master needs to have at his/her disposal a Decision Support System (DSS) to make his/her decisions on a rational basis instead of heuristics or his/her experience.

The information given to the master required by the international rules was proven to be inadequate especially during the sinking of Costa Concordia. After the grounding, it was not easy to even identify the breached compartments [1]. Besides, the mandatory onboard documentation regarding damage stability/control requires much time for consultation. Recently, some efforts have been made toward the digitalization of these documents. Nevertheless, most of them relate to a standard loading condition that is not likely met during navigation or to Safe Return to Port (SRtP) recovery actions and damage control in the final stage of flooding [2,3]. Besides, all modern ships are equipped with loading computers capable of carrying out damage stability calculations by applying the lost buoyancy method [4]. These tools might be capable of assessing the ship's survivability; however, they cannot consider physically consistent intermediate stages of flooding, which might lead to excessive heeling angles or to the ship capsizing. Moreover, they usually require the manual input of damaged compartments. The last problem has been overcome by the introduction of a mandatory flooding detection system on passenger vessels laid down after 1 July 2010 [5]. Furthermore, the installation of flooding sensors capable of measuring floodwater levels enabled the direct application of quasi-static progressive flooding simulation codes within onboard DSSs [6–8]. The flooding sensors shall be fit in each ship's internal space to permit the assessment of the damage dimension and location [9]. These data are then the input for the progressive flooding simulation, which allows the forecast of the damage consequences. The so improved situational awareness after damage occurrence can reduce the reaction time for damage control or ship abandonment, if required. However, up to now, most of the existing passenger ships are not equipped with flooding sensors and, thus, can suffer from the lack of emergency decision support. Besides, the flooding detection system retrofit required for the installation of the most advanced solutions available on the market is costly, discouraging the ship owners from adopting emergency DSSs based on time-domain simulations. A viable solution to overcome such a problem is the introduction of systems requiring a more basic set of sensors to make available the essential information during a flooding emergency. In this context, the time evolution of the damaged ship's floating position can be exploited instead of the floodwater levels [10], requiring only the measurement of the ship heel angle, trim angle, and draft during the progressive flooding of the ship. The required set of instruments is limited to inclinometers (usually fit on all the vessels) and one or more level radar(s) fit in still-pipes or below bridge wings to measure ship draught. Then, Machine Learning (ML) can be used to correlate the recorded floating position with the main flooding consequences [11].

In this context, Random Forests (RFs) have provided promising results [12]. RFs are trained using a database of progressive flooding simulations in the time domain. The training database is built according to a damage case generator that can be based on different mathematical formulations. Up to now, an extensive discussion of the effect of damage case generation on the classification and regression accuracy is lacking. To fill such a gap, the present work explores the effect of different damage-case-generation algorithms on the prediction of the progressive flooding consequences provided by RFs. In particular, four solutions were tested: a Parametric (P) one and three based on Monte Carlo (MC) sampling (according to probability distributions for damage dimensions used in the convention for Safety of Life at Sea (SOLAS), assuming a uniform distribution of the damage dimension or a uniform distribution of the damage area inverse). After a short overview of the progressive flooding consequences' prediction, the database generation algorithms are presented. The proposed methodology is, then, applied to a box-shaped barge using a large SOLAS-based database for validation purposes.

2. Prediction of Damage Consequences

When the hull integrity is compromised leading to the progressive flooding of the ship, a few pieces of information are essential to support the master's decision [11]. The most important one is the *final fate* of the ship, namely whether the ship will survive the damage scenario, reaching a new safe equilibrium position, or will sink, capsize, or shift towards an unsafe condition, e.g., excessive equilibrium heeling angle. Besides, the set of flooded watertight compartments should be known by the crew to promptly carry out the damage control procedures and prevent further spreading of floodwater towards intact watertight compartments. Finally, in a nonsurvival damage case requiring ship abandonment, it is vital to know the *time-to-flood* t_f, i.e., the time to reach the ship capsizing, sinking, or an unsafe condition, to manage the ship evacuation process. A viable method to assess this information from the floating position of the damaged ship employs ML. The process is sketched in Figure 1. More details about the applied methodology can be found in [11].

During the progressive flooding, the loading condition of the ship changes due to the embarked floodwater, leading to a variation of the floating position as well. The floodwater pouring among connected internal rooms is governed by the hydraulic laws, and thus, it can be predicted by applying progressive flooding simulation codes. Using RF, a link can be searched between the time records of sinkage s, heel ϕ and trim θ angles, i.e., the predictors, and the above-mentioned damage consequences, i.e., the responses [11]. Considering a time instant t^* during the progressive flooding, the past time evolution of the floating position is known. Therefore, ϕ, θ, and s can be sampled with a constant time step dt up to t^*, defining the predictor at t^*. The predictors are used within three specific learners that are trained to predict the main progressive damage consequences (*final fate*, *flooded compartments*, and *time-to-flood*). As time goes by, the predictor set's size increases, since new information about the floating position is available. Hence, specifically trained learners are produced at each time instant t^* by exploiting all the available information.

Among the ML algorithms present in the literature, RFs have shown good performances in addressing the classification problems (for *final fate* and *flooded compartments*) and the regression one (for the *time-to-flood*). This is why they were employed in the present work. The learners used in the DSS were trained during a preliminary preparation phase with a database of progressive flooding simulations defined according to a damage case generator. Moreover, to validate the trained learners, another independently generated database was utilized. To this end, the progressive flooding simulations included in the validation database provided the predictors' values up to the instant t^*, allowing statistically testing the accuracy of the responses. To ensure a reliable accuracy evaluation, the validation database shall be as much as possible representative of a real probability distribution of damage scenarios.

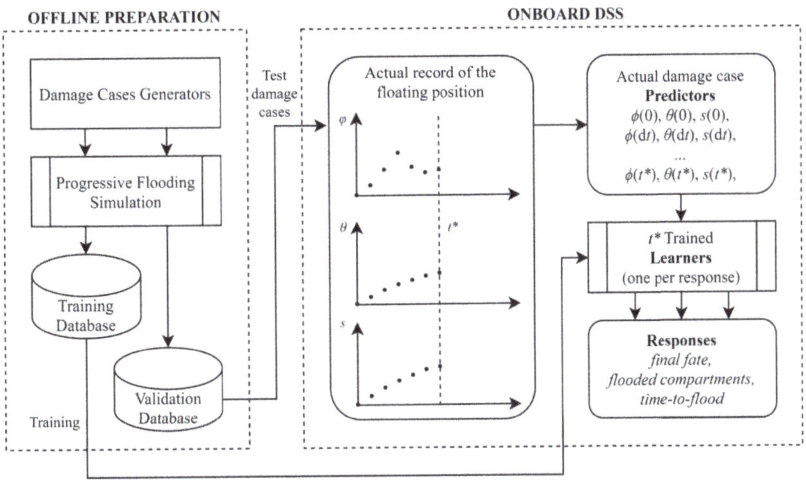

Figure 1. Flowchart of the classification process.

2.1. Random Forests

RFs are supervised ML algorithms belonging to the decision trees family. They can be employed in both classification and regression problems, where they provide a piecewise approximation of the response function. All the decision trees are based on binary decisions made according to the value of one predictor x_i at each node. Therefore, the process is shaped as a tree, starting from a root and reaching one leaf, i.e., the response, moving decision by decision (Figure 2). A single decision tree is trained with a database capable of describing the relationship between the predictors' values and a response [13]. RFs have been introduced to improve the accuracy of the prediction provided by standard decision trees [14]. As for bootstrap aggregation, the problem is decomposed into a set of "weak" trees trained with a partition of the original database instead of a single tree trained with the complete database. The response of the overall model is selected according to the vote of the multiple trees for classification problems and as the average of the responses for the regression one. In the present work, 30 weak learners were employed. In addition to the application of multiple learners, RFs also utilize a random selection of a predictors' subset at each split in a single tree to decorrelate the trees in the ensemble [15].

These features make RFs more resilient to noise/missing data and more capable of dealing with higher-dimensional data compared to standard decision trees and other ensemble methods. Hence, the choice of such an ML technique was initially considered for the studied problems, involving a large number of predictors for higher t^* and progressive flooding being affected by several uncertainties [16,17].

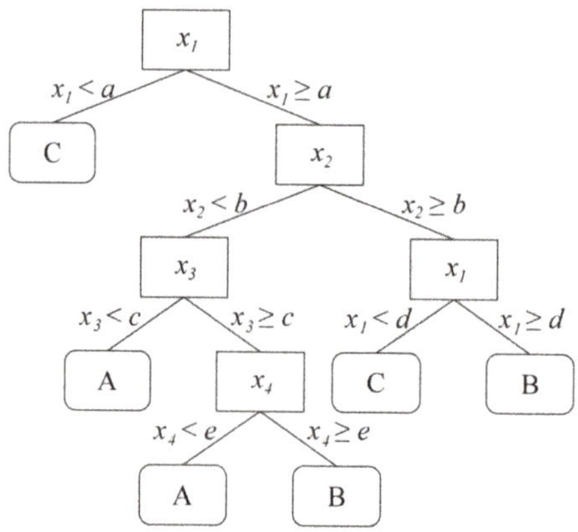

Figure 2. Structure of a decision tree.

2.2. Accuracy Estimation

The accuracy of trained learners can be estimated using a validation database independent of the training one. Considering a classification problem, the accuracy rate is usually defined as the capability of assigning a specific scenario from the validation database to the correct response class. Namely, given a time instant t^*, the accuracy of the related classifiers is defined as:

$$Acc(\%) = 100\frac{N_c}{N} \tag{1}$$

where N_c is the number of correctly classified damage scenarios and N the total number of scenarios induced in the validation database.

Aiming to predict the outcomes of the damage scenario, the so-called "ongoing" damage scenarios, i.e., the scenarios having $t_f > t^*$, are more interesting. Thus, the ongoing accuracy can also be defined as:

$$Acc^*(\%) = 100\frac{N_c^*}{N^*} \tag{2}$$

where N_c^* is the number of correctly classified ongoing damage scenarios and N^* the total number of ongoing scenarios induced in the validation database.

Regarding the regression problems, the accuracy can be checked by means of a proper statistical indicator. Here, the coefficient of determination R^2 was used:

$$R^2 = 1 - \frac{SSE}{SS_{tot}} \tag{3a}$$

$$SSE = \sum_{i=1}^{N}(y_i - y_i^*)^2 \tag{3b}$$

$$SS_{tot} = \sum_{i=1}^{N}(y_i - \overline{y})^2 \tag{3c}$$

where y_i are the known responses, $\overline{y}$ is their mean value, and y_i^* are the responses predicted by the model. Once again, an R^{2*} ongoing coefficient of determination was also defined based only on the N^* ongoing damage scenarios.

3. Database-Generation Methods

As mentioned, to assess the damage consequences by applying ML, a training dataset is needed, composed of progressive flooding simulations. The simulations are driven by a damage case generation algorithm. In the present section, several options are proposed to generate the training database according to a different characterization of the damages. All the progressive flooding simulations were carried out using a quasi-static technique based on the solution of a linearized differential-algebraic equation system [18,19]. The method represents a good compromise between accuracy and computational effort [20]; hence, it was considered adequate for the generation of large databases of progressive flooding simulations.

Usually, the damage is modeled as a parallelepiped box intersecting the hull [21–23]. With such an assumption, the surface of the hull shell enclosed in the damage box is removed to define the damage. Considering a collision case, the box-shaped damage is always crossing the waterline and can be completely defined by five parameters (Figure 3):

- Damage length L_d;
- Longitudinal position of the damage center X_d;
- Transversal damage penetration B_d measured from the shell side $(B/2)$;
- Vertical height Z_{max} of the highest tip of the damage above the Baseline (BL);
- Vertical height Z_{min} of the lowest tip of the damage above the BL.

In the present work, the damage penetration was neglected since all the internal structures were considered intact.

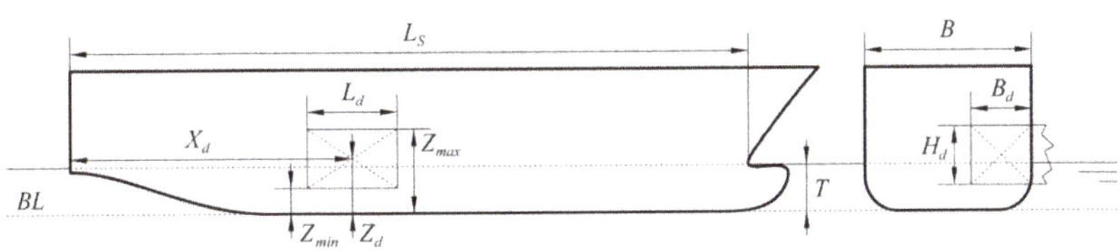

Figure 3. Bow-shaped damage parameters.

Here, two families of methods for the database generation of side damage cases in calm water were tested, one based on MC generation and the other based on a parametric generation aimed to cover all the possible damage scenarios involving multiple neighboring rooms. Applying MC sampling, the damage cases can be generated following the probability distribution of their parameters [24]. Here, three different options were explored. The first was based on the probability distributions embedded in SOLAS and the other two on two types of uniform distributions.

3.1. Monte Carlo with SOLAS Probability Distributions

The SOLAS probabilistic rule framework for ship damage stability is based on the statistical analysis of a database of side collision accidents [25]. In the SOLAS, the probability distributions are used to define a so-called zonal approach, so they are not explicitly defined. However, recent studies explored the so-called nonzonal approach, which directly applies the probability distributions on the damage parameters [26]. With this approach, the following damage parameter probability distributions can be taken from SOLAS: L_d, X_d, B_d, which lead to the definition of the p-factor, and Z_{max}, which is considered in the v-factor. The Z_{min} is not defined, since SOLAS adopts a worst-case approach in s-factor determination to consider horizontal subdivision below the waterline. However, the Z_{min} probability distribution can be taken from the statistical analysis of collision damage data available in the literature [22]. The adopted probability distributions for the SOLAS database generation are defined as follows:

Damage length was modeled with a bilinear probability density function, leading to the following cumulative distribution:

$$cdf(L_d) = \begin{cases} 0 & \text{if} \quad J \leq 0 \\ \frac{b_{11}}{2}J^2 + b_{12}J & \text{if} \quad 0 \leq J \leq J_k \\ \frac{b_{11}-b_{21}}{2}J_k^2 + (b_{12}-b_{22})J_k + \frac{b_{21}}{2}J^2 + b_{22}J & \text{if} \quad J_k < J \leq J_m \\ 1 & \text{if} \quad J > J_m \end{cases} \tag{4}$$

where $J = L_d/L_S$, and all the other parameters are defined as in SOLAS Ch.II-1 Part B-1 Regulation 7-1 [27].

The longitudinal position of the damage center is uniformly distributed along the ship subdivision length L_S:

$$cdf(X_d) = \begin{cases} 0 & \text{if} \quad X_d \leq 0 \\ \frac{X_d}{L_S} & \text{if} \quad 0 < X_d < L_S \\ 1 & \text{if} \quad X_d \geq L_S \end{cases} \tag{5}$$

The vertical height Z_{max} was modeled with a bilinear cumulative density function:

$$cdf(Z_{max}) = \begin{cases} 0 & \text{if} \quad Z_{max} - T \leq 0\,\text{m} \\ 0.8\frac{Z_{max}-T}{7.8} & \text{if} \quad 0\,\text{m} \leq Z_{max} - T \leq 7.8\,\text{m} \\ 0.8 + 0.2\frac{Z_{max}-T-7.8}{4.7} & \text{if} \quad 7.8\,\text{m} < Z_{max} - T \leq 12.5\,\text{m} \\ 1 & \text{if} \quad Z_{max} - T > 12.5\,\text{m} \end{cases} \tag{6}$$

The vertical height Z_{min} was modeled with a linear probability density function, leading to the following cumulative distribution:

$$cdf(Z_{min}) = \begin{cases} 0 & \text{if} \quad Z_{min} \leq 0 \\ 1.4\frac{Z_{min}}{T} - 0.4\left(\frac{Z_{min}}{T}\right)^2 & \text{if} \quad 0 < Z_{min} < T \\ 1 & \text{if} \quad Z_{min} \geq T \end{cases} \tag{7}$$

3.2. Monte Carlo with a Uniform Distribution of the Damage Dimensions

In this database-generation algorithm, the maximum damage dimensions were still taken from SOLAS. However, a uniform distribution was assumed for the damage length and height. The applied cumulative density functions are, then, defined as follows:

Damage length was assumed as uniformly distributed between zero and the maximum admissible nondimensional length according to SOLAS:

$$cdf(L_d) = \begin{cases} 0 & \text{if} \quad X_d \leq 0 \\ \frac{L_d}{J_m L_S} & \text{if} \quad 0 < X_d < J_m L_S \\ 1 & \text{if} \quad X_d \geq J_m L_S \end{cases} \tag{8}$$

The longitudinal position of the damage center is already uniformly distributed in SOLAS. Hence, Equation (5) can still be applied.

The damage height $H_d = Z_{max} - Z_{min}$ is uniformly distributed between zero and $T + 12.5$ m, i.e., the maximum value according to SOLAS:

$$cdf(Z_{max}) = \begin{cases} 0 & \text{if} \quad H_d \leq 0 \\ \dfrac{H_d}{T + 12.5\,\text{m}} & \text{if} \quad 0 < H_d < T + 12.5\,\text{m} \\ 1 & \text{if} \quad H_d \geq T + 12.5\,\text{m} \end{cases} \tag{9}$$

As the damage height is defined, the vertical position of the damage center Z_d is defined ensuring that the damage is crossing the waterline in compliance with SOLAS. Hence, Z_d was assumed as uniformly distributed in the interval:

$$\left[max\left(\frac{H_d}{2}, T - \frac{H_d}{2}\right), T + min\left(12.5\,(m) - \frac{H_d}{2}, T + \frac{H_d}{2}\right) \right] \tag{10}$$

3.3. Monte Carlo with a Uniform Distribution of the Damage Area Inverse

In this damage-case-generation algorithm, a uniform distribution of the inverse of the damage area $A_d = L_d(Z_{max} - Z_{min})$ was applied. The main objective of such a method is to generate a more uniform distribution of the *time-to-flood*, compared to the SOLAS one. SOLAS's probability distributions lead to many large-area damages having, consequently, a short *time-to-flood*. Such damages are not very interesting for decision support purposes since the events evolve too fast to gain any advantage from a DSS response. Moreover, a too-small number of long damage scenarios might affect the forecast accuracy of the learners due to the lack of training data.

The following cumulative probability function was applied to draw the damage areas:

$$cdf\left(\frac{1}{A_d}\right) = \begin{cases} 0 & \text{if} \quad \dfrac{1}{A} \leq \dfrac{1}{A_{max}} \\ \dfrac{\dfrac{1}{A} - \dfrac{1}{A_{max}}}{\dfrac{1}{A_{min}} - \dfrac{1}{A_{max}}} & \text{if} \quad \dfrac{1}{A_{max}} < \dfrac{1}{A} < \dfrac{1}{A_{min}} \\ 1 & \text{if} \quad \dfrac{1}{A} \geq \dfrac{1}{A_{min}} \end{cases} \tag{11}$$

where A_{min} and A_{max} are the minimum and maximum damage areas that can be defined for each different ship. In a real application, floodwater inflow due to very small damages can be controlled by the bilge system. Hence, it can be assessed considering the bilge pumps' capacity. The maximum area can be defined as the maximum damage area according to SOLAS: $A_{max} = (T + 12.5)\ J_m\ L_S$.

Given the damage area, the other parameters were defined. The longitudinal position of the damage center was here assessed as for SOLAS according to Equation (5). Two alternative procedures were then applied to half of the generated damage cases:

- The heights Z_{max} and Z_{min} were drawn following the SOLAS cdf provided in Equations (6) and (7), respectively. Then, defining the damage height $H_d = Z_{max} - Z_{min}$, the damage length was determined as $L_d = A_d / H_d$;
- The damage length was drawn following the SOLAS cdf provided in Equation (4), and damage height H_d was determined as $H_d = A_d / L_d$. Then, the vertical position of damage center Z_d was randomly drawn within the range defined in Equation (10) to have the damage crossing the waterline.

Damages having a height or length outside the ship boundaries were discarded and randomly generated again.

3.4. Parametric Method

The parametric generation of the damage cases was divided into two phases. In the first, the box-shaped damages were generated considering every single room laying on the hull shell. In the latter, the single-room damage cases were parametrically combined to define additional damage cases involving neighboring rooms.

In a previous study [17], it was observed that the longitudinal position of the damage center has only a limited impact on progressive flooding, whereas its vertical position has a relevant effect. Hence, all the single-room damages were applied at half of the room longitudinal extension assuming an L_d equal to the room longitudinal extension. On the other hand, at least three different vertical positions of damage center Z_d were considered for each room: the room's bottom, half-height, and top. If a room extends over multiple decks, additional intermediate positions were considered corresponding to the main decks' heights from the baseline. The damage center height was corrected for top and bottom damages, considering the applied damage area as shown in Figure 4, to avoid damages extending outside the room boundaries. For each possible damage location (X_d, Z_d), multiple damage sizes were considered. The area of the i-th damage in the j-th room, considering the k-th center, is evaluated as:

$$\frac{1}{A_{ijk}} = \frac{1}{A_{max_{jk}}} + \frac{k}{n_d}\left(\frac{1}{A_{max_{jk}}} - \frac{1}{A_{min_{jk}}}\right) \tag{12}$$

with $k = [1, 2, ..., n_d]$, where n_d is the so-called number of divisions, which is the main parameter governing the database size. Besides, A_{min} and A_{max} were defined for each room and each position of the damage center. The minimum area was defined as the one corresponding to an initial inflow equal to the bilge pumps' capacity. To define the maximum area, note that very large damages result in the almost instantaneous filling of the damaged room. Hence, the maximum damage area was defined as the area that causes the room filling in 15 s. Furthermore, every single room was assumed as lost at the beginning of progressive flooding (instantaneous flooding), defining an additional single-room damage case. After the definition of the single-room damage cases, they were combined with the ones related to the neighboring rooms. Namely, all the possible combinations of the damage areas of damages having the same center height (bottom, half-height, top) and sharing a boundary (watertight bulkhead, deck) were considered. For instance, at the intersection of a deck and a transverse bulkhead, the combinations were defined considering up to four rooms. Here, only one or two compartments' damages were generated. Nevertheless, the parametric generation technique can be easily extended to a higher number of contiguous compartments if required.

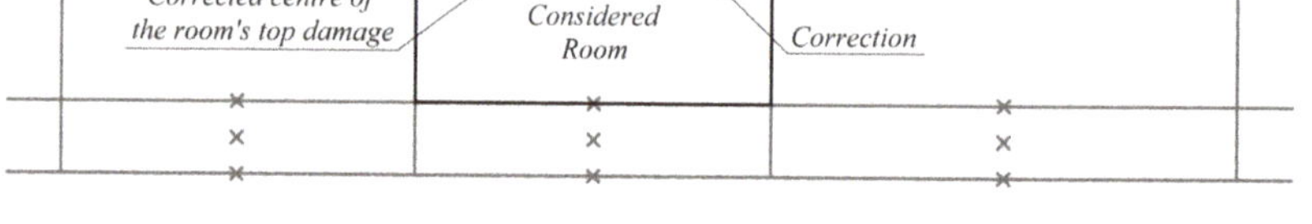

Figure 4. Correction of the vertical position of the damage center according to the damage area for a top damage.

4. Test Case

The multiple generation algorithms were tested on a simple test geometry to study their effect on damage consequences' assessment. In the present section, the test arrangement is described. Then, the tested training and validation databases are described.

4.1. Test Arrangement

The test arrangement was a box-shaped barge having the general arrangement shown in Figure 5, and the main particulars are provided in Table 1. The test barge had three decks: DK1 located near the water plane (at 5 m above the *BL*), DK2 (located at 10 m above the *BL*), and the bulkhead deck DK3 (located 15 m above the baseline). Beneath the bulkhead deck, the barge was divided into five main compartments. In first and third compartments, the lowest rooms extended from the barge bottom to DK2. In the lowest room within Compartments 1 and 3, longitudinal bulkheads were fit on the centerline. In addition, another longitudinal bulkhead was fit in Compartment 5 between DK2 and DK3 to create a long flooding chain when the lowest room was damaged.

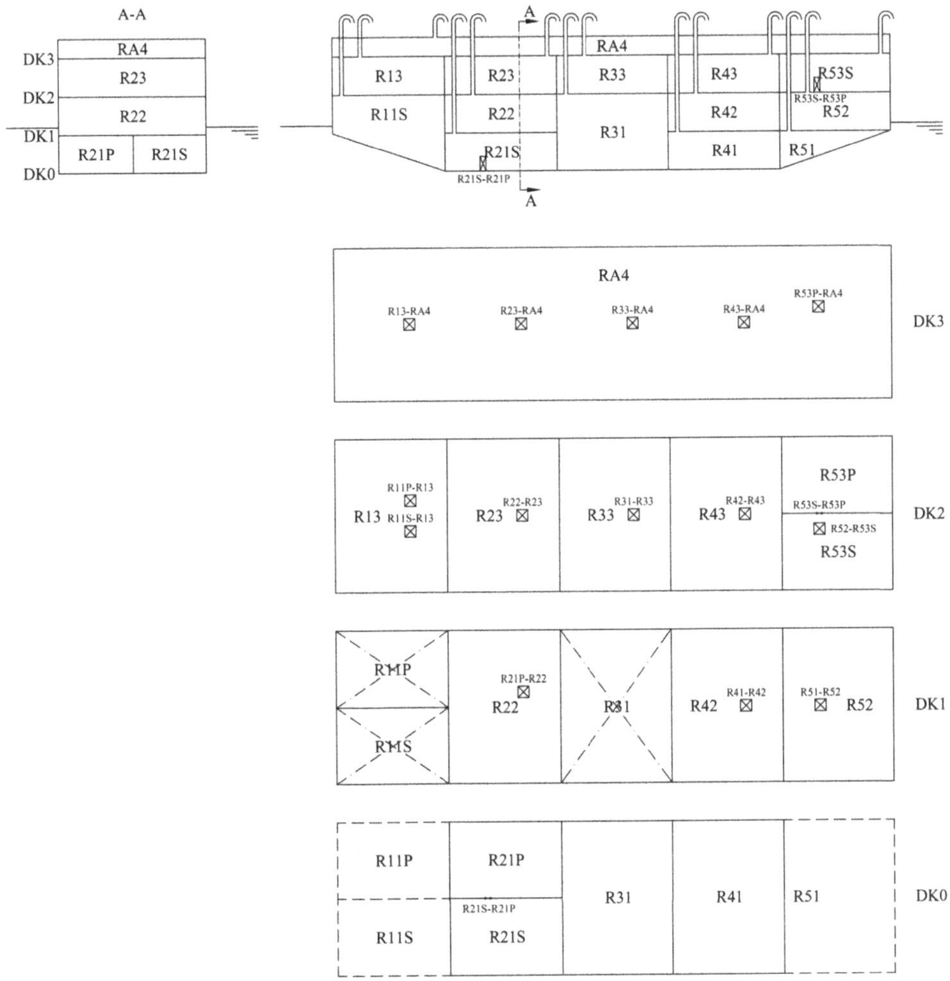

Figure 5. General arrangement of the test geometry.

Table 1. Main particulars of the test geometry.

Description	Symbol	Value
Length overall	L_{OA}	75 m
Breadth	B	20 m
Draught	T	6 m
Depth	D	17.5 m
Displacing volume	∇	7500 m^3
Metacentric height	GM	2.685 m

All the rooms were considered fully vented, had unitary permeability and were interconnected by a set of openings, as shown in Figure 5. The openings' size and position are provided in Table 2. To apply a hydrostatic code based on pressure integration technique [28], both the rooms and the openings were modeled with nonstructured triangular meshes having a maximum panel area of 3.125 m^2 and 0.25 m^2, respectively. A sketch of the 3D model of the test geometry is shown in Figure 6.

Table 2. Main characteristics of the test geometry openings. $C = (X_C, Y_C, Z_C)$ is the center of the opening in the ship-fixed reference system.

ID	Type	h (m)	w (m)	X_C (m)	Y_C (m)	Z_C (m)
R11S-R13	Horizontal	1.5	1.5	10	−2	10
R11P-R13	Horizontal	1.5	1.5	10	2	10
R13-RA4	Horizontal	1.5	1.5	10	0	15
R21P-R22	Horizontal	1.5	1.5	25	2	5
R22-R23	Horizontal	1.5	1.5	25	0	10
R23-RA4	Horizontal	1.5	1.5	25	0	15
R31-R33	Horizontal	1.5	1.5	40	0	10
R33-RA4	Horizontal	1.5	1.5	40	0	15
R41-R42	Horizontal	1.5	1.5	55	0	5
R42-R43	Horizontal	1.5	1.5	55	0	10
R43-RA4	Horizontal	1.5	1.5	55	0	15
R51-R52	Horizontal	1.5	1.5	65	0	5
R52-R53S	Horizontal	1.5	1.5	65	−2	10
R53P-RA4	Horizontal	1.5	1.5	65	2	15
R21S-R21P	Longitudinal	1.9	0.8	20	0	0.95
R53S-R53P	Longitudinal	1.9	0.8	65	0	10.95

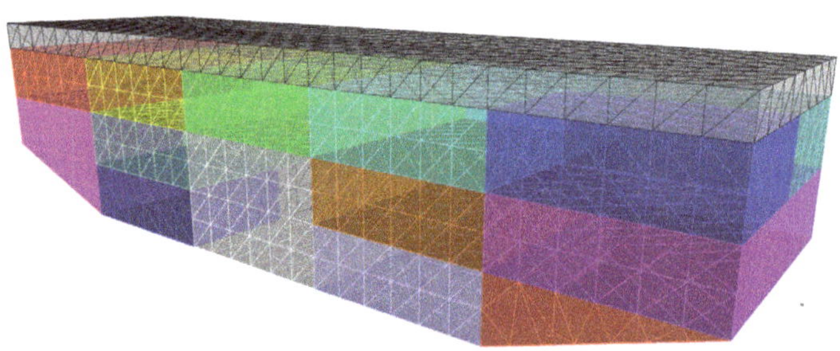

Figure 6. Mesh of the internal rooms of the test geometry.

4.2. Tested Databases

The different database-generation techniques defined in the previous section were applied to create five different training databases related to the test arrangement:

- MC20: based on the SOLAS probability distributions and being composed of 20,000 damage cases (3082 nonsurvival cases);
- MCD20: based on a uniform distribution of damage dimensions and being composed of 20,000 damage cases (8898 nonsurvival cases);
- MCA20: based on a uniform distribution of the damage area inverse and being composed of 20,000 damage cases (2318 nonsurvival cases);
- P8: a parametric database having $n_d = 8$ and being composed of 18,854 damage cases (15,388 nonsurvival cases).

The number of damage cases $N = 20{,}000$ was chosen to ensure a proper convergence of the accuracy [11]. For the same reason, 20,000 damage cases were included in the MCD20 and MCA20 databases, whereas $n_d = 8$ was chosen for the P8 database to have a similar number of damage cases and ensure a fair comparison of the methods. Figure 7 shows the resulting cumulative density functions of the damage dimensions, location, and *time-to-flood* of the damage cases in the four databases.

The MC-based methods provided a continuous distribution of all the parameters, whereas the parametric one led to scattered distributions except for t_f. As expected, an almost uniform distribution of t_f was associated with the MCA20 database. This means that many small damages were generated located near the free surface, while the number of nonsurvival scenarios was lower than that of MC20. On the contrary, both, the MCD20 and P8 databases contained a significantly higher percentage of nonsurvival scenarios compared to MC20. It is worth noticing that MC20 and MCD20 drive toward a comparable cumulative distribution of t_f, although driven by different probability distributions of damage parameters.

Besides the training databases, a single validation database was defined:

- MC50b: based on the SOLAS probability distributions and being composed of 50,000 damage cases (8059 nonsurvival cases).

The SOLAS probability distributions were chosen for the validation database since they were considered the most representative of the collision damages that might occur in an operative environment. Hence, aiming at the definition of a methodology to be employed during a real flooding emergency, such a choice was deemed appropriate.

For all the databases, the maximum simulation time was set to 2250 s.

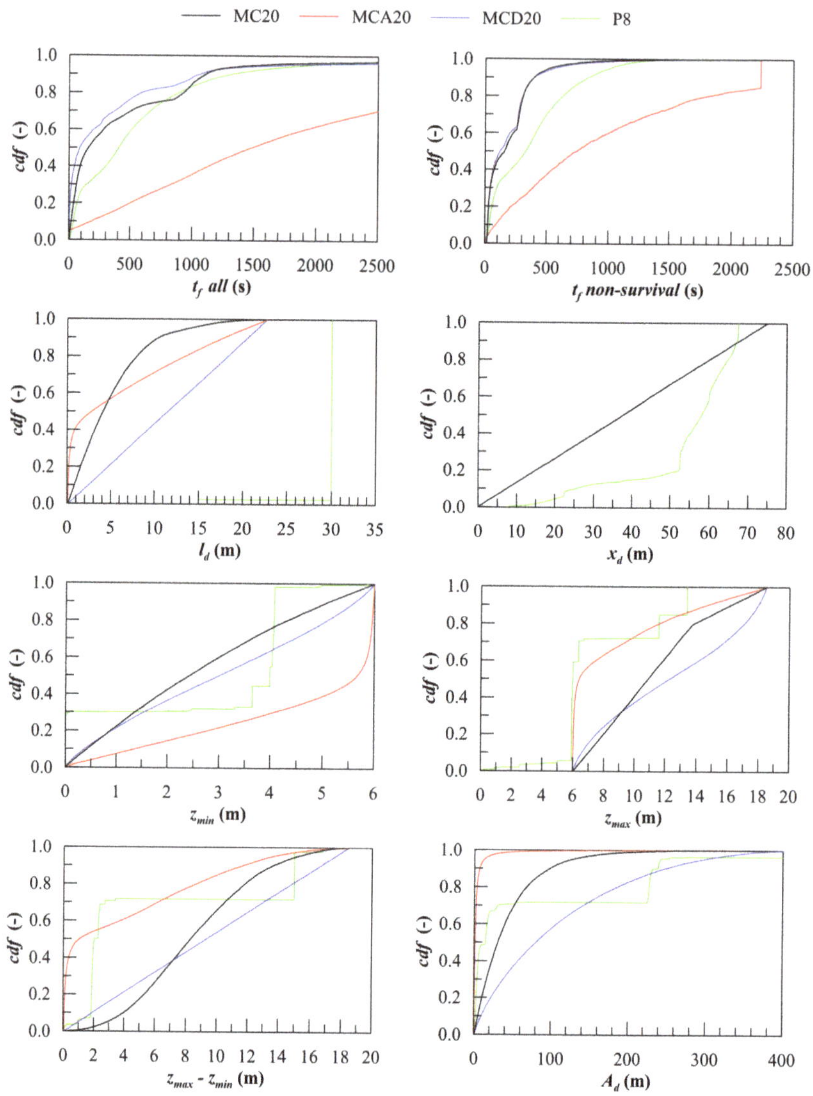

Figure 7. Cumulative density functions of damage dimensions, location, and *time-to-flood* related to the tested database-generation methods.

5. Results

In Figure 8, the overall performances obtained applying the different training databases are provided. It is worth noticing that the MC20 database employing the SOLAS probability distributions was not always associated with the best performances despite a SOLAS-based dataset always being employed for validation.

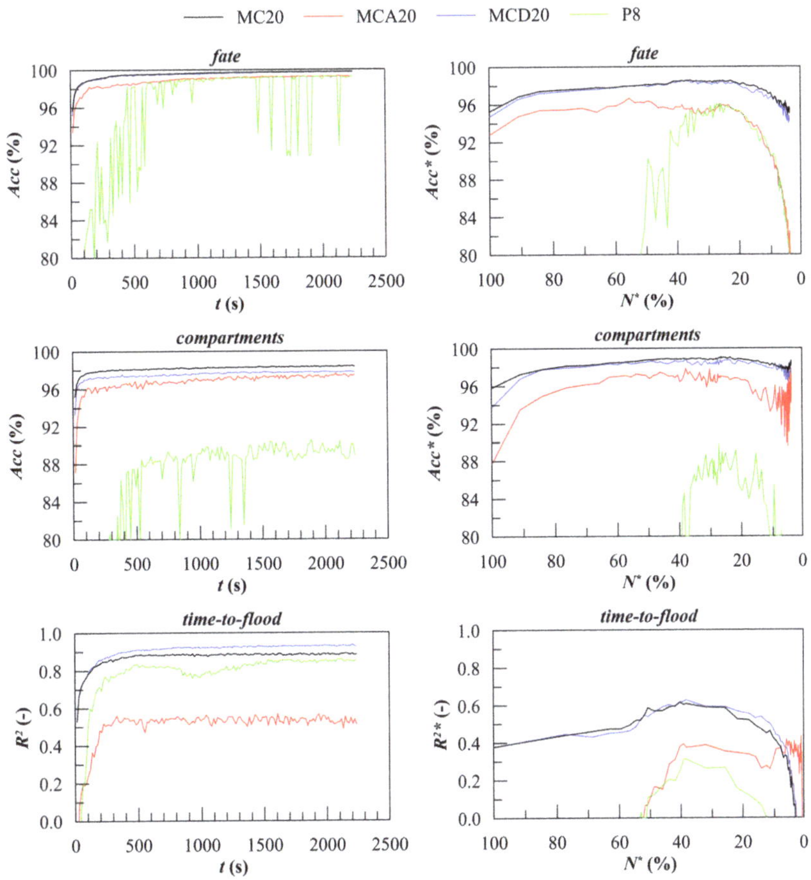

Figure 8. Comparison of the performances of the tested database-generation methods. Validation: MC50b; method: RFs.

Namely, the MCD20 database provided similar results for the *final fate* and *flooded compartments* classification problems compared to the MC20 one. Besides, the MCD20 dataset was more accurate in the prediction of t_f than MC20. The MCA20 database showed a lower accuracy than MC20 and MCD20 for all the studied problems. However, MCA20 showed a larger region of stable t_f forecast, although a lower maximum of R^2 and R^{2*} was reached. R^{2*} decayed under a null value at about $t = 1215$ s (corresponding to 0.5% of the ongoing damage scenarios in the validation database) instead of $t = 735$, which was related to the MCD20 database. The parametric generation method always led to lower accuracy compared to all the methods based on MC generation. Moreover, the P8 training database led to strong accuracy instability in both classification problems.

6. Discussion

Considering the *final fate* classification problem, the adoption of the MCD20 training database had a very limited effect on the overall and ongoing accuracy. In detail, a less skewed training dataset was obtained by the larger number of nonsurvival damage cases in MCD20. This reduced by about 0.05% the type I error (i.e., when a nonsurvival scenario was classified as a survival one by the learner), while increasing by about 0.15% the type II one (i.e., when a survival scenario was classified as a nonsurvival one by the learner).

The adoption of the MCA20 training database decreased the accuracy by about 2% in forecasting the *final fate* compared to MC20/MCD20. Furthermore, a strong decay of the ongoing accuracy for long damage scenarios was also observed. The main reason can be found in the larger number of damage scenarios exceeding the maximum simulation time compared to the other databases: in MCA20, 30% of damage scenarios had $t_f > 2250$ s. The adoption of the P8 training database led to a similar behavior, although the accuracy reached an almost stable value later (at about 500 s). Analyzing the results provided by the parametric generation, note that damage cases affected one or two adjoining compartments. On the other hand, the validation database (MC50b) was based on the SOLAS probability distributions. According to SOLAS, the maximum damage length for the test barge is 22.72 m, which exceeds the length of each watertight compartment (15 m). Hence, 7.4% of the damage scenarios in the validation database affected three adjoining compartments. The accuracy gap between MC20 and P8 for the *final fate* classification problem in the stable region was between 1% and 3%, being lower than 7.4%. This means that many of the three-compartment damage scenarios were still correctly classified by RFs trained with the P8 database.

On the contrary, the classification of *flooded compartments* led to an accuracy gap of about 9–10%, which was larger than the percentage of the three-compartment damage scenarios included in the validation database. For *flooded compartments* classification, the MCD20 and MCA20 databases had an overall accuracy 1% and 2% lower than the MC20 one, respectively. On the other end, the ongoing accuracy of MCD20 was comparable to the MC20 one, whereas the MCA20 one was again 2% worse and showed a greater instability for large t^* values.

The t_f regression related to nonsurvival scenarios based on the different techniques for training database generation led to more interesting results. In Figures 9–12, the predicted–observed plots evaluated at $t^* = 500$ s and $t^* = 1000$ s are provided.

As mentioned, the SOLAS-based training database was not always the best option. The MCD20 training database led to better results since it contained more than double the nonsurvival scenarios compared to MC20, while the two databases had almost the same distribution of the *time-to-flood*. Due to the increased density of the training database, the RFs better forecast the t_f up to $t^* = 1700$ s. Nevertheless, at larger t^* values, the ongoing accuracy still decayed. In such a region, better results could be reached by employing the MCA20 training database, which allowed a good forecast of the *time-to-flood* up to $t^* = 2000$ s. The main reason was the lower dimension of the generated damages, which led to a higher density of the training data in the most critical region. However, note that the model tended to overestimate t_f, highlighted by the clusters in the lower part in Figure 10.

Another criticality associated with the MCA20 training database was the poor capacity to deal with large damages having short t_f. This was mainly due to the limited number of nonsurvival damage scenarios. Bulky errors can be observed in the lower t_f region, leading to poor values of the overall R^2 (never greater than 0.55). Regarding the parametric generation method, it was not capable of ensuring good results for the t_f regression problem for the test geometry.

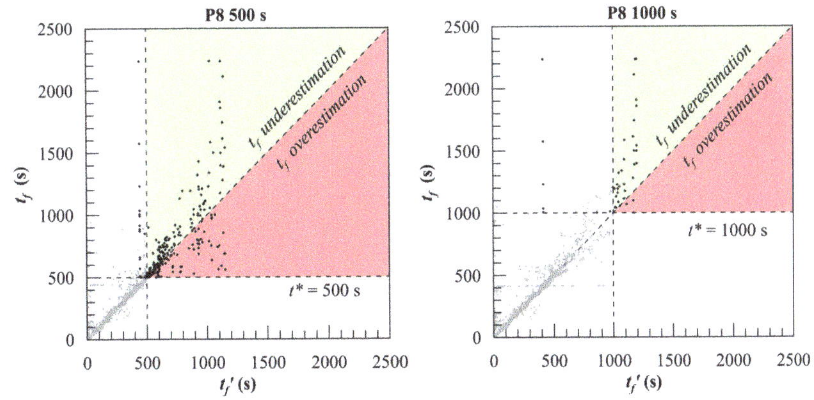

Figure 9. Predicted over observed values of t_f computed at $t^* = 500$ s and $t^* = 1000$ s. Training: MC20; validation: MC50b; method: RFs.

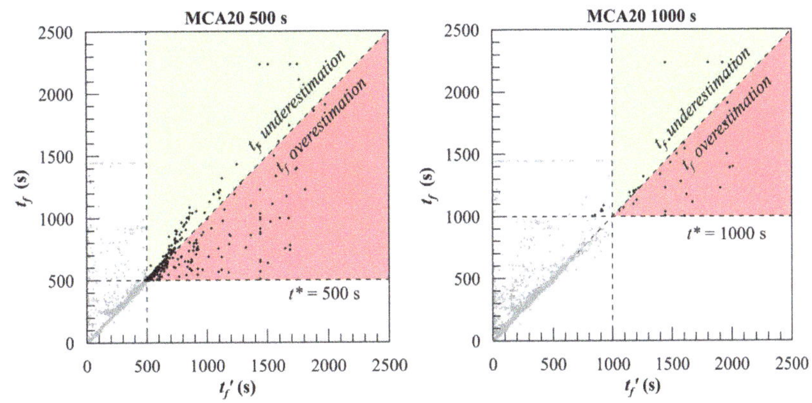

Figure 10. Predicted over observed values of the *time-to-flood* computed at $t^* = 500$ s and $t^* = 1000$ s. Training: MCA20; validation: MC50b; method: RFs.

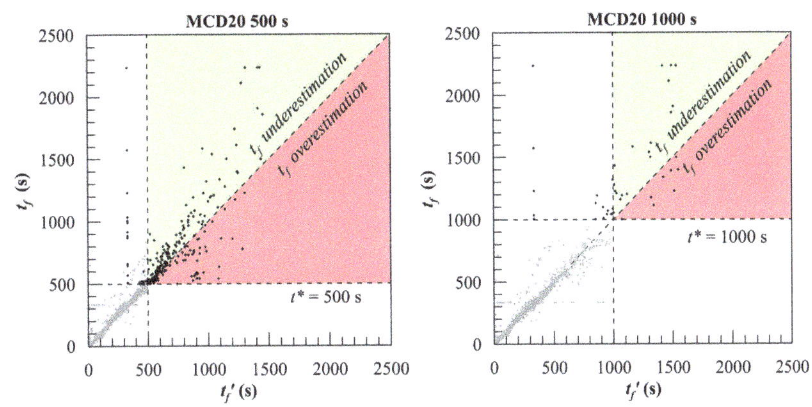

Figure 11. Predicted over observed values of the *time-to-flood* computed at $t^* = 500$ s and $t^* = 1000$ s. Training: MCD20; validation: MC50b; method: RFs.

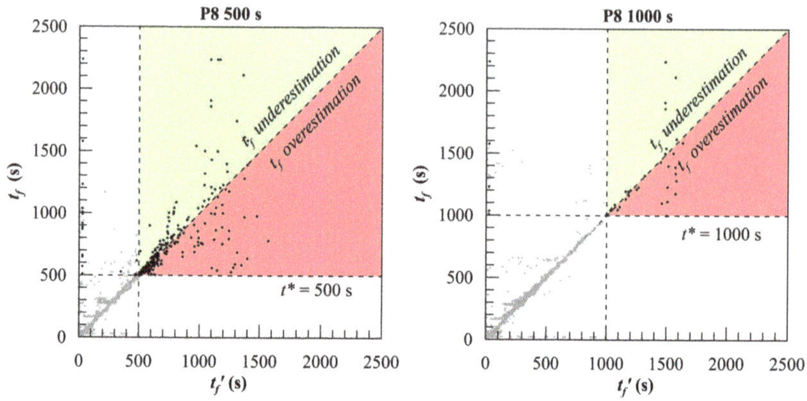

Figure 12. Predicted over observed values of the *time-to-flood* computed at $t^* = 500$ s and $t^* = 1000$ s. Training: P8; validation: MC50b; method: RFs.

7. Conclusions

The paper explored multiple options for training database generation on damage consequences' assessment. Four generation algorithms were tested on a box-shaped barge, demonstrating that the accuracy of prediction can be heavily affected by the database-generation method. An interesting result was that the training database generation based on the SOLAS probability distributions was not always the best option for the test geometry. Although the validation database was also based on the SOLAS probability distributions, different distributions applied to the damage parameters driving the MC sampling led to equal or better results.

Among the studied problems, the t_f regression problem was the most affected by the training database. Namely, the application of the uniform probability distribution of the damage dimensions or inverse area significantly improved the performances for short and large t^* values, respectively. For the two classification problems, the uniform probability distribution of the damage dimensions led to slightly better results compared to the SOLAS one.

The parametric generation method showed always poorer performances compared to the other tested options. However, this gap was mainly due to the assumption of a maximum of two adjoining compartments involved in a damage scenario valid for the parametrically generated database. Moreover, the parametric method showed quite good resilience in the *final fate* classification problem and led to a large percentage of long nonsurvival damage cases, thus to a less skewed training database. Hence, it can be concluded that the parametric generation is worthy of further investigation to assess its real effectiveness.

Future research should also focus on more complex geometries (such as a full-scale passenger vessel), the effect of the ship loading condition, waves, and internal openings' type and status (open/closed). All these issues shall be properly considered during the training database generation to move towards a real application of flooding-sensor-agnostic DSS in an operative environment. Nevertheless, the outcomes of the present study can help naval architects in addressing the issues that might affect the result of a flooding-sensor-agnostic DSS due to the adopted database-generation technique.

Author Contributions: Conceptualization, L.B., J.P.-O., and M.V.; methodology, L.B., J.P.-O. and M.V.; software, L.B.; validation, L.B. and M.V.; formal analysis, L.B. and M.V.; investigation, L.B.; resources, L.B.; data curation, L.B.; writing—original draft preparation, L.B.; writing—review and editing, J.P.-O. and M.V.; visualization, L.B., M.V. and J.P.-O.; supervision, J.P.-O. and M.V.; project administration, J.P.-O.; funding acquisition, J.P.-O. All authors have read and agreed to the published version of the manuscript.

Funding: This work was fully supported by the Croatian Science Foundation under the project IP-2018-01-3739.

Institutional Review Board Statement: Not applicable.

Informed Consent Statement: Not applicable.

Data Availability Statement: The data are contained within the article.

Acknowledgments: This work was also supported by the University of Rijeka (Project Nos. uniri-tehnic-18-18 1146 and uniri-tehnic-18-266 6469).

Conflicts of Interest: The authors declare no conflict of interest.

References

1. Marine Casualties Investigative Body. *Cruise Ship Costa Concordia, Marine Casualty on January 13, 2012, Report on the Safety Technical Investigation*; Technical Report; Ministry of Infrastructures and Transports: Rome, Italy, 2013.
2. Hu, L.; Ma, K. Genetic algorithm-based counter-flooding decision support system for damaged surface warship. *Int. Shipbuild. Prog.* **2008**, *55*, 301–315.
3. Kang, H.; Choi, J.; Yim, G.; Ahn, H. Time Domain Decision-Making Support Based on Ship Behavior Monitoring and Flooding Simulation Database for On-Board Damage Control. In Proceedings of the 27th International Ocean and Polar Engineering Conference, San Francisco, CA, USA, 25–30 June 2017.
4. Ruponen, P.; Pennanen, P.; Manderbacka, T. On the alternative approaches to stability analysisin decision support for damaged passenger ships. *WMU J. Marit. Aff.* **2019**, *18*, 477–494. [CrossRef]
5. IMO. *MSC.1/Circ.1291 Guidelines for Flooding Detection Systems on Passenger Ships*; Technical Report; International Maritime Organisation: London, UK, 2008.
6. Ruponen, P.; Larmela, M.; Pennanen, P. Flooding Prediction Onboard a Damage Ship. In Proceedings of the 11th International Conference on the Stability of Ships and Ocean Vehicles, Athens, Greece, 23–28 September 2012; pp. 391–400.
7. Dankowski, H.; Krüger, S. A Fast, Direct Approach for the Simulation of Damage Scenarios in the Time Domain. In Proceedings of the 11th International Marine Design Conference—IMDC 2012, Glasgow, UK, 11–14 June 2012.
8. Rodrigues, J.; Guedes Soares, C. A generalized adaptive mesh pressure integration technique applied to progressive flooding of floating bodies in still water. *Ocean Eng.* **2015**, *110*, 140–151. [CrossRef]
9. Ruponen, P.; Pulkkinen, A.; Laaksonen, J. A method for breach assessment onboard a damaged passenger ship. *Appl. Ocean Res.* **2017**, *64*, 236–248. [CrossRef]
10. Trincas, G.; Braidotti, L.; De Francesco, L. Risk-Based System to Control Safety Level of Flooded Passenger Ship. *Brodogradnja* **2017**, *68*, 31–60. [CrossRef]
11. Braidotti, L.; Valčić, M.; Prpić-Oršić, J. Exploring a Flooding-Sensors-Agnostic Prediction of the Damage Consequences Based on Machine Learning. *J. Mar. Sci. Eng.* **2021**, *9*, 271. [CrossRef]
12. Braidotti, L.; Prpić-Oršić, J.; Valčić, M. Application of decision trees to predict damage consequences during the progressive flooding. In Proceedings of the 8th International Conference on marine Structures, Trondheim, Norway, 7–9 June 2021.
13. Breiman, L.; Friedman, J.H.; Olshen, R.A.; Stone, C.J. *Classification and Regression Trees*; CRC Press: Boca Raton, FL, USA, 1984.
14. Breiman, L. Random Forests. *Mach. Learn.* **2001**, *45*, 5–32. [CrossRef]
15. James, G.; Witten, D.; Hastie, T.; Tibshirani, R. *An Introduction to Statistical Learning: With Applications in R*; Springer: New York, NY, USA, 2017.
16. Rodrigues, J.; Lavrov, A.; Hinostroza, M.; Guedes Soares, C. Experimental and numerical investigation of the partial flooding of a barge model. *Ocean Eng.* **2018**, *169*, 586–603. [CrossRef]
17. Braidotti, L.; Marinò, A.; Bucci, V. On the Effect of Uncertainties on Onboard Progressive Flooding Simulation. In Proceedings of the 3rd International Conference on Nautical and Maritime Culture—CNM 2019, Naples, Italy, 14–15 November 2019; pp. 21–30. [CrossRef]
18. Braidotti, L.; Mauro, F. A New Calculation Technique for Onboard Progressive Flooding Simulation. *Ship Technol. Res.* **2019**, *66*, 150–162. [CrossRef]
19. Braidotti, L.; Mauro, F. A Fast Algorithm for Onboard Progressive Flooding Simulation. *J. Marit. Sci. Eng.* **2020**, *8*, 369. [CrossRef]
20. Braidotti, L.; Degan, G.; Bertagna, S.; Bucci, V.; Marinò, A. A Comparison of Different Linearized Formulations for Progressive Flooding Simulations in Full-Scale. *Procedia Comput. Sci.* **2021**, *180*, 219–228. [CrossRef]
21. DNV GL. *Evaluation of Risk from Raking Damages Due to Grounding*; Technical Report PP090623/3-1; DNV GL: Høvik, Norway, 2014.
22. Bulian, G.; Cardinale, M.; Francescutto, A.; Zaraphonitis, G. Complementing SOLAS damage ship stability framework with a probabilistic description for the extent of collision damage below the waterline. *Ocean Eng.* **2019**, *186*, 106073. [CrossRef]
23. Ruponen, P.; Lindroth, D.; Routi, A.; Aartovaara, M. Simulation-based analysis method for damage survivability of passenger ships. *Ship Technol. Res.* **2019**, *66*, 180–192. [CrossRef]
24. Kruger, S.; Dankowsky, H. A Monte Carlo based simulation method for damage stability problems. In Proceedings of the 38th International Conference on Ocean, Offshore and Arctic Engineering—OMAE 2019, Glasgow, UK, 9–14 June 2019.

25. IMO. *Resolution MSC.281(85) Explanatory Notes to the SOLAS Chapter II-1 Subdivision and Damage Stability Regulations*; Technical Report; International Maritime Organisation: London, UK, 2008.
26. Bulian, G.; Lindroth, D.; Ruponen, P.; Zaraphonitis, G. Probabilistic assessment of damaged ship survivability in case of grounding: Development and testing of a direct non-zonal approach. *Ocean Eng.* **2016**, *120*, 331–338. [CrossRef]
27. IMO. Chapter Ch.II-1 Part B Subdivision and Stability. In *SOLAS 2018 Consolidated Edition*; International Maritime Organisation: London, UK, 2018.
28. Braidotti, L.; Trincas, G.; Bucci, V. Analysis of the Influence of Pressure Field on Accuracy for Onboard Stability Codes. In Proceedings of The 19th International Conference on Ships and Maritime Research—NAV 2018, Trieste, Italy, 20–22 June 2018; pp. 80–87. [CrossRef]

Journal of
Marine Science and Engineering

MDPI

Article

Numerical Evaluation of the Wave-Making Resistance of a Zero-Emission Fast Passenger Ferry Operating in Shallow Water by Using the Double-Body Approach

Suleyman Duman, Evangelos Boulougouris *, Myo Zin Aung, Xue Xu and Amin Nazemian

Department of Naval Architecture, Ocean and Marine Engineering, University of Strathclyde, Glasgow G4 0LZ, UK
* Correspondence: evangelos.boulougouris@strath.ac.uk

Abstract: The consideration of shallow water effects has gained in importance regarding inland operations. The interaction between the keel and the riverbed affects the hydrodynamic characteristics of marine vessels. The highly complex nature of the interference phenomenon in catamarans makes the shallow water problem more complicated as compared to monohulls. Hence, catamarans are very sensitive to speed changes, as well as to other parameters, such as the shallow water effects. This makes the design of catamarans more challenging than their monohull equivalents. At lower Froude numbers, the higher importance of the frictional resistance makes the greater wetted surface of the catamaran a disadvantage. However, at higher speeds, there is the potential to turn their twin hulls into an advantage. This study aims to investigate the wave-making resistance of a zero-carbon fast passenger ferry operating in shallow water. The URANS (unsteady Reynolds-averaged Navier–Stokes) method was employed for resistance simulations. Then, the double-body approach was followed to decompose the residual resistance into viscous pressure and wave-making resistance with the help of the form factors of the vessel calculated at each speed. The characteristics of the separated wave-making resistance components were obtained, covering low, medium, and high speeds. Significant findings have been reported that contribute to the field by providing insight into the resistance components of a fast catamaran operating in shallow waters.

Keywords: wave-making resistance; shallow water; form factor; double-body; wave interference; catamaran; URANS; CFD

Citation: Duman, S.; Boulougouris, E.; Aung, M.Z.; Xu, X.; Nazemian, A. Numerical Evaluation of the Wave-Making Resistance of a Zero-Emission Fast Passenger Ferry Operating in Shallow Water by Using the Double-Body Approach. *J. Mar. Sci. Eng.* **2023**, *11*, 187. https://doi.org/10.3390/jmse11010187

Academic Editors: Carlos Guedes Soares and Serge Sutulo

Received: 23 November 2022
Revised: 14 December 2022
Accepted: 2 January 2023
Published: 11 January 2023

1. Introduction

Fast and green transportation systems are becoming indispensable parts of our lives as they have the potential to provide a sustainable future. Some of them, such as ground vehicles, are already being used by a considerable number of people, and other types of vehicles are in a transition period or under feasibility assessment. Most of the metropolis and shore cities in the world rely on waterborne transportation to carry passengers on a daily basis from one port to the other. Providing large deck areas and having an immense potential to travel at high speeds, with relatively less power compared to equivalent monohull examples, catamarans are strong candidates for marine transportation.

The implementation of high-fidelity solution techniques is relatively new compared to the overall progress on fluid dynamics. RANS solvers emerged in the early 21st century and have become an indispensable part of the design, optimization, and simulation of engineering systems. CFD is now widely used in marine hydrodynamic problems, e.g., power predictions, manoeuvring simulations, and marine propeller design. Molland et al. (1994) investigated the resistance of symmetric catamarans [1]. Insel and Doctors (1995) presented a method to capture the wave patterns of monohulls and catamarans by linearized theory [2]. Van't Veer (1998) conducted experiments to evaluate the resistance and seakeeping performance characteristics of the Delft catamaran 372 (DC372) [3]. Castiglione et al. (2011)

investigated the seakeeping characteristics of DC372 in rough seas with URANS [4]. Milanov (2012) developed a manoeuvring model that is a function of advance speed and water depth for a water-jet-driven fast catamaran [5]. Broglia et al. (2014) investigated the wave interference phenomena between the demihulls of DC372 by conducting experiments at different demihull clearance arrangements [6]. Duman et al. (2020) utilized the double-body approach in CFD simulations to predict the wave interference of a trimaran surface combatant using form factors [7]. Basic et al. (2020) introduced an improved method of the original linear wave-making theory to predict the wave-making resistance of ships [8]. Papanikolaou et al. (2020) presented a study focusing on the hydrodynamic hull form optimization of a zero emission, battery driven, fast catamaran vessel [9]. Boulougouris et al. (2021) presented the implementation of state-of-the-art "Industry 4.0" methods and tools to enable fully electrical, fast, zero-emission waterborne urban transport [10]. Shi et al. (2021) conducted an extensive study to predict the shallow water effects acting on a zero-carbon fast passenger catamaran [11]. Duman and Bal (2022) applied the CFD method to DC372 to develop a non-linear manoeuvring mathematical model that can work at relatively high Froude numbers [12]. Ulgen and Dhanak (2022) analysed the hydrodynamic performance of an unmanned catamaran, WAM-V, in restricted waters by using OpenFOAM® [13].

This study was conducted as part of an ongoing H2020 project TrAM: Transport, Advanced and Modular to evaluate the hydrodynamic performance of the Stavanger Demonstrator of TrAM in shallow water conditions at a relatively low draught loading. The reason for choosing the presented operational conditions is to analyse the vessel's hydrodynamic performance in a different environment; in the present case, to keep the operational conditions as close to the Thames River as possible. First, the Stavanger Demonstrator was subjected to 2-DoF CFD simulations in deep and shallow water conditions to calculate the resistance performance and dynamic attitude. The computational analysis covers a wide range of Froude numbers between 0.20–0.66. Then, the catamaran hull was analysed by utilizing the double-body approach in CFD simulations. Single-phase simulation results have been used to calculate the form factors of the catamaran at all speeds. The residual resistance components that come from the 2-DoF resistance analysis have been decomposed into frictional and wave-making resistance components with the help of those form factors. Significant findings have been reported that contribute to the field of marine hydrodynamics by providing detailed insight into the resistance components of the subject vessel.

2. Mathematical Models

2.1. Geometric Properties of the Catamaran Hull

The catamaran hull form is designed for carrying passengers with fully electric power as part of TrAM, an ongoing EU-funded project [14]. The first demonstrator of the TrAM project, Stavanger Demonstrator, was launched and named MS Medstraum in Stavanger, Norway to provide zero-emission transportation with a maximum of 27 kn recorded speed (Figure 1). With these features, Stavanger Demonstrator is a special design as it is the world's first 100% battery-driven fast passenger ferry.

The catamaran hull form is characterized by a slender body, an axe-shaped bow, a flow-adopted propeller tunnel, a transom stern, and a bow thruster tunnel (Figure 2). The fully-appended Stavanger Demonstrator used in this study includes a twisted rudder, a shaft, and two symmetrical foil-shaped struts on both the port and starboard sides.

The main dimensions of the Stavanger Demonstrator are given in Table 1 in nondimensional form. Two different loading conditions (LC) have been analysed by the CFD method: $T' = 0.045$ (LC1) and $T' = 0.040$ (LC2). The displacement is changed by only re-arranging the draught position without any modification on the original Stavanger Demonstrator hull form to obtain main particulars for the LC2.

Figure 1. The Stavanger Demonstrator of the H2020 TrAM Project: MS Medstraum.

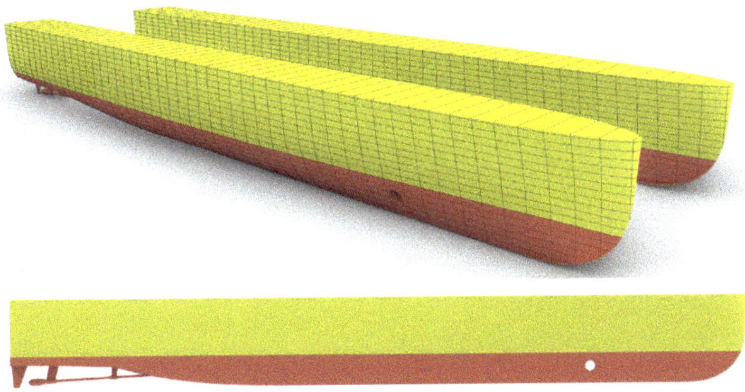

Figure 2. Hull surface in 3-D from the perspective (**upper**) and profile (**lower**) views.

Table 1. Principal dimensions of the Stavanger Demonstrator.

Dimension	Nondim.	Nondim. Value
Separation (s)	s/Lpp	0.227
Draught (T)	T/Lpp	0.040
KB	KB/Lpp	0.026
LCB	LCB/Lpp	0.460

2.2. Computational Method

All the hydrodynamic analyses of the Stavanger Demonstrator reported here have been done at model scale by using the commercial RANS solver software Star-CCM+ [15]. This chapter presents the field equations, boundary conditions, discretization of the computational domain, form factor calculation approach and the nondimensionalisation procedure followed in this study.

2.2.1. Governing Equations

Resistance simulations of the Stavanger Demonstrator are practised under 2-DoF conditions to allow the vessel to move freely on a vertical plane passing by the longitudinal symmetry centre. The flow field is built in two phases (water and air) by defining a free surface interaction level in a vertical direction (*z*-axis). The initial state of the free water surface is set to a flat plane where the ship floats at a laden draught. The volume of fluid (VoF) method is implemented to control the phase changes between the pre-described water and air cells [16]. As the vessel is released, any kind of motion or disturbance will cause gravity waves that will initiate a sequence of free surface waves in the vicinity of the vessel.

The fluid flow inside the computational domain is assumed to be Newtonian, incompressible, and turbulent. The fluid flow for both phases is handled by the Navier–Stokes equations in 3-dimensional space. Multi-phase flow simulations are achieved by solving the continuity equation (Equation (1)) and the Reynolds-averaged Navier–Stokes (RANS) equations (Equation (2)) in an unsteady way. As a result of averaging operation to the original Navier–Stokes (momentum) equations, Reynolds stresses (Equation (3)) appear that are modelled by the turbulence model. The near-wall treatment is activated within the turbulence model, Realizable k-epsilon, by setting the dimensionless wall distance (y+) to be between 30–300, which enables the wall function near the no-slip walls. Timestep, one of the key parameters for well-posed unsteady solutions, is determined according to the ITTC recommendations [17] following Equation (4) and is kept constant as all the computations are based on the implicit solution algorithm.

$$\frac{\partial(\overline{u}_i)}{\partial x_i} = 0 \tag{1}$$

$$\frac{\partial(\overline{u}_i)}{\partial t} + \overline{u}_j \frac{\partial(\overline{u}_i)}{\partial x_j} + \frac{\partial\left(\overline{u_i' u_j'}\right)}{\partial x_j} = -\frac{1}{\rho}\frac{\partial \overline{p}}{\partial x_j} + \frac{\partial \overline{\tau}_{ij}}{\partial x_j} \tag{2}$$

$$\overline{\tau}_{ij} = \nu\left(\frac{\partial \overline{u}_i}{\partial x_j} + \frac{\partial \overline{u}_j}{\partial x_i}\right) \tag{3}$$

$$\Delta t = 0.005 \div 0.01\frac{L}{V} \tag{4}$$

2.2.2. Computational Domain Boundaries and Grid Resolution

The sizes of the computational domain mainly depend on the type of the physical problem. Dimensions should provide adequate clearance for the initialization of the solution and for the developing flow without having an undesired numerical error. When previous CFD studies are searched, it is seen that at least 1.5 L upstream, 3 L downstream, and 2 L clearance are very common and give sufficient enough space to allow free surface waves to be developed for conventional resistance simulations. Bottom and top boundaries are generally assigned 1 L to 1.5 L far from the floating body. By considering previous experiences and recommendations and customize them to Stavanger Demonstrator, 2.37 L and 3.75 L distances are set for the CFD simulations in the upstream and downstream directions, respectively. Since there is no superstructure in the model geometry, the top boundary is located 1.68 L in height from the origin of the coordinate system, which is assigned at the intersection point of the aft perpendicular and keel of the vessel. The side boundaries are located approximately 2.25 L away from the centreline (Table 2). To avoid the bank shallow water effects in deep water simulations, an artificial infinite boundary effect is achieved by setting the top, bottom, and side faces as inlets where the fluid flows in the longitudinal direction, and yet no flux enters the domain from these faces. For the shallow water analysis, on the other hand, the bottom boundary is set to a pre-described depth and assigned as a no-slip moving wall, which has the same velocity as the free stream imposed at the inlet faces. This technique provides an identic solution to the physical operational conditions.

Table 2. Domain dimensions in CFD simulations.

Boundaries	Background (× L)	Overset (× L)
Upstream	2.370	0.072
Downstream	3.745	0.037
Top	1.685	0.169
Bottom	1.685	0.015
Side	2.247	0.187

The boundary conditions used in the CFD simulations are shown in Figure 3. The computational domain is divided into two main regions: background and overset. In the background domain, the front, top, bottom, and side faces are assigned as velocity inlets; the back face is set to outlet; and the boundary face on the centreline of the catamaran is set to symmetry conditions. In the overset domain: all faces except the boundary face on the centreline, which is set to symmetry, are assigned as overset boundaries. The Stavanger Demonstrator hull is assigned as a no-slip wall, where the tangential and normal velocities are zero on the hull surface. The Chimera or overset facilitates large amplitude motions without any deformation on the grid cells around the moving body. The flow features, such as velocity and pressure, are transferred between the overset boundaries and the background overlapping cells through the intersecting points [18].

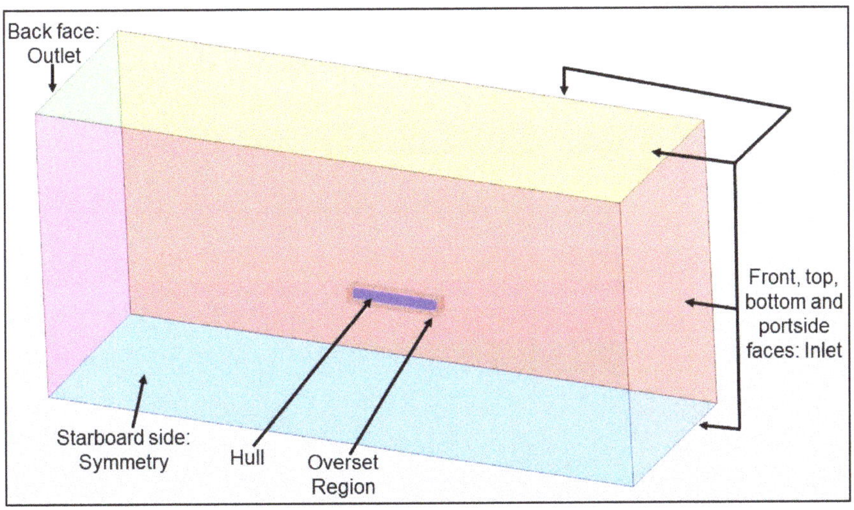

Figure 3. Boundary conditions in CFD simulations of the catamaran.

The Chimera or overset grid technique can provide high accurate dynamic solutions without any deformation on the grid cells around the moving body on that matter. The flow field is divided into two main regions, background and overset, and the flow information is transferred between the boundaries of the overset region and the background region overlapping cells through the intersecting points [18]. Coloured representations of the grid cells and information about the data transfer between those cells for catamarans in CFD simulations can be found in the literature [19,20].

In multiphase marine hydrodynamic problems, the interaction field should be handled in the first place rather than the far field. The mesh structure in this study is designed to start finer near the no-ship surfaces and to continue with a rate of change to get coarser in the relatively far points. The inner field between the demihulls and the overset boundary sites are discretized with smaller cells to capture the wave interference well and to provide a smooth transfer of the flow data between the background and overset regions (upper picture in Figure 4). The ideal way of discretizing the free water surface is to use very small and equally-sized cells on that horizontal plane. However, this approach will generate a very dense mesh and will be far more than practical and cost-effective due to the need for high computational power. Instead, a Kelvin-wave adopted grid has been applied in this study to capture the free surface deformations. The density of the grid resolution on the appendages and bow thruster tunnel of the Stavanger Demonstrator have been increased (Figure 5).

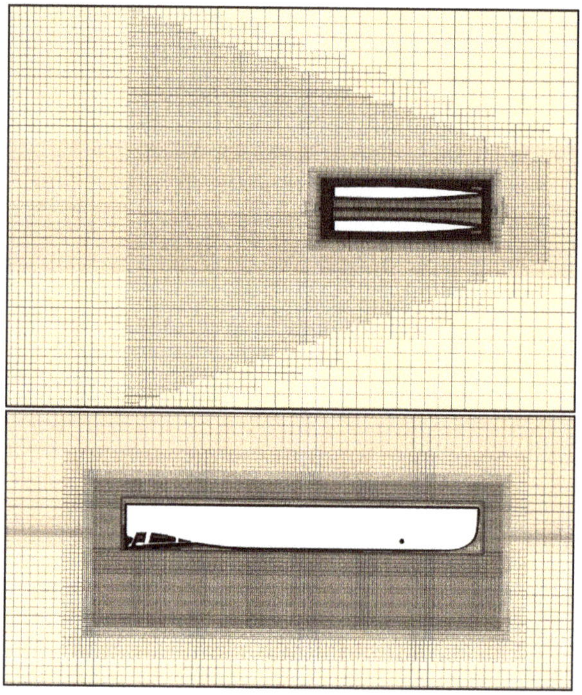

Figure 4. Grid structure on the free water surface (**upper**) and the symmetry plane of a demi-hull (**lower**) in the CFD simulations of the catamaran.

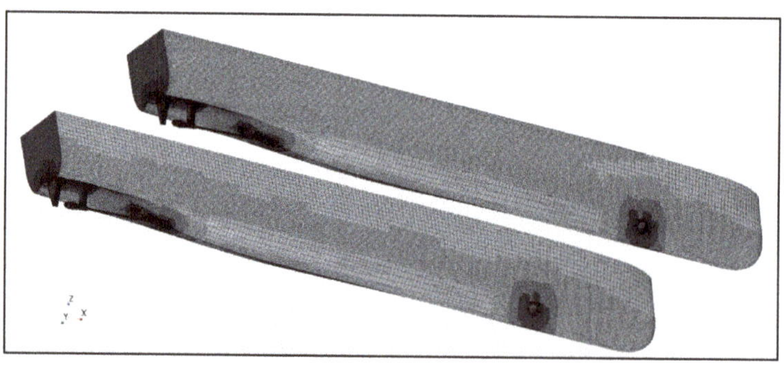

Figure 5. Grid structure on the fully-appended Stavanger Demonstrator.

2.2.3. Form Factor Prediction Method

There are two well-known and widely adopted ways of predicting the form factors of marine vessels: (1) Prohaska's method [21] and (2) the double-body geometric approach or the 1978 ITTC power prediction method [22]. The first method assumes that the wave-making resistance can be represented by the fourth order of the Froude number at low speeds (mostly between 0.10–0.20), and the total resistance coefficient can be written as in Equations (5) and (6). The principle is based on having a linear change at low speeds when the C_T/C_{F0} is plotted against Fr^4/C_{F0} where the intersection point of the linear curve and the C_T/C_{F0} axis gives the form factor (1 + k). Although it is a commonly adopted and well-defined method, Prohaska's method yields reliable results at low speeds.

$$C_T = (1+k)C_{F0} + C_w \tag{5}$$

$$\frac{C_T}{C_{F0}} = (1+k) + a\frac{Fr^4}{C_{F0}} \tag{6}$$

The second method, on the other hand, can even be utilized at high speeds. The methodology is simply cutting the hull form from the free water surface and eliminating the free surface effects by means of reducing the problem to a single-phase fluid dynamics case. The physical practice of this approach is testing the mirrored ship model in an adequate depth of water, like a submarine, and measuring the total resistance of the double-body. In this case, the total resistance is formed of viscous components only (Equation (7)). The form factor (1 + k) can be obtained by dividing the total resistance co-efficient by the frictional resistance coefficient (C_T/C_{F0}). Modelling this numerical problem is relatively less complicated compared to multi-phase simulations.

$$C_T = (1+k)C_{F0} \tag{7}$$

2.3. *Presentation of Data*

The force units are nondimensionalised by following the Equation (8), where it is necessary. X and Y represent and scalar force units, r is the fluid density, U is the ship advance speed and the S stands for the wetted surface area. The lengths are nondimensionalized by the ship length. All the calculations reported here have been done by using the fully appended catamaran's main particulars, i.e., S includes both hulls' and appendages' wetted surface areas unless the opposite is stated. The C_{F0} frictional resistance coefficient has been calculated by following the ITTC′57 frictional resistance coefficient formula given in Equation (9) [23]. The experimental resistance results have been decomposed by subtracting the C_{F0} from the C_T values that were measured in the experiments (Equation (10)). However, the frictional resistance C_{F0} has been calculated by the wall function that is implemented in the turbulence model in CFD simulations. The corresponding Froude numbers are also calculated using Equation (11) and provided in the tabulated results.

$$X', Y' = \frac{X, \; Y}{0.5\rho U^2 S} \tag{8}$$

$$C_{F0} = \frac{0.075}{(\log(Rn-2))^2} \tag{9}$$

$$C_T = C_{F0} + C_R \tag{10}$$

$$Fr = \frac{U}{\sqrt{gL_{WL}}} \tag{11}$$

3. Results and Discussion

3.1. *Verification and Validation Study*

The Stavanger Demonstrator hull form was tested in an HSVA experimental towing tank at a wide range of advance speeds [11]. The verification and validation study has been conducted for the LC1 loading condition in deep water due to the available experimental data.

3.1.1. Verification

Even though the comparison between the computational results and the experiments is a strong indication of the reliability of the method used, an uncertainty assessment is essential to draw the limits of the outputs and to observe the convergence of the method from a numerical analysis point of view. Therefore, the latest solution verification techniques, Grid Convergence Index (GCI), Correction Factor (CF) and Factors of Safety (FS) have been applied to calculate the uncertainties in the computations [24–27]. All those three methods are based on Richardson extrapolation. The total grid cell numbers are increased by changing the dimensions of the reference cell by $\sqrt{2}$ as recommended by ITTC [28]. Since the verification procedure is applied to a three-dimensional problem, the ratio between the total grid cell numbers of different qualities is defined as follows (Equation (12)):

$$r_{21} = \left(\frac{N_1}{N_2}\right)^{\frac{1}{3}} r_{32} = \left(\frac{N_2}{N_3}\right)^{\frac{1}{3}} \tag{12}$$

where N_1, N_2, and N_3 are the total cell numbers and $N_3 < N_2 < N_1$. The difference between any numerical scalar result, which, in this case, is the total resistance between two different grid cells, can be calculated as follows (Equation (13)):

$$E_{21} = X_2 - X_1 \; E_{32} = X_3 - X_2 \tag{13}$$

The convergence condition of the numerical solution can then be observed by dividing the constants calculated in Equation (13): $R = E_{21}/E_{32}$. The mathematical meaning of the ratio R is described in Equation (14) and the details can be found in the related reference [29].

$$\begin{array}{c} -1 < R < 0 \text{ Oscillatory convergence} \\ 0 < R < 1 \text{ Monotonic convergence} \\ R < -1 \text{ Oscillatory divergence} \\ 1 < R \text{ Monotonic divergence} \end{array} \tag{14}$$

Three different grid qualities have been tested at the 0.45 Froude number to calculate the total resistance of the Stavanger Demonstrator at LC1 (Table 3).

Table 3. Resistance values at Fr = 0.45 by different grid qualities.

Grid Quality	**Total Cell Numbers**	C_T **(1 × 10³)**
Fine	8.36×10^6	6.345
Medium	4.82×10^6	6.512
Coarse	2.92×10^6	6.648

The total resistance values gradually decrease as the grid resolution becomes dense. Although it seems to have monotonic convergence characteristics, it should be noted that the convergence condition R is the ultimate indication of this behaviour (Table 4). The highest uncertainty is calculated as 10% by the FS method, which is more up-to-date and already known as a conservative method in the literature [12,30].

Table 4. Uncertainty assessment results.

	GCI	CF	FS
r_{21}	1.20	1.20	1.20
r_{32}	1.18	1.18	1.18
R	0.73	0.73	0.73
P_{th}	2.00	2.00	2.00
P_{RE}	2.19	2.19	2.19
$C_{T\text{-}Ext}$ (1×10^3)	6.213	6.213	6.213
SF	1.25	1.12	1.10
Δ (%)	3.90	3.86	10.01

3.1.2. Validation

The validation of the numerical results has been achieved by comparing the resistance components and dynamic behaviour of the Stavanger Demonstrator with the experiments. Since the hull form and the tested conditions are symmetric about the centreline of the catamaran, the rest of the simulations, including double-body analyses have been carried out by adopting the half-body technique to save computational power and time. The experiments of the fully appended catamaran were conducted in HSVA towing tank for the T′ = 0.045 loading condition (LC1) [11].

Validation has been done through the resistance components and the dynamic behaviour of the Stavanger Demonstrator at model scale. The catamaran has been subjected to resistance simulations at a 0.24–0.63 Froude range with all appendages (rudder, shaft, and strut).

A significant match has been reported between the CFD results and those obtained by the experiments. The numerical total resistance values of the catamaran reflect the trend of the experimental results with good accuracy (Figure 6). The maximum relative error between the total resistance values is −3.32% at 0.39 Froude number (Table 5). The negative percentage means that the CFD gives higher values than the corresponding experimental results (see the formulation in Section 2.3). The experimental frictional resistance values have been calculated by using the ITTC′57 formula. The average relative percentage error between $C_{F\text{-}CFD}$ and $C_{F\text{-}Exp}$ has been calculated as approximately 5.6%, which can be concluded with the proven-functionality of the turbulence model (Table 5). The dynamic behaviour of the vessel, especially the extremum locations through the increasing advance speed, has been well-predicted with the CFD method (Figure 7).

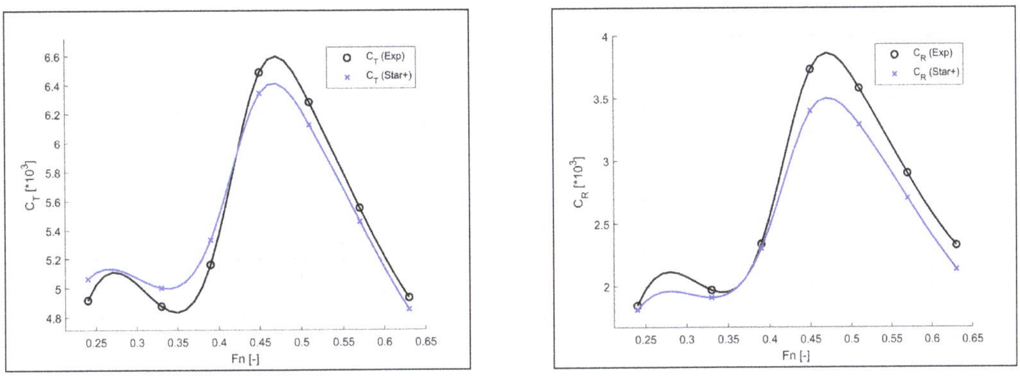

Figure 6. Numerical total (**left**) and residual (**right**) resistance values compared to the experiments.

Table 5. Resistance comparison for LC1.

Fr	$C_{T\text{-}CFD}$ (1 × 10³)	$C_{T\text{-}Exp}$ (1 × 10³)	e_{CT} (%)	$C_{F\text{-}CFD}$ (1 × 10³)	$C_{F\text{-}Exp}$ (1 × 10³)	e_{CF} (%)
0.24	5.064	4.917	−2.98%	3.248	3.071	−5.75%
0.33	5.002	4.876	−2.59%	3.089	2.906	−6.31%
0.39	5.332	5.161	−3.32%	3.028	2.825	−7.21%
0.45	6.345	6.488	2.21%	2.947	2.758	−6.87%
0.51	6.125	6.281	2.48%	2.837	2.701	−5.01%
0.57	5.461	5.552	1.65%	2.756	2.652	−3.91%
0.63	4.854	4.933	1.60%	2.720	2.610	−4.24%

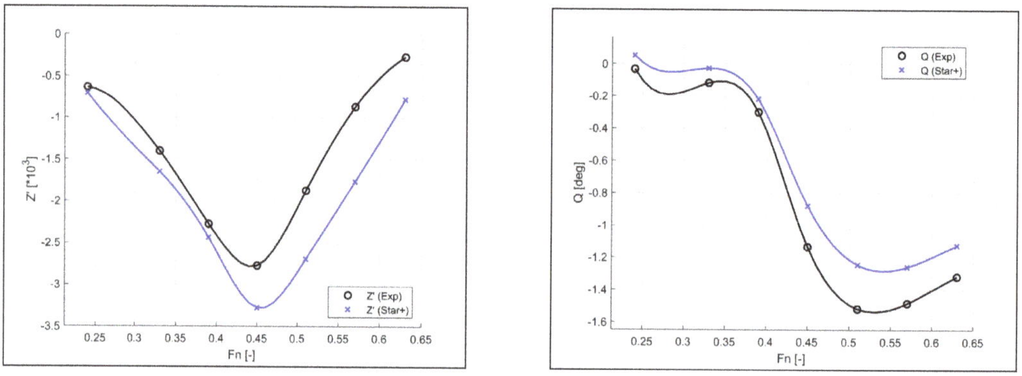

Figure 7. Numerical sinkage (**left**) and trim (**right**) values compared to the experiments.

The relative percentage between the numerical and experimental motion amplitudes are relatively high, yet these values are small enough to cause such high percentages. The difference in motion amplitudes can be the result of measuring points during the experiments, mounting conditions and, of course, the uncertainties in measurement systems and in computational analysis. The numerical uncertainty due to the selected grid system is reported in the previous section, yet no data was provided by the experimental facility regarding uncertainties in measurement systems. It can be noticed that the maximum sinkage values in both methods are obtained between 0.45–0.51 Froude numbers (Table 6). After this point, the catamaran enters the planning regime and the sinkage values decrease gradually (Figure 7).

Table 6. Full scale sinkage and trim comparison for LC1.

Fr	Z'_{CFD} (1 × 10³)	Q_{CFD} (deg)	Z'_{Exp} (1 × 10³)	Q_{Exp} (Deg)
0.24	−0.710	0.052	−0.635	−0.033
0.33	−1.649	−0.026	−1.404	−0.117
0.39	−2.436	−0.218	−2.273	−0.300
0.45	−3.276	−0.879	−2.774	−1.133
0.51	−2.694	−1.245	−1.872	−1.517
0.57	−1.763	−1.258	−0.869	−1.483
0.63	−0.779	−1.124	−0.267	−1.317

3.2. Deep and Shallow Water Resistance Analysis Results

The Stavanger Demonstrator has been analysed in deep and shallow water conditions at LC2 to observe the shallow water effects on the resistance and motion characteristics of the vessel. Deep water analyses have been performed in a 1.69 L depth body of water by assigning the bottom of the computational domain as the velocity inlet, which provides an identical numerical set-up to the actual physical conditions (see Section 2.2.2). In a real environment where the fluid is assumed to be still, with no other external effects such as wind and current, the ship travels at a certain speed. To imitate the physical conditions, the ship is captive (free to sink and trim) and the free stream passes through the surface of the vessel in the virtual environment (Figure 8, left caption). Seabed assumptions are extremely important when it comes to having an interaction with the floating body. A ship travelling in a finite depth water will be affected by the presence of the seabed to a certain level. When there is no current or wind, both the current and the seabed will have no speed but relative speed against the advancing vessel, i.e., the current and the seabed will have zero absolute speed. To have an identical numerical solution for the shallow water conditions, the vessel is kept captive (free to sink and trim) and the fluids enter the computational domain on the inlet faces and flow at the operational speed of the vessel; additionally, the seabed boundary condition is set to move in the longitudinal direction at the operational speed of the vessel by having no-slip surface properties (Figure 8, right caption).

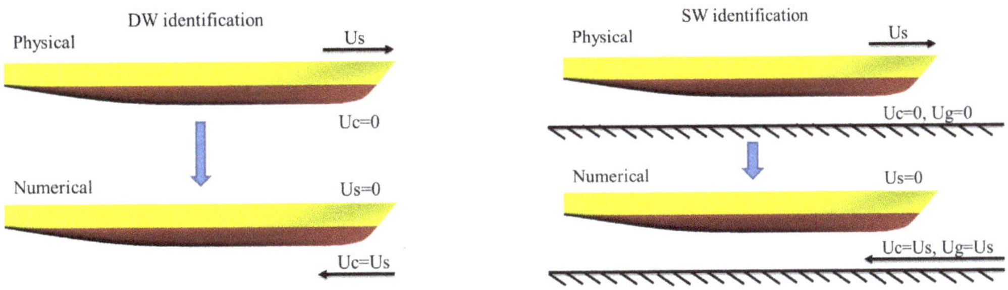

Figure 8. Illustrations for the identical numerical set-up for deep water (**left**) and shallow water (**right**).

The water depth for the shallow water analysis was set to 2.15 m to investigate the Stavanger Demonstrator's dynamic behaviour in potential operational conditions on the Thames River in London. The ratio between the water depth and the draught (H/T) becomes 1.79, which is a strong indication of the interaction between the free surface and the seabed, as practiced in many studies in the literature. The 2-DoF deep water resistance simulations have been performed at all speeds for comparison purpose. Resistance characteristics at LC2 have been reported to be in line with the LC1 validation case. Similarities between the sinkage and trim behaviours can be seen (Figures 7 and 9). The most striking result from the 2-DoF shallow water resistance analyses is the dramatic increase in the residual resistance between the 0.20–0.39 Froude numbers compared to the deep-water results (Figure 10, bottom caption). Interestingly, even though the advance speed increases, the total resistance coefficients remain around 4.6×10^{-3} in shallow water and start to increase after the 0.39 Froude number (Figure 10, top-left caption). The intersection point of two resistance curves has been plotted between 0.39–0.45 Froude numbers at where the shallow water effects turn in favour of the catamaran's resistance performance (Tables 7 and 8). The $Fr = 0.24$ has been identified as the critical speed where the shallow water effects reach the peak point. It should be noted that the highest impact has been detected due to pressure-based resistance components as no noticeable difference has been detected in the frictional resistance component (Figure 10, top-right caption). This was the authors' primary source of motivation in conducting this work to investigate the pressure-based resistance components by carrying on one step further and utilizing the double-body approach to investigate the wave-making resistance component.

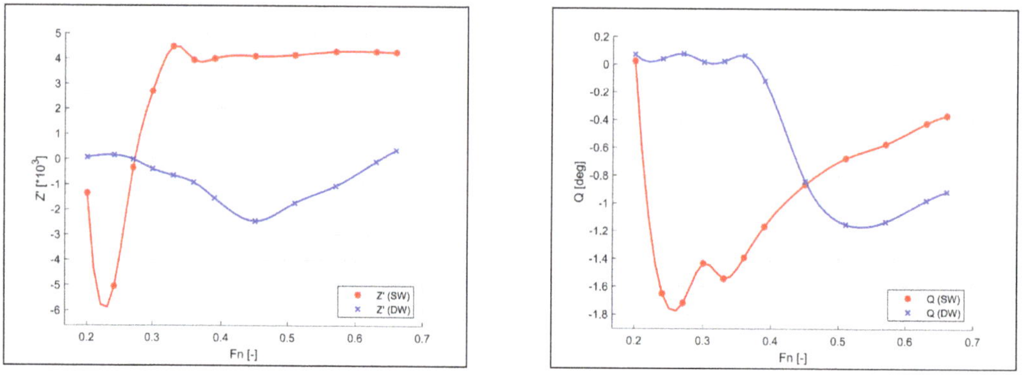

Figure 9. Sinkage (**left**) and trim (**right**) results of the catamaran in deep and shallow water.

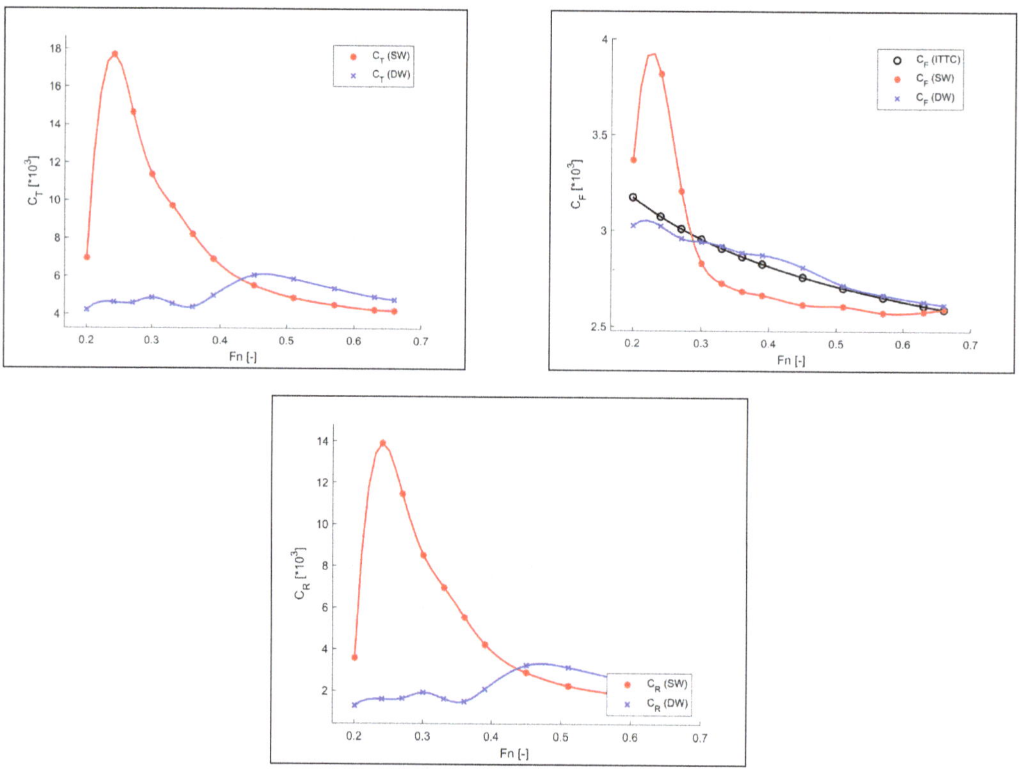

Figure 10. Total (**top-left**), frictional (**top-right**), and residual (**bottom**) resistance results of the catamaran in deep and shallow water at LC2.

Table 7. Resistance results of the catamaran in deep and shallow water at LC2.

Fn	$C_{T\text{-}DW}$ (1 × 10³)	$C_{F\text{-}DW}$ (1 × 10³)	$C_{T\text{-}SW}$ (1 × 10³)	$C_{F\text{-}SW}$ (1 × 10³)	e_{CT} (%)	e_{CF} (%)
0.20	4.233	3.026	6.952	3.369	−64.2%	−11.3%
0.24	4.634	3.024	17.720	3.817	−282.4	−26.2
0.27	4.601	2.959	14.657	3.206	−218.6	−8.3
0.30	4.875	2.943	11.362	2.830	−133.1	3.8
0.33	4.545	2.920	9.707	2.726	−113.6	6.6
0.36	4.377	2.886	8.212	2.684	−87.6	7.0
0.39	4.973	2.873	6.900	2.664	−38.8	7.3
0.45	6.066	2.810	5.515	2.615	9.1	7.0
0.51	5.872	2.717	4.864	2.607	17.2	4.1
0.57	5.368	2.665	4.492	2.574	16.3	3.4
0.63	4.924	2.630	4.244	2.580	13.8	1.9
0.66	4.773	2.614	4.189	2.592	12.2	0.8

Table 8. Sinkage and trim values of the catamaran in deep and shallow water.

Fn	Z'_{DW} (1 × 10³)	Z'_{SW} (1 × 10³)	Q_{DW} (Deg)	Q_{SW} (Deg)
0.20	0.0788	−1.3383	0.0709	0.0235
0.24	0.1690	−5.0447	0.0390	−1.6505
0.27	−0.0017	−0.3331	0.0759	−1.7179
0.30	−0.3753	2.6961	0.0177	−1.4326
0.33	−0.6346	4.4842	0.0222	−1.5421
0.36	−0.9065	3.9456	0.0637	−1.3896
0.39	−1.5303	4.0002	−0.1157	−1.1658
0.45	−2.4447	4.1026	−0.8402	−0.8632
0.51	−1.7202	4.1489	−1.1490	−0.6744
0.57	−1.0439	4.2890	−1.1295	−0.5704
0.63	−0.0822	4.2859	−0.9756	−0.4221
0.66	0.3753	4.2635	−0.9136	−0.3666

The CFD results have been further examined via visual media. The shock wave has been detected in the fore part of the catamaran at the 0.24 Froude number (Figure 11) and at other speeds up to the 0.45 Froude number. The wave contours have been formed more linear than deep water. The extremum points of the wave crest and wave trough ((ζ/L)E3) at model scale have been noted as 9.37 and −11.24 in deep water while 28.12 and −16.87 in shallow water, respectively. The phase distributions on the demihull at $Fr = 0.24$ shows that the numerical ventilation has not occurred in CFD simulations (Figure 12: blue represents the air phase on the catamaran hull and the water anywhere else). Squat effects in shallow water are strongly related to the velocity field beneath the hull. In the aft region, a noticeable increase in flow velocity has been observed (Figure 13). The decrease in flow speed in the fore region can be considered as a potential cause of the shock wave incidence.

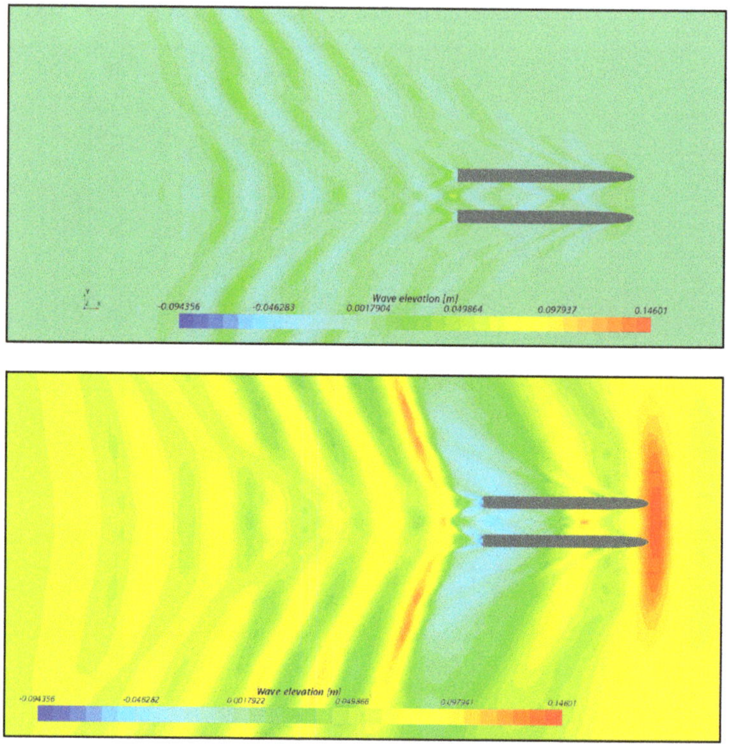

Figure 11. Free surface waves at Fr = 0.24 in deep (**top**) and shallow (**bottom**) water.

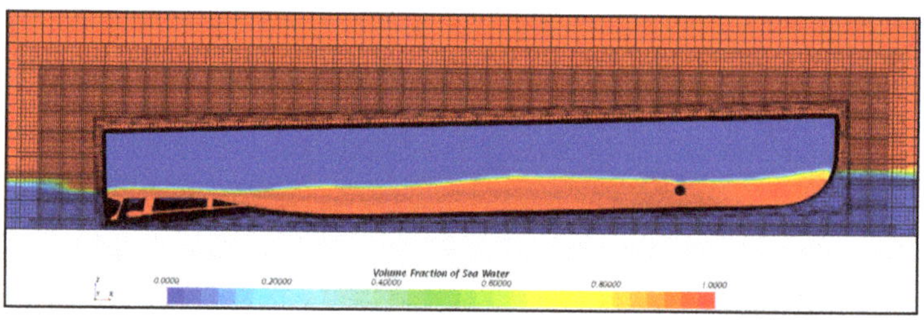

Figure 12. Phase distributions at Fr = 0.24 in shallow water.

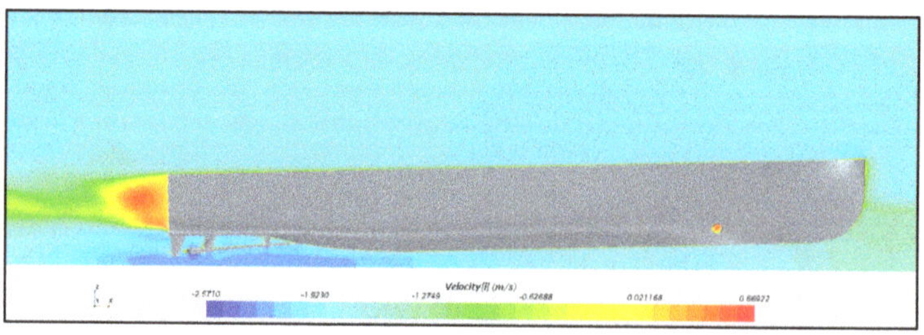

Figure 13. Velocity contours on the centre plane of the demihull at Fr = 0.24 in shallow water.

3.3. Double-Body Analysis Results

The form factors of the catamaran have been predicted by utilizing the double-body approach in CFD simulations. The hull form has been cut from the free surface level and subjected to a series of single-phase CFD simulations at all speeds in both deep and shallow water conditions. The form factors obtained in shallow water were found to be higher than those calculated in deep water for all speeds (Figure 14). The non-linear behaviour of the form factor values in deep water can be correlated by the steep change of the residual resistance component in 2-DoF simulations between 0.39–0.57 Froude numbers (see Figure 10, bottom caption). No matter which operational condition is chosen, form factors tend to have higher values with increasing speeds, which is in accordance with the literature findings. The maximum form factor has been calculated as 1.41 and 1.39 for the shallow and deep-water conditions at maximum speed, respectively.

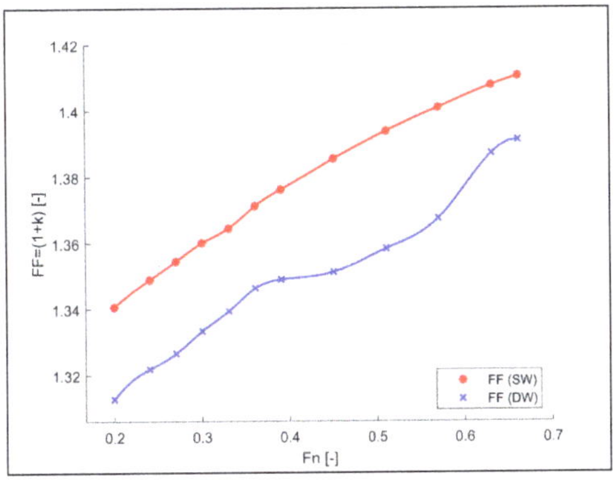

Figure 14. Form factors in deep and shallow water conditions.

The velocity distributions on the centre plane of the demihull at Fr −0.24 are presented in Figure 15 for the deep and shallow waters. Here, the submerged part of the catamaran is mirrored to obtain the double-body form, which has a similar 3-D shape to submarines. It can be noticed that the flow speed is higher when the fluid particles pass through the mid-body in the shallow water case (Figure 15, bottom caption).

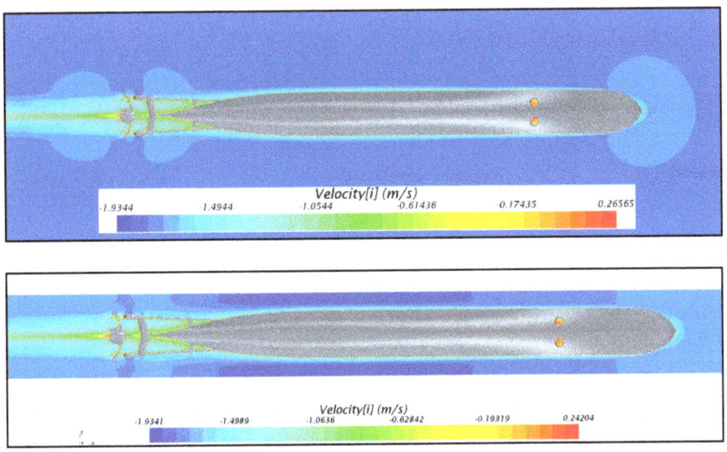

Figure 15. Velocity contours on the centre plane of the demihull at Fr = 0.24 in deep (**top**) and shallow (**bottom**) waters.

3.4. Wave-Making Resistance Calculations

The 2-DoF CFD simulations have provided the frictional (C_F) and residual resistance (C_R) components of the catamaran in deep and shallow waters. The output of the double-body analyses, form factors, have been used to decompose the pressure-based resistance into viscous pressure (C_{VP}) and gravitational waves (C_W) by following the 1978 ITTC power prediction method. The wave-making resistance coefficients are presented in Figure 16 and tabulated in Table 9. Nondimensional analysis of the wave-making resistance shows that the interaction between the hull and seabed becomes extremely important, even at moderate and relatively low speeds. The two peak points at Fr = 0.30 and Fr = 0.45 in deep water is no longer observed in shallow water results. Instead, gradually decreasing wave-making resistance values can be noticed with increasing speeds. Under the conditions defined in this study, it can be concluded that the shallow water effects are significantly dominant between 0.24–0.39. After passing the transition Froude number, which has been reported to be between 0.39–0.45, shallow water effects start to work in favour of the resistance performance of the catamaran.

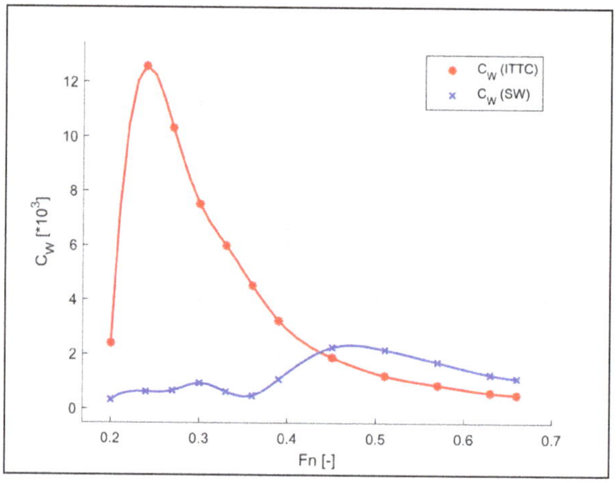

Figure 16. Wave-making resistance coefficients of the catamaran.

Table 9. Wave-making resistance values at LC2.

Fn	FF_{SW}	$C_{W\text{-}SW}$	FF_{DW}	$C_{W\text{-}DW}$
0.20	1.3406	2.41×10^3	1.3128	3.40×10^4
0.24	1.3489	1.26×10^2	1.3219	6.36×10^4
0.27	1.3545	1.03×10^2	1.3266	6.75×10^4
0.30	1.3601	7.51×10^3	1.3334	9.51×10^4
0.33	1.3644	5.99×10^3	1.3393	6.35×10^4
0.36	1.3712	4.53×10^3	1.3463	4.92×10^4
0.39	1.3762	3.23×10^3	1.3490	1.10×10^3
0.45	1.3853	1.89×10^3	1.3512	2.27×10^3
0.51	1.3936	1.23×10^3	1.3583	2.18×10^3
0.57	1.4008	8.86×10^4	1.3673	1.72×10^3
0.63	1.4076	6.12×10^4	1.3870	1.28×10^3
0.66	1.4104	5.33×10^4	1.3910	1.14×10^3

4. Conclusions

This study was conducted to look for insight into the resistance components, especially the wave-making resistance of a zero-emission fast catamaran ferry. The subject vessel, the Stavanger Demonstrator catamaran, was built and named MS Medstraum in Norway as a demonstrator of an ongoing H2020 project TrAM.

The catamaran has been analysed under free surface effects in deep and shallow water conditions to obtain resistance values and to observe the dynamic behaviour of the vessel. The 2-DoF CFD simulation results show a significantly good match with the experiments. It has been noted that the highest impact of the shallow water effects is on the residuary resistance component as no noticeable difference has been detected in the frictional resistance. The transition speed where the shallow water effects start to work in favour of the resistance performance of the catamaran has been recorded between the 0.39–0.45 Froude numbers. The critical speed, where the total resistance reaches the maximum point due to shallow water effects, has been recorded at $Fr = 0.24$. The catamaran travels at negligible trim angles up to $Fr = 0.39$; however, from this point on, the trim values increase exponentially towards the aft. On the other hand, significant trim-to-aft conditions occur in shallow water. The vertical motion of the catamaran is completely in the opposite direction in deep and shallow waters. Shock waves in the fore part of the catamaran have been captured at moderate and relatively small advance speeds in shallow water.

The study is followed by utilizing the double-body approach at all speeds in both deep and shallow water conditions. The form factors of the catamaran have been calculated for each speed and used to decompose the residual resistance (C_R) components into viscous pressure (C_{VP}) and wave-making resistance (C_W). Regardless of the operational condition, the form factors have been found to have higher values with increasing speeds, which complies with the literature. The analysis of the wave-making resistance shows that the interaction between the hull and seabed becomes extremely important even at moderate and relatively low speeds. In fact, the total resistance values of shallow water cases become lower than those obtained in deep water because of the significant reduction in wave-making resistance due to the interaction between the gravitational waves and the riverbed. This statement is not valid for low Froude numbers as the resistance of the catamaran increases dramatically at lower speeds.

The hydrodynamic performance of the Stavanger Demonstrator will be further investigated for changing depth, different loading conditions, and demihull clearance. The propeller effects will be taken into consideration in future studies.

Author Contributions: Conceptualization, S.D. and E.B.; methodology, S.D.; software, S.D. and M.Z.A.; validation, S.D., M.Z.A., X.X. and A.N.; formal analysis, M.Z.A. and X.X.; investigation, M.Z.A. and A.N.; resources, X.X.; data curation, A.N.; writing—original draft preparation, S.D.; writing—review and editing, E.B.; visualization, S.D. and A.N.; supervision, E.B.; project administration, E.B.; funding acquisition, E.B. All authors have read and agreed to the published version of the manuscript.

Funding: This work was funded by the H2020 European Union project "TrAM—Transport: Advanced and Modular" under the contract number 769303.

Institutional Review Board Statement: Not applicable.

Informed Consent Statement: Not applicable.

Data Availability Statement: Not applicable.

Acknowledgments: The authors affiliated with MSRC greatly acknowledge the funding from DNV and Royal Caribbean Group for the MSRC's establishment and operation. The opinions expressed herein are those of the authors and do not reflect the views of DNV and Royal Caribbean Group. CFD simulations run by the University of Strathclyde were performed on the ARCHIE-WeSt High Performance Computer (www.archie-west.ac.uk (accessed on 8 November 2022)).

Conflicts of Interest: The authors declare no conflict of interest.

Nomenclature

α	Constant
B, B_M	Beam moulded (m)
C_F, C_{F0}	Frictional drag coefficient
C_T	Total drag coefficient
$C_{T\text{-}CFD}$	Total drag coefficient by CFD
$C_{T\text{-}Exp}$	Total drag coefficient by experiments
$C_{T\text{-}Ext}$	Extrapolated total drag coefficient
ε	Relative Error (%)
λ	Model scale ratio
L, L_{PP}	Length between perpendiculars (m)
L_{WL}	Length of waterline (m)
υ	Kinematic viscosity (N s m^{-2})
∇	Displacement volume (m^3)
r	Density of water (kg m^{-3})
P	Pressure (N m^{-2})
Q	Pitch Amplitude (deg)
R	Convergence Condition
S	Wetted hull surface area (m^2)
T	Draught at Midship (m)
u, v, w	Fluid velocity components
$y+$	Dimensionless wall distance
Z	Heave Amplitude (m)
ζ	Wave height (m)
$(1 + k)$	Form Factor
BCs	Boundary Conditions
CF	Correction Factor
CFD	Computational Fluid Dynamics
DC372	Delft Catamaran 372
DoF	Degree of Freedom
EFD	Experimental Fluid Dynamics
Fr, Fn	Froude Number
FS	Factors of Safety
GCI	Grid Convergence Index
H2020	Horizon 2020
HSVA	Hamburg Ship Model Basin
ITTC	International Towing Tank Conference
KB	Vertical Centre of Buoyancy
LC	Loading Condition
LCB	Longitudinal Centre of Buoyancy
RANS	Reynolds-averaged Navier–Stokes
TrAM	Transport: Advanced and Modular
WAM-V	Wave adaptive modular vessel
VoF	Volume of Fluid

References

1. Molland, A.; Wellicome, J.; Couser, P. *Resistance Experiments on a Systematic Series of High Speed Displacement Catamaran Forms: Variation of Length-Displacement Ratio and Breadth-Draught Ratio*; Ship Science Report No. 71; University of Southampton: Iskandar Puteri, Malaysia, 1994.
2. Insel, M.; Doctors, L.J. Wave pattern prediction of monohulls and catamarans in a shallow-water canal by linearised theory. In Proceedings of the 12th Australian Fluid Mechanics Conference, Sydney, NSW, Australia, December 1995; pp. 259–262. Available online: https://www.afms.org.au/proceedings/12/Insel_and_Doctors_1995.pdf (accessed on 22 November 2022).
3. Veer, R.V. *Experimental Results of Motions, Hydrodynamic Coefficients and Wave Loads on the 372 Catamaran Model*; Technical Report 1129; Delft University of Technology: Delft, The Netherlands, 1998.

4. Castiglione, T.; He, W.; Stern, F.; Bova, S. Effects of shallow water on catamaran interference. In Proceedings of the 11th International Conference on FAST2011, Honolulu, HI, USA; 26–29 September 2011.
5. Milanov, E.; Chotukova, V.; Stern, F. System Based Simulation of Delft372 Catamaran Maneuvering Characteristics as Function of Water Depth and Approach Speed. In Proceedings of the 29th Symposium on Naval Hydrodynamics, Gothenburg, Sweden, 26–31 August 2012.
6. Broglia, R.; Jacob, B.; Zaghi, S.; Stern, F.; Olivieri, A. Experimental investigation of interference effects for high-speed catamarans. *Ocean. Eng.* **2014**, *76*, 75–85. [CrossRef]
7. Begovic, E. Wave Interference Prediction of a Trimaran Using Form Factor. In *HSMV 2020: Proceedings of the 12th Symposium on High Speed Marine Vehicles*; IOS Press: Amsterdam, The Netherlands; Volume 5, p. 99.
8. Bašić, J.; Blagojević, B.; Andrun, M. Improved estimation of ship wave-making resistance. *Ocean. Eng.* **2020**, *200*, 107079. [CrossRef]
9. Papanikolaou, A.; Xing-Kaeding, Y.; Strobel, J.; Kanellopoulou, A.; Zaraphonitis, G.; Tolo, E. Numerical and Experimental Optimization Study on a Fast, Zero Emission Catamaran. *JMSE* **2020**, *8*, 657. [CrossRef]
10. Boulougouris, E.; Papanikolaou, A.; Dahle, M.; Tolo, E.; Xing-Kaeding, Y.; Jürgenhake, C.; Seidenberg, T.; Sachs, C.; Brown, C.; Jenset, F. Implementation of Zero Emission Fast Shortsea Shipping. In Proceedings of the SNAME Maritime Convention; Online, 27–29 October 2021.
11. Shi, G. Numerical investigation of the resistance of a zero-emission full-scale fast catamaran in shallow water. *J. Mar. Sci. Eng.* **2021**, *9*, 563. [CrossRef]
12. Duman, S.; Bal, S. Turn and zigzag manoeuvres of Delft catamaran 372 using CFD-based system simulation method. *Ocean. Eng.* **2022**, *264*, 112265. [CrossRef]
13. Ulgen, K.; Dhanak, M.R. Hydrodynamic Performance of a Catamaran in Shallow Waters. *J. Mar. Sci. Eng.* **2022**, *10*, 1169. [CrossRef]
14. TrAM. Transport: Advanced and Modular (TrAM) Horizon2020 Project. 2022. Available online: https://tramproject.eu/ (accessed on 22 November 2022).
15. Simcenter STAR-CCM+. 2022. Available online: https://www.plm.automation.siemens.com/global/en/products/simcenter/STAR-CCM.html (accessed on 22 November 2022).
16. Hirt, C.W.; Nichols, B.D. Volume of fluid (VOF) method for the dynamics of free boundaries. *J. Comput. Phys.* **1981**, *39*, 201–225. [CrossRef]
17. ITTC. Practical Guidelines for Ship CFD Applications. In Proceedings of the 26th ITTC, Rio de Janerio, Brazil. 2011. Available online: https://ittc.info/media/1357/75-03-02-03.pdf (accessed on 22 November 2022).
18. Benek, J.A.; Steger, J.L.; Dougherty, F.C.; Buning, P.G. Chimera: A Grid-Embedding Technique. AEDC-TR-85-64; Arnold Engineering Development Center: Fort Belvoir, VA, USA, 1986.
19. Begovic, E. Pure Yaw Simulations of Fast Delft Catamaran 372 in Deep Water. In *HSMV 2020: Proceedings of the 12th Symposium on High Speed Marine Vehicles*; IOS Press: Amsterdam, The Netherlands; Volume 5, p. 109.
20. Duman, S.; Bal, S. Prediction of maneuvering coefficients of Delft catamaran 372 hull form. In *Sustainable Development and Innovations in Marine Technologies*, 1st ed.; Georgiev, P., Soares, C.G., Eds.; CRC Press: Boca Raton, FL, USA, 2019; pp. 167–174. [CrossRef]
21. Prohaska, C. A simple method for the evaluation of the form factor and the low speed wave resistance. In *Hydro-and Aerodynamics Laboratory, Lyngby, Denmark, Hydrodynamics Section, Proceedings of the 11th International Towing Tank Conference, ITTC'66*; Resistance Committee: Tokyo, Japan, 1966; pp. 65–66.
22. Hughes, G. Friction and form resistance in turbulent flow, and a proposed formulation for use in model and ship correlation. In National Physical Laboratory, NPL, Ship Division, Presented at the Institution of Naval Architects, Paper no. 7, RINA Transactions, London; April 1954. Available online: https://repository.tudelft.nl/islandora/object/uuid%3A9a642c53-27f0-45fe-a5e4-5e1296b62af1 (accessed on 22 November 2022).
23. ITTC. Report of 8th ITTC Resistance Committee. In Proceedings of the 8th International Towing Tank Conference, Madrid, Spain, September 1957. Available online: https://repository.tudelft.nl/islandora/object/uuid%3A4aa44007-881c-441c-82dd-70f45cc70ff2 (accessed on 22 November 2022).
24. Roache, P.J. Perspective: A Method for Uniform Reporting of Grid Refinement Studies. *J. Fluids Eng.* **1994**, *116*, 405. [CrossRef]
25. Stern, F.; Wilson, R.V.; Coleman, H.W.; Paterson, E.G. Comprehensive Approach to Verification and Validation of CFD Simulations—Part 1: Methodology and Procedures. *J. Fluids Eng.* **2001**, *123*, 793. [CrossRef]
26. Celik, I.B.; Ghia, U.; Roache, P.J.; Freitas, C.J. Procedure for Estimation and Reporting of Uncertainty Due to Discretization in CFD Applications. *J. Fluids Eng.* **2008**, *130*, 078001. [CrossRef]
27. Xing, T.; Stern, F. Factors of Safety for Richardson Extrapolation. *J. Fluids Eng.* **2010**, *132*, 061403. [CrossRef]
28. ITTC. Resistance, Uncertainty Analysis, Example for Resistance Test. In Proceedings of the International Towing Tank Conference. 2002. Available online: https://ittc.info/media/1818/75-02-02-02.pdf (accessed on 22 November 2022).

29. Phillips, T.S.; Roy, C.J. Richardson Extrapolation-Based Discretization Uncertainty Estimation for Computational Fluid Dynamics. *J. Fluids Eng.* **2014**, *136*, 121401. [CrossRef]
30. Korkmaz, K.B. CFD based form factor determination method. *Ocean. Eng.* **2021**, *220*, 108451. [CrossRef]

Journal of
Marine Science and Engineering

MDPI

Article

Research on the Karhunen–Loève Transform Method and Its Application to Hull Form Optimization

Haichao Chang [1,2], Chengjun Wang [1,2,*], Zuyuan Liu [1,2,*], Baiwei Feng [1,2], Chengsheng Zhan [1,2] and Xide Cheng [1,2]

1 Key Laboratory of High Performance Ship Technology, Wuhan University of Technology, Ministry of Education, Wuhan 430063, China
2 School of Naval Architecture, Ocean and Energy Power Engineering, Wuhan University of Technology, Wuhan 430063, China
* Correspondence: 15924090997@163.com (C.W.); wtulzy@whut.edu.cn (Z.L.)

Abstract: Hull form optimization becomes prone to the curse of dimensionality as the number of design variables increases. The traditional sensitivity analysis method requires massive computational fluid dynamics (CFD) computations and analyzing the effects of all variables on the output; thus, it is extremely time-consuming. Considering this, the development of a rapid and effective dimensionality reduction method is particularly important. The Karhunen–Loève (K–L) transform method projects data from a high-dimensional space onto a low-dimensional space in the direction of the eigenvectors corresponding to large-variance eigenvalues. It extracts the principal components that represent the hull offset information to represent the hull geometric characteristics by analyzing the relationship between the variables in the sample offset matrix. The geometric information matrices of new hull forms can be rapidly reconstructed from the principal components. Compared with direct optimization methods, fewer variables are used to control the deformation of the hull form from the perspective of geometric deformation, avoid a large number of CFD calculations, and improve the efficiency of optimization. This study examined the relevant K–L matrix solution methods and the corresponding hull form reconstruction methods and proposed eigenvalue-based hull form reconstruction equations. The K–L transform method was combined with a previously developed multidisciplinary platform for a comprehensive optimization of ship hydrodynamic performance for hull form optimization, and its effectiveness was verified by using it to optimize DTMB 5415. The results showed that the K–L transform-based dimensionality reduction method significantly reduces the time consumption of optimization while maintaining an acceptable optimization performance.

Keywords: Karhunen–Loève transform; hull form optimization; dimensionality reduction reconstruction method

Citation: Chang, H.; Wang, C.; Liu, Z.; Feng, B.; Zhan, C.; Cheng, X. Research on the Karhunen–Loève Transform Method and Its Application to Hull Form Optimization. *J. Mar. Sci. Eng.* **2023**, *11*, 230. https://doi.org/10.3390/jmse11010230

Academic Editors: Carlos Guedes Soares and Serge Sutulo

Received: 1 November 2022
Revised: 8 January 2023
Accepted: 8 January 2023
Published: 16 January 2023

1. Introduction

With the rapid development of computer technology, computational fluid dynamics (CFD) has been significantly advancing. Ship designers have introduced optimization technology into ship design and have combined it with CFD technology to develop a CFD-based method for automatic hull form optimization. The CFD method substantially reduces the time and economic costs compared with a ship model test. However, for complex optimization problems or scenarios with many design variables, it can have a very high computation cost and even lead to the curse of dimensionality. Therefore, effectively reducing the number of design variables and the dimensionality of the design space is a major problem in hull form design optimization.

Currently, the above problem is solved mainly using the sensitivity analysis (SA) method. The SA method can qualitatively or quantitatively assess the effects of the input variables on the output objective function. Cheng et al. [1] proposed an uncertainty and sensitivity analysis framework for ship motion data, in which the sensitivity and uncertainty of the samples or weights generated by an artificial-neural-network-based surrogate model are analyzed. Hu et al. [2] adjusted an airfoil curve using control points and a free-form deformation technique and reduced the dimensionality of control variables by SA. Guan et al. [3] used an SA procedure to estimate the effects of the values assigned to the key calculation parameters of a fully parametric hull form optimization algorithm on the calculation process and results. Hamed [4] numerically optimized a trimaran hull form using an optimization platform, with the total resistance and the propeller intake flow defined as the study objectives. Moreover, information about highly significant regions of a hull was obtained by surface SA based on an adjoint solver. Jung [5] introduced examples of SA of the effects on the added resistance and speed loss in waves caused by the variations in ship dimensions. SA has been proven to be a feasible method for dimensionality reduction, and variance-based SA has been widely used to determine variables that are unimportant for this purpose. Therefore, Liu et al. [6] implemented variance-based SA in a hull form optimization model and identified a variable that did not affect dimensionality reduction. Guan et al. [7] developed an automatic calculation framework for the resistance of a small-waterplane-area twin hull at different speeds, for which a fully parametric model was established to realize automatic geometric transformation. They also performed an SA of all parameters to the resistance and a correlation analysis between the design variables and the resistance, including the calculation and examination of the correlation coefficients. Jeon et al. [8] identified important hull form parameters and estimated their optimal values that satisfied the required intercept time of mission competence and the turning rate of a specific underwater vehicle based on SA. Coppedè et al. [9] proposed a computational framework for the hydrodynamic shape optimization of complex hull forms and applied it to improve the calm-water performance of the KRISO Container Ship. They also performed a preliminary SA of the mesh size to reduce the computational burden of the CFD solver. Fu et al. [10] proposed a multiobjective optimization method for autonomous underwater gliders based on the fast elitist nondominated sorting genetic algorithm (NSGA-II) and performed an SA of the variables. In hull form optimization, models of the input variables and performance indicators are established to compare the effects of the variables on the ship performance. This enables identification of the important variables and elimination of the variables with the minimum effects on the performance, to realize dimensionality reduction.

SA requires calculation of the effects of all variables on the output and, thus, has a high computational cost. Different from SA, the Karhunen–Loève (K–L) transform reduces data dimensionality by projecting data in the direction of the eigenvectors corresponding to large eigenvalues based on an analysis of the data matrix. K–L transform–based dimensionality reduction needs to consider only the relationships between the variables in the sample data matrix and does not require calculating the effects of the variables on the objective function.

Considering that K–L and directional seismic wave methods can extract target signals and suppress random noise, Yue et al. [11] used these methods to solve the problem of difficulty in the separation of weak effective signals. Fan et al. [12] developed a spatiotemporal model based on the K–L decomposition, a multilayer perceptron, and a long short-term memory network for estimating temperature distributions by a three-step procedure. Ahuja et al. [13] suggested that a large-sized discrete Fourier transform of a colored random process approximates the behavior of the K–L transform in terms of achieving approximate decorrelation between frequency-domain terms. Trudu et al. [14] proposed a new method for computing the K–L transform for application to complex voltage data for the detection and noise reduction of astronomical signals. Siripatana et al. [15] applied a generalized K–L expansion that incorporated the construction of a reference basis of spatial modes and a coordinate transformation to the prior field. Reed and Roberts [16] applied the K–L transform to a 44-group version of the 2-D C5G7 benchmark problem. Allaix and Carbone [17] proposed a method for the discretization of random fields by combining a K–L series expansion and the finite element method. Azevedo et al. [18] studied the numerical approximation of a homogeneous Fredholm integral equation of the second kind associated with the K–L expansion of Gaussian random fields. Liu et al. [19] proposed to encode the third spectral information with an adaptive K–L transform and mathematically proved that interspectral correlations are preserved in two-dimensional randomly encoded spatial information. Ai [20] obtained the K–L expansion and distribution identity for a demeaned stationary Ornstein–Uhlenbeck process and applied them to the small-deviation asymptotic behavior of the L2 norm and Laplace transform of the process. Chowdhary and Najm [21] introduced a Bayesian K–L expansion procedure that allows developing a probabilistic model based on the principal components and accounts for the inaccuracies due to a limited sample size. Feng et al. [22] suggested that a K–L representation of a target random medium can be efficiently estimated by projecting reconstructed samples onto the K–L basis. Diez et al. [23–25] elaborated the mathematical principle behind the K–L transform and applied it to the optimization of the David Taylor Model Basin (DTMB) 5415 hull form for the first time. Using control points as the design variables based on a free-form deformation technique, they reduced the initial number of design variables from 29 to 4 using the K–L transform. Concurrently, 95% variability of the original design space was retained. In addition, the convergence speed was improved while achieving optimization results approximate to those without dimensionality reduction.

In hull form optimization, the dimensionality of the hull form geometric information matrix can be reduced using the K–L transform to extract several principal components to represent the geometric characteristics of the full shape. This allows for rapid reconstruction of the geometric information matrix of a new hull form and the control of hull form variability using a few variables from the perspective of geometric transformation. Consequently, complex CFD computations are avoided, and the time cost is reduced.

The remainder of this paper is structured as follows: Section 1 presents the mathematical principle, derivation, and geometric implications of the K–L transform and the detailed steps and procedure of using it to reduce dimensionality. Section 2 describes the methods for solving the eigenvalues and eigenvectors of the covariance matrices. Section 3 discusses the methods for obtaining the eigenvalues of a matrix. Section 4 proposes a K–L transform–based method for hull form reconstruction, which serves as the basis for applying the K–L transform to the geometric transformation of a hull form. Section 5 presents the multidisciplinary platform developed for a comprehensive optimization of ship hydrodynamic performance and its application to the optimization of the DTMB 5415 hull form. Section 6 summarizes this study and provides an outlook for future research.

2. K–L Transform Method

2.1. Basic Principle of K–L Transform

Harold [26–28] proposed the K–L transform, by which discrete signals are transformed into a series of uncorrelated coefficients. The K–L transform describes a sample using fewer features, thereby reducing the dimensionality of the feature space. Its most salient advantage is decorrelation, and it is the best transform in terms of the mean square error (MSE). Its mathematical definition is as follows.

For a discrete case, a uniform sampling is performed on $x(t)$ in interval $a \le t \le b$; thus, x can be expressed using the following vector form:

$$x = [x(t_1), x(t_2), \ldots, x(t_n)]^T \quad (1)$$

x is expanded using a system of orthonormal vectors $u_j, j = 1, 2, \ldots, \infty$ as follows:

$$x = \sum_{j=1}^{\infty} c_j u_j \quad (2)$$

x is estimated with a finite series d as follows:

$$\hat{x} = \sum_{j=1}^{d} c_j u_j \quad (3)$$

Thus, the MSE is

$$\varepsilon = E[(x-\hat{x})^T(x-\hat{x})] \tag{4}$$

because $u_i{}^T u_j = \begin{cases} 1, j=i \\ 0, j \neq i \end{cases}, c_j = u_j{}^T x$,

$$\varepsilon = E[\sum_{j=d+1}^{\infty} u_j{}^T x x^T u_j] = \sum_{j=d+1}^{\infty} u_j{}^T E[xx^T] u_j \tag{5}$$

Setting $\Psi = E[xx^T]$ yields

$$\varepsilon = \sum_{j=d+1}^{\infty} u_j{}^T \Psi u_j \tag{6}$$

When the orthogonality condition is satisfied, the following equations can be obtained using the Lagrange multiplier method:

$$g(u_j) = \sum_{j=d+1}^{\infty} u_j{}^T \Psi u_j - \sum_{j=d+1}^{\infty} \lambda_j [u_j{}^T u_j - 1] \tag{7}$$

$$\frac{d}{du_j} g(u_j) = 0,\ j = d+1, \ldots, \infty \tag{8}$$

$$(\Psi - \lambda_j I) u_j = 0,\ j = d+1, \ldots, \infty \tag{9}$$

Thus, it can be concluded that when x is expanded in the eigenvector direction of matrix Ψ, the MSE of truncation reaches the minimum value. When x is approximated with d number of u_j, the MSE of truncation is

$$\varepsilon = \sum_{j=d+1}^{\infty} \lambda_j \tag{10}$$

$$\beta = \frac{\sum_{j=1}^{d} \lambda_j}{\sum_{j=1}^{\infty} \lambda_j} \tag{11}$$

where λ_j represents the eigenvalues of matrix Ψ.

Thus, the MSE of truncation when x is expanded using the eigenvectors corresponding to the largest d number of eigenvalues of the matrix is smaller than that from the expansion of x using d coordinates in any other orthogonal coordinate system. The percentage ratio of the sum of the d largest eigenvalues to the sum of all eigenvalues, β, is referred to as the feature contribution rate, to which a threshold is assigned during dimensionality reduction. The orthogonal coordinate system consisting of the d eigenvectors used in the expansion is referred to as the d-dimensional K–L transform coordinate system in the D-dimensional space where x is located. The coefficient for the expansion of x in the K–L coordinate system is referred to as the K–L transform of x.

2.2. K–L Transform–Based Data Dimensionality Reduction Process

The K–L transform represents an orthogonal transform $A \in R^{n \times n}$, where vectors $X \in R^n$ are mapped onto vectors $Y \in R^n$ and the components of vector Y are uncorrelated, that is,

$$C = E\left\{(X - \overline{X})(X - \overline{X})^T\right\} \tag{12}$$

where $\overline{X}$ is the mean of the original data and C is the covariance matrix of the demeaned data.

The covariance matrix is subjected to eigendecomposition, and the resulting eigenvalues are arranged in a descending order as follows:

$$Cz_i = \lambda_i z_i \tag{13}$$

where λ_i represents the eigenvalues and Z_i represents the eigenvectors corresponding to λ_i.

The dimensionality after dimensionality reduction depends on the threshold assigned to β, with a larger threshold leading to a larger retained variance and a smaller MSE. If the predefined threshold is satisfied and the dimensionality of the original matrix can be reduced to K, the threshold of β is defined as

$$\beta = 1 - \varepsilon = \frac{\sum\limits_{i=K+1}^{\infty} \lambda_i}{\sum\limits_{i=1}^{\infty} \lambda_i} \tag{14}$$

For a transformation matrix A consisting of the eigenvectors corresponding to the largest K eigenvalues, the mapping relationship from X to Y is

$$Y = A\left(X - \overline{X}\right) \tag{15}$$

Figure 1 shows a flowchart of the dimensionality reduction process using the K–L transform. The detailed steps are as follows:

(1) An $M \times N$ sample dataset (where M and N are the numbers of samples and feature dimensions, respectively) is demeaned.
(2) The covariance matrix of the demeaned sample data matrix is calculated.
(3) The eigenvalues and eigenvectors of the resulting $N \times N$ covariance matrix are calculated.
(4) The eigenvalues are arranged in a descending order, and their corresponding eigenvectors are arranged in a corresponding order.
(5) A threshold is assigned to β. If the ratio of the sum of the K ($K < N$) largest eigenvalues to the sum of all eigenvalues (feature contribution rate) is larger than the threshold, the dimensionality of the original data can be reduced to K. An $N \times K$ transformation matrix is established using the eigenvectors corresponding to the largest K eigenvalues.
(6) An $M \times K$ matrix is obtained by multiplying the normalized $M \times N$ sample data matrix with the $N \times K$ transformation matrix, thereby reducing the dimensionality of the sample data matrix from N to K.

As shown in Figure 1, the core of dimensionality reduction using the K–L transform is obtaining the eigenvalues of the covariance matrix. In the following, we discuss and present the investigation on the methods for solving the eigenvalues and eigenvectors of the covariance matrix and reducing the solution time.

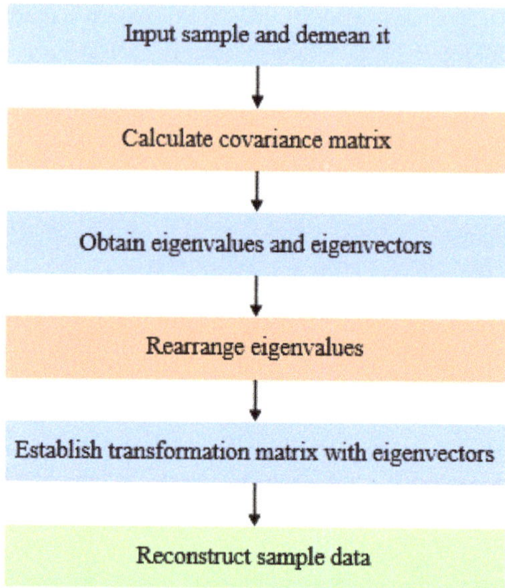

Figure 1. Flowchart of dimensionality reduction using the K–L transform.

3. Methods for Eigenvalue of Matrix

The core step of the K–L transform method is solving the eigenvalues and eigenvectors of the covariance matrix of the sample data matrix. The methods for obtaining the eigenvalues of matrices mainly include the Jacobi iteration and rapid principal component analysis (PCA) methods, as described below.

3.1. Jacobi Iteration Method

Low-order linear equations, such as $AX = B$, are typically solved using the pivot elimination method. However, engineering problems frequently involve computer-aided solutions of large sparse matrix equations, which need to be obtained using an iterative method. Its major principle is as follows:

Consider the following equations:

$$\begin{cases} a_{11}x_1 + a_{12}x_2 + \ldots + a_{1n}x_n = b_1 \\ a_{21}x_1 + a_{22}x_2 + \ldots + a_{2n}x_n = b_2 \\ \cdots \\ a_{n1}x_1 + a_{n2}x_2 + \ldots + a_{nn}x_n = b_n \end{cases} \tag{16}$$

where coefficient matrix A is nonsingular and $a_{ii} \neq 0$. These equations can be transformed to obtain the following iteration equations:

$$\begin{cases} x^{(0)} = \left(x_1{}^{(0)}, x_2{}^{(0)}, \ldots, x_n{}^{(0)}\right)^T \\ x_i{}^{(k+1)} = \frac{1}{a_{ii}}\Big(b_i - \sum\limits_{\substack{i=1 \\ j \neq i}}^{n} a_{ij}x_j{}^{(k)}\Big) \end{cases} \tag{17}$$

where $x^{(0)}$ represents the initial vectors and $x^{(k)}$ represents the vectors at the k-th iteration.

The matrix form is as follows: First, the coefficient matrix, A, of the equations is decomposed into three components, that is, $A = L + D + U$:

$$D = \begin{bmatrix} a_{11} & & & \\ & a_{22} & & \\ & & \ddots & \\ & & & a_{nn} \end{bmatrix} L = \begin{bmatrix} 0 & & & \\ a_{21} & 0 & & \\ \vdots & & \ddots & \\ a_{n1} & a_{n2} & \cdots & 0 \end{bmatrix} U = \begin{bmatrix} 0 & a_{12} & \cdots & a_{1n} \\ & 0 & \cdots & a_{2n} \\ & & \ddots & \vdots \\ & & & 0 \end{bmatrix} \tag{18}$$

where D, L, and U are the diagonal, lower triangular, and upper triangular matrices, respectively.

Thus, the following matrix formulation can be derived:

$$Dx = -(L+U)x + b \tag{19}$$

which can be rearranged as

$$x = -D^{-1}(L+U)x + D^{-1}b \tag{20}$$

which can be simplified as

$$x = Bx + d \tag{21}$$

Thus, the Jacobi iteration equations can be expressed in the following matrix form:

$$\begin{cases} x^{(0)} = \left(x_1{}^{(0)}, x_2{}^{(0)}, \ldots, x_n{}^{(0)}\right)^T \\ x^{(k+1)} = Bx^{(k)} + d \end{cases} \tag{22}$$

From the above process, the Jacobi iteration method involves a simple calculation process, requiring only the multiplication of matrices in each iteration. However, it has remarkable disadvantages, including a low convergence speed and a large computer memory requirement.

3.2. Rapid PCA Method

PCA requires the calculation and eigendecomposition of the covariance matrix of the input sample data to obtain its eigenvalues and eigenvectors. For a sample data matrix X of size $n \times d$ (where n and d are the numbers of samples and feature dimensions, respectively), its covariance matrix S is a $d \times d$ square matrix. Extremely complex calculations are required when d is large. The rapid PCA algorithm solves the transposed matrix of the covariance matrix, instead of directly solving the covariance matrix. The dimensionality of the covariance matrix of the transposed matrix depends on the number of samples. When the number of feature dimensions is much larger than the number of samples, the transposed matrix can be very easily solved using the rapid PCA algorithm.

The derivation of the rapid PCA algorithm is as follows: For an $n \times d$ matrix of a sample dataset Z with the number of samples, n, much smaller than the number of feature dimensions, d, the covariance matrix is

$$S = Z^T Z \tag{23}$$

The transposed matrix of the covariance matrix is

$$R = ZZ^T \tag{24}$$

S and R have the same nonzero eigenvalues:

$$(ZZ^T)v = \lambda\, v \tag{25}$$

Multiplying both sides of the above equation with Z^T yields

$$(Z^T Z)Z^T v = \lambda\, Z^T v \tag{26}$$

Thus, the eigenvectors of S are $Z^T v$.

From the above equations, the eigenvalues and eigenvectors of the original covariance matrix can be obtained by solving those of its transposed matrix.

From the principles of the two methods described above, the Jacobi iteration method is typically suitable for obtaining computer-aided eigenvalues of matrices. It can deal with moderate computational complexity; however, it requires a long solution time in cases of massive computations. The rapid PCA algorithm is suitable when the number of feature dimensions is much larger than the number of samples. By solving the feature matrix of the transposed matrix, it transforms a high-dimensional matrix to be solved into a low-dimensional matrix, thereby improving the computational efficiency.

We previously developed a multidisciplinary platform for a comprehensive optimization of ship hydrodynamic performance (SHIPMDO-WUT) using a radial basis function (RBF)– based interpolation method [29–33] for hull form modification. In this method, the number of discrete points on a hull surface usually reaches thousands, resulting in tens of thousands of dimensions of covariance matrix, making the direct eigendecomposition of the covariance matrix considerably difficult. Because the number of feature dimensions is much larger than the number of samples, the covariance matrix was solved using the rapid PCA method, thereby substantially reducing the dimensionality by eigendecomposition and improving the solution efficiency.

4. K–L Transform–Based Hull Form Reconstruction Method

The core of the K–L transform–based hull form reconstruction method is solving the problem that the design variables in the projected dimensionality-reduced matrix are not points on the hull surface and do not have physical relevance. This problem makes obtaining a new hull form directly by RBF modification impossible. This study developed a transformation-matrix-based method for offset matrix reconstruction that can ensure surface smoothness of the new reconstructed hull form. In the following, the hull form reconstruction process is described, including the presentation of the derivation of its equations and the application of the process to actual cases.

4.1. Hull Form Reconstruction Process

In the K–L transform–based dimensionality reduction, a sample offset matrix of hull forms obtained by varying the original hull form is subjected to dimensionality reduction to obtain projection and eigenvector matrices. Each column of the projection matrix following the dimensionality reduction represents a new design variable, which is a combination of the initial variables and does not correspond to a discrete point on the hull surface. Because the new design variables do not have physical relevance, a new hull form cannot be obtained by surface modification, thus necessitating another method to generate it. The process of generating a new hull form using the new design variables is referred to as hull form reconstruction. Figure 2 shows a flowchart of the hull form reconstruction process. The steps are detailed as follows:

(1) Using SHIPMDO-WUT, a hull surface is discretized [29–33], and a certain number of variable points are selected as the design variables.
(2) A certain number of samples are collected using the uniform experimental design method to generate a sample offset matrix and the corresponding hull forms using the RBF-based shape modification method.
(3) For each sample hull form generated, M offset points are selected, and their x-, y-, and z-coordinates are arranged in a geometric information matrix.
(4) The covariance matrix of the demeaned geometric information matrix is calculated using the K–L transform method to obtain its eigenvalues and eigenvectors. The eigenvectors are arranged according to the eigenvalues, and a threshold of β is assigned to generate a transformation matrix. The sample offset matrix is normalized and multiplied with the transformation matrix to produce a projection matrix.

(5) The new design variables in the dimensionality-reduced projection matrix and their respective variation ranges are defined.
(6) Hull form reconstruction is performed using the hull form reconstruction equations to obtain the hull offset matrix and generate the corresponding hull form.

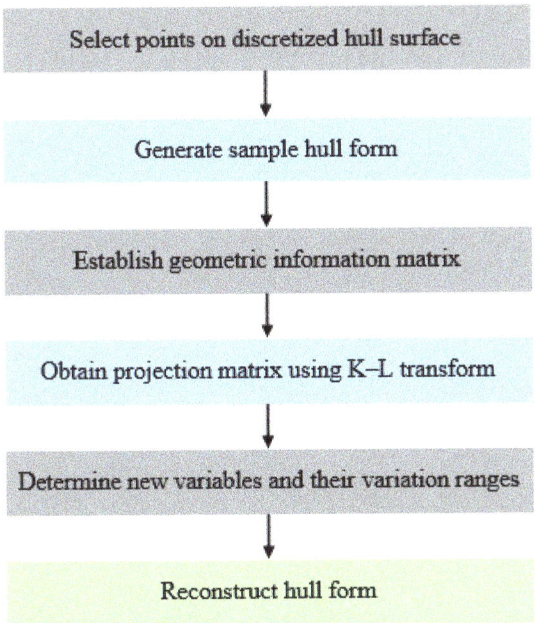

Figure 2. Flowchart of hull form reconstruction.

In the hull form reconstruction method, the core problem is the reconstruction of the new offset matrix from the dimensionality-reduced projection matrix. In the following, this core problem is discussed.

N sample hull forms are sampled. For each hull form, M offset points are selected, and their x-, y-, and z-coordinates are arranged as a geometric information matrix, which can be expressed as

$$x_i{}^T = (x_1, y_1, z_1, x_2, y_2, z_2, \ldots, x_M, y_M, z_M); \tag{27}$$

The geometric information matrices of the N sample hull forms are used as the basis function. Subsequently, the offset matrix of each hull form is decomposed into a linear combination of several sample hull forms as follows:

$$x_i = \overline{x} + \sum_{j=1}^{N} a_{ij} u_j \tag{28}$$

The mean of each column is subtracted from the elements in the corresponding column. The geometric characteristics of each demeaned sample hull form can be expressed as

$$\widetilde{x}_i = x_i - \overline{x}_i \tag{29}$$

The demeaned geometric information matrices of all sample hull forms can be expressed as

$$\widetilde{X} = (\widetilde{x}_1, \widetilde{x}_2, \ldots, \widetilde{x}_N) \tag{30}$$

The covariance matrix of the geometric information matrices can be obtained as

$$S = \widetilde{X}\widetilde{X}^T = \sum_{i=1}^{N} \widetilde{x}_i \widetilde{x}_i^T \tag{31}$$

The covariance matrix is subjected to eigendecomposition to obtain the eigenvalues and the corresponding eigenvectors as follows:

$$XX^T u_j = \lambda_j u_j \tag{32}$$

where λ_j represents the eigenvalues and u_j represents the eigenvectors.

The threshold of β, which determines the dimensionality after dimensionality reduction, is calculated as

$$\beta = \frac{\sum_{j=1}^{M_P} \lambda_j}{\sum_{j=1}^{3M} \lambda_j} \tag{33}$$

The eigenvalues are arranged in a descending order. The eigenvectors corresponding to the first M_p eigenvalues constitute the transformation matrix,

$$U = (u_1, u_2, \ldots, u_{M_P}) \tag{34}$$

Each geometric information matrix is reconstructed to obtain the offsets of the corresponding reconstructed hull form as follows:

$$x_i^{rec} = \overline{x} + \sum_{j=1}^{M_P} a_{i\,j} u_j \tag{35}$$

where $a_{ij} = \widetilde{x}_i^T u_j$ represents the new design variables, u_j represents the retained eigenvectors, and $\overline{x}$ is the mean geometric information matrix, that is, a matrix consisting of the means of the geometric information of all sample hull forms.

In summary, in the K–L transform/dimensionality reduction–based hull form reconstruction, the offset matrix of a hull form can be obtained using Equation (35). This is possible when the mean geometric information matrix, values of the new design variables, and retained eigenvector matrix (i.e., transformation matrix) are known, thereby generating a new hull form.

4.2. Examples of Hull Form Reconstruction

As shown in Figure 3, the Series 60 hull form was optimized using SHIPMDO-WUT. First, the Series 60 hull surface was discretized. The model consisted of two surfaces, and each surface was discretized in the U and V directions into a 40 × 30 point cloud. As shown in Figure 4, 10 variable points are selected and defined to be variables in the Y-direction. Table 1 lists their respective variation ranges. Using a uniform experimental design, 50 samples were collected, and the surface was varied using the RBF method; consequently, a sample offset matrix of 50 sample hull forms was obtained, with each hull surface represented as a 40 × 30 point cloud. The offset matrix was subjected to dimensionality reduction using the K–L transform procedure, with the threshold of β set as 95%. Accordingly, a total of 180,000 eigenvalues were obtained, which were arranged in a descending order. Except for the first 50 eigenvalues, all eigenvalues were approximately zero. Based on Table 2, the sum of the first 5 eigenvalues is 10.17 and that of the first 50 eigenvalues is 10.61. The first 10 eigenvalues are significant in magnitude, based on Table 3. Figure 5 shows the distribution of the 10 largest eigenvalues. The percentage ratio of the sum of the first 5 eigenvalues to the sum of all eigenvalues is 95.9%, which is higher than the preset threshold (95%), as shown in Figure 6. Thus, the eigenvectors corresponding to the first 5 eigenvalues were selected to constitute the transformation matrix.

Figure 3. Schematic of a 3D model of Series 60.

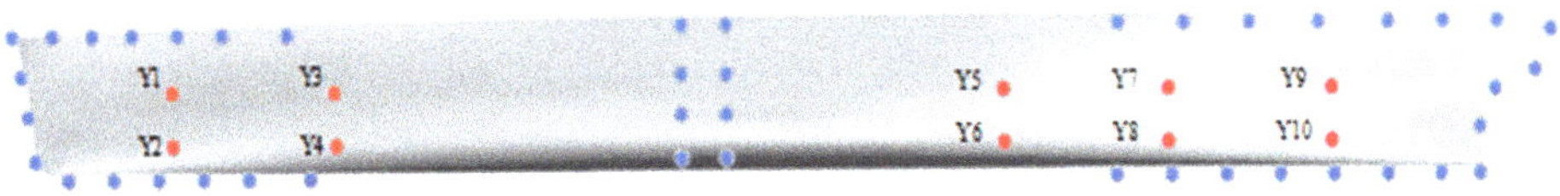

Figure 4. Distributions of variable and fixed points on the Series 60 hull surface (blue: fixed points; red: variable points).

Table 1. S60 variation ranges of variables.

Serial Number	Y_1	Y_2	Y_3	Y_4	Y_5	Y_6	Y_7	Y_8	Y_9	Y_{10}
Upper limit	1.243	1.885	2.396	3.22	3.371	4.203	2.743	3.369	1.058	1.273
Lower limit	1.051	1.577	2.055	2.795	2.963	3.751	2.270	3.010	0.908	1.075

Table 2. Five largest eigenvalues.

Serial Number	1	2	3	4	5
Eigenvalue	6.49	1.95	1.05	0.46	0.22

Table 3. Ten largest eigenvalues of S60.

Serial Number	1	2	3	4	5	6	7	8	9	10
Eigenvalue	6.49	1.95	1.05	0.46	0.22	0.16	0.13	0.08	0.06	0.01

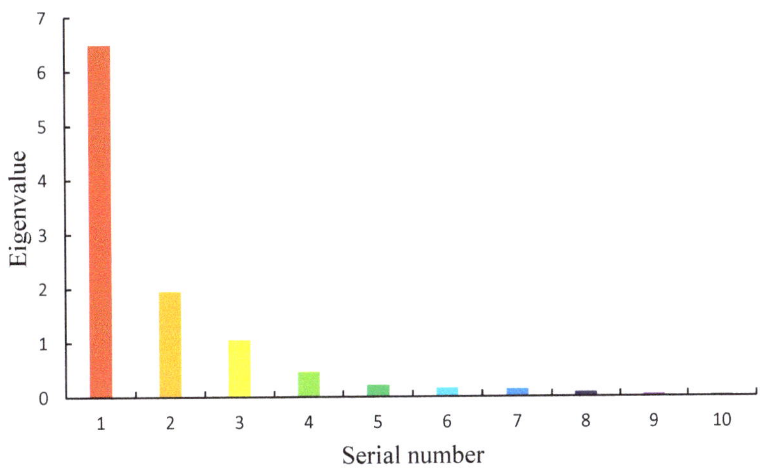

Figure 5. Distributions of the 10 largest eigenvalues.

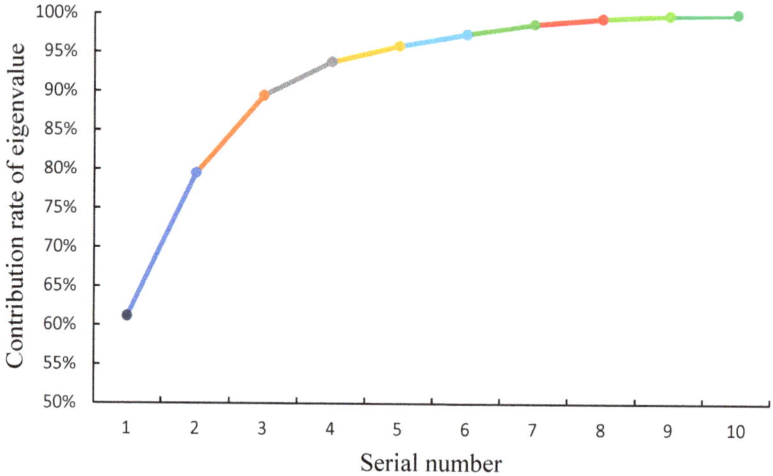

Figure 6. Contribution rates of the 10 largest eigenvalues.

The hull surface was reconstructed by feeding a random set of new variables a_{ij} into Equation (36). Figures 7–9 show the line plans of the reconstructed hull forms.

$$x_i^{rec} = \overline{x} + \sum_{j=1}^{5} a_{ij} u_j \tag{36}$$

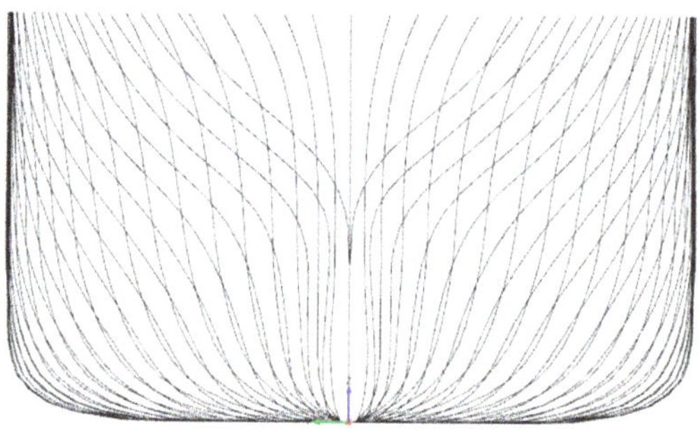

Figure 7. Line plan of reconstructed hull form KLE-5.

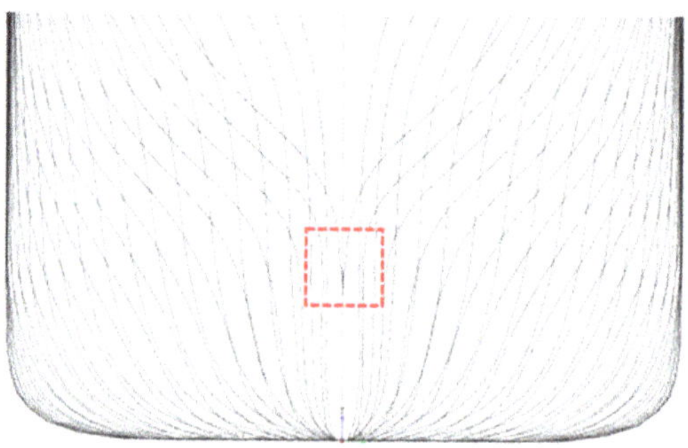

Figure 8. Line plan of reconstructed hull form KLE-4.

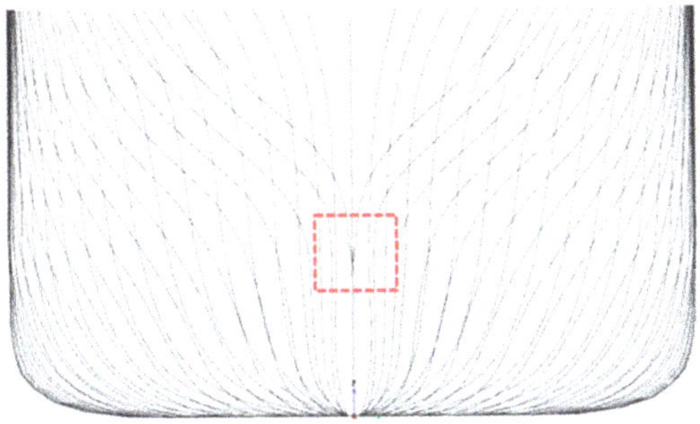

Figure 9. Line plan of reconstructed hull form KLE-3.

First, the threshold of β for dimensionality reduction was set as 95% by default. The eigenvectors corresponding to the 5 largest eigenvalues were used to constitute the transformation matrix. The reconstructed hull form (KLE-5) had smooth molded lines, indicating that the 5 principal components were adequate to retain most of the geometric information in the original space and, thus, could replace the initial 10 design variables.

When the threshold of β was set as 90%, the eigenvectors corresponding to the 4 largest eigenvalues were used to constitute the transformation matrix. The percentage ratio of the sum of the 4 largest eigenvalues to the sum of all eigenvalues was 94.1%. Figure 8 shows the line plan of the reconstructed hull surface (KLE-4).

When the threshold of β was changed to 85%, the number of eigenvectors retained in the transformation matrix decreased to 3, because the percentage ratio of the sum of the 3 largest eigenvalues to the sum of all eigenvalues was 89.8%. Figure 9 shows the line plan of the hull form reconstructed after dimensionality reduction (KLE-3).

The body plans of KLE-4 and KLE-3 exhibited unsmooth molded lines compared with that of KLE-5. Based on the principle of the K–L transform, a smaller threshold of β leads to a smaller number of retained eigenvalues, a smaller number of corresponding eigenvectors, and a lower dimensionality of the transformation matrix. Consequently, a lower dimensionality of the design space is achieved after dimensionality reduction. In addition, exclusion of eigenvectors corresponding to large eigenvalues leads to an increased MSE and loss of information and, thus, an unsmooth surface of the reconstructed hull form. Thus, an excessively small threshold of β for dimensionality reduction using the K–L transform will lead to retention of an inadequate number of principal components, loss of information, and finally, an unsmooth surface of the reconstructed hull form. For hull form optimization, a 95% threshold of β for dimensionality reduction is safe to ensure that the major geometric information in the design space is retained after dimensionality reduction.

5. Optimization of Wave-Making Resistance Coefficient for DTMB 5415

5.1. Optimized Object

Figure 10 shows a three-dimensional (3D) model of DTMB 5415. Table 4 lists the major parameters of the model. L_{wl} is the waterline length, B_{wl} is the maximum waterplane width, L_{cb} is the longitudinal center of buoyancy (LCB), and T is the draft. C_b is the block coefficient, C_m is the fullest cross-section area, ∇ is the volume of displacement, and S_{wet} is the wetted surface area.

Figure 10. Schematic of a 3D model of DTMB 5415.

Table 4. Major parameters of DTMB 5415.

L_{wl}	L_{cb}	B_{wl}	T	C_b	C_m	∇	S_{wet}
5.699 m	3.293 m	0.679 m	0.248 m	0.339	0.548	0.550 m^3	4.904 m^2

5.2. Description of Optimization Problem

(1) Objective of optimization

The objective of optimization was to minimize the wave-making resistance (WMR) coefficient C_w at Froude number $Fr = 0.22$.

$$\min f_{obj} = C_w s.\ t.\ Fr = 0.22$$

(2) Setting of optimization variables

The DTMB 5415 hull surface was discretized into a 40 × 30 point cloud. Some points on the bow, centerline of the stern, and deck side line were fixed. Figure 11 shows the locations of the fixed points (blue dots) on the hull surface. Ten points on the bow are selected as variable points Y_1–Y_{10}. The red dots in Figure 11 represent the positions of the variable points on the hull surface. The variable points are defined as variable in the Y-direction; that is, their Y-coordinates are the design variables to be optimized. Table 5 summarizes the variation ranges of the variable points.

Figure 11. Distributions of variable and fixed points (blue: fixed points; red: variable points).

Table 5. DTMB variation ranges of variables.

Serial Number	Y_1	Y_2	Y_3	Y_4	Y_5	Y_6	Y_7	Y_8	Y_9	Y_{10}
Upper limit	0.1325	0.1933	0.0955	0.0404	0.0629	0.1157	0.0152	0.0258	0.2100	0.2836
Lower limit	0.1104	0.1581	0.0801	0.0331	0.0514	0.0946	0.0125	0.0211	0.1860	0.2502

(3) Constraint conditions.

Hydrostatic constraint: Constraint on the volume of displacement: $\frac{|\nabla - \nabla_{opti}|}{\nabla} \leq 1\%$.

Constraint on the LCB: $\frac{|L_{cb} - L_{cbopti}|}{L_{cb}} \leq 1\%$.

Geometric constraints: Some points on the stern, deck side line, and centerline were fixed to ensure the ship length, width, and molded depth did not vary.

5.3. CFD Simulation

This study used the commercial software SHIPFLOW as CFD solver. Solve the wave-making resistance coefficient based on potential flow theory. Create a surface mesh using the XMESH module in SHIPFLOW. Then the wave-making resistance coefficient is obtained by surface mesh with fine precision. The mesh distribution is shown in Figure 12.

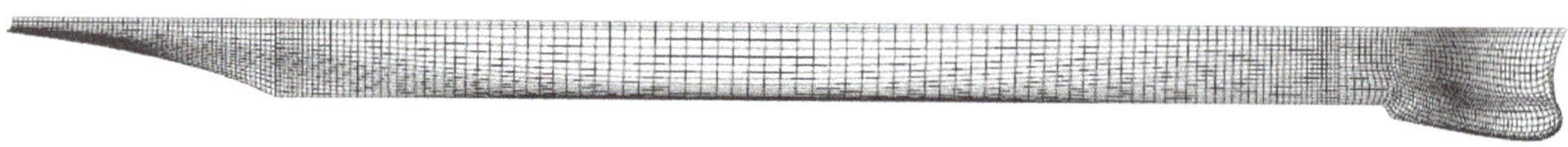

Figure 12. SHIPFLOW meshes of DTMB 5415.

The wave resistance coefficient C_w and the friction coefficient C_f are calculated by the formulas of SHIPFLOW and the International Towing Tank Conference (ITTC). The results are then compared with the experimental data. The specific calculation formula is as follows:

$$C_f = \frac{0.075}{(\log_{10}\mathrm{Re} - 2)^2} \tag{37}$$

Re is the Reynolds number.

The resistance coefficients calculated by SHIPFLOW and ITTC formulas are in good agreement with the experimental data. Therefore, SHIPFLOW meets the requirements of simulation optimization. The data comparison is shown in Figure 13. The relevant SHIPFLOW calculation results are verified in [34,35].

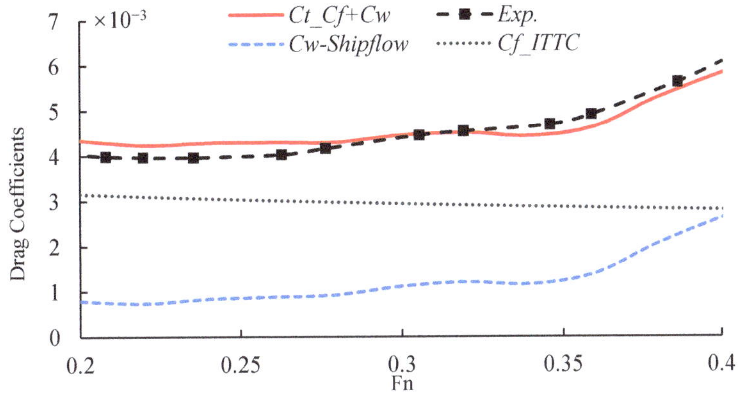

Figure 13. Data comparison chart of DTMB 5415.

5.4. Optimization Process

The whole optimization process is based on SHIPMDO-WUT, a comprehensive optimization platform for ship hydrodynamic performance developed by our team. The optimization process is shown in Figure 14.

(1) Direct optimization process

First, the original ship is discretized into 40 × 30 spatial point clouds, and the invariant points and variable points are selected according to the ship optimization experience. Then, based on the hull surface deformation method of radial basis function interpolation, the parametric deformation is realized, and the deformed hull is subjected to hydrostatic calculation to select the hull that meets the hydrostatic constraints. Finally, the hull that meets the hydrostatic constraints is imported into the SHIPFLOW software to calculate the wave-making resistance, and the iterative optimization is carried out based on the multiobjective particle swarm optimization (MOPSO) of the optimization algorithm module. If the convergence condition is satisfied, the optimized ship type is obtained, and if the convergence condition is not satisfied, the optimization variables need to be modified.

(2) Dimensionality reduction optimization process

The overall optimization process is consistent with the above. The difference is that the optimization variable in the direct optimization process is the variable point Y-direction coordinate value. However, the optimization variables in the dimension reduction optimization process are based on the optimization variables in the direct optimization process, which are obtained by the K–L transformation method.

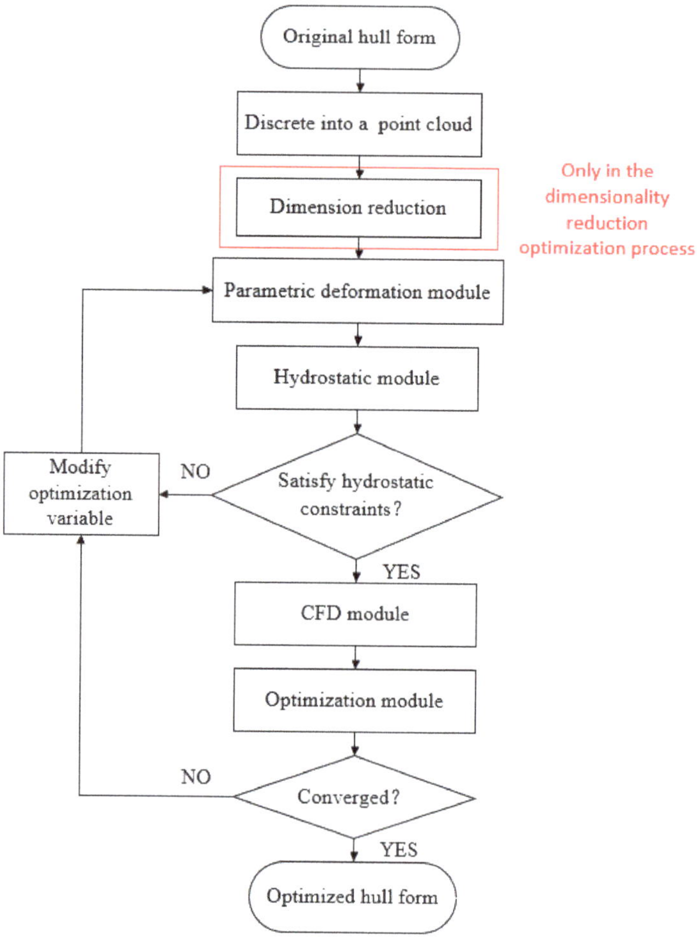

Figure 14. Optimization process based on the SHIPMDO-WUT platform.

5.5. Optimization Results

Direct optimization and optimization using the K–L transform-based dimensionality reduction reconstruction method were performed to compare their optimization performances and efficiencies. The results are discussed below.

(1) Results of direct optimization

Using the hydrodynamic performance optimization platform, the Y-coordinates of the 10 variable points were defined as the design variables, and the hull surface was modified by varying the coordinates using the RBF method. The WMR was calculated using SHIPFLOW. A particle swarm optimization algorithm was used, with both the number of particles and iterations set as 50. The hull surface was directly optimized.

The resulting optimal hull form had a WMR coefficient of $C_w = 4.90 \times 10^{-4}$ and a WMR of 3.328 N, whereas the original shape hull had a WMR coefficient of $C_w = 6.83 \times 10^{-4}$ and a WMR of 4.647 N. Thus, direct optimization decreased the WMR coefficient and WMR by 28.25% and 28.38%, respectively. The optimized hull form had a larger volume of displacement and smaller LCB and wetted surface area than the original hull form. Figure 15 shows the optimization convergence curve. The objective function is observed to converge after approximately 3000 iterations. Tables 6 and 7 summarize the optimization results and the variations in the hydrostatic parameters, respectively. Figures 16 and 17 compare the line plans of the optimized and original hull forms. The shape of the bulbous bow of the optimized hull form is not markedly different from that of the original hull form. The optimized hull form is concave near the bow part of the design waterline, making the design waterline appear slenderer and changing the geometric shape under the design waterline from V- to U-shaped compared with the original hull form.

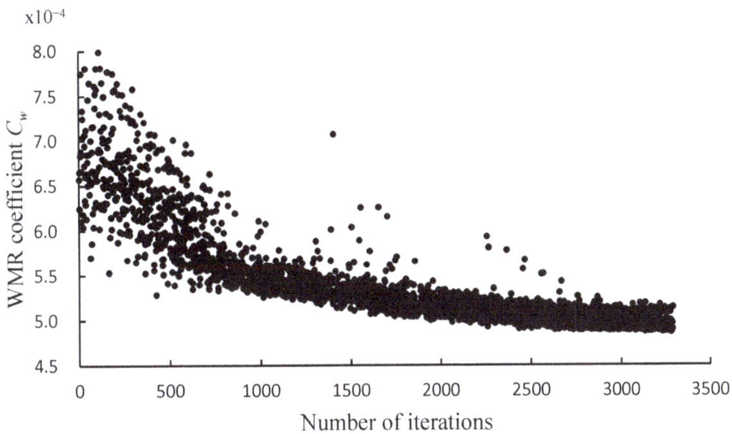

Figure 15. Optimization convergence curve of direct optimization.

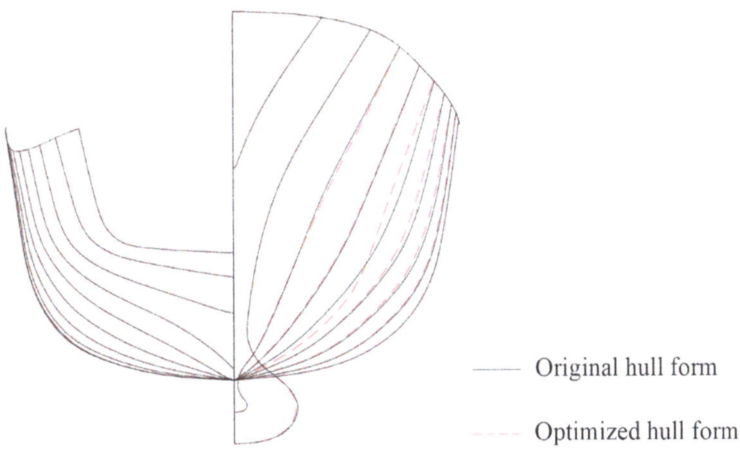

Figure 16. Comparison of body plan views of direct optimized and original hull forms.

Figure 17. Comparison of profile views of direct optimized and original hull forms.

Table 6. Optimization results of direct optimization.

	Number of Iterations for Convergence	WMR Coefficient C_w (Variation %)	WMR R_w (Variation %)
Original hull form	-	6.83×10^{-4}	4.647 N
Direct optimization	3000	4.90×10^{-4} (−28.25%)	3.328 N (−28.38%)

Table 7. Comparison of hydrostatic parameters of direct optimization.

	Volume of Displacement ∇ (Variation %)	LCB L_{cb} (Variation %)	Wetted Surface Area S_{wet} (Variation %)
Original hull form	0.550 m^3	3.293 m	4.904 m^2
Direct optimization	0.553 m^3 (+0.55%)	3.286 m (−0.21%)	4.903 m^2 (−0.2%)

(2) Results of dimensionality reduction optimization

The uniform experimental design method was used to sample the 10 design variables (Y_1–Y_{10}) described above to obtain 200 samples. The hull form was varied using the RBF method to generate 200 sample hull forms, which were screened to retain only those shapes that satisfied the predefined constraints of the volume of displacement and LCB. The offset information of the sample hull forms that satisfied the constraints was used to generate the offset matrix. A total of 113 sample hull forms satisfied the constraint conditions. The 3D model of DTMB 5415 was divided into three surfaces. Each surface was discretized into a 40 × 30 point cloud, generating a total of 3600 points. Each point was divided into three dimensions of x, y, and z. Thus, the offset matrix for a sample hull form had a size of 113 × 10,800.

The dimensionality was reduced using the K−L transform procedure, with the threshold of β set as 95%. A total of 406,800 eigenvalues were obtained and arranged in a descending order. Except for the first 113 eigenvalues, all eigenvalues were approximately zero. The sum of the first 6 eigenvalues was 2.95 (Table 8), and the sum of the first 113 eigenvalues was 3.065. The first 10 eigenvalues were significant in magnitude (Table 9). The ratio of the sum of the first 6 eigenvalues to the sum of the total eigenvalues was 96.24%, which was larger than the preset threshold of β (95%); this can be seen from Figure 18. Figure 19 shows the distribution of the first 6 eigenvalues. Thus, the eigenvectors corresponding to the first 6 eigenvalues after dimensionality reduction constituted the transformation matrix, which was projected to obtain new design variables P_1–P_6.

The offsets of the reconstructed hull forms were obtained using Equation (35). The original hull form was optimized using the new design variables, P_1–P_6. The WMR coefficient was calculated using SHIPFLOW. A particle swarm optimization algorithm was used, with both the number of particles and iterations set as 50.

Table 8. Six largest eigenvalues.

Serial Number	1	2	3	4	5	6
Eigenvalue	1.75	0.52	0.23	0.19	0.15	0.11

Table 9. Ten largest eigenvalues.

Serial Number	1	2	3	4	5	6	7	8	9	10
Eigenvalue	1.75	0.52	0.23	0.19	0.15	0.11	0.06	0.03	0.01	0.003

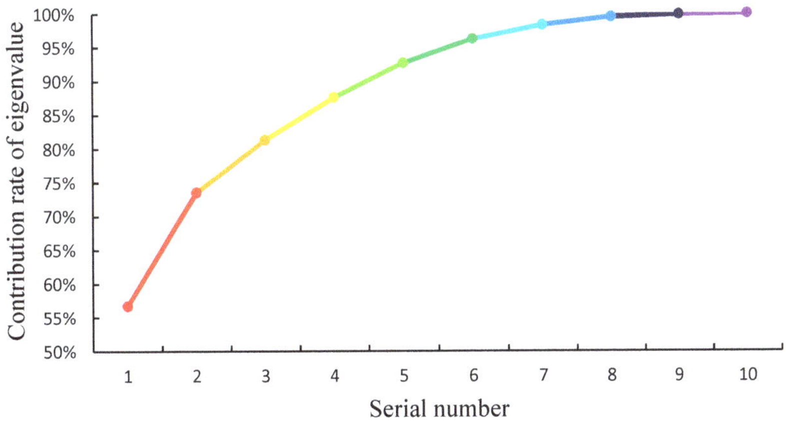

Figure 18. Contribution rates of eigenvalues.

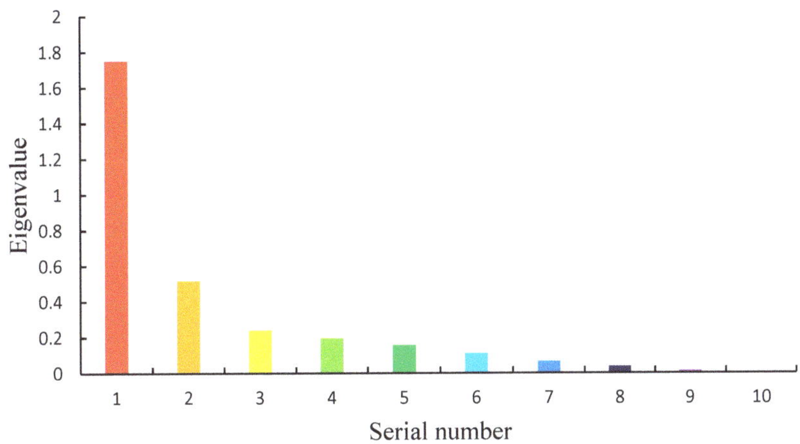

Figure 19. Distribution of eigenvalues.

The optimized hull form had a WMR coefficient of $C_w = 4.98 \times 10^{-4}$ and a WMR of 3.394 N, representing decrease of 27.07% and 26.96% compared with those of the original hull form, respectively. The optimized hull form had a larger volume of displacement, larger wetted surface area, and slightly smaller LCB than the original hull form. Figure 20 shows the convergence curve for the WMR optimization. The objective function is observed to converge after approximately 1000 iterations. Table 10 shows the optimization results, and Table 11 compares the hydrostatic parameters of the original and optimized hull forms. Figures 21 and 22 compare the original and optimized hull forms. The bulbous bow shape of the optimized hull form does not vary markedly from that of the original hull form. The optimized hull form is concave near the design waterline at the bow, making the design waterline appear slenderer and changing the geometric shape under the design waterline from V- to U-shaped compared with the original hull form.

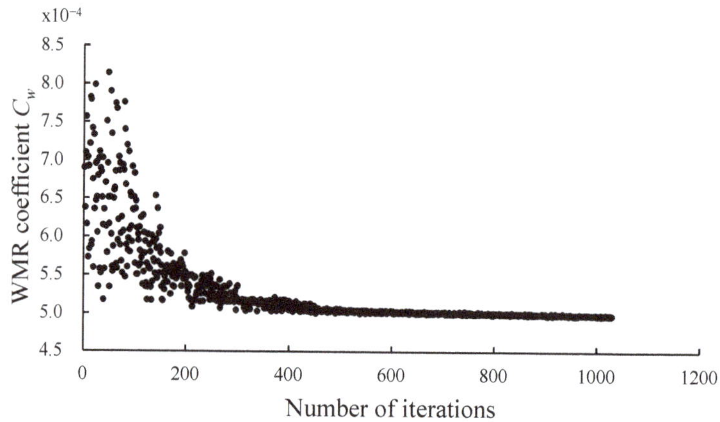

Figure 20. Optimization convergence curve of dimensionality reduction optimization.

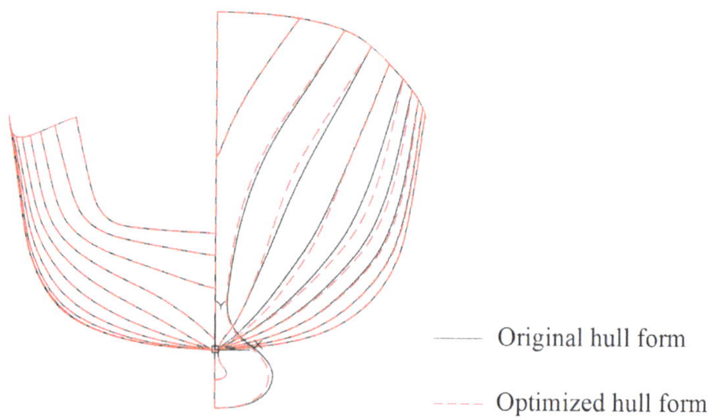

Figure 21. Comparison of body plan views of dimensionality reduction optimized and original hull forms.

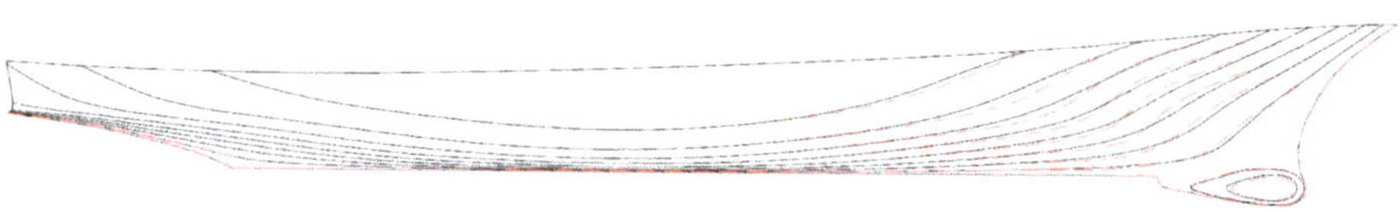

Figure 22. Comparison of profile views of dimensionality reduction optimized and original hull forms.

Table 10. Optimization results of dimensionality reduction optimization.

	Number of Iterations for Convergence	WMR Coefficient C_w (Variation %)	WMR R_w (Variation %)
Original hull form		6.83×10^{-4}	4.647 N
Dimensionality reduction optimization	1000	4.98×10^{-4} (−27.07%)	3.394 N (−26.96%)

Table 11. Comparison of hydrostatic parameters of dimensionality reduction optimization.

	Volume of Displacement ∇ (Variation %)	LCB L_{cb} (Variation %)	Wetted Surface Area S_{wet} (Variation %)
Original hull form	0.550 m^3	3.293 m	4.904 m^2
Dimensionality reduction optimization	0.554 m^3 (+0.73%)	3.283 m (−0.30%)	4.920 m^2 (+0.33%)

(3) Comparison of direct and dimensionality reduction optimization results.

Figures 23 and 24 compare the line plans of the hull forms obtained using the two different optimization methods. The bulbous bow shape obtained from the dimensionality reduction optimization is slightly slenderer than that obtained from the direct optimization. The hull form is concave in the design waterline near the bow, making the design waterline appear slender. Moreover, it is convex in that near the midship, making the design waterline appear fuller and changing the geometric shape under the design waterline from V- to U-shaped. Figures 25–27 compare the wave disturbances of the hull forms obtained using the two optimization methods and the original hull form. The original hull form has markedly more scattered waves than the hull forms obtained by both the direct and dimensionality reduction optimizations. However, the hull form obtained by the dimensionality reduction optimization has fewer scattered waves than that by the direct optimization, as indicated by the red ellipses. Figure 28 compares the transverse cross sections of the longitudinal waves at $y/lpp = 0.12$. Owing to the optimized bow shape, the hull forms obtained by the direct and dimensionality reduction optimizations have low-magnitude wave disturbances around the hull, which are favorable, and markedly low waves at the bow and midship, which decrease the WMR. Tables 12 and 13 compare the results and time consumptions and the hydrostatic parameters of the hull forms obtained by the two optimizations, respectively. The hull forms obtained by the two optimizations have larger volumes of displacement and LCBs closer to the stern than the original hull form. The optimal hull forms obtained by the two optimizations and the original hull form do not have significantly different wetted surface areas. The hull form obtained by the dimensionality reduction optimization has a larger wetted surface area, a larger volume of displacement, and an LCB closer to the stern than that obtained by the direct optimization. Compared with the WMR coefficient of the original hull form ($C_w = 6.833 \times 10^{-4}$), those of the hull forms obtained by the direct and dimensionality reduction optimizations decrease by 28.25% and 27.07%, respectively. Compared with the WMR of the original hull form, those of the hull forms obtained by the direct and dimensionality reduction optimizations decrease by 28.38% and 26.96%, respectively. Thus, the results of the direct and dimensionality reduction optimizations do not differ significantly.

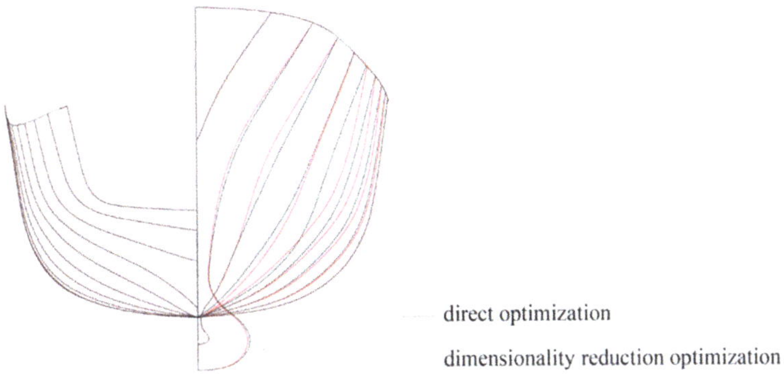

Figure 23. Comparison of body plan views of hull forms obtained by direct and dimensionality reduction optimizations.

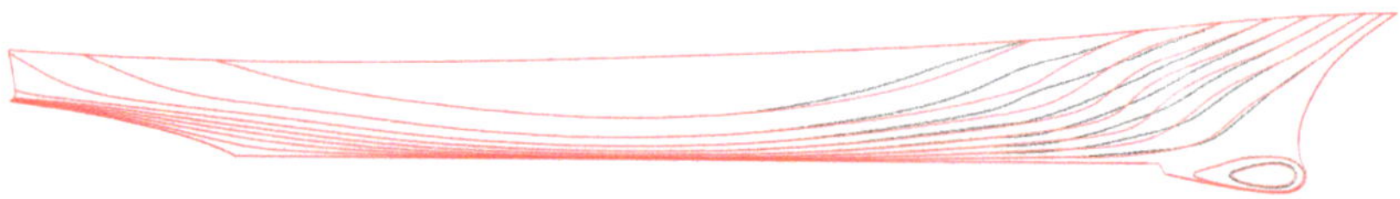

Figure 24. Comparison of profile views of hull forms obtained by direct and dimensionality reduction optimizations.

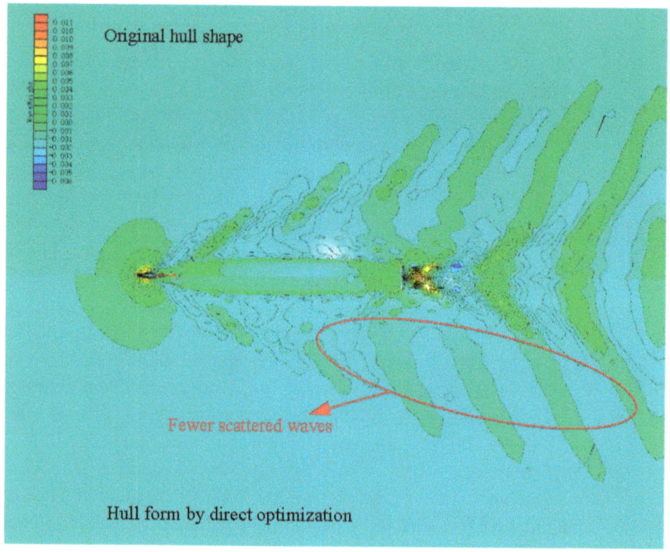

Figure 25. Comparison of wave disturbances of original hull form and hull form obtained by direct optimization.

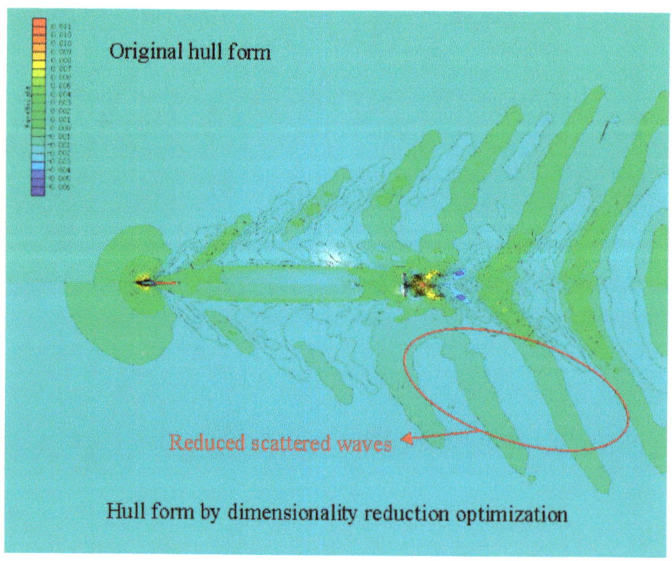

Figure 26. Comparison of wave disturbances of original hull form and hull form obtained by dimensionality reduction optimization.

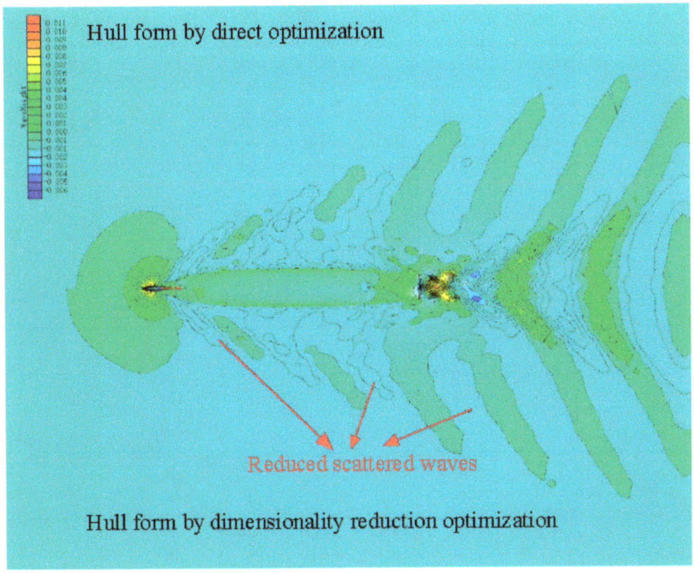

Figure 27. Comparison of wave disturbances of hull forms obtained by direct and dimensionality reduction optimizations.

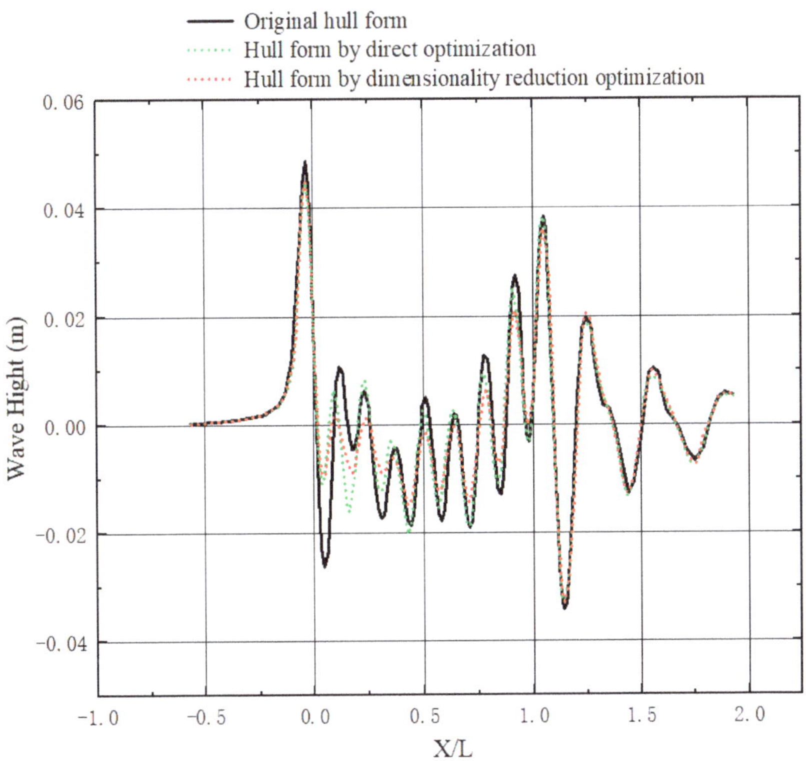

Figure 28. Comparison of transverse cross sections of longitudinal waves at $y/lpp = 0.12$.

Table 12. Comparison of optimization results.

	Number of Iterations for Convergence	WMR Coefficient C_w (Variation %)	WMR R_w (Variation %)	Time Consumption of Optimization
Original hull form		6.83×10^{-4}	4.647 N	
Direct optimization	3000	4.98×10^{-4} (−28.25%)	3.328 N (−28.38%)	15 h
Dimensionality reduction optimization	1000	4.98×10^{-4} (−27.07%)	3.394 N (−26.96%)	5 h

Table 13. Comparison of hydrostatic parameters of three hull forms.

	Volume of Displacement ∇ (Variation %)	LCB L_{cb} (Variation %)	Wetted Surface Area S_{wet} (Variation %)
Original hull form	0.550 m^3	3.293 m	4.904 m^2
Direct optimization	0.553 m^3 (+0.55%)	3.286 m (−0.21%)	4.903 m^2 (−0.2%)
Dimensionality reduction optimization	0.554 m^3 (+0.73%)	3.283 m (−0.30%)	4.920 m^2 (+0.33%)

Both optimizations were performed using a computer equipped with a 6-core 12-thread i7-8700 CPU, 16 G memory, and 64-bit operating system. The direct optimization converged after approximately 3000 iterations, consuming 15 h. The dimensionality reduction optimization reduced the number of design variables from 10 to 6 and converged after 1000 iterations, consuming 5 h, representing a 75% time saving. This suggests that the K–L transform–based hull form reconstruction method effectively reduces the number of design variables, considerably reduces the time consumption of optimization, and improves the efficiency of optimization while yielding results that differ insignificantly from those by the direct optimization.

6. Conclusions

This study derived the equations for a K–L transform–based hull form reconstruction method and optimized the DTMB 5415 hull form by dimensionality reduction and direct optimizations. The following can be concluded from the results. The dimensionality reduction reduced the number of design variables, markedly reduced the number of iterations at optimization convergence, and significantly reduced the computation time compared with the direct optimization. The results yielded by the dimensionality reduction and direct optimizations differed insignificantly. This suggests that the K–L transform–based dimensionality reduction method reduces the time consumption and improves the optimization efficiency while retaining the variability of the design space. This proves the feasibility of the developed hull form optimization method combining the dimensionality reduction method and SHIPMDO-WUT.

During the study, several aspects could not be investigated thoroughly due to time constraints, and these are summarized as follows:

(1) The K–L transform is a linear dimensionality reduction method and assumes the presence of a linear relationship between data. Thus, its dimensionality reduction performance could be limited on nonlinear data. This study did not investigate the application of nonlinear dimensionality reduction methods, which will be tested in the future.

(2) To reduce the computation time, the sample data used in the dimensionality reduction optimization performed in this study only included the geometric information of the hull and excluded its physical information. The consideration of physical information requires a detailed computation of the ship performance in the early stage, and the

physical information of a new hull form can be directly obtained by dimensionality reduction reconstruction. This is a major direction for future research.

Author Contributions: Conceptualization: C.W., H.C. and Z.L.; methodology: C.W., H.C. and Z.L.; software: B.F., C.Z. and X.C.; validation: C.W. and H.C.; formal analysis: C.W., H.C. and Z.L.; data curation: C.W. All authors have read and agreed to the published version of the manuscript.

Funding: This research was funded by the National Natural Science Foundation of China (grant numbers 51979211, 52271327, 52271330), Key Research and Development Plan of Hubei Province (2021BID008), and 111 Project (BP0820028).

Institutional Review Board Statement: Not applicable.

Informed Consent Statement: Not applicable.

Data Availability Statement: The data presented in this study are available in this article (tables and figures).

Conflicts of Interest: The authors declare no conflict of interest.

Nomenclature

L_{wl}	m	the waterline length
L_{cb}	m	the longitudinal center of buoyancy
B_{wl}	m	the maximum waterplane width
T	m	
C_b	/	the block coefficient
C_m	/	the fullest cross-section area
∇	m^3	the volume of displacement
S_{wet}	m^2	the wetted surface area
F_r	/	Froude number
R_w	N	the wave-making resistance
C_w	/	the wave-making resistance coefficient

References

1. Cheng, X.; Li, G.; Skulstad, R. Data-driven uncertainty and sensitivity analysis for ship motion modeling in offshore operations. *Ocean Eng.* **2019**, *179*, 261–272. [CrossRef]
2. Hu, H.; Zhang, G.; Li, D. Shape optimization of airfoil in ground effect based on free-form deformation utilizing sensitivity analysis and surrogate model of artificial neural network. *Ocean Eng.* **2022**, *257*, 514–527. [CrossRef]
3. Guan, G.; Yang, Q.; Gu, W. A new method for parametric design and optimization of ship inner shell based on the improved particle swarm optimization algorithm. *Ocean Eng.* **2018**, *169*, 551–566. [CrossRef]
4. Hamed, A. Multi-objective optimization method of trimaran hull form for resistance reduction and propeller intake flow improvement. *Ocean Eng.* **2022**, *244*, 352–366. [CrossRef]
5. Jung, Y.-W.; Kim, Y. Hull form optimization in the conceptual design stage considering operational efficiency in waves. *Proc. Inst. Mech. Eng. Part M J. Eng. Marit. Environ.* **2019**, *233*, 745–759. [CrossRef]
6. Liu, Q.; Feng, B.; Liu, Z.; Zhang, H. The improvement of a variance-based sensitivity analysis method and its application to a ship hull optimization model. *J. Mar. Sci. Technol.* **2017**, *22*, 694–709. [CrossRef]
7. Guan, G.; Zhuang, Z.; Yang, Q. Design parameter sensitivity analysis for SWATH with minimum resistance at design and service speeds. *Ocean Eng.* **2021**, *240*, 961–972. [CrossRef]
8. Jeon, M.; Yoon, H.K.; Hwang, J. Analysis of the dynamic characteristics for the change of design parameters of an underwater vehicle using sensitivity analysis. *Int. J. Nav. Archit. Ocean Eng.* **2018**, *10*, 508–519. [CrossRef]
9. Coppedè, A.; Gaggero, S.; Vernengo, G. Hydrodynamic shape optimization by high fidelity CFD solver and Gaussian process based response surface method. *Appl. Ocean Res.* **2019**, *90*, 841–851. [CrossRef]
10. Fu, X.; Lei, L.; Yang, G. Multi-objective shape optimization of autonomous underwater glider based on fast elitist non-dominated sorting genetic algorithm. *Ocean Eng.* **2018**, *157*, 339–349. [CrossRef]
11. Yue, Y.; Han, C.; Cui, Y. TSP wavefield separation and noise suppression based on the LC-KL-DSW method. *J. Appl. Geophys.* **2022**, *197*, 552–560. [CrossRef]
12. Fan, Y.; Xu, K.; Wu, H. Spatiotemporal Modeling for Nonlinear Distributed Thermal Processes Based on KL Decomposition, MLP and LSTM Network. *IEEE Access* **2020**, *8*, 11–21. [CrossRef]
13. Ahuja, R.; Lall, B.; Prasad, S. On impropriety for a large-sized discrete fourier transform of a real-valued stationary process. *Signal Process.* **2022**, *193*, 397–402. [CrossRef]

14. Trudu, M.; Pilia, M.; Hellbourg, G.; Pari, P.; Antonietti, N.; Maccone, C.; Melis, A.; Perrodin, D.; Trois, A. Performance analysis of the Karhunen–Loève Transform for artificial and astrophysical transmissions: Denoizing and detection. *Mon. Not. R. Astron. Soc.* **2020**, *494*, 69–83. [CrossRef]
15. Siripatana, A.; Le Maitre, O.; Knio, O. Bayesian inference of spatially varying Manning's n coefficients in an idealized coastal ocean model using a generalized Karhunen-Loève expansion and polynomial chaos. *Ocean Dyn.* **2020**, *70*, 1103–1127. [CrossRef]
16. Reed, R.L.; Roberts, J.A. Application of the Karhunen-Loève Transform to the C5G7 benchmark in the response matrix method. *Ann. Nucl. Energy* **2017**, *103*, 350–355. [CrossRef]
17. Allaix, D.L.; Carbone, V.I. An efficient coupling of FORM and Karhunen–Loève series expansion. *Eng. Comput.* **2016**, *32*, 1–13. [CrossRef]
18. Azevedo, J.S.; Wisniewski, F.; Oliveira, S.P. A Galerkin method with two-dimensional Haar basis functions for the computation of the Karhunen–Loève expansion. *Comput. Appl. Math.* **2018**, *37*, 1825–1846. [CrossRef]
19. Liu, L.; Yan, J.; Zheng, X.; Peng, H.; Guo, D.; Qu, X. Karhunen-Loève transform for compressive sampling hyperspectral images. *Opt. Eng.* **2015**, *54*, 014106. [CrossRef]
20. Ai, X. A note on Karhunen–Loève expansions for the demeaned stationary Ornstein–Uhlenbeck process. *Stat. Probab. Lett.* **2016**, *117*, 113–117. [CrossRef]
21. Chowdhary, K.; Najm, H.N. Bayesian estimation of Karhunen–Loève expansions; A random subspace approach. *J. Comput. Phys.* **2016**, *319*, 280–293. [CrossRef]
22. Feng, J.W.; Cen, S.; Li, C.F.; Owen, D.R.J. Statistical reconstruction and Karhunen-Loève expansion for multiphase random media. *Int. J. Numer. Methods Eng.* **2016**, *105*, 3–32. [CrossRef]
23. Diez, M.; Campana, E.F.; Andstern, F. Design-space dimensionality reduction in shape optimization by Karhunen–Loève expansion. *Comput. Methods Appl. Mech. Eng.* **2015**, *283*, 1525–1544. [CrossRef]
24. Diez, M.; Serani, A.; Campana, E.F. Design-Space Dimensionality Reduction for Single- and Multi-Disciplinary Shape Optimization. In Proceedings of the 17th AIAA/ISSMO Multidisciplinary Analysis and Optimization Conference, Washington, DC, USA, 13–17 June 2016.
25. Diez, M.; Serani, A.; Stern, F. Combined Geometry and Physics Based Method for Design-space Dimensionality Reduction in Hydrodynamic Shape Optimization. In Proceedings of the 31st Symposium on Naval Hydrodynamics, Monterey, CA, USA, 11–16 September 2016.
26. Hotelling, H. Note on Edgeworth's Taxation Phenomenon and Professor Garver's Additional Condition on Demand Functions. *Econometrica* **1933**, *1*, 408–409. [CrossRef]
27. Hotelling, H.; Hotelling, F. A new analysis of duration of pregnancy data. *Am. J. Obstet. Gynecol.* **1932**, *23*, 643–657. [CrossRef]
28. Hotelling, H. Demand Functions with Limited Budgets. *Econometrica* **1935**, *3*, 66–78. [CrossRef]
29. Feng, B.; Zhan, C.; Liu, Z. Application of Basis Functions for Hull Form Surface Modification. *J. Mar. Sci. Eng.* **2021**, *9*, 1005. [CrossRef]
30. Chang, H.; Zhan, C.; Liu, Z.; Cheng, X.; Feng, B. Dynamic sampling method for ship resistance performance optimisation based on approximated model. *Ships Offshore Struct.* **2021**, *16*, 386–396. [CrossRef]
31. Cheng, X.; Feng, B.; Liu, Z. Hull surface modification for ship resistance performance optimization based on Delaunay triangulation. *Ocean Eng.* **2018**, *153*, 333–344. [CrossRef]
32. Cheng, X.; Feng, B.; Chang, H. Multi-objective optimisation of ship resistance performance based on CFD. *J. Mar. Sci. Technol.* **2019**, *24*, 152–165. [CrossRef]
33. Ouyang, X.; Chang, H.; Feng, B. Information Matrix-Based Adaptive Sampling in Hull Form Optimisation. *J. Mar. Sci. Eng.* **2021**, *9*, 973. [CrossRef]
34. Qiang, Z.; Bai-Wei, F.; Zu-Yuan, L.; Hai-Chao, C.; Xiao, W. Optimization method for hierarchical space reduction method and its application in hull form optimization. *Ocean Eng.* **2022**, *262*, 112–129. [CrossRef]
35. Qiang, Z.; Hai-Chao, C.; Bai-Wei, F.; Zu-Yuan, L.; Cheng-Sheng, Z. Research on knowledge-extraction technology in optimisation of ship-resistance performance. *Ocean Eng.* **2019**, *179*, 325–336. [CrossRef]

Journal of
Marine Science and Engineering

MDPI

Article

An Approach to Determine Optimal Bow Configuration of Polar Ships under Combined Ice and Calm-Water Conditions

Hui Li [1], Yan Feng [1,*], Muk Chen Ong [2], Xin Zhao [1] and Li Zhou [3]

[1] College of Shipbuilding Engineering, Harbin Engineering University, Harbin 150001, China; huili@hrbeu.edu.cn (H.L.); zhaoxinyi@hrbeu.edu.cn (X.Z.)
[2] Department of Mechanical and Structural Engineering and Materials Science, University of Stavanger, NO-4036 Stavanger, Norway; muk.c.ong@uis.no
[3] School of Naval Architecture and Ocean Engineering, Jiangsu University of Science and Technology, Zhenjiang 212003, China; zhouli209@hotmail.com
* Correspondence: fengyan@hrbeu.edu.cn

Abstract: Selecting an optimal bow configuration is critical to the preliminary design of polar ships. This paper proposes an approach to determine the optimal bow of polar ships based on present numerical simulation and available published experimental studies. Unlike conventional methods, the present approach integrates both ice resistance and calm-water resistance with the navigating time. A numerical simulation method of an icebreaking vessel going straight ahead in level ice is developed using SPH (smoothed particle hydrodynamics) numerical technique of LS-DYNA. The present numerical results for the ice resistance in level ice are in satisfactory agreement with the available published experimental data. The bow configurations with superior icebreaking capability are obtained by analyzing the sensitivities due to the buttock angle γ, the frame angle β and the waterline angle α. The calm-water resistance is calculated using FVM (finite volume method). Finally, an overall resistance index devised from the ship resistance in ice/water weighted by their corresponding weighted navigation time is proposed. The present approach can be used for evaluating the integrated resistance performance of the polar ships operating in both a water route and ice route.

Keywords: bow configuration; level ice; ice resistance; sensitivity analysis; integrated evaluation

Citation: Li, H.; Feng, Y.; Ong, M.C.; Zhao, X.; Zhou, L. An Approach to Determine Optimal Bow Configuration of Polar Ships under Combined Ice and Calm-Water Conditions. *J. Mar. Sci. Eng.* **2021**, *9*, 680. https://doi.org/10.3390/jmse9060680

Academic Editors: Carlos Guedes Soares and Serge Sutulo

Received: 9 June 2021
Accepted: 17 June 2021
Published: 21 June 2021

1. Introduction

The bow configuration of a polar ship is determined by a basic idea, i.e., the minimum power to move forward in the ice. From the design point of view, level ice resistance is also one of the most significant factors. There are many cases of polar ships with self-icebreaking ability sailing in level ice, for example, PUGNAX (polar deck carrier) [1], MV Xue Long (polar scientific research vessel) and MV Xue Long II. At the current stage of polar exploration, this kind of ship sailing in level ice is also considered as an important condition. The primary requirement for these polar ships is a good performance in level ice. Good performance means low ice resistance, high propulsive efficiency and guaranteed continuity in icebreaking [2]. Some researches [3–5] suggested that ice resistance is the main contribution to the total resistance in level ice. Apart from the ice properties and the power of polar ships, the ice resistance is significantly influenced by the bow configurations. In the evolution of the bow configuration, the frame angle has been increasing, while the waterline angle has decreased, and the buttock angle has reduced from about 30° to 20° in the early years [6]. From the perspective of ship–ice interaction, a description about the changes of the bow angles is explained below. When the ship is operating in ice region, the ice is crushed at the stem post of the bow. The buttock angle has been reduced because a smaller buttock angle induces more bending moment and less horizontal force. The waterline angle mainly affects the crushed ice-pushing capability of the bow. The decrease in waterline angle leads to the decrease in longitudinal crushed ice force. However, the bow needs more contact area and extrusion force to bend the ice with the decrease in frame angle. Therefore, the purpose of these changes of the bow angles is to achieve a superior icebreaking capability of a polar ship. It seems that the bow configurations of modern polar ships are quite similar. The ice resistance may nevertheless show notable differences. Consequently, all the changes of the bow configuration aim to obtain a superior icebreaking capability of polar ships navigating in ice region.

Several methods for ice resistance prediction have been proposed, which take into account the influence of geometric variations of the hull. In practical ship design, empirical or semi-empirical prediction methods may be applied to estimate the resistance in the preliminary design phase. Erceg and Ehlers [7] reviewed six semi-empirical level ice resistance prediction methods (Vance [8]; Lewis et al. [9]; Zahn and Phillips [10]; Keinonen [11,12]; Lindqvist [13]; Riska et al. [14]). Ice resistances of four ships with different sizes were calculated and compared based on the six semi-empirical prediction methods. The methods show significant discrepancies for larger ships, with differences in the predicted ice resistance as large as 100–200%. In the study of Lindström [15], most of the relevant ice parameters and hull configurations including the bow angles were regarded in the ice resistance prediction. Based on the study of Lindqvist [13], Riska et al. [14] proposed a formula for calculating the level ice resistance. This formula takes into account the particulars of the ship and the ice thickness. The empirical coefficients are derived from full-scale tests performed in the Baltic Sea. Jeong [16] decomposed the ice resistance into ice buoyancy resistance, ice-clearing resistance and icebreaking resistance. Using the component method, Jeong [16] conducted an ice model test of MOERI standard icebreaker model, and non-dimensional coefficients in ice resistance prediction formula are obtained based on the data of model test. Lubbad and Løset [17] established new analytical closed-form solutions to represent the icebreaking process of ship–ice interaction in real-time. These works have offered good insights into the method of estimating the ship performance accounting for ship–ice interaction.

Because the experimental research can reveal real physical effects of ship–ice interaction, some detailed analyses have been carried out to study the bow configurations based on the full-scale experiment as well as the ice model tests. Varges [18] investigated the bow shape of an icebreaker based on the full-scale experiment. The vessel of concern was the Soviet polar icebreaker MUDYUG, which was built in 1972 and given a new Thyssen/Waas forebody in 1986. The conventional angled bow was replaced with a form more resembling a sled. The sleigh-shaped bow has better performance and less vibration in the icebreaking process. Based on the ice model test results of HSVA's database from 1996 to 2014, Myland and Ehlers [19] focused in particular on the icebreaking patterns in level ice with regard to relevant hull shape parameters. The ship models with similar scales were chosen for analysis, so as to have model ice conditions, model speed values and ship model dimensions in the same range.

More recently, numerical simulations, e.g., discrete element method (DEM) and finite element method (FEM), have been widely used to analyze the influence of ship hull forms on the ice performance. Lu et al. [20] applied FEM to calculate the ice load acting on conical structures colliding with level ice. The mesh size dependency of the bending failure of level ice was eliminated by taking into account the random spatial distribution of material properties. Wang and Zou [21] used FEM to investigate the effects of the wedge stem and the shovel stem on the damage of the ice layer and the resistance during the icebreaking process. The result of wedge-shaped bow and scoop bow in the process of icebreaking were compared, and some suggestions for the design of icebreaker hull lines were given. Li et al. [22] estimated the ice load acting on the shoulder and midship of a polar ship by applying the extended finite element method (XFEM). Long et al. [23] adopted DEM with the bond and failure model to simulate the breaking process of sea ice colliding with a conical structure and vertical pile. The results showed that the bending failure occurs as the ice cover moves upward or downward, depending on the water level. Wang et al. [24] combined an elastoplastic softening constitutive model with cohesive element method (CEM) to simulate the continuous icebreaking process in level ice. This method captured well the main features, including the local crushing and the bending failure, of the ship–ice interactions.

In the past few years, the smoothed particle hydrodynamics (SPH) method has been used to effectively model large deformation and failure behavior of solids, including ice. Das and Ehlers [25] carried out a numerical simulation of crushing and bending failure of ice using SPH. In their study, the numerical bending results for force, displacement and failure time were compared with earlier simulations of in situ four-point bending test results and finite element simulations. The crushing failure was compared with experiments conducted by Häusler [26], and it was found that the SPH approach was in good agreement with the experiment. Furthermore, the SPH method is extended so as to simulate the actual behavior of sea ice as ships progress through level ice. Zhang et al. [27] reported that the SPH results agreed well with the experimental data of three-point bending test. This implies that the present SPH model can produce accurate results for simulating the ice failure problem. Zhang et al. [27] also showed a good agreement between the results of the SPH and the experimental data for the ice–ship interaction. They indicated that the numerical accuracy and stability of SPH were satisfactory.

Usually, polar ships also navigate in water ways, but the calm-water performance is ignored in the researches above. Especially for polar ships converted from the merchant vessel, which, most of the time, are operated in both water ways and ice regions. It is necessary to conduct integrated analyses that account for the resistance both in ice and in water. However, the literature on the optimization of bow configuration of polar ships navigating in combined routes is limited. Polach et al. [28] proposed a method which includes the ship performance evaluation based on ship merit factor (SMF). This method combined SMF with a route-specific ship and allows one to compare the technical and economic performance of ships operating in open water and ice. The performance of different ship designs that operated along the route Rotterdam to Yokohama through the Suez Canal and the Northern Sea Route was studied.

This study aims at devising an approach to determine the optimal bow of polar ships operating in both water route and ice route. In the paper, a numerical simulation of an icebreaking vessel going straight ahead in level ice is performed using the SPH numerical technique. The present numerical results are compared with the experimental data by Zhou et al. [29]. A polar research vessel, MV Xue Long, is used as a reference ship to demonstrate the usage of the proposed approach. Comparative analyses are carried out to evaluate the sensitivities of the buttock angle γ, the frame angle β and the waterline angle α on the icebreaking capability. Integrated numerical simulations are conducted to study the icebreaking capability and calm-water performance of the polar ship with different bow configurations. An overall resistance index C_r is proposed to determine the optimal bow configuration of polar ships navigating in a combined route.

2. Validation of Ice Material Model

According to the mechanical properties of ice, the constitutive model of ice is established using the SPH method. The numerical model is validated by comparing the results with the experiment by Kim et al. [30]. Furthermore, a ship–ice–water interaction model is proposed for the simulation of a ship moving in level ice. The present numerical results for the icebreaking patterns and the ice resistance are then compared with the experimental data by Zhou et al. [29].

2.1. Verification of Ice Material Model

During the process of deformation to failure of ice, the ductile–brittle transition occurs due to the effect of the loading rate. The behavior of ice at high strain rates (greater than 10^{-3} s^{-1}) is similar to the linear elastic material with brittle failure [31]. The material model takes into account high strain rates and is widely used to model the dynamic behavior of ice including elastic deformation and brittle fracture. Strain rate (s^{-1}) is the change in strain of a material with respect to time. In this study, the strain rate effect was transformed to the strain effect included in the material model. An isotropic elastic-plastic material model with a von Mises yield surface was used in the numerical simulation.

The von Mises yield condition is given by:

$$\phi = J_2 - \frac{\sigma_y^2}{3} \tag{1}$$

where the second stress invariant, J_2, is defined in terms of the deviatoric stress components as

$$J_2 = \frac{1}{2} s_{ij} s_{ij} \tag{2}$$

and the yield stress, σ_y, is a function of the effective plastic strain, ε_{eff}^p, and the plastic hardening modulus, E_p:

$$\sigma_y = \sigma_0 + E_p \varepsilon_{eff}^p \tag{3}$$

Pressure is given by the expression

$$p^{n+1} = K\left(\frac{1}{V^{n+1}} - 1\right) \tag{4}$$

where n is the number of time steps, K is the bulk modulus, and V is the relative volume defined as the ratio of the current volume over the initial volume.

When an element reaches the plasticity phase under the action of compression, it follows the yield surface until failure. An element is removed by erosion if the pressure surpasses the failure pressure. Meanwhile, failure by element erosion is activated if the equivalent plastic strain reaches the plastic failure strain. The values including plastic failure strain and failure pressure for the cone-shaped ice were also used in the ice sheet simulations. When the ice breaks, compression failure acquirement is met, as shown in Table 1.

Table 1. The parameters used in the numerical simulation.

Materials	Items	Values
Cone-shaped ice	Mass density	920 kg/m^3
	Yield stress	2.12 MPa
	Shear modulus	2.20 GPa
	Plastic failure strain	0.35
	Failure pressure	−4 MPa
Ice sheet	Mass density	906 kg/m^3
	Thickness	0.77 m/0.76 m
	Bending strength	724 kPa/844 kPa
	Compressive strength	1748 kPa/2192 kPa
	Elastic modulus	929 MPa/984 MPa
Water	Mass density	998 kg/m^3
	Acoustic velocity	1650 m/s
	Cut-off pressure	10 N/m^2
	Dynamic viscosity coefficient	1.0×10^{-3} N·s/m^2
	Volume correction	1.39
	Gruneisen gamma	0.49
	Internal energy	0.25 Kg/m^{2}s^2

In order to verify the accuracy of the above constitutive model of ice, a numerical simulation was carried out on the compressive cone-shaped ice experiments [30]. The models of cone-shaped ice and steel plate for the present numerical simulation are shown in Figure 1a. The SPH method is used to simulate the cone-shaped ice with a diameter of

100 mm and a cone angle of 120 degrees. In the bottom plane of the numerical cone-shaped ice, the movement of SPH particles is constrained to six degrees of freedom to simulate the ice fixed on the experimental device. A rigid body is used to simulate the steel plate. The cone-shaped ice is vertically compressed by the steel plate at the speed of 1 mm/s and 100 mm/s. The brittle fracture was observed in the numerical calculations, as shown in Figure 1b,c.

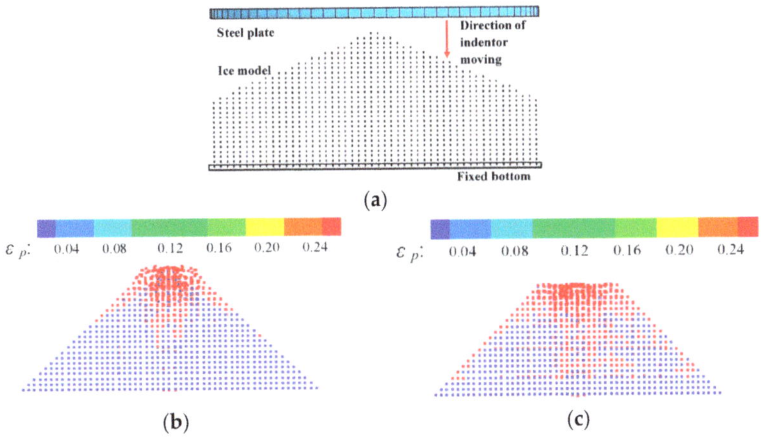

Figure 1. The compressive cone-shaped ice, where (**a**) is the present numerical models of the cone-shaped ice and the steel plate; and the brittle fracture of the ice cone in the numerical simulation at different time (compression speed v = 100 mm/s): (**b**) t = 0.04 s; and (**c**) t = 0.08 s (contours of plastic strain).

Figure 2a gives the convergence tests on compression force, in which N is the total particle number. Three particle numbers of 27,974; 37,196 and 51,192 are tested in the compressive cone-shaped ice. The relative error Er_f is defined as the errors between experimental results [30] and SPH results, which are calculated by:

$$Er_f = \sqrt{\sum_{i=1}^{n} (f_i - f_{0i})^2} \Big/ \sqrt{\sum_{i=1}^{n} f_{0i}^2} \tag{5}$$

where f_i is the compression force of SPH with different particle numbers at $t = t_i$; and f_0 is the compression force of experimental data at $t = t_i$. The results of Figure 2a indicate that the errors of the compression force decrease with the increasing particle number and demonstrate a roughly first-order convergence ratio. Therefore, 37,196 particles are deemed adequate for the simulation of ice cone. The force-displacement curves of the numerical simulation at the compression speed of 1 mm/s and 100 mm/s are presented in Figure 2b,c. Numerical results are compared with the experimental data [30]. It is found that, in both the high-speed loading and the low-speed loading, the ice force increases with the displacement. As shown in Figure 2b,c, the force pattern is captured by the numerical simulation with satisfactory accuracy. The numerical method agrees better with the experiment in the amplitude of ice force for lower loading speeds. It also can be seen that more fluctuations are present in the solution for higher loading speed. When the ice is broken to a certain length, the ice force does not increase but decrease due to the unloading of the pressure. In Figure 2b,c, the strain rate is 10^{-2} s^{-1} for v = 1 mm/s, and 1 s^{-1} for v = 100 mm/s. Both fall into the range of high strain rates (>10^{-3} s^{-1}), where the ice has the behavior as brittle materials. Moreover, the same cut-off stress value is used in the present numerical ice model. In fact, the cut-off stress decreases slightly with the increase in loading speed in the brittle range. This is why the peak contact forces are somewhat close (35 and 38 kN, respectively), and the present numerical method shows a little overestimation of the amplitude for a higher loading speed. The comparisons show that the present numerical ice model can give a similar trend of the compression process of ice with that of the experiment, although there are some differences in details.

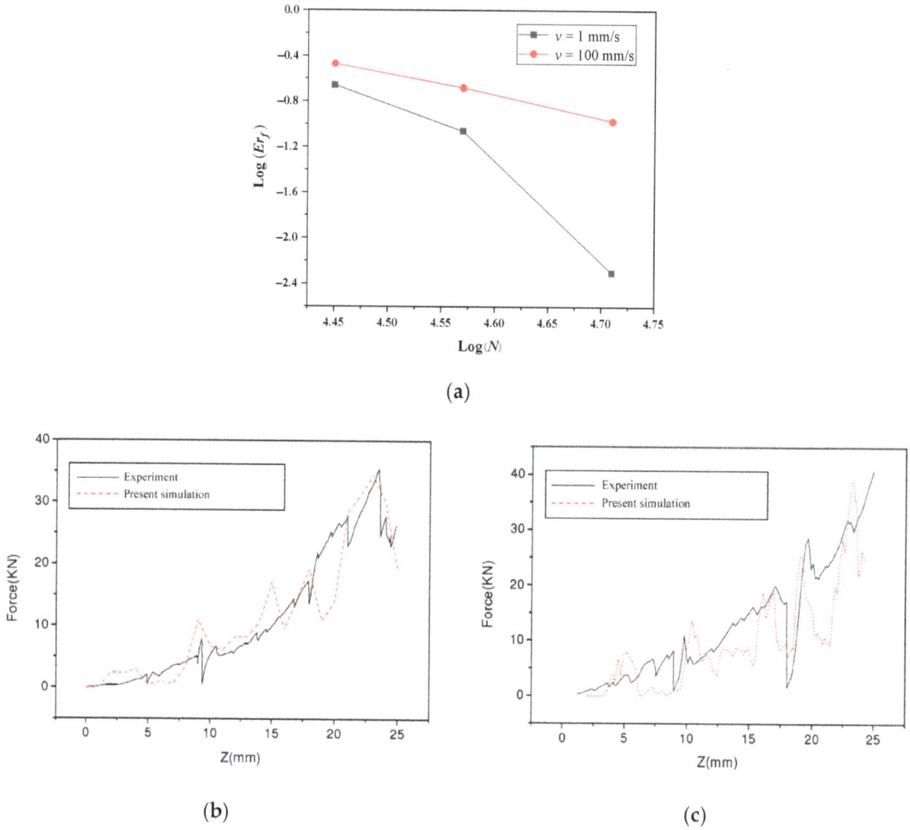

Figure 2. Convergence tests of force by different compression speed (**a**), and comparisons of the force-displacement curves at different compression speed (**b**) $v = 1$ mm/s and (**c**) $v = 100$ mm/s.

2.2. Description of Model Test

Zhou et al. [29] conducted a series of ice resistance tests where the model of an icebreaking vessel, MT Uikku, was towed in an ice basin. A scale model of 1:31.6 was built without any appendices such as rudders and propellers. The full-scale primary dimensions of MT Uikku are given in Table 2. The model was towed with the carriage to simulate the ice–hull interaction process. In the model tests, the ice drift angle and the hull heading angle were constant at 0°. The ice force was measured and digitally sampled on a computer and recorded as an analogue signal as a back-up. The measured data were sampled at a frequency of 107 Hz. The test program and measured ice properties are summarized in Table 3, where h_i is ice thickness; σ_b and σ_c are bending strength and compressive strength of ice; E_i is the elastic modulus of ice; and V_i is ship speed.

Table 2. Full-scale primary dimensions of MT Uikku.

Items	Values
Length	150 m
Length of parallel	65 m
Breadth	21 m
Tested draft	9.50 m

Table 3. Test matrix with the model speed and ice properties.

Test No.	h_i (m)	σ_b (kPa)	σ_c (kPa)	E_i (MPa)	V_i (m/s)
1	0.77	724	1748	929	0.2
2	0.76	844	2192	984	0.5

2.3. Numerical Simulation

A finite element ship model of the MT Uikku is created in the numerical simulation. The ship form parameters are presented in Table 2. The icebreaking pattern and the ice resistance are investigated regardless of the response of ship structures. Therefore, only the hull of the ship is established and simplified to a rigid body in the present numerical simulation. The model is going straight ahead in the x-direction at predefined constant speeds. Based on the SPH numerical technique of LS-DYNA hydrocode, the problem of a ship moving in level ice is simulated. The dimensions of domain are 600 m $\times$ 150 m $\times$ 75 m. The level ice floats on water, and its two long sides are constrained in the x-direction.

In the mathematical description of fluid, the equation of state describes the relationship between volume deformation and pressure. The null material is used to describe the water flow. The Gruneisen equation is adopted to describe the state of water. The parameters used in the numerical simulation are shown in Table 1. The force data are sampled every 0.01 s. Figure 3 shows the typical configuration of the numerical model applied to calculate the ice resistance.

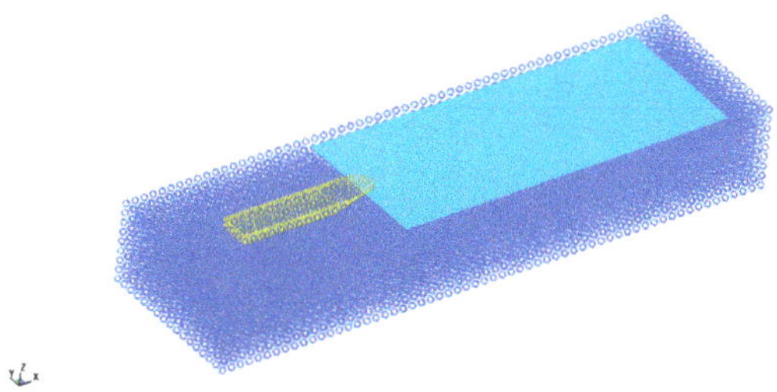

Figure 3. A typical configuration of the present numerical model.

2.4. Comparisons and Discussions of Results

Numerical simulations are performed to simulate the Test 1 and Test 2 in Table 3. Figure 4 shows the icebreaking stress and the icebreaking pattern in the numerical simulation of Test 2. As shown in Figure 4a, the stress is generated by the deformation of the ice, which mainly occurs in the contact area between ice and hull. A concentrated area of high stress is present near the bow, and larger ice pieces are observed in the bow and shoulder area. As the hull goes straight ahead in level ice, the open channel is generated in Figure 4b. As can be seen in Figure 4a,b, the width of the channel is approximately equal to the beam of the model ship. The icebreaking patterns obtained in the present numerical simulation is similar to that in Zhou et al. [29].

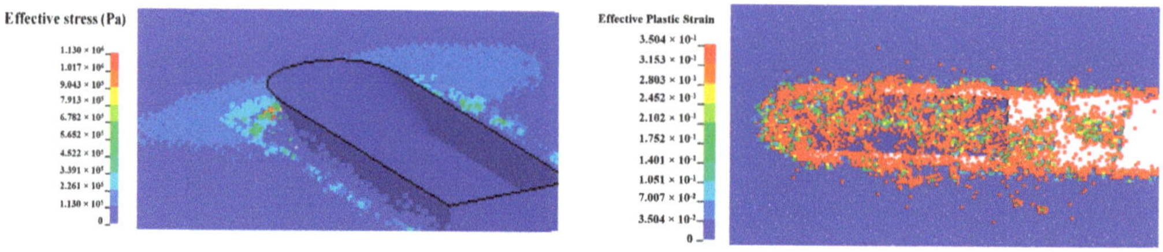

(a) (b)

Figure 4. Icebreaking pattern of the present numerical simulation, where (**a**) is the stress distribution of level ice and (**b**) is the open channel in the icebreaking process (contours of plastic strain).

Figure 5a gives the convergence tests on ice resistance, in which M is the total particle number. Three particle numbers of 210,555; 262,709 and 307,601 are tested in the icebreaking simulations. The relative error Er_r is defined as the errors between experimental results [29] and SPH results, which are calculated by:

$$Er_r = |r_0 - r| / r_0 \tag{6}$$

where r is the ice resistance of SPH with different particle numbers from t = 0 s to t = 30 s, r_0 is the ice resistance of experimental data from t = 0 s to t = 30 s. The results of Figure 5a indicate that the errors of the ice resistance decrease with the increasing particle number and demonstrate a roughly first-order convergence ratio. Hence, 262,709 particles are deemed adequate for the simulation of icebreaking. Time histories of ice forces in the longitudinal direction of Test 1 and Test 2 (Table 3) from the present simulation and the experiment [29] are presented in Figure 5b,c. Since the ice resistance commonly refers to the time average of all longitudinal forces due to ice acting on the ship, a comparison of results in a quantitative way is conducted. The mean values of the ice forces at the steady stage are also presented in Figure 5b,c. The differences between numerical results and model test data are identified to a certain extent. The ice force fluctuates during the icebreaking process while the force is never zero. This is because that there are always multiple ice contacts along the hull as the ship moves forward. These contacts are a series of events during the process, and the ice forces are the summation of all the contact forces. In addition, ice submergence resistance is a significant contribution. Some previous numerical simulation, e.g., [3], shows that submersion plays a significant role even at slow speed. The existence of submersion is also one of the reasons that the time history of ice forces rarely drops to zero. The mean values of the present simulation results for the ice forces, i.e., ice resistance, approach those of the experimental data from [29]. The fluctuation amplitudes of the simulated ice force are slightly higher than those of the experimental results. Figure 4b shows that the pieces of large ice cusps generated in the present numerical simulation are more than those observed in the experiment [29]. Data processing methods in Wang et al. [24] are referenced in Figure 5b,c. In order to eliminate the noise from the results, the numerical force curves are plotted by processing the original data through the filter used in the experiment. Higher fluctuation amplitudes of calculated ice resistance have been reduced. The results of the present numerical simulations are in better agreement with the experimental data. The ice resistances of numerical simulations and experiments are denoted by $R_{i\,num}$ and $R_{i\,exp}$ in Table 4, respectively. The standard deviations (StDev) of $R_{i\,num}$ are 156 kN for Test 1 and 99 kN for Test 2. The standard deviations of ice resistance versus the mean value (CV) for $R_{i\,num}$ are 0.31 for Test 1 and 0.16 for Test 2. The relative errors (E_r) between $R_{i\,num}$ and $R_{i\,exp}$ are 9.04% for Test 1 and 10.56% for Test 2. Although this observation indicates that the present numerical simulation cannot capture all the details of the broken ice floes around the hull, the approach adopted here, as shown in the comparative analyses above, is appropriate for evaluating the ice resistance performance. Thus, the present approach can serve as a tool to study the bow configuration of the polar ship.

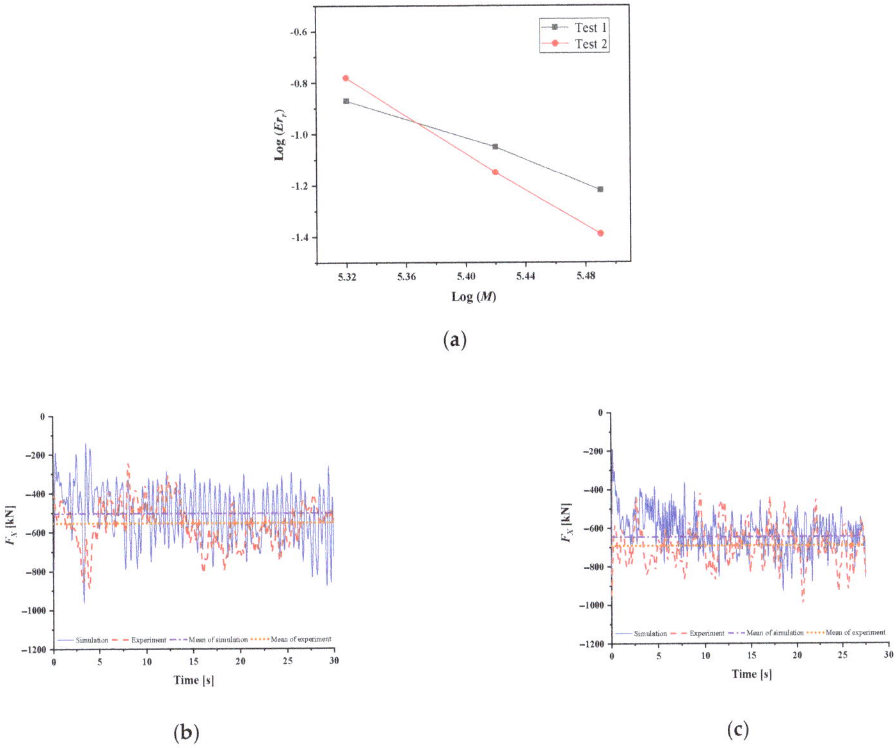

Figure 5. Convergence tests of ice resistance (**a**) and comparisons of longitudinal force between experiments and simulations, where (**b**) is the result of Test 1 and (**c**) is the result of Test 2, respectively.

Table 4. Ice resistances of numerical simulations and experiments.

Test No.	$R_{i\,num}$ (kN)	StDev of $R_{i\,num}$ (kN)	CV of $R_{i\,num}$	$R_{i\,exp}$ (kN)	E_r
1	503	156	0.31	553	9.04%
2	618	99	0.16	691	10.56%

3. Approach to Determine the Optimal Bow Configuration of Polar Ships

Based on the dimensions of MV Xue Long, a full-scale hull model is established. The full-scale model is taken as the master model. Nine hull models are created with different bow configurations so as to analyze the icebreaking capability. Then, the models are moved in the calm water (without ice) at a speed of 15 knots. The workflow of the approach is shown in Figure 6.

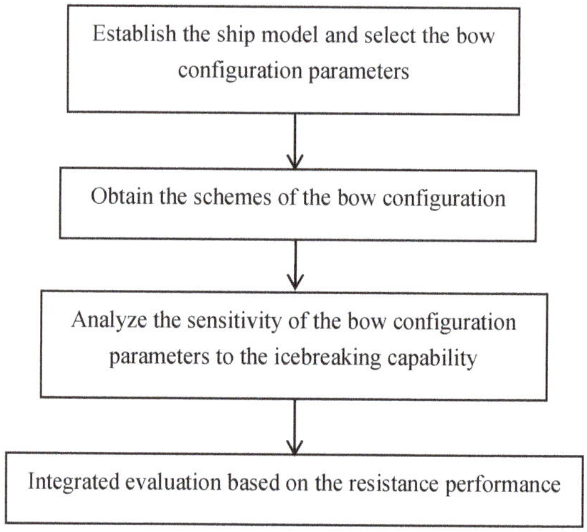

Figure 6. General workflow for the approach to determine the optimal bow configuration.

3.1. Ship Model and Bow Configuration Parameters

An icebreaking scientific research vessel, MV Xue Long, is modeled for the present study. MV Xue Long is an icebreaking research vessel owned by the Polar Research Institute of China. It was built at the Kherson Shipyard in Ukraine in 1993. It started as an icebreaking cargo and supply ship designed for the Russian Arctic. It was then converted from an Arctic cargo ship to a polar research and resupplying vessel by the mid-1990s. The ice class of MV Xue Long, assigned by the China Classification Society (CCS), is B1 [32].

The primary dimensions of MV Xue Long are presented in Table 5. As the present study focuses on the bow configuration instead of the structural response, the ship is modeled as a rigid body without any hull structures. The finite element model of the hull is shown in Figure 7. The bow configuration can be characterized by the buttock angle γ, the frame angle β and the waterline angle α [33], as illustrated in Figure 8.

Table 5. Primary dimensions of MV Xue Long.

Items	Values
Length overall	167.0 m
Length between perpendiculars	147.2 m
Breadth	22.6 m
Draught	9.0 m
Displacement	21,025 tons
Deadweight	8759 tons
Propulsion power	13.2 MW
Bollard pull	1650 kN
Maximum calm-water speed.	18 knots

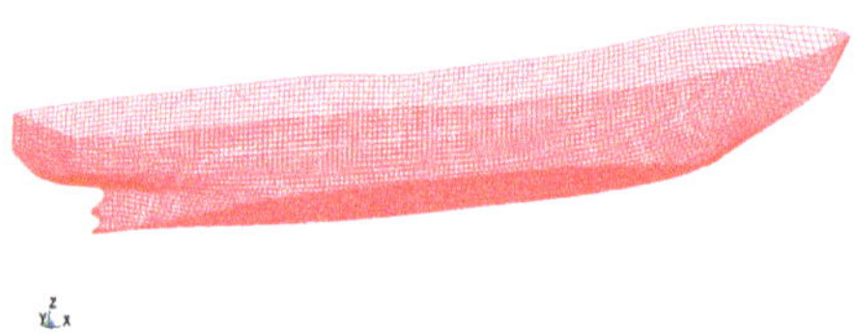

Figure 7. The numerical model used as a reference ship in the icebreaking simulations.

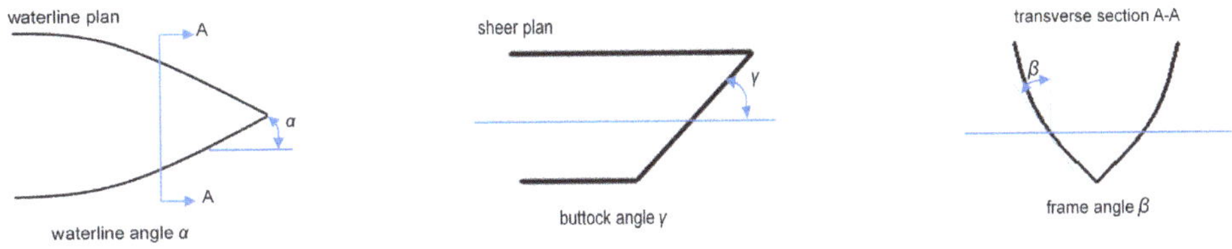

(**a**) (**b**) (**c**)

Note: α = upper ice waterline angle at the one-quarter of ship breadth
γ = buttock angle at upper ice waterline (angle of buttock line measured from horizontal)
β = frame angle at upper ice waterline
A-A is the transverse section at the one-quarter of ship breadth

Figure 8. Definition of bow configuration parameters, where (**a**) is α, (**b**) is γ and (**c**) is β, respectively.

3.2. Schemes of Bow Configuration

An analysis is conducted to analyze the effect of γ on the icebreaking capability, which is done by varying γ with α and β being fixed. In the simulations, the ship model is going straight ahead in level ice (h_{ice} = 0.5 m). The speed the ship can attain in level ice is calculated based on the present numerical simulation. The ice resistance equals the extra thrust available at different power levels and speeds. The net thrust curves used in the analysis are based on the theoretical calculation given by Equation (7) [34] and are considered as a preliminary curve. The average longitudinal speed the ship achieves in the designated ice conditions is determined from the intersection of the net thrust curve and the ice resistance curve.

$$T_{net} = T_B\left[1 - \frac{1}{3}\cdot\frac{u}{v_{ow}} - \frac{2}{3}\cdot\left(\frac{u}{v_{ow}}\right)^2\right] \tag{7}$$

where T_B is the bollard pull; u is the initial ship speed; and v_{ow} is the maximum calm-water speed.

Figure 9a shows the effect of γ on the average longitudinal ship speed. It can be seen that the relationship between the longitudinal ship speed in the continuous icebreaking process and γ is approximately a downward curve. The main reason is that a small γ induces a large bending moment and a small horizontal force. The effects of α and β on the average longitudinal ship speed are also analyzed using in the same manner. Figure 9b shows the average longitudinal ship speed for α = 17°, 21°, 25°, 29° and 33° with γ fixed to 23° and β to 53°. It is observed that the longitudinal ship speed decreases slightly with α increases. The reason is that the longitudinal crushed ice force F_x increases with α and results in the decrease in the longitudinal ship speed, as shown in Figure 9d. Unlike γ and α, β has a non-monotonic effect on the longitudinal ship speed when γ is fixed to 23° and α to 25°, as shown in Figure 9c. A fluctuating pattern of the influence of β on the average longitudinal speed is observed. However, the difference between the maximum and the minimum ship speed is less than 4%. It suggests that β has little effect on the total ice resistance and the ship speed.

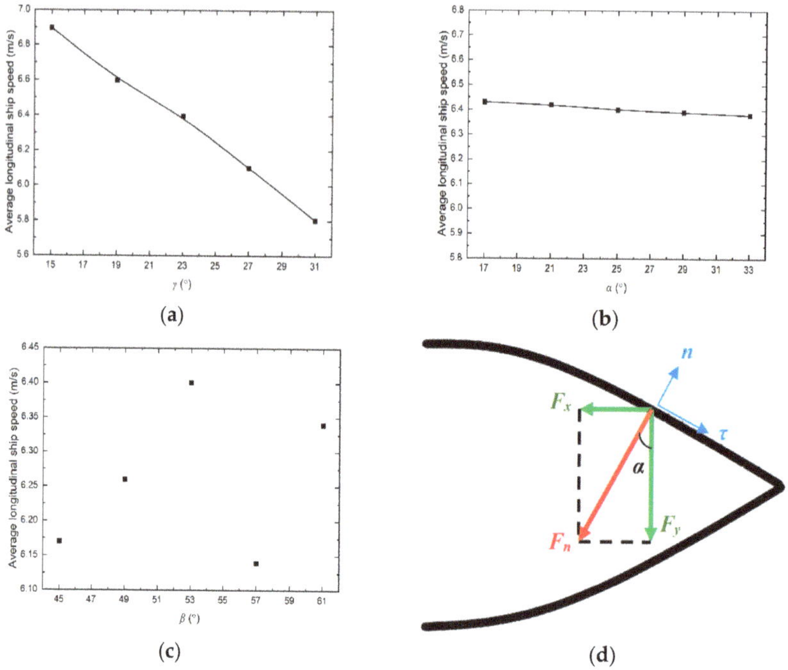

Figure 9. Average longitudinal ship speeds at different bow configuration parameters with (**a**) fixed $\alpha = 25°$ and $\beta = 53°$, (**b**) fixed $\gamma = 23°$ and $\beta = 53°$ and (**c**) fixed $\gamma = 23°$ and $\alpha = 25°$, respectively, and (**d**) illustrates the ice forces on the stem side.

To determine the optimal bow configuration, schemes of combinations of α, γ and β are obtained using the orthogonal design method [35]. A comprehensive research with three factors (α, γ and β) at three levels (i.e., three values of each factor) requires twenty-seven different cases. By using the orthogonal design method, there are only nine non-repetitive cases. As shown in Figure 10, the nine cases are evenly distributed in the entire research domain. The bow configuration schemes and parameters used in the orthogonal design are listed in Table 6.

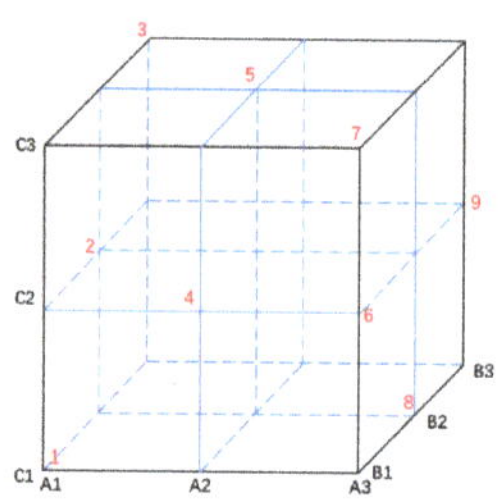

Figure 10. Distribution of the nine cases.

Table 6. Analysis of variance for the ice resistance.

Case	α	γ	β	Indicator (kN)
1	24°	22°	52°	895
2	24°	24°	54°	945
3	24°	26°	56°	1065
4	26°	22°	54°	865
5	26°	24°	56°	1025
6	26°	26°	52°	1155
7	28°	22°	56°	795
8	28°	24°	52°	1025
9	28°	26°	54°	1095
K_1	2905	2555	3075	T = 8865
K_2	3045	2995	2905	T^2 = 78,588,225
K_3	2915	3315	2885	P = 8,732,025
${K_1}^2$	8,439,025	6,528,025	9,455,625	Q_T = 8,842,825
${K_2}^2$	9,272,025	8,970,025	8,439,025	—
${K_3}^2$	8,497,225	10,989,225	8,323,225	—
Q_A	8,736,091.67	882,9091.67	8,739,291.67	—
S_A	4066.67	97,066.67	7266.67	—

3.3. Sensitivity of the Icebreaking Capability to the Bow Configuration Parameters

Sensitivity studies are carried out to analyze the influence of bow configurations on the icebreaking capability. Analysis of variance is employed to identify the critical affecting factor in the significance tests. The variance of the test results originates from the test errors and the variations of different factors:

$$S_T = Q_T - P \tag{8}$$

where S_T denotes the total variance of the test results. Q_T is the sum of squares of the indicators, and P is the revised value.

$$Q_T = \sum_{k=1}^{n} x_k^2 \tag{9}$$

$$P = \frac{1}{n}\left(\sum_{k=1}^{n} x_k\right)^2 \tag{10}$$

where n is the total number of tests, and x_k is the indicator for each test.

The variance of each factor can be expressed as

$$S_A = Q_A - P \tag{11}$$

where

$$Q_A = \frac{1}{a}\sum_{i=1}^{n_a} K_i^2 \tag{12}$$

$$K_i = \sum_{j=1}^{a} x_{ij} \tag{13}$$

where K_i denotes the sum of test results for factor A at level i. a is the number of tests for each level, and x_{ij} is the indicator of j-th test for factor A at level i.

Taking the ice resistance as the indicator of the icebreaking capability, Table 6 presents the variance analysis results for the significance test, where T is the sum of indicators. The variance of factors at three levels are ranked in the order of S_γ, S_β, and S_α.

The total variance of the test results $S_T = Q_T - P$ = 110,800, and the variance of the test error $S_E = S_T - S_\alpha - S_\gamma - S_\beta$ = 2400. Table 7 shows the significance test results of the effects of α, γ and β on the ice resistance, where d.f. denotes the degrees of freedom. The total d.f. of the test $f_T = n - 1 = 8$, the d.f. of each factor $f_A = n_a - 1 = 2$, and the d.f. of the test error $f_E = f_T - f_\alpha - f_\gamma - f_\beta = 2$.

Table 7. Significance test of the bow configuration parameters to the ice resistance.

Source	α	γ	β	Error
Sum Sq.	4066.67	97,066.67	7266.67	2400
d.f.	2	2	2	2
Mean Sq.	2033.33	48,533.33	3633.33	1200
F Value	1.70	40.50 *	3.01	—
$F_{0.05}(2,2)$	19.00	19.00	19.00	—
$F_{0.01}(2,2)$	99.01	99.01	99.01	—

Note: * is the significant factor affecting the indicator.

F is defined as the mean squared deviation (Mean Sq.) of each factor divided by the mean squared deviation of test error, which represents the magnitude of the sensitivity of each factor to the test results. Here, the F values of α, γ and β are 1.70, 40.5 and 3.01, respectively. Compared with the F-critical value, i.e., $F_{0.05}(2, 2) = 19.0$, which is obtained by a joint hypotheses test, the F value of γ is greater, indicating that γ has a significant effect on the test results. In other words, γ is the most important factor to the ice resistance, marked with *. In contrast, the F values of α and β are much less than 19.0, suggesting that the effect of α and β on the ice resistance is insignificant. Therefore, the bow configurations from Case 1, Case 4 and Case 7 have an excellent icebreaking capability, which is achieved with the minimum value of γ (i.e., $\gamma = 22°$).

3.4. Integrated Evaluation Based on the Resistance Performance

For polar ships sailing in both water and ice regions, it is economically beneficial to consider both water resistance and ice resistance by using a typical route for conceptual design. In the present study, the calm-water resistance of Case 1, Case 4 and Case 7 is calculated using the FVM-based commercial code of STAR-CCM+. The fluid domain and mesh of the present numerical model are shown in Figure 11a. The ship speed is 15 knots for the estimation of calm water resistance, which represents the design speed during operation. A grid convergence study and time-step study are performed in order to verify the present numerical model. Table 8 indicates the grid information and the resulting average calm-water resistance. Three grid numbers of 1,357,071; 3,088,921, and 4,401,879 are tested, respectively. Figure 11b shows the plot of average calm-water resistance with varying grid numbers. As the grid number increases, the average calm-water resistances approach an asymptotic infinite-grid number value. Three simulations (coarse, medium and fine) are completed with a constant refinement ratio $r = 2$. The order of convergence, p, is calculated using:

$$p = \ln[(193.875 - 188.520)/188.520 - 188.385]/\ln(2) = 5.31$$

Richardson extrapolation is performed to predict an estimate of the value of the average calm-water resistance at infinite grid number,

$$f_{n\to\infty} = 188.385 + (188.385 - 188.520)/\left(2^{5.31} - 1\right) = 188.382\,(\mathrm{kN})$$

This value is also plotted on Figure 11b. The grid convergence index for the fine grid solution can now be computed. A factor of safety of $F_S = 1.25$ is used since three grids were used to estimate the average calm-water resistance. The grid convergence index (GCI) for the medium and fine refinement levels is:

$$GCI_{23} = 1.25|(188.520 - 188.385)/188.385|/\left(2^{5.31} - 1\right) \cdot 100\% = 0.002316\%$$

The grid convergence index (GCI) for the coarse and medium refinement levels is:

$$GCI_{12} = 1.25|(193.875 - 188.520)/188.520|/\left(2^{5.31} - 1\right) \cdot 100\% = 0.091826\%$$

Grids are ensured in the asymptotic range of convergence by checking:

$$0.091826\%/\left(2^{5.31}0.002316\%\right) = 0.99938$$

which is approximately one and indicates that the solution is well within the asymptotic range of convergence. Therefore, 3,088,921 grids are sufficient to produce accurate results, which are used to calculate the calm-water resistance.

For the time-step convergence study, the time step is selected based on the numerical simulations in which a variety of Courant numbers, C = 0.075, 0.1, 0.25, and 0.5 are carried out. The Courant number describes the relationship between the time step and the space step. Intuitively speaking, the Courant number is the number of grids that a fluid particle can pass through in a time step. As shown in Figure 11c, when the Courant numbers of 0.075 and 0.1, the calm-water resistance obtained with Courant number being 0.075 and 0.1 is insignificant. Therefore, the Courant number C of 0.1 is used in this present study.

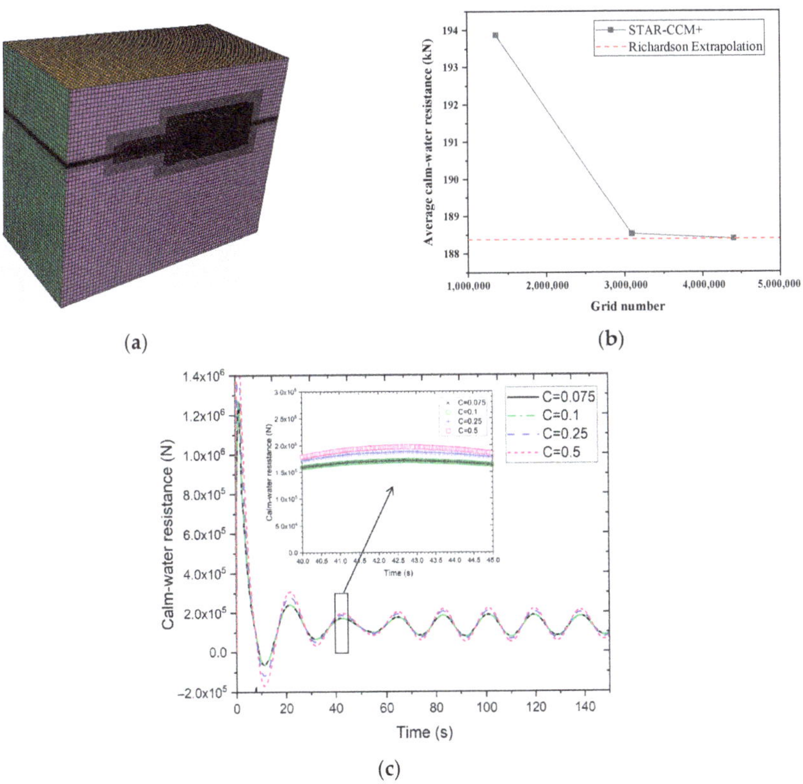

Figure 11. Numerical model and convergence study in the calculation of calm-water resistance, where (**a**) is the fluid domain extents and grid partition, (**b**) is the relationship of grid numbers to the average calm–water resistance and (**c**) is the time-step convergence study of the numerical calculations carried out with the Courant numbers, where C = 0.075, 0.1, 0.25, and 0.5, respectively.

Table 8. Grid information and the resulting average calm-water resistance.

Grid	Grid Number	Average Calm-Water Resistance (kN)
Coarse (3)	1,357,071	193.875
Medium (2)	3,088,921	188.520
Fine (1)	4,401,879	188.385

Figure 12 shows the time histories of the calm-water resistance for Case 1, Case 4 and Case 7. The calm-water resistance amplitude in the stable range, from 80 to 200 s, is used to calculate the mean value, i.e., the average calm-water resistance. Comparison of the numerical results shows that Case 7 has the maximum calm-water resistance. As can be seen in Table 6, Case 7 has the fullest bow, corresponding to the largest α and β, resulting in the maximum calm-water resistance. Results for the average calm-water resistance (R_{water}) and the ice resistance (R_{ice}) are presented in Table 9. It was found that the average calm-water resistance for Case 4 is about 3.87% larger than Case 1, and the average calm-water resistance for Case 7 is 4.79% larger than Case 4. The ice resistance for Case 4 is 3.35% smaller than for Case 1, and the ice resistance for Case 7 is 8.09% smaller than for Case 4. R_{water} and R_{ice} are significantly influenced by the bow configuration. The present results indicate that full bows are favorable for the performance in the ice but not beneficial to the performance in the calm-water. In addition, Table 9 shows that total resistance in the ice region is more than four times higher than in water region, because it is assumed that the ship speed is fixed to 15 knots in the water route and the invariant ice thickness is 0.5 m in the ice region. In this case, the ice resistance is approximately four times higher than the water resistance. However, the speed changes when the ship navigates along the ice routes according to the ice condition, such as the variety of ice thickness. In addition, the ship probably does not sail at the same speed in the water region. Therefore, the ratio of the ice resistance to the water resistance is not constant.

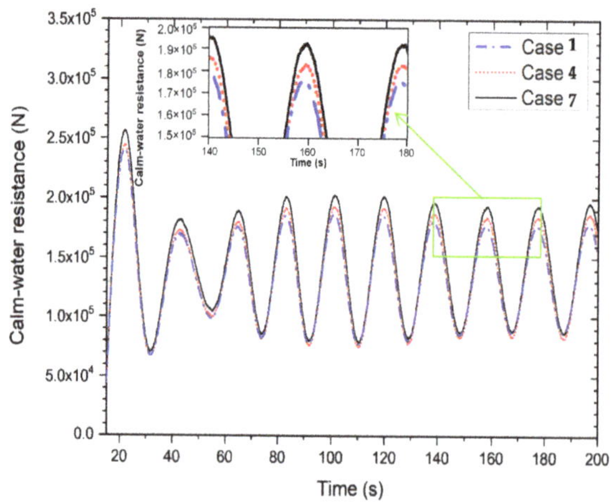

Figure 12. Calm-water resistance of Case 1, Case 4 and Case 7.

Table 9. C_r of the polar ship navigating in a combined route.

Case	R_{ice} (N)	R_{water} (N)	k_{ice}	k_{water}	C_r (N)
1	8.95×10^5	1.81×10^5	0.2	0.8	3.24×10^5
4	8.65×10^5	1.88×10^5	0.2	0.8	3.23×10^5
7	7.95×10^5	1.97×10^5	0.2	0.8	3.17×10^5

In most ice regions, the ship speed changes when a ship navigates along the ice routes according to the ice condition, such as the variety of ice thickness. If the ship captain wants to achieve a certain speed in the ice region, the ship needs to overcome resistance in ice. For that purpose, propeller must provide sufficient thrust, i.e., the ship engine is selected based on this. Ship probably does not sail at the same speed in ice region and in region without ice due to significantly different total resistances. In this case, the entire route of a polar ship should be divided by different segments based on ship speed from the harbor to the ice region. Thus, an overall resistance index C_r comprising the ship resistance and the corresponding weighted navigation time is proposed as follows,

$$C_r = \sum_{i=1}^{m} k_{water\,i} R_{water\,i} + \sum_{j=1}^{n} k_{ice\,j} R_{ice\,j} \tag{14}$$

where $k_{water\,i}$ and $k_{ice\,j}$ are the weight factors of the i-th route segment in the water region and the j-th route segment in the ice region, respectively; $R_{water\,i}$ and $R_{ice\,j}$ are the resistance of the i-th route segment in the water region and the j-th route segment in the ice region, correspondingly.

In the present study, C_r is calculated by only one uniform water segment and one uniform ice segment. It is assumed that the speeds of a polar ship in calm water and in level ice are both constant. For MV Xue Long, the weighted factor k_{water} is assumed to be 0.8, and k_{ice} is 0.2. C_r of Case 1, Case 4 and Case 7 is shown in Table 9, where C_r of Case 7 is the smallest. Therefore, Case 7 is selected as the optimal bow configuration for overall performance in both water and ice. It should be noted that k_{water} and k_{ice} are consistent with the actual navigation time of the polar ship of concern. For example, in August 2017, the Russian icebreaking LNG carrier Christophe de Margerie [36] passed through the Northern Sea Route (NSR) in six days and completed the entire 19-day journey from Hammerfest, Norway, to Boryeong, South Korea. In this case, k_{water} = 0.68 and k_{ice} = 0.32 can be used according to the article posted by Rachael [36]. The present method can be used to improve and optimize the bow configuration by assessing the resistance performance. The optimal route of a polar ship from the harbor to the ice region can also be studied using Equation (14). For the operation of vessels in the different routes, the mileage is considered in k_{water} and k_{ice}.

4. Conclusions

An approach to determine the optimal bow of polar ships based on present numerical simulation and available published experimental studies was proposed. In order to validate the constitutive model of ice, numerical simulations were performed using SPH, and the results were compared with the experiment by Kim et al. [21]. Then, a ship–ice–water interaction model was proposed for the simulation of a ship moving in level ice. The present numerical results for the ice resistance in level ice were in satisfactory agreement with the experimental data by Zhou et al. [20]. This method was then used to analyze the ice resistance of a polar ships with various bow configuration parameters, including the buttock angle γ, the frame angle β and the waterline angle α. Sensitivity analyses of the ice resistance to these parameters were evaluated by performing the analysis of variance. It was found that the effect of γ on the ice resistance is much more significant than that of α and β; thus, small γ values are desirable.

To assess the overall resistance performance of a ship that is operated in both ice and waters regions, the resistance in water was calculated using the FVM-based method. Then, an overall resistance index C_r devised from the ship resistance in ice/water weighted by their corresponding weighted navigation time was proposed. The formula to calculate the index takes a simple form but is particularly useful in practice as the resistances and weights can be easily obtained. Since the bow geometry is the most important factor to the overall resistance performance, the proposed formula provides a convenient approach to determine the optimal bow configuration of polar ships. A polar research vessel, MV Xue

Long, was used as a reference ship to demonstrate the usage of the proposed formula. In most cases, the entire route of a polar ship should be divided by different segments based on ship speed from the harbor to the ice region. In the present study, C_r was calculated by one uniform water segment and one uniform ice segment. It was assumed that the speeds of a polar ship in calm water and in level ice are both constant. More calculations by Equation (14) are needed when the route is divided by different segments based on ship speed. In addition, the present approach is applied to the evaluation based on the resistance performance of polar ships moving in level ice, which is a typical condition for polar ships with self-icebreaking ability. For merchant polar ships that may be operated in pack ice, improvement and validation of the approach shall be made for the desired accuracy.

Author Contributions: Conceptualization and methodology, H.L., Y.F. and M.C.O.; software, validation and formal analysis, Y.F. and X.Z.; investigation, H.L., Y.F., M.C.O. and L.Z.; resources, H.L., M.C.O. and L.Z.; writing—original draft preparation, H.L. and Y.F.; writing—review and editing, H.L., Y.F., M.C.O., X.Z. and L.Z.; visualization, H.L., Y.F. and X.Z.; supervision, H.L. and M.C.O.; project administration and funding acquisition, H.L. All authors have read and agreed to the published version of the manuscript.

Funding: This research was funded by National Natural Science Foundation of China, grant number 52071110.

Institutional Review Board Statement: Not applicable.

Informed Consent Statement: Not applicable.

Data Availability Statement: Not applicable.

Acknowledgments: The authors would like to acknowledgment College of Shipbuilding Engineering, Harbin Engineering University in purchasing and supporting the use of LS-DYNA and STAR-CCM+ software. The authors would also like to thank X. Q. Zhou in Harbin Engineering University for his discussions and suggestions.

Conflicts of Interest: The authors declare no conflict of interest.

References

1. van Lievenoogen, R.; Hovilainen, M.; Ville, V. AUDAX and PUGNAX, Unique PC-3 Class Heavy Transport Vessels Designed for Module Transportation for LNG Liquefaction Projects in Arctic Remote Locations. In Proceedings of the OTC Arctic Technology Conference, Houston, TX, USA, 5–7 November 2018.
2. Myland, D.; Ehlers, S. Theoretical investigation on ice resistance prediction methods for ships in level ice. In Proceedings of the 33rd International Conference on Ocean, Offshore and Arctic Engineering, OMAE 2014-23304, San Francisco, CA, USA, 8–13 June 2014.
3. Valanto, P. On the cause and distribution of resistance forces on ship hulls moving in level ice. In Proceedings of the 16th International Conference on Port and Ocean Engineering Conference under Arctic Conditions, POAC'01, Ottawa, ON, Canada, 12–17 August 2001.
4. Puntigliano, F. Experimental and Numerical Research on the Interaction between Ice Floes and a Ship's Hull during Icebreaking. Ph.D. Dissertation, Technical University Hamburg-Harburg, TUHH, Hamburg, Germany, 2003.
5. Riska, K. *Ship-Ice Interaction in Ship Design: Theory and Practice*; Cold Regions Science and Marine Technology: Amsterdam, The Netherlands, 2006; p. 15.
6. Dicke, R.A. An empirical review of the design and performance of icebreakers. *Mar. Technol.* **1989**, *4*, 145–159.
7. Erceg, S.; Ehlers, S. Semi-empirical level ice resistance prediction methods. *Ship Technol. Res.* **2017**, *64*, 1–14. [CrossRef]
8. Vance, G.P. *Analysis of the Performance of a 140-Ft Great Lakes Icebreaker: USCGC Katmai Bay*; Report No. 80-8; Army CRREL: Hanover, NH, USA, 1980.
9. Lewis, J.W.; Debord, F.W.; Bulat, V.A. Resistance and propulsion of iceworthy ships. *Trans. Soc. Nav. Archit. Mar. Eng.* **1982**, *90*, 249–276.
10. Zahn, P.B.; Phillips, L. *Towed Resistance Trials in Ice of the USCGC Mobile Bay (WTGB103)*; Transport Canada, Transportation Development Centre: Montreal, QC, Canada, 1987.
11. Keinonen, A.J.; Browne, R.P.; Revill, C.R. *Icebreaker Design Synthesis-Phase 2-Analysis of Contemporary Icebreaker Performance*; Report for Transportation Development Centre, Transport Canada, TP 10923E; AKAC Inc.: Victoria, BC, Canada, September 1991.
12. Keinonen, A.J.; Browne, R.P.; Revill, C.R.; Reynolds, A. *Icebreaker Characteristics Synthesis*; Report for Transportation Development Centre, Transport Canada, TP 12812E; AKAC Inc.: Victoria, BC, Canada, July 1996.
13. Lindqvist, G. A straightforward method for calculation of ice resistance of ships. In Proceedings of the 10th International Conference on Port and Ocean Engineering under Arctic Conditions, POAC'89, Lulea, Sweden, 12–16 June 1989; pp. 722–735.

14. Riska, K.; Wilhelmson, M.; Englund, K.; Leiviskä, T. *Performance of Merchant Vessels in the Baltic*; Research report no 52; Winter Navigation Research Board: Norrkoping, Sweden; Helsinki University of Technology, Ship Laboratory: Espoo, Finland, 1998.
15. Lindström, C.-A. Numerical estimation of ice forces acting on inclined structures and ships in level ice. In Proceedings of the 22nd International Offshore Technology Conference, OTC 6445, Houston, TX, USA, 7–10 May 1990.
16. Jeong, S.Y.; Lee, C.J.; Cho, S.R. Ice Resistance Prediction for Standard Icebreaker Model Ship. In Proceedings of the Twentieth (2010) International Offshore and Polar Engineering Conference, Beijing, China, 20–25 June 2010; pp. 1300–1304.
17. Lubbad, R.; Løset, S. A numerical model for real-time simulation of ship-ice interaction. *Cold Reg. Sci. Technol.* **2011**, *65*, 111–127. [CrossRef]
18. Varges, G. Advances in icebreaker design-the conversion of the soviet polar icebreaker MUDYUG into a Thyssen/Waas ship. *Schiff und Hafen* **1987**, *39*, 11.
19. Myland, D.; Ehlers, S. Influence of bow design on ice breaking resistance. *Ocean Eng.* **2016**, *119*, 217–232. [CrossRef]
20. Lu, W.; Løset, S.; Lubbad, R. Simulating ice-sloping structure interactions with the cohesive element method. *J. Offshore Mech. Arct. Eng.* **2014**, *136*, 031501. [CrossRef]
21. Wang, J.; Zou, Z. Ship's structural response during its collision with level ice based on nonlinear finite element method. *J. Vib. Shock* **2015**, *34*, 125–130.
22. Li, F.; Korgesaar, M.; Kujala, P.; Goerlandt, F. Finite element based meta-modeling of ship-ice interaction at shoulder and midship areas for performance simulation. *Mar. Struct.* **2020**, *71*, 102736. [CrossRef]
23. Long, X.; Liu, S.; Ji, S. Breaking characteristics of ice cover and dynamic ice load on upward-downward conical structure based on DEM simulations. *Comput. Part. Mech.* **2020**, *1*, 1–17. [CrossRef]
24. Wang, F.; Li, Z.; Zou, Z.; Song, M.; Wang, Y. Study of continuous icebreaking process with cohesive element method. *Brodogradnja Teor. I praksa Brodogr. I Pomor. Teh.* **2019**, *70*, 93–114. [CrossRef]
25. Das, J.; Ehlers, S. Numerical Simulation of Crushing and Bending Failure of Ice Using SPH. In Proceedings of the ASME 2015 34th International Conference on Ocean, Offshore and Arctic Engineering, Volume 8: Ian Jordaan Honoring Symposium on Ice Engineering, St. John's, NL, Canada, 31 May–5 June 2015; p. V008T07A019.
26. Häusler, F.U. Multiaxial compressive strength tests on saline ice with brush-type loading platens. In Proceedings of the IAHR. International Association for Hydraulic Research. International Symposium on Ice, Quebec, QC, Canada, 27–31 July 1981; Volume 2, pp. 526–539.
27. Zhang, N.; Zheng, X.; Ma, Q.; Hu, Z. A numerical study on ice failure process and ice-ship interactions by Smoothed Particle Hydrodynamics. *Int. J. Nav. Archit. Ocean Eng.* **2019**, *11*, 796–808. [CrossRef]
28. Polach, R.; Ehlers, S.; Erikstad, S. A Decision-based Design Approach for Ships Operating in Open Water and Ice. *J. Ship Prod. Des.* **2015**, *30*, 209–219. [CrossRef]
29. Zhou, L.; Riska, K.; und Polach, R.V.B.; Moan, T.; Su, B. Experiments on level ice loading on an icebreaking tanker with different ice drift angles. *Cold Reg. Sci. Technol.* **2013**, *85*, 79–93. [CrossRef]
30. Kim, H.; Daley, C.; Colbourne, B. A numerical model for ice crushing on concave surfaces. *Ocean Eng.* **2015**, *106*, 289–297. [CrossRef]
31. Schulson, E.M. Brittle failure of ice. *Eng. Fract. Mech.* **2001**, *68*, 1839–1887. [CrossRef]
32. Chinese Arctic and Antarctic Administration. A Brief Introduction of R/V Xuelong. 2012. Available online: http://www.chinare.gov.cn/en/index.html?pid=stations&st=xuelong (accessed on 20 May 2021).
33. International Association of Classification Societies (IACS). *Requirements Concerning POLAR CLASS*; IACS: London, UK, 2016.
34. Bertram, V. *Practical Ship Hydrodynamics*; Butterworth-Heinemann: Oxford, UK, 2000.
35. Logothetis, N.; Haigh, A. Characterising and optimising multi-response processes by the Taguchi method. *Qual. Reliab. Eng. Int.* **1988**, *4*, 159–169. [CrossRef]
36. Rachael, G. The Complexities of Arctic Maritime Traffic. Center for Circumpolar Security Studies. Available online: https://www.thearcticinstitute.org/complexities-arctic-maritime-traffic/?cn-reloaded=1 (accessed on 30 January 2018).

Journal of
Marine Science and Engineering

MDPI

Article

Effect of Propeller Cup on the Reduction of Fuel Consumption in Realistic Weather Conditions

Mina Tadros [1,2,*], Roberto Vettor [1], Manuel Ventura [1] and C. Guedes Soares [1]

1 Centre for Marine Technology and Ocean Engineering (CENTEC), Instituto Superior Técnico, Universidade de Lisboa, Av. Rovisco Pais 1, 1049-001 Lisboa, Portugal; roberto.vettor@centec.tecnico.ulisboa.pt (R.V.); manuel.ventura@centec.tecnico.ulisboa.pt (M.V.); c.guedes.soares@centec.tecnico.ulisboa.pt (C.G.S.)

2 Department of Naval Architecture and Marine Engineering, Faculty of Engineering, Alexandria University, Alexandria 21544, Egypt

* Correspondence: mina.tadros@centec.tecnico.ulisboa.pt

Abstract: This paper presents the effect of a propeller cup on the propeller cavitation and the fuel consumption of a bulk carrier in both calm water and different weather conditions towards improving the energy efficiency of the ship and reducing the level of emissions in terms of design and operation. Based on the propeller optimization model, previously developed that couples NavCad and a Matlab code to select the geometry and the operating point of the propeller at the engine operating point with minimum fuel consumption, the optimized propeller performance is evaluated for different percentages of the cup; light, medium and heavy and compared with the performance of the propeller without a cup in both calm water and several sea states. By evaluating the cavitation criteria, it is concluded that increasing the percentage of cupping reduces the occurrence of cavitation based on the Keller and Burrill methods; moreover, the fuel consumption is reduced by up to 5.4% and 6.6% at the propeller with a higher percentage of cup compared with the uncapped propeller in calm water and among the ship route, respectively.

Keywords: bulk carrier; propeller cup; fuel consumption; decarbonization; weather condition; cavitation

Citation: Tadros, M.; Vettor, R.; Ventura, M.; Guedes Soares, C. Effect of Propeller Cup on the Reduction of Fuel Consumption in Realistic Weather Conditions. *J. Mar. Sci. Eng.* **2022**, *10*, 1039. https://doi.org/10.3390/jmse10081039

Academic Editor: Kostas Belibassakis

Received: 5 July 2022
Accepted: 27 July 2022
Published: 28 July 2022

1. Introduction

The International Maritime Organization (IMO) has applied stringent regulations to reduce the amount of exhaust emissions, especially the level of greenhouse gas (GHG) in the maritime industry; these regulations aim to achieve a reduction in the carbon intensity by at least 40% by 2030 and the total annual GHG by at least 50% by 2050 compared to 2008 [1]. The level of emissions is evaluated using several types of indices, which can be applied to the new ship, such as the Energy Efficiency Design Index (EEDI). At the same time, the Energy Efficiency Existing Ship index (EEXI) and the Carbon Intensity Indicator (CII) take into account the design and operational points of view that will enter into force at the beginning of 2023 for the existing ship.

Several solutions are proposed to improve the energy efficiency of the ships, mainly consisting of reducing the amount of fuel consumption and thus, exhaust emissions as well as increasing the amount of cargo held at a suitable ship speed [2]; these solutions mainly focus on optimizing engine performance [3] to support fuel conversion techniques to replace the fossil fuel used in ship operation with new clean fuels with much fewer carbon emissions, such as biofuel, natural gas, methanol and ammonia [4–8]; moreover, installing after-treatment systems, especially the carbon capture and storage (CCS), will be an effective solution to capture the carbon emissions before entering the atmosphere [9,10]. Other solutions consider the design of the ship hull [11,12] and improving the propulsive coefficients [13,14] combined with the use of wind energy [15] and managing the ballast system [16,17], as well as finding the optimal route [18,19] and reducing speed [20–22], which based on simulations, becomes essential to assist the shipmaster during the voyage to reduce the resistance on the hull and thus use lower energy to operate the ship; this is mainly thanks to the online availability of wind and wave weather data worldwide [23].

The propeller selection, which is the main focus of this paper, is another area of interest in reducing fuel consumption and ensuring the sustainability of the ship along the trips. Selecting the propeller at the maximum efficiency is important to ensure a high propeller performance during the design [24–26] and operation, especially in sea states [27,28] and comply with the cavitation limits. Furthermore, the vibration and noise can be reduced, and the propeller efficiency can increase by considering the propeller boss cap fins [29]; moreover, the selection of propellers by minimizing the fuel consumption is performed in Refs. [30–33], which has been extended in Ref. [34] to ensure both objectives; maximum propeller efficiency and minimum fuel consumption.

Considering the propeller cup while designing the propeller can improve the propeller performance by increasing the thrust loading and cavitation margin. A propeller cup is the deformation of a propeller's trailing edge toward the pressure face, as shown in Figure 1 acts as a propeller with a higher pitch. As a result, the pressure distribution is changed along the blade, which increases the lift force toward the trailing edge. According to MacPherson [35], the propeller cup is computed as a percentage of the propeller diameter and varies from 0.5% for the light cup to 1.5% for the heavy cup.

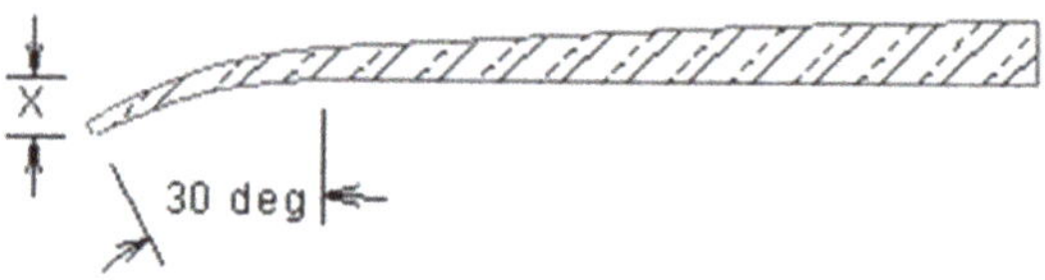

Figure 1. Section of propeller with a cup [35].

Few papers have investigated the effect of the propeller cup on the overall propeller performance. Hwang et al. [36] compared the measured data to show the effect of the cup on a series of three-blade Gawn-Burrill propellers. The value of cavitation is computed, and the cup percentage can be estimated. Tsai [37] studied the performance of a propeller with a cup, and the author showed an increase in the lift force and suggested an improvement in cavitation. Yari and Moghadam [38] used the boundary element method (BEM) to predict the performance of a partially submerged propeller (PSP) with and without a cup. The computational results are validated using experimental data; they concluded an increase in the static pressure of the face side, an increase in thrust coefficient of more than 50% and a decrease in torque coefficient with the same percentage as the thrust coefficient among all advance coefficients. Samsul [39] used the computational fluid dynamics (CFD) method to study the effect of a propeller cup on cavitation. The author achieved lower cavitation in the case of the propeller cup than in the uncupped propeller.

From that point of view, the contribution of this paper focuses on presenting the effect of propeller cupping in terms of fuel consumption toward ship decarbonization and the assessment of cavitation on propeller blades for different sea states during the trip of a bulk carrier from Europe to the USA.

The remainder of this paper is organized as follows. The numerical model used to perform the simulation is presented in Section 2. The computed results and the evaluation of the propeller performance in both calm water and weather conditions are presented in Section 3. Finally, a summary of the main findings and future recommendations are presented in Section 4.

2. Numerical Model

The numerical model that is used to perform the simulation in this study is a propeller optimization model that was previously developed by Tadros et al. [31], coupling NavCad software [40] and a nonlinear optimizer. The main objective of this model is to benefit from the operation research technique to select the propeller geometry at the engine operating point with minimum fuel consumption and verify the proposed constraints. More details about the developed numerical tool that is consisted of two modules to optimize the engine and propeller performance can be found in Refs. [3,31,34,41–44].

In this study, a bulk carrier 154 m in length is chosen as a case study to perform the numerical simulation. The ship's characteristics and the main engine installed are given in Tables 1 and 2, respectively.

Table 1. Main characteristics of the bulk carrier.

Item	Unit	Value
Length waterline	m	154.00
Breadth	m	23.11
Draft	m	10.00
Displacement	tonne	27,690
Service speed	knot	14.5
Maximum speed	knot	16.0
Number of propellers	-	1
Type of propellers	-	FPP
Rated power	kW	7140

Table 2. Main characteristics of diesel engine [45].

Item	Unit	Value
Engine builder	-	MAN Energy Solutions
Brand name	-	MAN
Bore	mm	320
Stroke	mm	440
Displacement	liter	4954
Number of cylinders	-	14
Rated speed	rpm	750
Rated power	kW	7140

Once the ship is chosen, a systematic procedure is applied to define all the inputs required in NavCad as described in Ref. [31] to select the propeller geometry and the operating point based on the gearbox ratio at the design speed. The computed parameters are exported from Maxsurf [46] based on the defined 3D hull of the ship and then defined in NavCad. A range of ship speeds, the series of the propeller (Wageningen B-series [47]) and the number of blades (five blades), the engine load diagram and the efficiencies of the propulsion system are introduced. After filling all the required inputs, the Holtrop method is presented in Refs. [48,49] is selected to compute the ship resistance associated with the speed range, while the method presented in Holtrop and Mennen [50] is considered for computing the propulsive coefficients such as wake fraction, thrust deduction factor and relative rotative efficiency and these methods show a good agreement with CFD results [51].

Other methods can be used during the resistance and propulsion computation and integrated into the software, while these two selected methods are ranked first based on the expert ranking provided in NavCad. Due to the uncertainty of these methods, as they are based on regression models and require validation with real tests or CFD models [51], a design margin of 10% is considered during the computation of ship resistance. The same procedures are considered by changing the value of the propeller cup from light to medium to heavy presented in mm, and as a function of propeller diameter as in the following equation:

$$Prop_{Cup\text{-}mm} = Prop_{Cup\text{-}percentage} \times Prop_{Dia\text{-}mm} \quad (1)$$

Then, the effective propeller pitch is computed using the following expression as presented by MacPherson [35]:

$$P_{Eff} = P_{Geo} + 21\left(Prop_{Cup\text{-}mm}\right) \quad (2)$$

where P_{Eff} is the effective pitch and P_{Geo} is the uncupped face pitch.

Once the propeller characteristics are computed from NavCad, the propeller performance is integrated into the engine load diagram, previously developed in Refs. [3,52], as functions of engine speed and brake power to compute the brake-specific fuel consumption (BSFC) and the exhaust emissions.

After that, the propeller is optimized, and the performance of the system is recalculated. Figure 2 shows the schematic diagram of the optimization model to select the propeller in calm water.

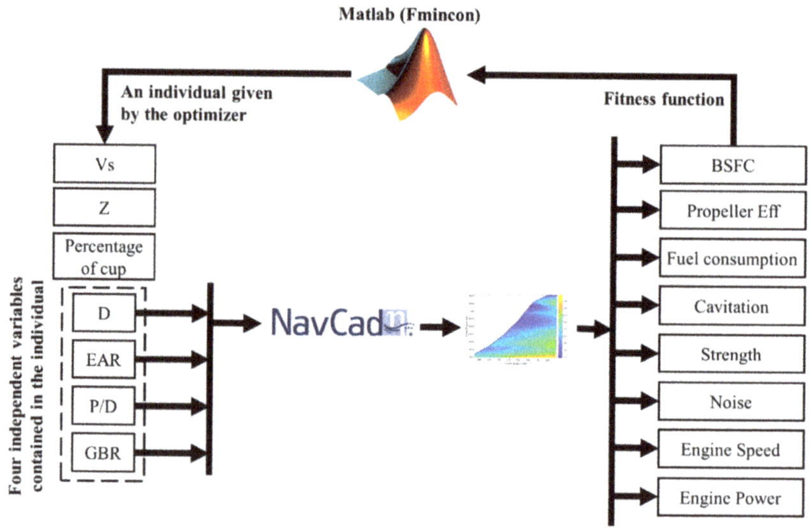

Figure 2. Schematic diagram of the propeller optimization model.

The objective and the constraints of the optimization model are combined into a developed fitness function as in the following equation, where the main objective is the minimization of fuel consumption and the constraints are related to cavitation, noise and strength issues.

$$Fitness\ Function = FC + R\sum_{i=1}^{j} max(g_i(x), 0) \quad (3)$$

where FC is the fuel consumption, $g(x)$ is the static penalty function, x is the number of variables, j is the number of constraints, and R is a penalty function.

After that, the fitness function is evaluated by the nonlinear optimizer (fmincon) integrated into Matlab [53]. Although the fmincon is considered a local optimizer, several initial starts have been performed to ensure the accuracy of the results and achieve the minimum value of fuel consumption along all the trials performed.

After selecting the optimal propeller geometry and the operational point in calm water, the performance of the ship, engine and propeller are computed along the shipping route with different sea states. The considered sea states represent the expected wave climate in the northern route from the British Channel to the west coast of the US [54], as shown in Figure 3; it also accounts for the effect of storm avoidance on the probability distribution of rough weather conditions [55], where the significant wave height (H_W) varies between

0 and 10 m, and the modal wave period (T_P) varies between 4 and 18 s; moreover, the most occurrent sea state varies between 1 to 2 m and 6 to 8 s. The scatter diagram of the selected route is shown Figure 4.

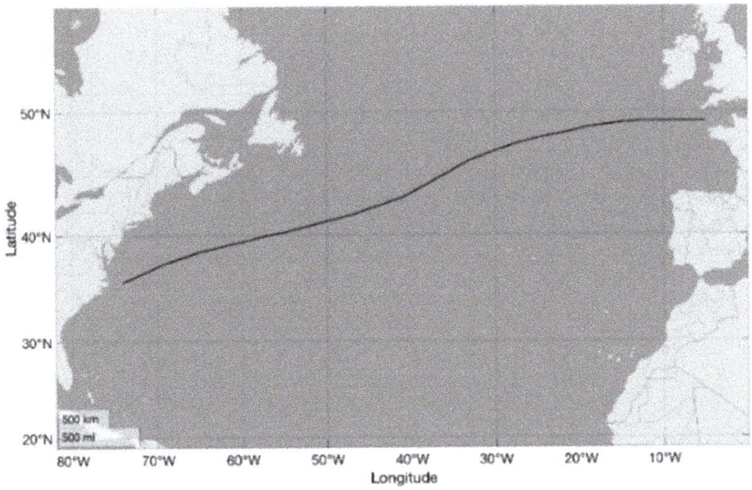

Figure 3. Northern route from the British Channel to the west coast of the US.

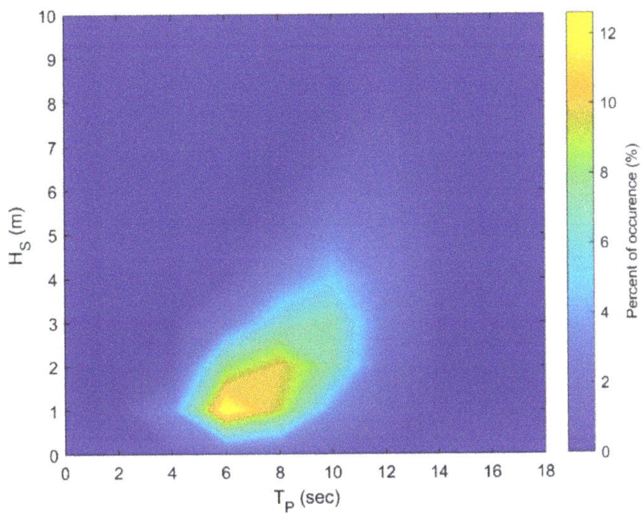

Figure 4. Scatter diagram of the northern route from the British Channel to the west coast of the USA.

Based on the collected weather data, the added resistance due to the existing waves in each sea state is computed using Aertssen [56] method; this method is integrated into NavCad, does not account for ship type and can be easily used by defining the significant wave height and modal wave period of each sea state; it is adapted by the team of HydroComp [40] to compute the added resistance due to the waves only while extracting the wind resistance.

Using an application programming interface (API), which can allow the connection between NavCad and Matlab as a third party, the simulation of all sea states is performed in a simple loop; then, the computed results are easily exported. A schematic diagram of all the simulation and optimization processes is presented in Figure 5. For a meaningful comparison among the different configurations, the different parameters are computed for each defined sea state and then averaged, considering each weather condition occurrence using the following expression:

$$P_W = \frac{\sum_{i=1}^{n} P_i \times SS_{Occ,i}}{\sum_{i=1}^{n} SS_{Occ,i}} \quad (4)$$

where P_W is the weighted average parameter, P is the computed parameter, SS_{Occ} is the occurrence of the sea state.

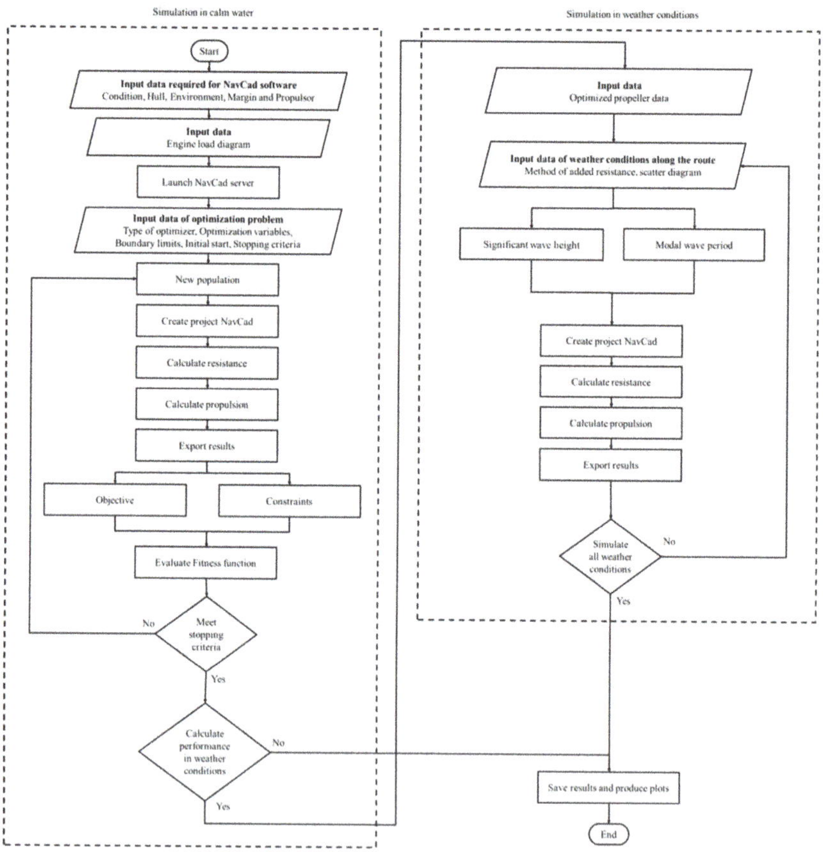

Figure 5. Schematic diagram of optimization tool in calm water and weather conditions.

3. Results

3.1. Simulation in Calm Water

First, the simulation is performed to select the optimum propeller by minimizing the fuel consumed at the design speed and in calm water conditions for several percentages of cupping, which varies from no cup to heavy cup. Table 3 shows the performance of the optimum propeller for each simulated case and the expanded view of the propeller blade is shown in Figure 6, showing the propeller cupping at the trailing edge along the four simulated cases. The propeller geometry is selected, including propeller diameter (D), expanded area ratio (*EAR*), pitch diameter ratio (*P/D*) and the operating point computed by the relation between propeller speed (*N*) and gearbox ratio (*GBR*).

Table 3. Propeller characteristics for different percentages of cupping.

	Parameters	Unit	No Cup	Light Cup	Medium Cup	Heavy Cup
Propeller Type				Wageningen B-Series		
Ship Speed	V_s	[kn]	14.5	14.5	14.5	14.5
Propeller characteristics	Cup	[%]	0.0	0.5	1.0	1.5
	D	[m]	6.00	6.00	6.00	6.00
	EAR	[-]	0.47	0.78	0.81	0.45
	P	[m]	6.58	6.38	6.26	5.59
	N	[RPM]	75	71	68	68
	Thrust	[kN]	576.49	576.49	576.49	576.49
	Torque	[kN.m]	573.30	600.90	620.70	613.2
	η_o	[%]	0.59	0.60	0.61	0.61
	J	[-]	0.62	0.65	0.68	0.68
	K_T	[-]	0.28	0.31	0.34	0.33
	K_Q	[-]	0.05	0.05	0.06	0.06
	w	[-]	0.38	0.38	0.38	0.38
	t	[-]	0.19	0.19	0.19	0.19
Cavitation	Tip Speed	[m/s]	23.61	22.37	21.32	21.48
	EAR_{min}	[-]	0.47	0.42	0.38	0.34
	Average loading pressure	[kPa]	43.56	21.15	16.40	23.21
	Back Cavitation	[%]	7.40	2.00	2.00	2.00
	$Pitch_{min}$	[m]	4.97	5.25	5.51	5.47
Gearbox characteristics	GBR	[-]	9.50	9.63	10.06	9.88
Engine characteristics	Speed	[RPM]	714	687	682	676
	Brake Power	[kW]	4682.30	4735.80	4671.00	4552.20
	Loading ratio	[%]	65.6	66.3	65.4	63.8
	BSFC	[g/kW.h]	192	189	188	187
	Fuel consumption	[l/nm]	74.17	73.94	72.47	70.20
Exhaust emissions	CO_2	[g/kW.h]	607.99	599.27	595.47	591.98
	NO_x	[g/kW.h]	6.68	7.40	7.31	6.95
	SO_x	[g/kW.h]	9.59	9.45	9.39	9.34

The propeller is always selected at the maximum propeller diameter (6 m). At the same time, the other parameters are changed to provide a suitable thrust, which is equal among the cases and complies with the cavitation limits; it has been found that the EAR is minimized in the case of a no- and heavy-cup propeller compared to the other propellers that have a percentage of cupping. The propeller speed among the cases is almost close; it has been found among the four cases that the propeller pitch decreases when increases the cupping percentage and reaches its lowest value at the heavy cup.

Therefore, the tip speed has been reduced by up to 2.3 m/s (10%) for the cupped propeller than the normal one. The minimum area required to avoid cavitation, as suggested by Keller [57], is minimized while increasing the cupping percentage by up to 27%, reducing cavitation occurrence in different weather conditions; moreover, the value of average loading pressure suggested by Burrill and Emerson [58] is reduced while increasing the cupping percentage by up to 62.4% compared to the uncupped propeller. The back cavitation shows its lowest value among all cases compared to the limit value suggested by HydroComp [40]; moreover, face cavitation is considered by identifying the minimum value of pitch to avoid this type of cavitation, as presented in Ref. [59]; it has been found that the design pitch values are higher than the minimum required to avoid face cavitation.

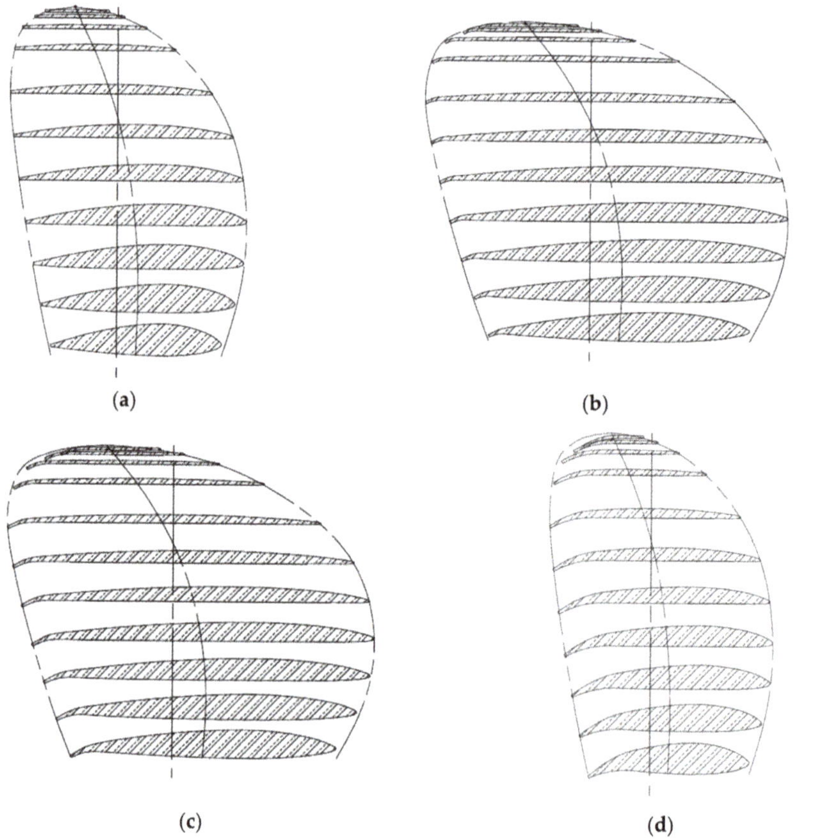

Figure 6. Expanded view of the optimized propeller. (**a**) No cup, (**b**) Light cup, (**c**) Medium cup, (**d**) Heavy cup.

Regarding the engine operating point, the propeller is selected to operate at the minimum loading conditions, which shows its lowest level at the higher cupping propeller with a reduction of 2.1% than the uncupped propeller; this achieves a reduction in fuel consumption by around 0.4%, 2.3% and 5.4% at the propeller with a light, medium and heavy cup, respectively, compared to the uncupped propeller, thus reducing the amount of exhaust emissions produced from the engine, namely the CO_2 and SO_x emissions, which has been computed based on emission factors and from the polynomial equations presented in Ref. [52].

3.2. Simulation in Different Weather Conditions

After the propeller is selected using the developed optimization model in calm water, the propeller's performance is simulated for several sea states among a selected ship route. The added resistance for each sea state is computed based on the significant wave height and modal wave period, which is then added to the calm water resistance to compute the total resistance of the ship. After that, the thrust required is computed as well as the power needed to drive the ship. Finally, the propulsive coefficients presented by the wake fraction and the thrust deduction factor are computed using empirical formulas based on the average value, and it is kept constant in both calm water and sea state conditions.

Despite the method's limitation in estimating the variation in the values of propulsive coefficients, this method shows agreement with the concept presented in Ref. [60], which is based on experimental tests and shows that the values of propeller coefficients fluctuate around the average value computed in calm water conditions. Thus, this method shows its suitability for estimating the propeller performance in a preliminary stage for a given operating area with a defined sea state.

After simulation of the ship and propeller performance along the ship route with several sea states, it has been found that the brake power is significantly increasing so that the ship can operate at the same design speed; this increment in the value of brake power can vary between 3% at the lower conditions of sea states and about to reach 100% at high sea states conditions, as shown in Figure 7; more of an increment in brake power is achieved when the propeller is cupped than the uncupped propeller, mainly based on the propeller's initial design and operational point.

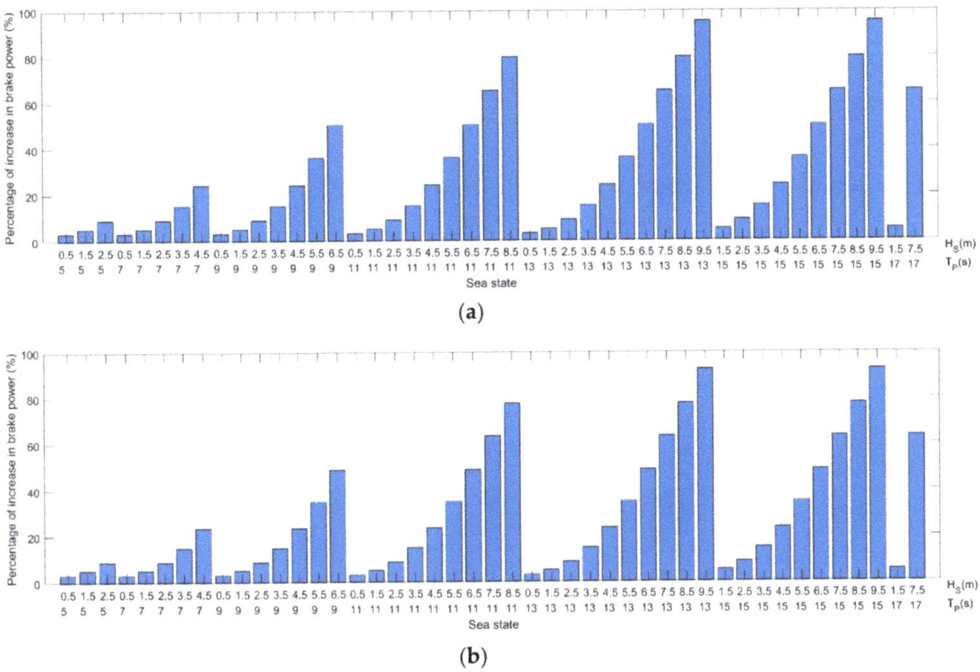

Figure 7. Percentage of increase in brake power for different sea states. (**a**) no cup, (**b**) heavy cup.

Regarding the cavitation criteria, the four techniques, as previously presented and suggested by Keller [61] and Burrill and Emerson [58] methods, are considered to quantify the limits of cavitation; these cavitation criteria are valid for calm water and under exposure to weather conditions. To consider the safety aspects during sailing in rough weather conditions, the speed reduction is limited so that the operational points of the propeller that exceed 90% of the rated power will operate at the 90% of the rated power, which is the maximum value of the continuous engine operational area.

Based on the Keller method, it has been found that the cavitation can occur in the case of the no-cup propeller in all sea states while increasing the wave height because the minimum value of EAR to avoid cavitation is equal to the designed EAR, as shown in Figure 8a. Therefore, any increase in the added resistance requires more loads, and thus the probability of occurrence of cavitation increases. A big deviation between the selected EAR and the minimum required EAR to avoid cavitation is noticed in the cupped propeller except in the higher sea states. Therefore, the probability of occurrence of cavitation is significantly reduced as the minimum required EAR to avoid cavitation is lower than the selected EAR among most of the sea states; this difference between the two values increases while increasing the percentage of cupping shown in Figure 8b.

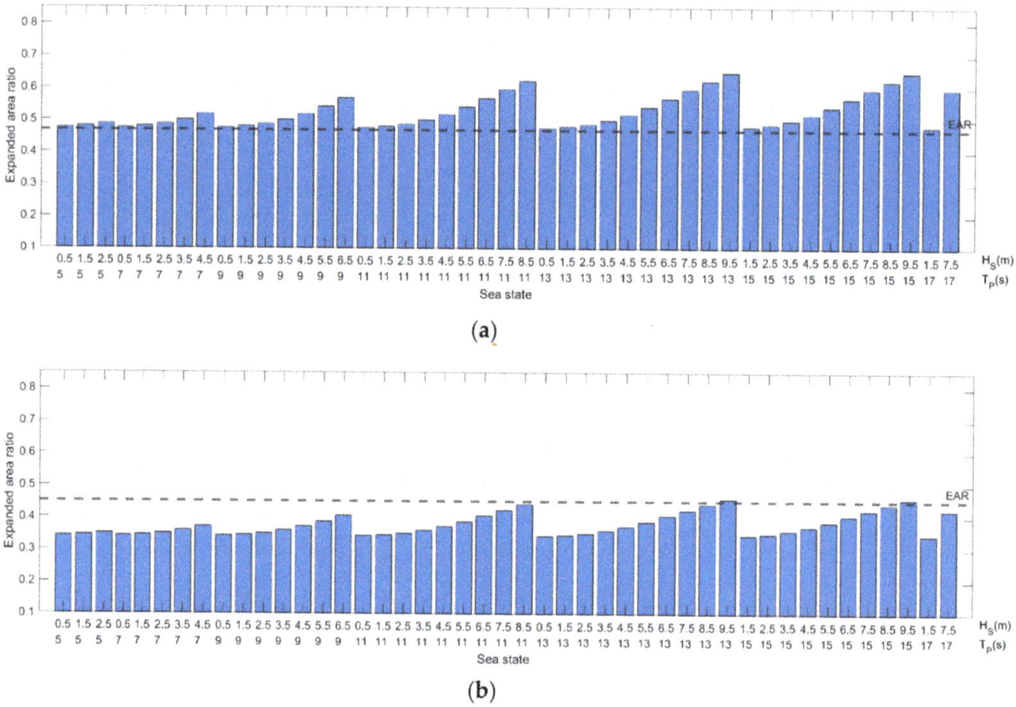

Figure 8. Variation of the minimum EAR for different sea states. (**a**) no cup, (**b**) heavy cup.

Based on the Burrill method, the average propeller loading pressure is computed and presented in Figure 9. Figure 9a shows the average propeller loading pressure for the uncupped propeller; it shows that the cavitation can occur at the higher sea state with more than 8.5 m significant wave height as the average propeller loading pressure values exceed the maximum limit required to avoid cavitation (65 kPa) suggested by HydroComp [40]; however, the values of the average propeller loading pressure decrease while increasing the cupping percentage. Thus, the occurrence of cavitation is significantly reduced among the sea states, as shown in Figure 9b.

The average predicted back cavitation percentage follows the same trend as the average propeller loading pressure. The cavitation can occur for uncupped propeller and in case of significant wave height with more than 8.5 m as the average predicted back cavitation percentage values exceed the 15% limit suggested by HydroComp [40] shown in Figure 10a. While increasing the cupping percentage, the average predicted back cavitation percentage values significantly decreased and reached their lowest level, as shown in Figure 10b.

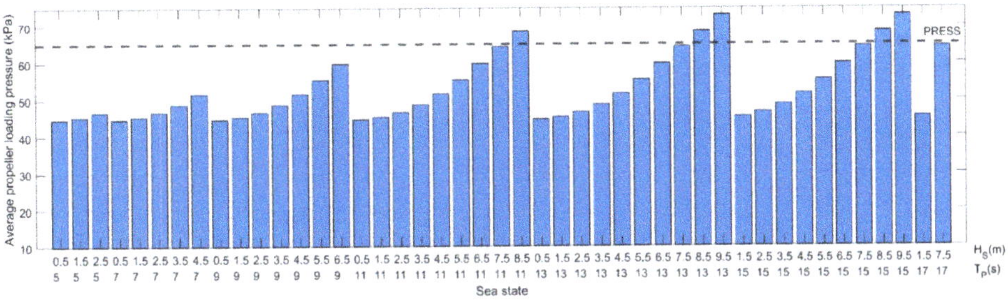

(**a**)

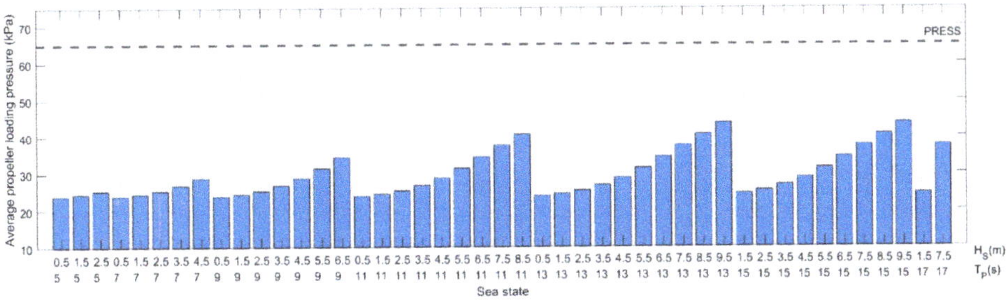

(**b**)

Figure 9. Variation of average propeller loading pressure for different sea states. (**a**) no cup, (**b**) heavy cup.

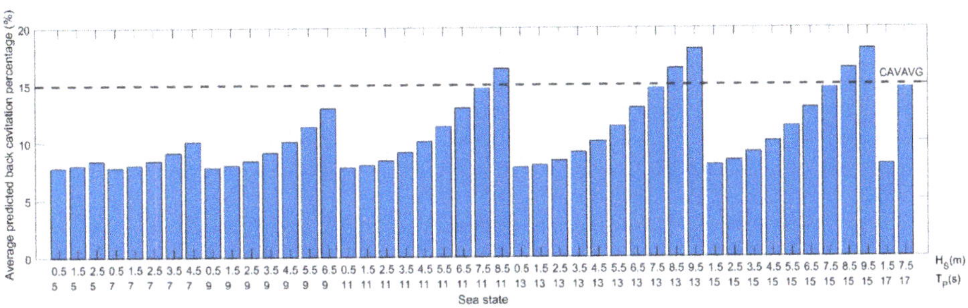

(**a**)

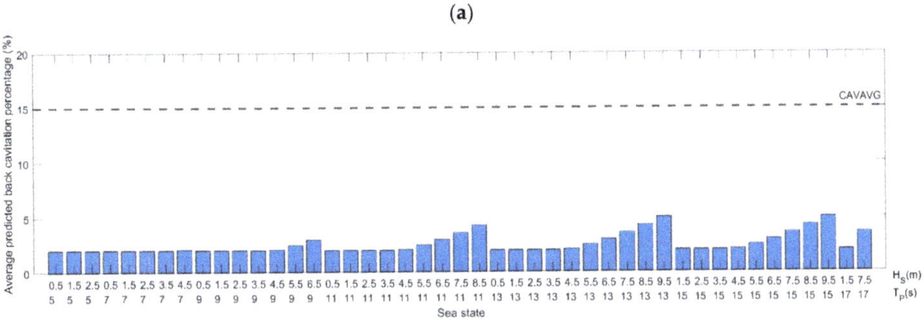

(**b**)

Figure 10. Variation of average back cavitation percentage for different sea states. (**a**) no cup, (**b**) heavy cup.

Regarding the noise level evaluated by the tip speed, all the optimized propellers, either with or without a cup, comply with the maximum noise level equal to 46 m/s and show the same trend, as shown in Figure 11.

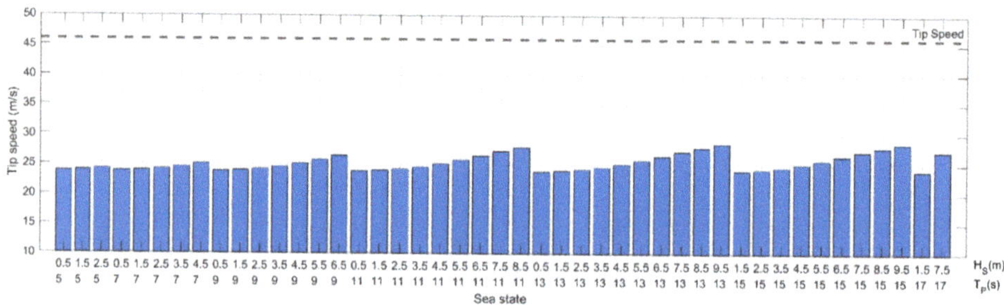

(a)

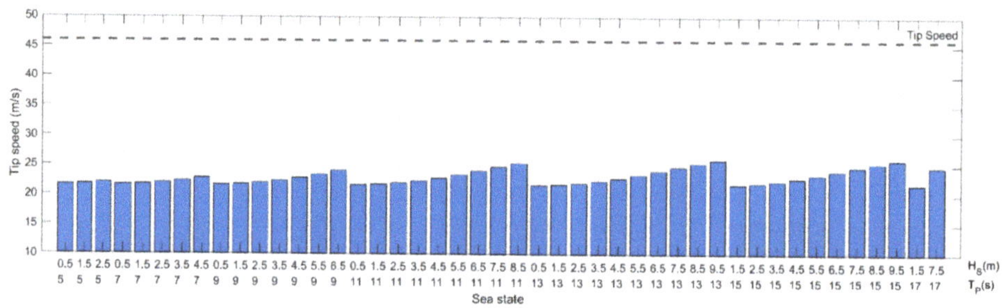

(b)

Figure 11. Variation of tip speed for different sea states. (**a**) no cup, (**b**) heavy cup.

Figure 12 shows the propeller performance inside the engine load diagram where no speed reduction is performed when the ship operates in sea states with a significant wave height below 6 m. With more than 6 m significant wave height, a reduction technique is considered to avoid the engine from overloading with no more than 90% of the rated power, where the brake power and engine speed are corrected and the reduction in ship speed is computed. By comparing the two sub-figures, there is no significant difference in speed reduction between the cases (up to 0.2 kn).

Finally, Table 4 presents the weighted average of the main important parameters based on the number of occurrences of each sea state to evaluate the effect of cupping percentage among the ship route; it has been found the weighted average engine speed is reduced while increasing the cupping percentage by 4.0%, 4.5% and 5.3% for light, medium and heavy cups, respectively. The weight average brake power is increased by 1% in the case of a light cup compared to no cup, following the same concept in calm water; however, the weight average brake power is reduced by 0.3% and 2.8% in the case of medium and heavy cups, respectively, compared to the uncapped propeller. The weighted average ship speed is almost the same among the four cases at around 14.47 kn.

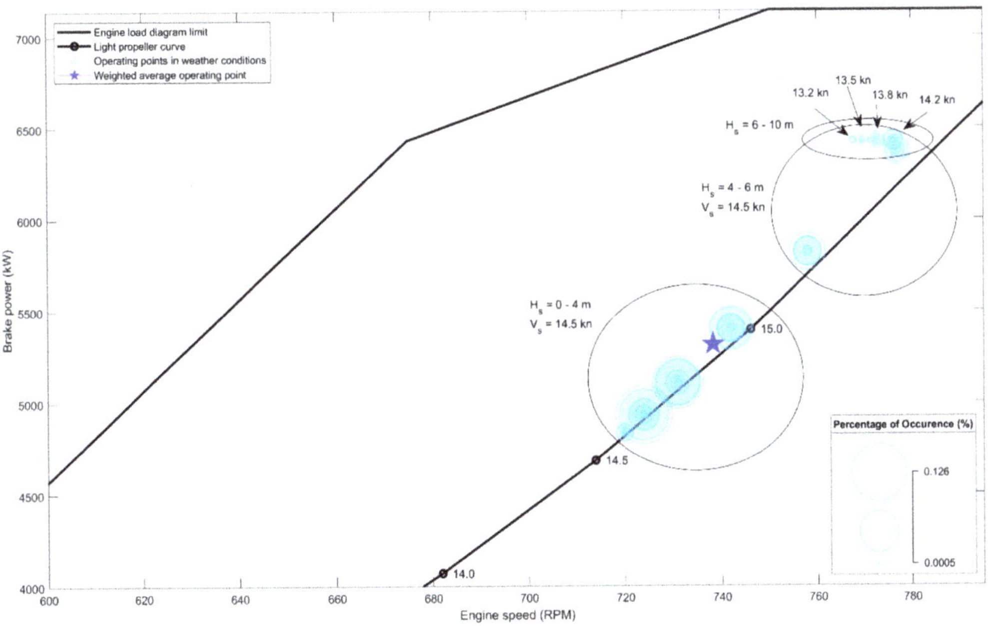

(**a**)

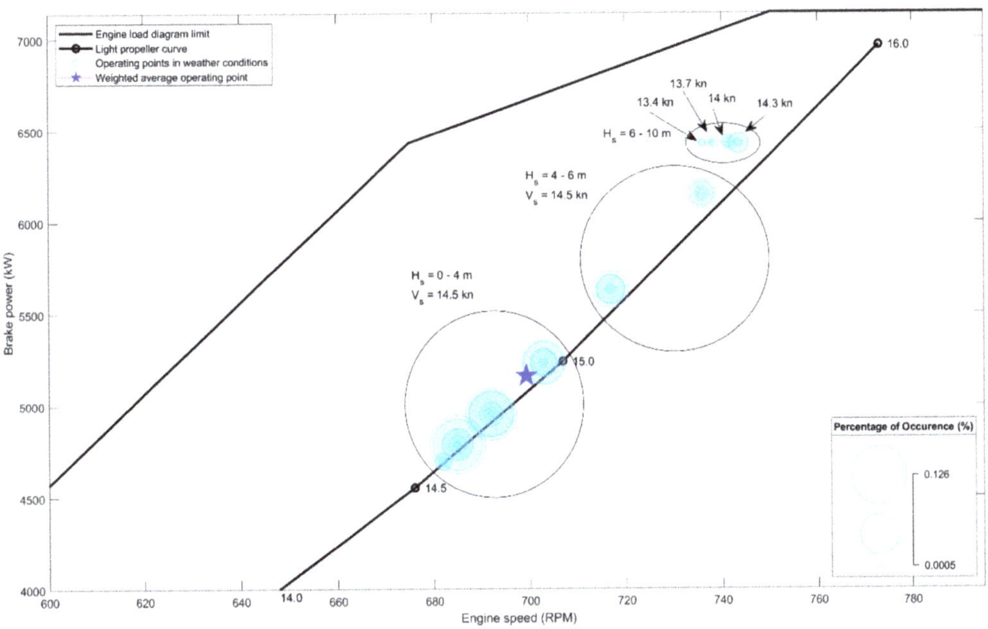

(**b**)

Figure 12. Propeller performance inside engine load diagram. (**a**) no cup, (**b**) heavy cup.

Table 4. Comparison between the weighted average results of the simulated cases.

Item (Weighted Average)	Unit	No Cup	Light Cup	Medium Cup	Heavy Cup
Engine speed	rpm	738	708	705	699
Brake power	kW	5314	5370	5298	5161
Ship speed	Knot	14.47	14.47	14.47	14.48
Fuel consumption	l/nm	89.54	88.41	87.03	83.61

The weighted average fuel consumption is computed among the four cases, as the main parameter used to evaluate the propeller performance; it has been shown that the fuel consumption in the case of a light cup is reduced by 1.3% compared to the uncapped propeller; this percentage value is increased while considering the medium cup to achieve a fuel consumption reduction of 2.8%. The maximum reduction in fuel consumption is achieved by the propeller with a heavy cup with a 6.6% reduction compared to the case of the uncupped propeller.

4. Conclusions

This paper presents the effect of the propeller cup on propeller performance as a solution toward increasing the ship's energy efficiency by reducing the level of fuel consumption and thus reducing the level of emissions. Different cupping percentages are considered in this study, varying from no cup, passing through the light and medium cup to the heavy cup percentage. A propeller optimization model previously developed coupling NavCad and Matlab is used to perform the simulation using empirical formulas and optimize the geometry and operating point of the propeller for each cupping percentage. The main objective of this optimization model is to select the propeller at the minimum operating point with fuel consumption while complying with the cavitation and noise limits.

The propeller is first optimized at the ship design speed in calm water conditions; then, the propeller performance is evaluated among several sea state conditions based on the computed ship resistance. The added resistance due to wave is computed based on the known significant wave height and modal wave period as suggested by the Aertssen method. Next, a speed reduction technique is considered to ensure the propeller operation inside the engine load diagram. Any operation point exceeding 90% of rated power will be reduced to operate at the defined limit of the continuous engine operation area (90% of the rated power). Finally, a weighted average technique is used to evaluate the overall performance of the ship route.

In terms of cavitation, it has been found that the propeller can be able to avoid cavitation while increasing the cupping percentage in both calm water and weather conditions based on the evaluation criteria suggested by Keller and Burrill.

In terms of fuel consumption, a significant reduction in fuel consumption is achieved by up to 5.4% in calm water and by up to 6.6% among the ship route.

This work presents a preliminary estimation of propeller performance with cupping percentage in both calm water and weather conditions. Validation procedures using CFD methods can be considered for future work to estimate the propeller performance as well as the values of propulsive coefficients in different sea conditions.

Author Contributions: The concept of the problem is developed by M.T. The analysis is performed by M.T. and the writing of the original draft manuscript is done by M.T., R.V., M.V. and C.G.S. All authors have read and agreed to the published version of the manuscript.

Funding: This work was performed within the scope of the Strategic Research Plan of the Centre for Marine Technology and Ocean Engineering (CENTEC), which is financed by the Portuguese Foundation for Science and Technology (Fundação para a Ciência e Tecnologia-FCT) under contract UIDB/UIDP/00134/2020.

Institutional Review Board Statement: Not applicable.

Informed Consent Statement: Not applicable.

Data Availability Statement: The data presented in this study are available on request from the corresponding author.

Conflicts of Interest: The authors declare no conflict of interest.

References

1. IMO-MEPC. Reduction of GHG emissions from ships. In *Fourth IMO GHG Study 2020*; IMO: London, UK, 2020; Volume 53, pp. 1689–1699.
2. Green Ship of the Future. 2019 Retrofit Project. Available online: https://greenship.org/project/2019-retrofit-series/ (accessed on 8 December 2021).
3. Tadros, M.; Ventura, M.; Guedes Soares, C. Optimization procedure to minimize fuel consumption of a four-stroke marine turbocharged diesel engine. *Energy* **2019**, *168*, 897–908. [CrossRef]
4. Korberg, A.D.; Brynolf, S.; Grahn, M.; Skov, I.R. Techno-economic assessment of advanced fuels and propulsion systems in future fossil-free ships. *Renew. Sustain. Energy Rev.* **2021**, *142*, 110861. [CrossRef]
5. Chiong, M.-C.; Kang, H.-S.; Shaharuddin, N.M.R.; Mat, S.; Quen, L.K.; Ten, K.-H.; Ong, M.C. Challenges and opportunities of marine propulsion with alternative fuels. *Renew. Sustain. Energy Rev.* **2021**, *149*, 111397. [CrossRef]
6. Altosole, M.; Benvenuto, G.; Campora, U.; Laviola, M.; Zaccone, R. Simulation and performance comparison between diesel and natural gas engines for marine applications. *Proc. Inst. Mech. Eng. Part M J. Eng. Marit. Environ.* **2017**, *231*, 690–704. [CrossRef]
7. Tadros, M.; Ventura, M.; Guedes Soares, C. A Review of the Use of Biodiesel as a Green Fuel for Diesel Engines. In *Developments in Maritime Technology and Engineering*; Guedes Soares, C., Santos, T., Eds.; Taylor & Francis Group: London, UK, 2021; Volume 2, pp. 481–490.
8. Elkafas, A.G.; Elgohary, M.M.; Shouman, M.R. Numerical analysis of economic and environmental benefits of marine fuel conversion from diesel oil to natural gas for container ships. *Environ. Sci. Pollut. Res.* **2021**, *28*, 15210–15222. [CrossRef]
9. Wärtsilä. Wärtsilä Advances Carbon Capture and Storage in Maritime as Part of LINCCS Consortium. Available online: https://www.wartsila.com/media/news/08-09-2021-wartsila-advances-carbon-capture-and-storage-in-maritime-as-part-of-linccs-consortium-2972116 (accessed on 10 January 2022).
10. Irena, K.; Ernst, W.; Alexandros, C.G. The cost-effectiveness of CO_2 mitigation measures for the decarbonisation of shipping. The case study of a globally operating ship-management company. *J. Clean. Prod.* **2021**, *316*, 128094. [CrossRef]
11. Ventura, M.; Guedes Soares, C. Integration of a Voyage Model Concept into a Ship Design Optimization Procedure. In *Towards Green Marine Technology and Transport*; Guedes Soares, C., Dejhalla, R., Pavletic, D., Eds.; Taylor & Francis Group: London, UK, 2015; pp. 539–548.
12. Zha, L.; Zhu, R.; Hong, L.; Huang, S. Hull form optimization for reduced calm-water resistance and improved vertical motion performance in irregular head waves. *Ocean. Eng.* **2021**, *233*, 109208. [CrossRef]
13. Feng, Y.; el Moctar, O.; Schellin, T.E. Parametric Hull Form Optimization of Containerships for Minimum Resistance in Calm Water and in Waves. *J. Mar. Sci. Appl.* **2021**, *20*, 670–693. [CrossRef]
14. Farkas, A.; Degiuli, N.; Martić, I.; Dejhalla, R. Numerical and experimental assessment of nominal wake for a bulk carrier. *J. Mar. Sci. Technol.* **2019**, *24*, 1092–1104. [CrossRef]
15. Ammar, N.R.; Seddiek, I.S. Wind assisted propulsion system onboard ships: Case study Flettner rotors. *Ships Offshore Struct.* **2021**, *1*, 1–12. [CrossRef]
16. Lakshmi, E.; Priya, M.; Achari, V.S. An overview on the treatment of ballast water in ships. *Ocean Coast. Manag.* **2021**, *199*, 105296. [CrossRef]
17. Lu, K.-T.; Lui, H.-K.; Chen, C.-T.A.; Liu, L.-L.; Yang, L.; Dong, C.-D.; Chen, C.-W. Using Onboard-Produced Drinking Water to Achieve Ballast-Free Management. *Sustainability* **2021**, *13*, 7648. [CrossRef]
18. Vettor, R.; Guedes Soares, C. Development of a ship weather routing system. *Ocean Eng.* **2016**, *123*, 1–14. [CrossRef]
19. Zaccone, R.; Ottaviani, E.; Figari, M.; Altosole, M. Ship voyage optimization for safe and energy-efficient navigation: A dynamic programming approach. *Ocean Eng.* **2018**, *153*, 215–224. [CrossRef]
20. Prpić-Oršić, J.; Vettor, R.; Faltinsen, O.M.; Guedes Soares, C. The influence of route choice and operating conditions on fuel consumption and CO_2 emission of ships. *J. Mar. Sci. Technol.* **2016**, *21*, 434–457. [CrossRef]
21. Vettor, R.; Tadros, M.; Ventura, M.; Guedes Soares, C. Route Planning of a Fishing Vessel in Coastal Waters with Fuel Consumption Restraint. In *Maritime Technology and Engineering 3*; Guedes Soares, C., Santos, T.A., Eds.; Taylor & Francis Group: London, UK, 2016; pp. 167–173.
22. Tadros, M.; Vettor, R.; Ventura, M.; Guedes Soares, C. Effect of Different Speed Reduction Strategies on Ship Fuel Consumption in Realistic Weather Conditions. In *Trends in Maritime Technology and Engineering*; Guedes Soares, C., Santos, T.A., Eds.; Taylor & Francis Group: London, UK, 2022; Volume 1, pp. 553–561.
23. Vinayak, P.P.; Prabu, C.S.K.; Vishwanath, N.; Prakash, S.O. Numerical simulation of ship navigation in rough seas based on ecmwf data. *Brodogradnja* **2021**, *72*, 19–58. [CrossRef]

24. Tadros, M.; Ventura, M.; Guedes Soares, C. Optimization Scheme for the Selection of the Propeller in Ship Concept Design. In *Progress in Maritime Technology and Engineering*; Guedes Soares, C., Santos, T.A., Eds.; Taylor & Francis Group: London, UK, 2018; pp. 233–239.
25. Vlašić, D.; Degiuli, N.; Farkas, A.; Martić, I. The preliminary design of a screw propeller by means of computational fluid dynamics. *Brodogradnja* **2018**, *69*, 129–147. [CrossRef]
26. Bacciaglia, A.; Ceruti, A.; Liverani, A. Controllable pitch propeller optimization through meta-heuristic algorithm. *Eng. Comput.* **2020**, *37*, 2257–2271. [CrossRef]
27. Ghaemi, M.H.; Zeraatgar, H. Analysis of hull, propeller and engine interactions in regular waves by a combination of experiment and simulation. *J. Mar. Sci. Technol.* **2021**, *26*, 257–272. [CrossRef]
28. Dai, K.; Li, Y.; Gong, J.; Fu, Z.; Li, A.; Zhang, D. Numerical study on propulsive factors in regular head and oblique waves. *Brodogradnja* **2022**, *73*, 37–56. [CrossRef]
29. Sun, Y.; Wu, T.; Su, Y.; Peng, H. Numerical prediction on vibration and noise reduction effects of propeller boss cap fins on a propulsion system. *Brodogradnja* **2020**, *71*, 1–18. [CrossRef]
30. Tadros, M.; Ventura, M.; Guedes Soares, C. Optimum Design of a Container Ship's Propeller from Wageningen B-Series at the Minimum BSFC. In *Sustainable Development and Innovations in Marine Technologies*; Georgiev, P., Guedes Soares, C., Eds.; Taylor & Francis Group: London, UK, 2020; pp. 269–274.
31. Tadros, M.; Vettor, R.; Ventura, M.; Guedes Soares, C. Coupled Engine-Propeller Selection Procedure to Minimize Fuel Consumption at a Specified Speed. *J. Mar. Sci. Eng.* **2021**, *9*, 59. [CrossRef]
32. Tillig, F.; Ringsberg, J.; Mao, W.; Ramne, B. A generic energy systems model for efficient ship design and operation. *Proc. Inst. Mech. Eng. Part M J. Eng. Marit. Environ.* **2017**, *231*, 649–666. [CrossRef]
33. Marques, C.H.; Belchior, C.R.P.; Caprace, J.D. Optimising the engine-propeller matching for a liquefied natural gas carrier under rough weather. *Appl. Energy* **2018**, *232*, 187–196. [CrossRef]
34. Tadros, M.; Ventura, M.; Guedes Soares, C. Design of Propeller Series Optimizing Fuel Consumption and Propeller Efficiency. *J. Mar. Sci. Eng.* **2021**, *9*, 1226. [CrossRef]
35. MacPherson, D.M. Small Propeller Cup: A Proposed Geometry Standard and a New Performance Model. In Proceedings of the 8th Propeller and Shafting Symposium, Virginia Beach, VA, USA, 23–24 September 1997.
36. Hwang, J.-L.; Tsai, J.-F.; Li, C.-Y. Cupped propeller test and analysis. *Ship Technol. Res.* **1995**, *42*, 186–192.
37. Tsai, J.-F. Study on the cavitation characteristics of cupped foils. *J. Mar. Sci. Technol.* **1997**, *2*, 123–134. [CrossRef]
38. Yari, E.; Moghadam, A.B. BEM applied to the cup effect on the partially submerged propeller performance prediction and ventilation pattern. *J. Mar. Eng. Technol.* **2020**, *21*, 159–177. [CrossRef]
39. Samsul, M.B. Blade Cup Method for Cavitation Reduction in Marine Propellers. *Pol. Marit. Res.* **2021**, *28*, 54–62. [CrossRef]
40. HydroComp. NavCad: Reliable and Confident Performance Prediction. Available online: https://www.hydrocompinc.com/solutions/navcad/ (accessed on 30 January 2019).
41. Tadros, M.; Ventura, M.; Guedes Soares, C. Data Driven In-Cylinder Pressure Diagram Based Optimization Procedure. *J. Mar. Sci. Eng.* **2020**, *8*, 294. [CrossRef]
42. Tadros, M.; Ventura, M.; Guedes Soares, C. A nonlinear optimization tool to simulate a marine propulsion system for ship conceptual design. *Ocean Eng.* **2020**, *210*, 107417. [CrossRef]
43. Tadros, M.; Ventura, M.; Guedes Soares, C. Optimization of the performance of marine diesel engines to minimize the formation of SOx emissions. *J. Mar. Sci. Appl.* **2020**, *19*, 473–484. [CrossRef]
44. Tadros, M.; Ventura, M.; Guedes Soares, C. Simulation of the Performance of Marine Genset Based on Double-Wiebe Function. In *Sustainable Development and Innovations in Marine Technologies*; Georgiev, P., Guedes Soares, C., Eds.; Taylor & Francis Group: London, UK, 2020; pp. 292–299.
45. MAN Diesel & Turbo. Four-Stroke Project Guides. Available online: https://www.man-es.com/marine/products/planning-tools-and-downloads/project-guides/four-stroke (accessed on 22 July 2022).
46. Bentley. MAXSURF: Maximize Vessel Performance. Available online: https://www.bentley.com/en/products/product-line/offshore-structural-analysis-software/maxsurf (accessed on 4 April 2020).
47. Oosterveld, M.; Van Oossanen, P. Further Computer-Analyzed Data of the Wageningen B-Screw Series. *Int. Shipbuild. Prog.* **1975**, *22*, 251–262. [CrossRef]
48. Holtrop, J. A statistical re-analysis of resistance and propulsion data. *Int. Shipbuild. Prog.* **1984**, *31*, 272–276.
49. Holtrop, J. A Statistical Resistance Prediction Method With a Speed Dependent Form Factor. In Proceedings of the Scientific and Methodological Seminar on Ship Hydrodynamics (SMSSH '88), Varna, Bulgaria, 17–22 October 1988; Bulgarian Ship Hydrodynamics Centre: Varna, Bulgaria, 1988; pp. 1–7.
50. Holtrop, J.; Mennen, G.G.J. An approximate power prediction method. *Int. Shipbuild. Prog.* **1982**, *29*, 166–170. [CrossRef]
51. Islam, H.; Ventura, M.; Guedes Soares, C.; Tadros, M.; Abdelwahab, H.S. Comparison between Empirical and CFD Based Methods for Ship Resistance and Power Prediction. In *Trends in Maritime Technology and Engineering*; Guedes Soares, C., Santos, T.A., Eds.; Taylor & Francis Group: London, UK, 2022; Volume 1, pp. 347–357.
52. Tadros, M.; Ventura, M.; Guedes Soares, C. Surrogate Models of the Performance and Exhaust Emissions of Marine Diesel Engines for Ship Conceptual Design. In *Maritime Transportation and Harvesting of Sea Resources*; Guedes Soares, C., Teixeira, A.P., Eds.; Taylor & Francis Group: London, UK, 2018; pp. 105–112.

53. The MathWorks Inc. Fmincon. Available online: https://www.mathworks.com/help/optim/ug/fmincon.html (accessed on 2 June 2017).
54. Vettor, R.; Guedes Soares, C. Detection and Analysis of the Main Routes of Voluntary Observing Ships in the North Atlantic. *J. Navig.* **2015**, *68*, 397–410. [CrossRef]
55. Vettor, R.; Guedes Soares, C. Assessment of the storm avoidance effect on the wave climate along the main North Atlantic routes. *J. Navig.* **2016**, *69*, 127–144. [CrossRef]
56. Aertssen, G. The Effect of Weather on Two Classes of Container Ships in the North Atlantic. *Nav. Archit.* **1975**, *1*, 11–13.
57. Carlton, J. *Marine Propellers and Propulsion*, 2nd ed.; Butterworth-Heinemann: Oxford, UK, 2012.
58. Burrill, L.C.; Emerson, A. Propeller cavitation: Further tests on 16in. propeller models in the King's College cavitation tunnel. *Int. Shipbuild. Prog.* **1963**, *10*, 119–131. [CrossRef]
59. MacPherson, D.M. Reliable Propeller Selection for Work Boats and Pleasure Craft: Techniques Using a Personal Computer. In *Fourth Biennial Power Boat Symposium*; SNAME: Alexandria, VA, USA, 1991.
60. Saettone, S.; Taskar, B.; Steen, S.; Andersen, P. Experimental measurements of propulsive factors in following and head waves. *Appl. Ocean Res.* **2021**, *111*, 102639. [CrossRef]
61. Keller, W.H. Extended Diagrams for Determining the Resistance and Required Power for Single-Screw Ships. *Int. Shipbuild. Prog.* **1973**, *20*, 133–142. [CrossRef]

Journal of
Marine Science and Engineering

MDPI

Article

Towards Fuel Consumption Reduction Based on the Optimum Contra-Rotating Propeller

Mina Tadros [1,2,*], Manuel Ventura [1] and C. Guedes Soares [1]

[1] Centre for Marine Technology and Ocean Engineering (CENTEC), Instituto Superior Técnico, Universidade de Lisboa, Av. Rovisco Pais 1, 1049-001 Lisboa, Portugal
[2] Department of Naval Architecture and Marine Engineering, Faculty of Engineering, Alexandria University, Alexandria 21544, Egypt
* Correspondence: mina.tadros@centec.tecnico.ulisboa.pt

Abstract: This paper presents the effect of selecting a contra-rotating propeller (CRP) for a bulk carrier at the engine operating point with minimum fuel consumption, as well as ensuring the safety of the propeller in terms of cavitation and noise. Using a developed optimization model, the geometry of a CRP was selected for different propeller diameters, the same propeller diameter as that of a fixed pitch propeller (FPP) installed on the bulk carrier, and at 90% of the FPP diameter. Additionally, each case was optimized with both no-cup and heavy-cup configurations. In general, the CRP showed better performance than the FPP in terms of efficiency, cavitation, and fuel economy. At the same time, the level of performance was increased when considering the CRP cupping percentage. It was concluded that the CRP can achieve a gain in fuel economy of up to 6.2% in a no-cup configuration when compared to an FPP, and up to 11.7% with a cupped configuration.

Keywords: bulk carrier; contra-rotating propeller; fuel consumption; decarbonization; cavitation; cupping CRP

Citation: Tadros, M.; Ventura, M.; Guedes Soares, C. Towards Fuel Consumption Reduction Based on the Optimum Contra-Rotating Propeller. *J. Mar. Sci. Eng.* **2022**, *10*, 1657. https://doi.org/10.3390/jmse10111657

Academic Editor: Decheng Wan

Received: 4 October 2022
Accepted: 3 November 2022
Published: 4 November 2022

1. Introduction

Due to the need to improve the energy efficiency of ships by reducing fuel consumption and thus carbon dioxide (CO_2) emissions, several solutions have been proposed to minimize the level of energy consumption and achieve the lowest level of exhaust emissions [1]. However, due to the complexity of ships, it has been concluded that no single solution greatly affects the decarbonization process [2,3], and there are many challenges to reducing energy demand even with equipment with lower consumption. However, every retrofitting process from the design or operation concept helps save energy and improve fuel economy.

The propulsion system is the most important system that directly affects the reduction of fuel consumption and exhaust emissions, starting from diesel engines as the prime movers and source of emissions, up to the type of propeller. Therefore, the main interest of industrial companies such as MAN Energy Solutions and Wärtsilä is to improve engine performance by replacing fossil fuels with alternative ones, mainly methanol and ammonia, to be further used in the near future according to strategic plans that present the vision of international organizations [4].

The propulsor, or the propeller, is the second component of the marine propulsion system that directly affects the amount of power transmitted from the engine to the ship hull. The selection of an effective propeller must be carefully performed to ensure high technical efficiency, the required safety during the operation of the ship in calm waters and in severe weather conditions, and to achieve a high level of economic benefit [5,6]. Therefore, optimizing hull forms [7] as well as ship transoms [8] are essential solutions to achieve an appropriate propeller inflow while placing energy-saving devices [9] forward of the propeller, such as pre-swirl ducts [10], pre-swirl fins [11], and vortex generator fins (VGFs) [12], and can be effective solutions for improved energy efficiency.

Besides the previous points, the selection of the marine propeller(s), including the type and the size, coupled with the engine performance, is essential to improve propeller efficiency and reduce fuel consumption [13–15]. This process-based optimization can be performed by comparing several types of propellers from different series and with different propeller shapes, either ducted or non-ducted, to find the optimal fixed pitch propeller (FPP) performance. Additionally, the right choice of the number of blades would be a solution to help reduce fuel consumption [16], especially when reducing the ship speed [17]. The same concept can be considered to achieve a 5% reduction in the fuel consumed, while choosing to utilize a controllable-pitch propeller (CPP) at the engine operating point with minimum fuel consumption compared to only maximizing propeller efficiency [18,19]. Furthermore, increased fuel savings can be achieved by replacing the normal propeller with a cupping type; in addition, an extreme decrease in cavitation values will be observed [20].

The contra-rotating propeller (CRP) is another concept that was developed by Wagner [21] to reduce engine loads and thus increase fuel economy. This concept is based on adding a smaller propeller that rotates in the opposite direction, positioned aft of the main one. As the FPP causes water circulation, placing two different propellers in front of each other assist in neutralizing the water circulation. Therefore, the energy losses from the sideways forces due to water circulation are recovered by the small propeller (aft propeller) and force the water to flow in a horizontal direction parallel to the thrust direction. This system will create a larger thrust force than in the case of a single FPP and increase the propeller efficiency. Figure 1 shows the configuration of a CRP in a model ship.

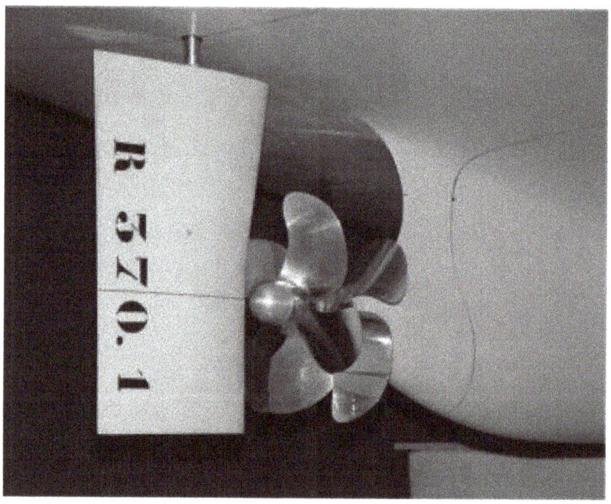

Figure 1. The CRP system at the stern of the model ship [22].

In terms of propeller performance, van Manen and Oosterveld [23] presented the results of the performance of a CRP in open-water conditions compared to a conventional propeller. They predicted higher efficiency with the CRP, accompanied by a reduction in the delivered horsepower (DHP). Min et al. [22] performed model tests of several CRP configurations and concluded that the propulsion efficiency was improved compared to that with a conventional configuration. Koronowicz et al. [24] developed a numerical model to design and generate the two propellers in a 3D mode and computed the hydrodynamic performance of each one for a given ship speed. Additionally, Ghassemi and Taherinasab [25] validated the numerical results from a 3D model developed with real data. They concluded that the hydrodynamic coefficient of the rear propeller was higher than that of the front one due to the wake. The cavitation was improved due to the distribution of loads between the two propellers, and a pressure distribution more uniform than that obtained with the single propeller was observed. Nouri et al. [26] used optimization procedures coupled with computational fluid dynamics (CFD) techniques to find both propellers' optimal geometries.

In terms of maneuverability, Kayano et al. [27] showed that using a CRP with the tandem arrangement of pod propulsor improves the ship's maneuverability by almost 50% compared to the international standards. Torneman [28] concluded that by using a CRP, the maneuverability as well as the fuel efficiency of a wind farm support vessel (WFSV) on the high seas were increased during transportation from shore to the offshore site.

In terms of operation, Hou et al. [29] used CFD methods to find the operational speed of both propellers. They concluded that the speed of the forward propeller must be higher than that of the rear one to achieve higher propulsive efficiency and avoid any negative effects related to the net torque of the CRP on transverse stability.

In terms of emissions, Minami and Kano [30] showed that the use of a super marine gas turbine (SMGT) coupled with a CRP could reduce CO_2 emissions by 25%, nitrogen oxide (NO_x) emissions by 92%, and sulfur oxide (SO_x) emissions by 73%

From that point of view, this paper contributes to the selection of a coaxial CRP using optimization procedures for a given vessel at the engine operating point with minimum fuel consumption. Based on the literature review, simulations were performed for several propeller diameters. Furthermore, the results were compared to a conventional propeller installed on the ship, as mentioned in the literature, while considering different levels of cupping.

The remainder of this paper is organized as follows. The numerical model used to perform the simulations is presented in Section 2. Next, the computed results and the evaluation of the propeller performance for the design concept are presented in Section 3. Finally, a summary of the main findings and future recommendations are presented in Section 4.

2. Numerical Model

This study considers the selection of a CRP for a bulk carrier of 154 m in length by performing optimization procedures to identify the optimum propeller geometry. The characteristics of the bulk carrier and the main engine installed are given in Table 1.

Table 1. Main characteristics of the bulk carrier.

	Characteristics	Unit	Value
Ship characteristics	Length waterline	m	154.00
	Breadth	m	23.11
	Draft	m	10.00
	Displacement	tonnes	27,690
	Service speed	knots	14.5
	Maximum speed	knots	16.0
	Number of propellers	-	1
	Type of propeller	-	FPP
	Rated power	kW	7140
Engine characteristics	Engine builder	-	MAN Energy Solutions [31]
	Brand name	-	MAN
	Bore	mm	320
	Stroke	mm	440
	Displacement	liters	4954
	Number of cylinders	-	14
	Rated speed	rpm	750
	Rated power	kW	7140

The optimization model used in this study was previously developed by Tadros et al. [20]. In the current paper, the model was adapted to select a contra-rotating propeller and to compute its performance coupling NavCad software [32], an engine load diagram computed from a 1D engine model [33–35], and a nonlinear optimizer integrated into Matlab [36] using an application programming interface (API) that allows interaction between NavCad and other third parties. A schematic diagram is presented in Figure 2, showing the processing of data through the optimization model. The model was developed to find the optimal propeller parameters and the operational point in order to minimize the fuel consumption of the bulk carrier under different input parameters such as the ship design speed (V_s), number of propeller blades (Z), type of propeller series and the percentage of propeller cupping. The model complies with the limitation of noise and cavitation methods applied.

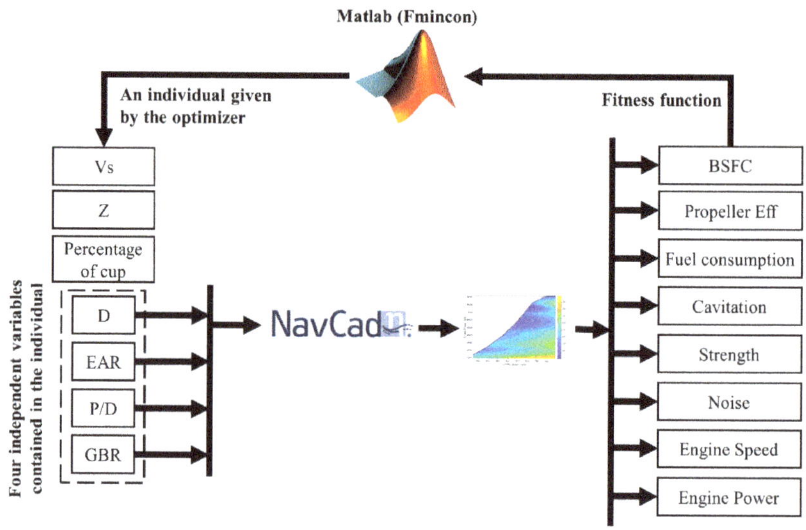

Figure 2. Schematic diagram of the propeller optimization model [20].

In order to perform the simulation, the ship data were exported from the 3D hull and input into NavCad to first compute the ship resistance. Then, the methods presented in [37,38] were used according to the expert ranking tab provided in NavCad to provide the priority of the suitable method among all methods considered in the software, with the total resistance coefficient (C_T) s computed using the following equation:

$$C_T = (1 + k)\, C_F + C_R + C_A \tag{1}$$

where k is the form factor computed based on the equations of the International Towing Tank Conference (ITTC) [39], C_F is the frictional coefficient, C_R is the residuary coefficient, and C_A is the correlation allowance computed according to the recommendations of ITTC 78 [40].

These methods to compute the ship resistance have been validated by comparing the empirical results of several ships with the ones exported from CFD computation, showing good agreement and high accuracy [41]. Figure 3 shows the calculated ship resistance for several ship speeds.

Once the resistance was computed, the propulsive coefficient, including wake fraction (w), thrust deduction factor (t) and relative rotative efficiency (η_{RR}), were estimated using the methods presented in [37,42]. Then, the propeller geometry and gearbox ratio were computed based on the defined propeller series (Wageningen B-series [43] was used in this study), as well as the efficiencies of the propeller shaft and gearbox.

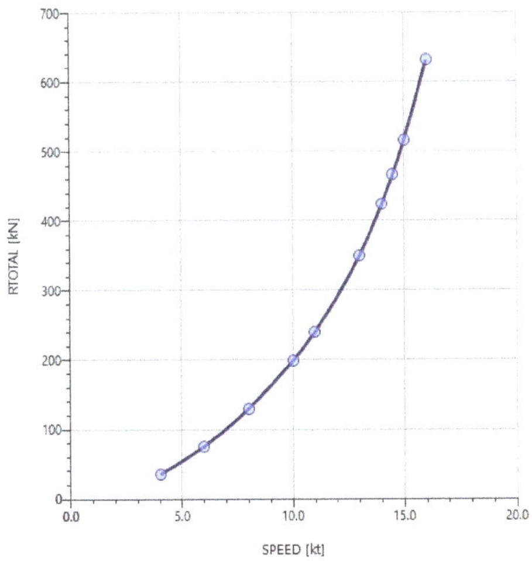

Figure 3. Total ship resistance at different speeds.

The optimization procedure started by defining the initial conditions of propeller geometry, including propeller diameter (D), expanded area ratio (EAR), pitch diameter ratio (P/D) and gearbox ratio (GBR). All of this information was processed in NavCad, and the propeller's performance was computed. Then, the data were exported and passed through the engine load diagram using interpolation to find the interaction between engine and propeller performance. The engine performance at this point could be computed by estimating the brake-specific fuel consumption (BSFC) and the different exhaust emissions (CO_2, NO_x and SO_x). After that, the fuel consumption (FC) in liters per nautical mile was computed using the following equation:

$$FC_{l/nm} = \frac{BSFC \times P_B \times 1000}{\rho_{fuel} \times V_S} \quad (2)$$

where P_B is the brake power, and ρ_{fuel} is the fuel density.

The FC, as the study's main objective, and the different constraints, including cavitation, noise and effective speed-power area inside the engine load diagram presented as penalty functions, were implemented in a fitness function and evaluated using the nonlinear optimizer. More information about the detailed fitness function presented in the following equation can be found in [16]. Then, the process of optimization was repeated until the stopping criteria were met, as shown in Figure 4.

$$Fitness\ Function = FC_{l/nm} + R\sum_{i=1}^{j} max(g_i(x), 0) \quad (3)$$

where R is a constant, g is the penalty function, and j is the number of constraints.

The contra-rotating propeller performance developed in NavCad can be adapted to consider any type of propeller series. At the same time, some corrections have been applied using empirical formulas based on propeller performance in [23,44,45], comparing the FPP to the CRP. Therefore, it is a simplified model that can predict the performance of the CRPs with the definition of the forward propeller only.

This model divides equally the thrust and torque of the total propulsor in half between the blade rows and has lower blade loading than without the other blade row. It considers the increased induced velocity that manifests as a reduction of thrust coefficient (K_T) and torque coefficient (K_Q). Additionally, the recovery of rotational losses manifests as a reduction in propeller torque for the after-blade row, thus increasing the relative rotative efficiency, which is a measure of the torque. Therefore, based on the propeller model test and self-propulsion tests described in the previous references, some internal corrections in NavCad were applied, without user interaction, to compute the corrected open-water propeller efficiency (η_o), hull efficiency (η_H), and relative rotative efficiency η_{RR} [32] compared to a single FPP. As this model is based on a single propeller, the computed thrust and torque coefficient were for the total CRP unit, while the cavitation parameters were for the forward propeller only.

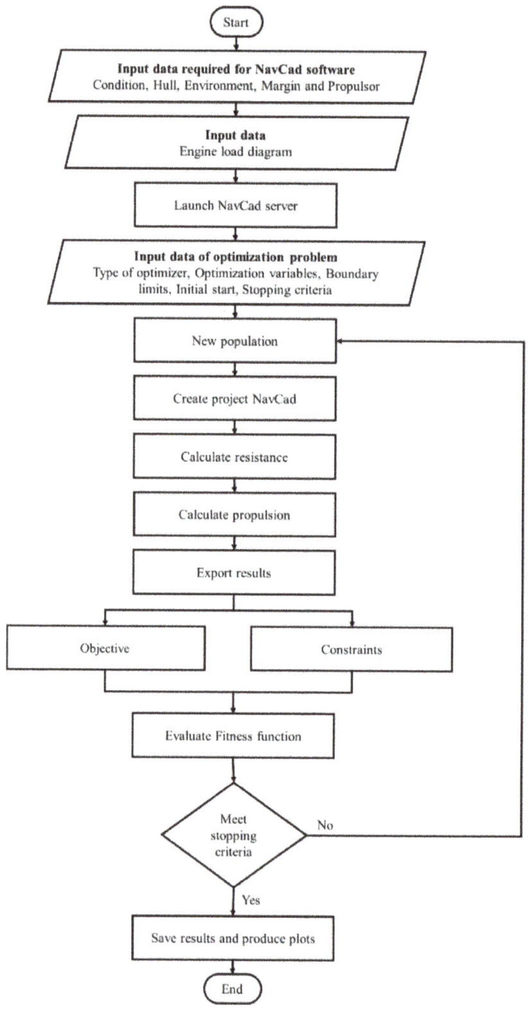

Figure 4. Schematic diagram of optimization tool in calm water.

3. Results and Discussion

After establishing the numerical model, the propeller type, CRP, was selected to perform the optimization procedure. Different upper boundary conditions of propeller diameter were selected to compare the selected CRP with the FPP. The first case was for an upper boundary equal to the FPP (6 m), while the second one equaled 90% of the diameter of the FPP (5.4 m) compared to the propeller mentioned by van Lammeren et al. [46]. Then, each propeller was simulated in two cases; no cup and a heavy cup. All simulations were performed for a five-blade propeller at a 14.5 knot designed speed. The results are presented in normalized data in Figures 5 and 6, while the real data are presented in Table A1 in Appendix A.

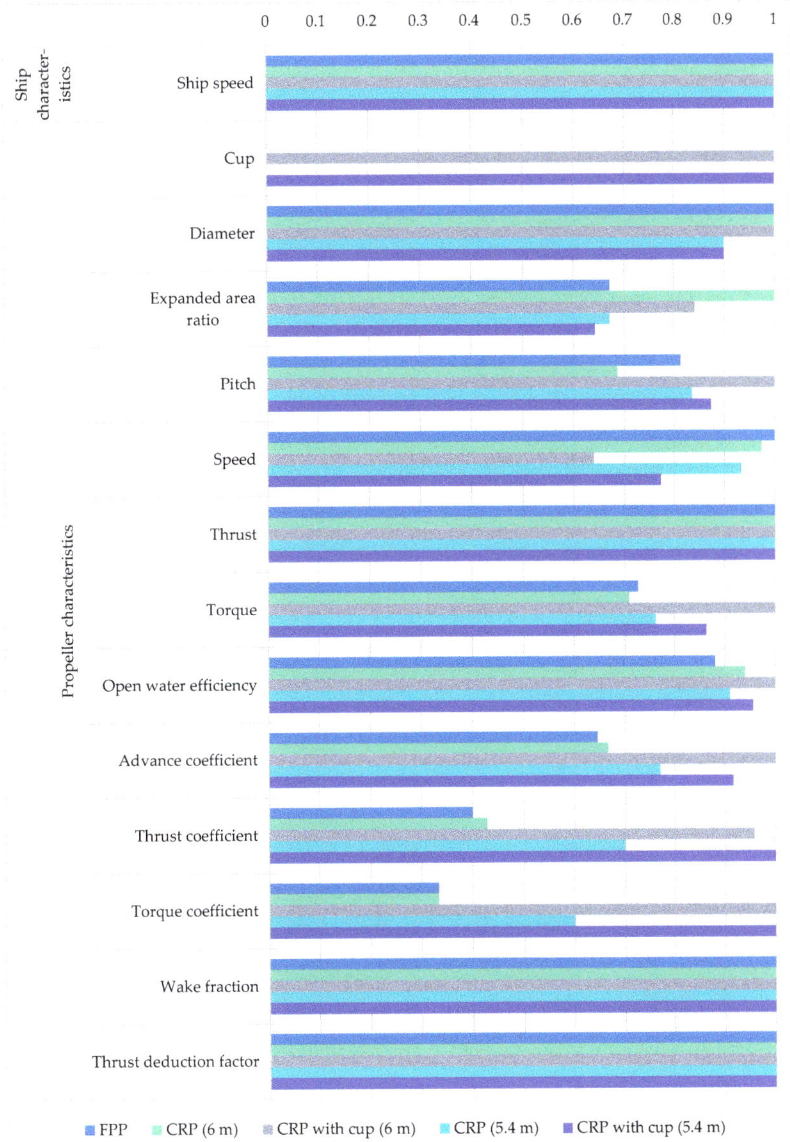

Figure 5. Normalized propeller characteristics for different configurations.

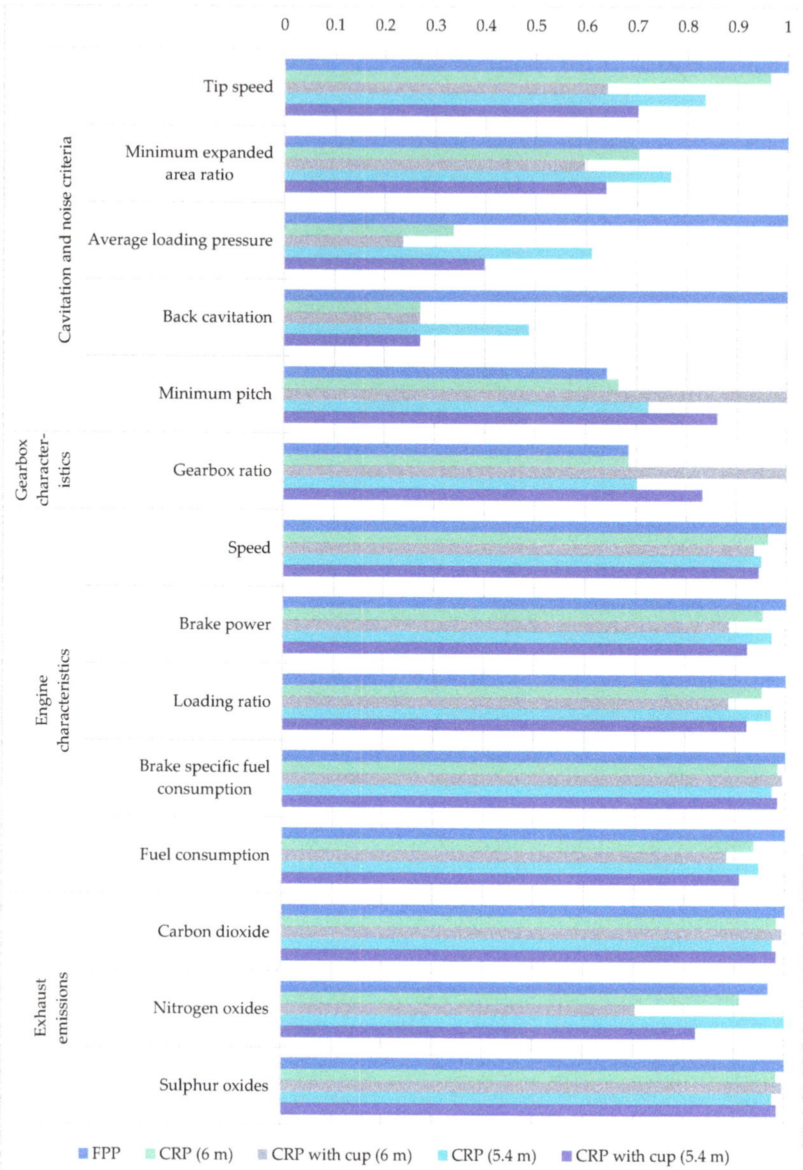

Figure 6. Normalized cavitation criteria, gearbox and engine characteristics and exhaust emissions for different configurations.

After that, the optimization procedures were performed, and the propeller geometry and the operational point were selected to minimize fuel consumption. The model complied with the limitations of the different criteria of cavitation that were defined as model constraints: (1) minimum EAR [43], where the computed EAR must be greater than or equal to minimum EAR based on the Keller method, (2) average loading pressure [47] computed on the basis of the Burrill chart, with values that cannot exceed 65 kPa, (3) the average predicted back cavitation percentage [48], generated from the high power that cannot be handled due to lower blade area, and that cannot exceed 15%, and (4) minimum pitch to avoid face cavitation [49], where the lower angle of attack causes the blade to generate lift on both blade sides, leading to face cavitation; therefore, the propeller pitch must be greater or equal to the value of the minimum pitch.

Further, the tip speed limitation [32] and minimum blade thickness [43] were verified. The propeller diameter for each case was selected at the upper boundary of the defined parameter. Then, the other parameters, such as EAR and pitch, were selected accordingly by considering the cavitation criteria.

In contrast, the GBR was selected such that the propeller speed did not exceed the level of noise as well as the operating range defined in accordance with the engine load diagram, where the engine could operate smoothly with a high combustion process. All of the selected propellers produced the same thrust required to operate the ship at the designed speed. It was found that the open water efficiency in the case of CRP was increased relative to the FPP and achieved a further increment in the case of a cupping propeller. In addition, the larger diameter selected, the higher the efficiencies achieved relative to the FPP.

The advance coefficient (J_A) was increased in the cases of cupping propellers, where there was a reduction in propeller speed accompanied by an increase in K_T and K_Q. However, there were no changes in the wake fraction and thrust deduction factor values in the five different cases, as confirmed in some studies, such as in [50,51].

Due to the reduction in propeller speed, the tip speed was reduced in the case of CRP and achieved an increment in speed when the propeller was cupped by 33% in the case of a 6 m propeller and 16% in the case of a 5.4 m propeller. The EAR was always greater than the minimum EAR to avoid cavitation, and the minimum EAR in the CRP was smaller than that of the FPP by 30%, while the minimum EAR was reduced in the case of a cupping propeller compared to a normal CRP, by 15% and 17%, respectively, in the case of 6 and 5.4 m propellers, because the propeller was operating at a higher pitch, and therefore, the pressure on the blades was reduced. The same concept applied when computing the average loading pressure and back cavitation criteria, where the first criterion was reduced by 76% compared to the FPP, and the second criterion was reduced by 73% compared to the FPP.

The GBR was computed as a ratio between the propeller and engine speeds, and some increase was achieved in the case of a 6 m propeller with a heavy cup, because the propeller speed was lower.

The engine speed was reduced in the case of a CRP because it could produce the same amount of thrust at lower speeds. Additionally, the brake power was reduced by 3 to 5% compared to that of the FPP. By considering the percentage of cupping, the CRP could operate at speeds 5 to 7% lower than those of an FPP.

There were no significant differences in BSFC for all cases, but because of the lower brake power, fuel consumption was reduced. The no-cup CRP could reduce fuel consumption, by 6.2% in the case of a 6 m propeller and 5.2% in the case of a 5.4 m propeller, compared to an FPP. Moreover, with a cupping propeller, fuel consumption was significantly reduced, by 11.7% in the case of a 6 m and 9.2% in the case of a 5.4 m propeller, compared to an FPP. As fuel consumption was reduced, the different exhaust emissions were also reduced, to levels corresponding to the reduced levels of fuel consumption.

4. Conclusions

This paper presents the selection of CRPs with different propeller diameters corresponding to the ship selected and the cases presented in a literature review. Further, the cupping percentage was considered to select a more effective propeller.

The propeller was selected for the designed speed using a developed optimization model coupling NavCad, an optimizer in the Matlab environment, and a surrogate model presenting the engine performance. The optimization model was able to select the propeller geometry and the operational point in order to minimize fuel consumption. Additionally, the model complied with stipulated limitations on cavitation and noise.

The propeller selection process is very fast, as it depends on the propeller series to perform a preliminary investigation in propeller selection for the concept of ship design. With some simplifications, the geometry of the forward propeller was selected, while the model was able to select the performance of the CRP unit.

In terms of propeller performance, it was found that the performance of the CRP unit, in general, was higher than that of an FPP and increased when the cupping percentage was considered.

In terms of cavitation and noise, a CRP unit is more resistant to cavitation, as the amount of thrust is divided between the two rows; therefore, the probability of the occurrence of cavitation was reduced when the different cavitation criteria were evaluated. In addition, the noise level was reduced, as the tip speed was reduced in comparison with that of an FPP.

In terms of fuel consumption, the CRP showed better fuel economy, as the propeller was operated at a lower loading ratio than that of an FPP. This percentage was increased when the model considered a cupping CRP. Compared to an FPP, a no-cup CRP unit could achieve a reduction in fuel consumption by up to 6.2%, while a cupped CRP could achieve a reduction of up to 11.7%.

This work presents a preliminary estimation of the performance of a CRP in comparison with an FPP, with and without cupping. More investigations can be performed for other propeller series that are able to reach higher values of fuel savings and propeller performance than the B series. Furthermore, it would be interesting to consider CFD computations for the same propeller selected to validate the calculation of the wake field, which could provide additional improvements to the wake calculation using empirical formulas.

Author Contributions: The study concept was developed by M.T. The analysis was performed by M.T., and the original draft manuscript was written by M.T., M.V. and C.G.S. All authors have read and agreed to the published version of the manuscript.

Funding: This work was performed within the scope of the Strategic Research Plan of the Centre for Marine Technology and Ocean Engineering (CENTEC), which is financed by the Portuguese Foundation for Science and Technology (Fundação para a Ciência e Tecnologia-FCT) under contract UIDB/UIDP/00134/2020.

Institutional Review Board Statement: Not applicable.

Informed Consent Statement: Not applicable.

Data Availability Statement: The data presented in this study are available on request from the corresponding author.

Conflicts of Interest: The authors declare no conflict of interest.

Abbreviations

η_H	Hull efficiency
η_o	Open-water propeller efficiency
η_{RR}	Relative-rotative efficiency
3D	Three dimensional
API	Application programming interface
BSFC	Brake-specific fuel consumption
C_A	Correlation allowance
CAV_{AVG}	Back cavitation
C_F	Frictional coefficient
CFD	Computational fluid dynamics
CO_2	Carbon dioxide
CPP	Controllable pitch propeller
C_R	Residuary coefficient
CRP	Contra-rotating propeller
C_T	Total resistance coefficient
D	Propeller diameter

DHP	Delivered horsepower
EAR	Expanded area ratio
EAR_{min}	Minimum expanded area ratio
FC	Fuel consumption
FPP	Fixed pitch propeller
g	Penalty function
GBR	Gearbox ratio
ITTC	International Towing Tank Conference
j	Number of constraints
J_A	Advance coefficient
k	Form factor
K_Q	Torque coefficient
K_T	Thrust coefficient
LR	Loading ratio
N	Propeller speed
NO_x	Nitrogen oxides
P/D	Pitch diameter ratio
P_B	Brake power
P_{FC}	Minimum pitch
PRESS	Average loading pressure
Q	Propeller torque
R	Constant
RPM	Engine speed
SMGT	Super marine gas turbine
SO_x	Sulphur oxides
t	Thrust deduction factor
T	Propeller thrust
VGFs	Vortex generator fins
V_s	Ship design speed
V_{tip}	Tip speed
w	Wake fraction
WFSV	Wind farm support vessel
Z	Number of propeller blades
ρ_{fuel}	Fuel density

Appendix A

Table A1. Optimum results for different configurations.

Main Characteristics	Parameters	Symbol	Unit					
Propeller type			[-]	FPP	CRP (6 m)	CRP with cup (6 m)	CRP (5.4 m)	CRP with cup (5.4 m)
Ship characteristics	Ship speed	V_s	[kn]	14.5	14.5	14.5	14.5	14.5
Propeller characteristics	Series	[-]	[-]	Wageningen B-series				
	Cup	[-]	[%]	0.00	0.00	1.50	0.00	1.50
	Diameter	D	[m]	6.00	6.00	6.00	5.40	5.40
	Expanded area ratio	EAR	[-]	0.47	0.70	0.59	0.47	0.45
	Pitch	P	[m]	6.58	5.55	8.08	6.76	7.06
	Speed	N	[RPM]	75	73	48	70	58
	Thrust	T	[kN]	576.49	576.49	576.49	576.49	576.49
	Torque	Q	[kN·m]	573.30	559.20	788.50	600.90	680.70
	Open water efficiency	η_o	[%]	59	63	67	61	64
	Advance coefficient	J_A	[-]	0.62	0.64	0.96	0.74	0.88
	Thrust coefficient	K_T	[-]	0.28	0.30	0.67	0.49	0.70
	Torque coefficient	K_Q	[-]	0.05	0.05	0.15	0.09	0.15
	Wake fraction	w	[-]	0.38	0.38	0.38	0.38	0.38
	Thrust deduction factor	t	[-]	0.19	0.19	0.19	0.19	0.19

Table A1. *Cont.*

Main Characteristics	Parameters	Symbol	Unit					
Cavitation and noise criteria	Tip Speed	V_{tip}	[m/s]	23.61	22.78	15.12	19.70	16.53
	Minimum expanded area ratio	EAR_{min}	[-]	0.47	0.33	0.28	0.36	0.30
	Average loading pressure	PRESS	[kPa]	43.56	14.62	10.30	26.57	17.32
	Back cavitation	CAV_{AVG}	[%]	7.40	2.00	2.00	3.60	2.00
	Minimum pitch	P_{FC}	[m]	4.98	5.16	7.77	5.62	6.69
Gearbox characteristics	Gearbox ratio	GBR	[-]	9.50	9.50	13.88	9.73	11.54
Engine characteristics	Speed	RPM	[RPM]	714	688	668	678	675
	Brake power	P_B	[kW]	4682	4465	4151	4551	4321
	Loading ratio	LR	[%]	65.6	62.5	58.1	63.7	60.5
	BSFC	BSFC	[g/kW·h]	192	189	191	187	189
	Fuel consumption	FC	[l/nm]	74.17	69.56	65.48	70.28	67.38
Exhaust emissions	Carbon dioxide	CO_2	[g/kW·h]	608	598	605	593	598
	Nitrogen oxides	NO_x	[g/kW·h]	6.68	6.28	4.85	6.90	5.68
	Sulphur oxides	SO_x	[g/kW·h]	9.59	9.43	9.55	9.35	9.44

References

1. Bouman, E.A.; Lindstad, E.; Rialland, A.I.; Strømman, A.H. State-of-the-art technologies, measures, and potential for reducing GHG emissions from shipping—A review. *Transp. Res. D Transp. Environ.* **2017**, *52*, 408–421. [CrossRef]
2. Green Ship of the Future. 2019 Retrofit Project. Available online: https://greenship.org/project/2019-retrofit-series/ (accessed on 8 December 2021).
3. Karatuğ, Ç.; Arslanoğlu, Y.; Guedes Soares, C. Evaluation of decarbonization strategies for existing ships. In *Trends in Maritime Technology and Engineering*; Guedes Soares, C., Santos, T.A., Eds.; Taylor & Francis Group: London, UK, 2022; pp. 45–54.
4. DNV. *Maritime Forecast to 2050: Energy Transition Outlook 2020*; DNV: Bærum, Norway, 2020.
5. Tadros, M.; Vettor, R.; Ventura, M.; Guedes Soares, C. Effect of different speed reduction strategies on ship fuel consumption in realistic weather conditions. In *Trends in Maritime Technology and Engineering*; Guedes Soares, C., Santos, T.A., Eds.; Taylor & Francis Group: London, UK, 2022; pp. 553–561.
6. Vettor, R.; Tadros, M.; Ventura, M.; Guedes Soares, C. Route planning of a fishing vessel in coastal waters with fuel consumption restraint. In *Maritime Technology and Engineering 3*; Guedes Soares, C., Santos, T.A., Eds.; Taylor & Francis Group: London, UK, 2016; pp. 167–173.
7. Ventura, M. Ship dimensioning in the initial design. In *Developments in Maritime Transportation and Exploitation of Sea Resources*; Guedes Soares, C., López Peña, F., Eds.; Taylor & Francis Group: London, UK, 2014; pp. 531–539.
8. Feng, Y.; el Moctar, O.; Schellin, T.E. Parametric Hull Form Optimization of Containerships for Minimum Resistance in Calm Water and in Waves. *J. Mar. Sci. Appl.* **2021**, *20*, 670–693. [CrossRef]
9. Stark, C.; Xu, Y.; Zhang, M.; Yuan, Z.; Tao, L.; Shi, W. Study on Applicability of Energy-Saving Devices to Hydrogen Fuel Cell-Powered Ships. *J. Mar. Sci. Eng.* **2022**, *10*, 388. [CrossRef]
10. Andersson, J.; Shiri, A.A.; Bensow, R.E.; Yixing, J.; Chengsheng, W.; Gengyao, Q.; Deng, G.; Queutey, P.; Xing-Kaeding, Y.; Horn, P.; et al. Ship-scale CFD benchmark study of a pre-swirl duct on KVLCC2. *Appl. Ocean Res.* **2022**, *123*, 103134. [CrossRef]
11. Gaggero, S.; Martinelli, M. Pre-swirl fins design for improved propulsive performances: Application to fast twin-screw passenger ships. *J. Ocean Eng. Mar. Energy* **2022**. [CrossRef]
12. Seol, H. Virtue and Function. *The Naval Architect*. Available online: https://www.rina.org.uk/Virtue_and_function.html (accessed on 8 September 2022).
13. Tadros, M.; Ventura, M.; Guedes Soares, C. Design of Propeller Series Optimizing Fuel Consumption and Propeller Efficiency. *J. Mar. Sci. Eng.* **2021**, *9*, 1226. [CrossRef]
14. Nelson, M.; Temple, D.W.; Hwang, J.T.; Young, Y.L.; Martins, J.R.R.A.; Collette, M. Simultaneous optimization of propeller–hull systems to minimize lifetime fuel consumption. *Appl. Ocean Res.* **2013**, *43*, 46–52. [CrossRef]
15. Tadros, M.; Ventura, M.; Guedes Soares, C. Optimum design of a container ship's propeller from Wageningen B-series at the minimum BSFC. In *Sustainable Development and Innovations in Marine Technologies*; Georgiev, P., Guedes Soares, C., Eds.; Taylor & Francis Group: London, UK, 2020; pp. 269–274.
16. Tadros, M.; Vettor, R.; Ventura, M.; Guedes Soares, C. Coupled Engine-Propeller Selection Procedure to Minimize Fuel Consumption at a Specified Speed. *J. Mar. Sci. Eng.* **2021**, *9*, 59. [CrossRef]
17. Jaurola, M.; Hedin, A.; Tikkanen, S.; Huhtala, K. A TOpti simulation for finding fuel saving by optimising propulsion control and power management. *J. Mar. Sci. Technol.* **2020**, *25*, 411–425. [CrossRef]
18. Tadros, M.; Ventura, M.; Guedes Soares, C. Optimization procedures for a twin controllable pitch propeller of a ROPAX ship at minimum fuel consumption. *J. Mar. Eng. Technol.* **2022**. [CrossRef]

19. Makino, H.; Umeda, N.; Ohtsuka, T.; Ikejima, S.; Sekiguchi, H.; Tanizawa, K.; Suzuki, J.; Fukazawa, M. Energy savings for ship propulsion in waves based on real-time optimal control of propeller pitch and electric propulsion. *J. Mar. Sci. Technol.* **2017**, *22*, 546–558. [CrossRef]
20. Tadros, M.; Vettor, R.; Ventura, M.; Guedes Soares, C. Effect of propeller cup on the reduction of fuel consumption in realistic weather conditions. *J. Mar. Sci. Eng.* **2022**, *10*, 1039. [CrossRef]
21. Wagner, R. Rückblick und Ausblick auf die Entwicklung des Contrapropellers. In *Jahrbuch der Schiffbautechnischen Gesellschaft: 30. Band*; Springer: Berlin/Heidelberg, Germany, 1929; pp. 195–256.
22. Min, K.-S.; Chang, B.-J.; Seo, H.-W. Study on the Contra-Rotating Propeller system design and full-scale performance prediction method. *Int. J. Nav. Arch. Ocean Eng.* **2009**, *1*, 29–38. [CrossRef]
23. van Manen, J.D.; Oosterveld, M.W.C. Model Tests on Contra-Rotating Propellers. *Int. Shipbuild. Prog.* **1969**, *15*, 401–417. [CrossRef]
24. Koronowicz, T.; Krzemianowski, Z.; Tuszkowska, T.; Szantyr, J. A complete design of contra-rotating propellers using the new computer system. *Pol. Marit. Res.* **2010**, *17*, 14–24. [CrossRef]
25. Ghassemi, H.; Taherinasab, M. Numerical calculations of the hydrodynamic performance of the contra-rotating propeller (CRP) for high speed vehicle. *Pol. Marit. Res.* **2013**, *20*, 13–20. [CrossRef]
26. Nouri, N.M.; Mohammadi, S.; Zarezadeh, M. Optimization of a marine contra-rotating propellers set. *Ocean Eng.* **2018**, *167*, 397–404. [CrossRef]
27. Kayano, J.; Haraguchi, T.; Tsukada, Y.; Kano, T. On the ship maneuverability of tandem arrangement CRP pod propulsion system. In *Maritime Transportation and Exploitation of Ocean and Coastal Resources*; Guedes Soares, C., Garbatov, Y., Fonseca, N., Eds.; Taylor and Fransis: London, UK, 2005; pp. 189–193.
28. Torneman, G. Multiple pod units for efficient vessel handling in wind farm operations. In *Design & Operation of Offshore Wind Farm Support Vessels*; Royal Institution of Naval Architects: London, UK, 2015.
29. Hou, L.; Yin, L.; Hu, A.; Chang, X.; Lin, Y.; Wang, S. Optimal matching investigation of marine contra-rotating propellers for energy consumption minimization. *J. Mar. Sci. Technol.* **2021**, *26*, 1184–1197. [CrossRef]
30. Minami, Y.; Kano, T. Evaluation of the emissions from the super eco-ship and the corresponding conventional ship. In *Maritime Transportation and Exploitation of Ocean and Coastal Resources*; Guedes Soares, C., Garbatov, Y., Fonseca, N., Eds.; Taylor and Fransis: London, UK, 2005; pp. 1721–1727.
31. MAN Diesel & Turbo. *32/44CR Project Guide—Marine Four-Stroke Diesel Engines Compliant with IMO Tier II*; MAN Diesel & Turbo: Augsburg, Germany, 2017.
32. HydroComp. NavCad: Reliable and Confident Performance Prediction. *HydroComp Inc.* Available online: https://www.hydrocompinc.com/solutions/navcad/ (accessed on 30 January 2019).
33. Tadros, M.; Ventura, M.; Guedes Soares, C. Surrogate models of the performance and exhaust emissions of marine diesel engines for ship conceptual design. In *Maritime Transportation and Harvesting of Sea Resources*; Guedes Soares, C., Teixeira, A.P., Eds.; Taylor & Francis Group: London, UK, 2018; pp. 105–112.
34. Tadros, M.; Ventura, M.; Guedes Soares, C. Optimization procedure to minimize fuel consumption of a four-stroke marine turbocharged diesel engine. *Energy* **2019**, *168*, 897–908. [CrossRef]
35. Tadros, M.; Ventura, M.; Guedes Soares, C. Simulation of the performance of marine Genset based on double-Wiebe function. In *Sustainable Development and Innovations in Marine Technologies*; Georgiev, P., Guedes Soares, C., Eds.; Taylor & Francis Group: London, UK, 2020; pp. 292–299.
36. The MathWorks Inc. Fmincon. Available online: https://www.mathworks.com/help/optim/ug/fmincon.html (accessed on 2 June 2017).
37. Holtrop, J. A statistical re-analysis of resistance and propulsion data. *Int. Shipbuild. Prog.* **1984**, *31*, 272–276.
38. Holtrop, J. A Statistical Resistance Prediction Method With a Speed Dependent Form Factor. In *Proceedings of Scientific and Methodological Seminar on Ship Hydrodynamics (SMSSH'88)*; Bulgarian Ship Hydrodynamics Centre: Varna, Bulgaria, 1988; pp. 1–7.
39. ITTC. Skin Friction and Turbulence Stimulation. In Proceedings of the 8th ITTC, Madrid, Spain, 15–23 September 1957.
40. ITTC. 1978 ITTC Performance Prediction Method. In Proceedings of the 28th ITTC, Wuxi, China, 27 November–2 December 2017.
41. Islam, H.; Ventura, M.; Guedes Soares, C.; Tadros, M.; Abdelwahab, H.S. Comparison between empirical and CFD based methods for ship resistance and power prediction. In *Trends in Maritime Technology and Engineering*; Guedes Soares, C., Santos, T.A., Eds.; Taylor & Francis Group: London, UK, 2022; pp. 347–357.
42. Holtrop, J.; Mennen, G.G.J. An approximate power prediction method. *Int. Shipbuild. Prog.* **1982**, *29*, 166–170. [CrossRef]
43. Oosterveld, M.; Van Oossanen, P. Further computer-analyzed data of the Wageningen B-screw series. *Int. Shipbuild. Prog.* **1975**, *22*, 251–262. [CrossRef]
44. Lindgren, H.; Johnsson, C.-A.; Dyne, G. Studies of the Application of Ducted and Contrarotating Propellers on Merchant Ships. In *Seventh ONR Symposium on Naval Hydrodynamics*; Cooper, R.D., Doroff, S.W., Eds.; Office of Naval Research: Arlington, VA, USA, 1968.
45. Bjarne, E. Systematic Studies of Contra-rotating Propellers for Merchant Ships. In *Proceedings International Maritime and Shipping Conference (IMAS)*; Institute of Marine Engineers: London, UK, 1973.
46. van Lammeren, W.P.A.; van Manen, J.D.; Oosterveld, M.W.C. The Wageningen B-screw series. *Trans. SNAME* **1969**, *77*, 269–317.
47. Burrill, L.C.; Emerson, A. Propeller cavitation: Further tests on 16in. propeller models in the King's College cavitation tunnel. *Int. Shipbuild. Prog.* **1963**, *10*, 119–131. [CrossRef]

48. Blount, D.L.; Fox, D.L. Design Considerations for Propellers in a Cavitating Environment. *Mar. Technol.* **1978**, *15*, 144–178. [CrossRef]
49. MacPherson, D.M. Reliable Propeller Selection for Work Boats and Pleasure Craft: Techniques Using a Personal Computer. In *SNAME Fourth Biennial Power Boat Symposium*; SNAME: Alexandria, VA, USA, 1991.
50. Sasaki, N.; Kuroda, M.; Fujisawa, J.; Imoto, T.; Masaharu, S. On the Model Tests and Design Method of Hybrid CRP Podded Propulsion System of a Feeder Container Ship. In *Proceedings of the First International Symposium on Marine Propulsors (SMP'09)*; Koushan, K., Steen, S., Eds.; Norwegian Marine Technology Research Institute (MARINTEK): Trondheim, Norway, 2009.
51. Sasaki, N.; Murakami, M.; Nozawa, K.; Soejima, S.; Shikaki, A.; Aono, T. Design system for optimum contra-rotating propellers. *J. Mar. Sci. Technol.* **1998**, *3*, 3–21. [CrossRef]

Journal of
Marine Science and Engineering

MDPI

Article

Mitigation of Hub Vortex Cavitation with Application of Roughness

Savas Sezen [1,2,*] and Mehmet Atlar [1]

1 Department of Naval Architecture, Ocean & Marine Engineering, University of Strathclyde, Glasgow G1 1XQ, UK
2 Lloyd's Register EMEA, Technical Investigation Department (TID), Southampton SO16 7QF, UK
* Correspondence: savas.sezen@strath.ac.uk

Abstract: This study investigates the influence of roughness on hydrodynamic performance, especially for the hub vortex—and, hence, hub vortex cavitation—of a benchmark propeller operating under uniform flow conditions using the RANS method. The Schnerr–Sauer cavitation model is also used for modelling the cavitation on and off the propeller blades. In order to include the effects of roughness in the numerical calculations, the experimentally obtained roughness functions were incorporated with the wall function of the CFD solver. The applicability and effectiveness of the roughness application applied on the propeller hub as a novel concept were explored to mitigate hub vortex cavitation. The results are first validated with experimental data on smooth conditions through the propeller hydrodynamic performance characteristics and cavitation extension. Then, the propeller hub is covered with four different sizes of roughness. The results show that the degradation effects of roughness applied to the hub on propeller performance are negligible, and the maximum efficiency loss is around 0.25% with respect to the smooth condition when the propeller hub was roughened. Favourable impacts of roughness are found for the hub vortex, and hence, hub vortex mitigation. Applying the roughness on the propeller changed the flow properties (e.g., pressure, velocity and turbulent kinetic energy) inside the vortex, enabling the early breakdown of the extension of hub vortices. These flow changes in the presence of roughness result in a mitigation of hub vortex cavitation up to 50% depending on the roughness size with respect to the smooth condition. Thus, this proposed novel concept, application of roughness to the propeller hub, can be used to mitigate hub vortex cavitation, rudder erosion and propeller URN for both newly designed and retrofitted projects by keeping the efficiency loss as minimum as possible.

Keywords: CFD; cavitation; mitigation; roughness; hub vortex cavitation

Citation: Sezen, S.; Atlar, M. Mitigation of Hub Vortex Cavitation with Application of Roughness. *J. Mar. Sci. Eng.* **2022**, *10*, 1426. https://doi.org/10.3390/jmse10101426

Academic Editor: Alon Gany

Received: 4 August 2022
Accepted: 22 September 2022
Published: 4 October 2022

1. Introduction

Shipping is a significant contributor to the underwater radiated noise levels (URN) in the world's oceans, particularly in the low-frequency range of the noise spectrum. These escalated URN levels have caused several detrimental impacts on marine fauna. This is because marine animals use a certain frequency range for various fundamental living activities such as communication, interaction and feeding. For this reason, the sudden increase in URN levels may disorient them, destroy their ability to communicate with each other and even cause their local extinction. Thus, the mitigation of URN levels caused by the shipping industry is of great importance, although this subject is currently of a low priority compared to other sustainability concerns within the shipping industry (e.g., Greenhouse Gas Emissions (GHG)). In addition, the lack of mandatory international regulations and noise limits makes progress slow for URN mitigation investigations. Nevertheless, the effort in this field is getting more attention to highlight the possible short- and long-term detrimental impacts of ship URN on marine mammals. Recently, the Marine Environment Protection Committee of the IMO has accepted the proposal from Australia, Canada and the United States to review the existing 2014 Guidelines [1] for URN mitigation radiated by commercial vessels [2–4]

The main source of ship URN is propeller cavitation, which dominates the other noise sources such as machinery, flow noise, etc. Despite the many side effects of cavitation (e.g., performance degradation, noise, vibration and material damage), inevitably, commercial ship propellers will always operate under cavitating conditions above a certain speed limit. Depending on the operating conditions and propeller action, several cavitation formations (e.g., sheet, vortex, bubble and cloud cavitation) can be present on and off the propeller blades. Amongst those, propellers generally operate in a condition where sheet and vortex cavitation are present. Vortex cavitation, which appears at the blade tips, leading edge and propeller hub, is not erosive unless it is present excessively, but it is a source of noise. Vortex cavitation is formed by the low-pressure core of the shed vortices. Tip vortex cavitation (TVC), created by the vortex at the blade tips and observed as the first type of cavitation on well-designed propellers, is an important noise source and leads to a substantial increase in noise levels when bursting or collapsing phenomena occurs [5,6]. Another type of vortex cavitation is hub vortex cavitation, which appears around the propeller hub; it is formed through the contribution of different vortices, which are unlikely to cavitate individually, shed from the blade root. Under the impact of the propeller's converging cone, the combination of blade-root vortices tends to cavitate. When the vortices start to cavitate, the resulting cavitation (i.e., hub vortex cavitation) becomes very stable and appears like a rope with strands corresponding to the number of propeller blades [7].

Similar to TVC, hub vortex cavitation, which is the main interest of this study, is also of practical importance. First of all, the waste energy is spent by the excessive swirl of the flow in the propeller slipstream near the propeller's rotational axis. This swirl will result in a pressure reduction with respect to ambient pressure, and hence, it will generate an undesirable drag force on the propeller. In addition, when the swirl is large, the delivery of energy to the fluid near the hub will not produce axial thrust, and the mixing of turbulence will dissipate it. Secondly, hub vortex cavitation is a large cavity. Thus, if the rudder or any control surfaces are positioned in line with the propeller-shaft-system axis, they can lose the lift force which is aimed to be produced. Eventually, the vortices around the hub reduce the propeller's efficiency by increasing energy loss depending on the axial load distribution on the propeller and hub geometry. Lastly, hub vortex cavitation can also be associated with undesirable vibration, noise and, in some cases, erosion on the rudder [7–9].

In order to mitigate the cavitation and associated URN, several active and passive noise-control methods can be applied for marine propellers, especially for retrofit projects [10] The most common passive noise-control methods, mainly implemented for TVC mitigation, are blade-geometry modification, reduction of the propeller tip's circulation, the inclusion of additional geometry at the tip, drilling holes, leading-edge tubercle modifications and application of roughness to propeller blades (e.g., [11–16]) In addition to these methods, propeller boss cap fins (PBCF), primarily designed to improve a propeller's performance characteristics, have also been utilised as an Energy Saving Device (ESD). Moreover, with this concept, the strength of the hub vortex is weakened, and the kinetic energy of the flow around the boss can be recovered by applying a number of fins corresponding to the propeller-blade number. Thus, it can also be used as a passive noise-control concept to mitigate hub vortex cavitation, associated URN and rudder erosion. The PBCF investigations were carried out by [17–19]; the authors found out that with the application of PBCF, thrust increases while the torque reduces, resulting in an efficiency increase. Since then, there have been several experimental and numerical investigations conducted using different PBCF concepts to investigate its effects on propeller hydrodynamic performance, including cavitation and URN (e.g., [20–24]).

Another new passive noise-control concept, namely, roughness application on the propeller's hub—which is the main interest of this study—can also be used as an alternative to PBCF to mitigate hub vortex cavitation, associated URN and rudder erosion. With roughness, near-wall flow structures can change significantly, and turbulence transition can be stimulated in the laminar boundary layer. The roughness elements interact with the vortices around the hub, and it enables early breakdown; hence, hub vortex cavitation mitigation can be achieved. However, unlike the PBCF, the roughness application on the propeller's hub may cause performance degradation, resulting in efficiency loss. Nevertheless, this less efficiency loss can be further minimised by applying the roughness on a strategical section of the hub, similar to the strategic application of roughness on propeller blades to keep the efficiency loss as minimal as possible [15] In this regard, the authors recently explored the influence of roughness on the hydrodynamic performance of a model-scale INSEAN E779A propeller under non-cavitating and cavitating conditions. With the application of roughness to the propeller blades, the cavitation volume decreased, which is mainly due to the TVC; this resulted in URN mitigation up to 10dB at a certain frequency range, whereas the propeller efficiency loss was found to be 25% [16]. Later on, the authors extended their research to find a compromise between cavitation-volume reduction and efficiency loss by applying the roughness heterogeneously on strategic areas on the blades for model- and full-scale benchmark propellers in a wide range of operating conditions (i.e., uniform, inclined and non-uniform flow conditions). The results showed that the detrimental effects of roughness on propeller hydrodynamic performance could be significantly reduced by applying the roughness on the back side (or suction side) of the propeller blades between the radius of 0.9 and 1 [15].

In light of previous findings regarding propeller hydrodynamic performance—including cavitation volume reduction and URN—with a roughness application on propeller blades, the authors have further extended their research with this study by applying a similar type of roughness on the propeller's hub. This study aims to explore hub vortex cavitation mitigation with this alternative new concept, roughness application on a propeller's hub for a model-scale propeller under uniform flow conditions, for the first time in the literature. This study is a pioneer in investigating propeller URN, propeller–hull–rudder interaction and rudder erosion by mitigating hub vortex cavitation using a roughness application both in model- and full-scale propellers under non-uniform flow conditions.

In this study, the standard RANS method with the k-ω SST turbulence model was used to solve the cavitating flow around the propeller. Sheet and hub vortex cavitation were modelled using the mass transfer cavitation model, Schnerr–Sauer, within the commercial CFD solver, Star CCM. The propeller's open-water characteristics (i.e., thrust and torque coefficients) were validated with the available experimental data when the propeller blades and hub were clean (i.e., in smooth condition). In addition, the cavitation extensions were compared between numerical calculations and experimental observation. Then, the propeller hub was covered with a particular type of biofouling roughness. The four different roughness heights were used in the calculations to investigate the effects of roughness on propeller hydrodynamic performance, including cavitation. The results obtained in the presence of roughness were compared with respect to the smooth condition to explore the impact of roughness on propeller hydrodynamic performance, particularly for hub vortex and hub vortex cavitation mitigation.

The structure of the paper is as follows. The hydrodynamic model and details of the roughness are given in Section 2. The test case and numerical modelling details are given in Section 3. The numerical results are given in Section 4, and the results are consequently presented in Section 5.

2. Theoretical Background

2.1. Hydrodynamic Model

The governing flow equations are solved using the computational fluid dynamics (CFD) solver [25]. The CFD solver uses a finite volume method to discretise the Navier–Stokes equations. The unsteady RANS method with k-ω SST turbulence model was used to solve cavitating flow around the propeller.

Cavitation is modelled using a homogenous seed-based approach. The Schnerr–Sauer cavitation model based on the reduced Rayleigh–Plesset equation was used with the VOF

(Volume of Fluid) approach to model the liquid and vapor phases. The adapted cavitation model neglects the influence of the bubble growth and collapse rates, viscous and surface-tension effects, and is the simplified version of the full Rayleigh–Plesset equation. The detailed information about the governing equations and cavitation model can be found in the user guide of CFD solver [25], and as such is not repeated here.

Cavitation and roughness models were incorporated to account for effects of roughness on propeller hub vortex cavitation. The wall function of CFD solver was utilised for the roughness modelling, whereas a mass transfer model, the Schnerr–Sauer cavitation model, was used for the cavitation modelling at the same time.

2.2. Roughness Model

Biofouling accumulation increases surface roughness, affecting the flow around the ship and propeller. This effect can be described as a downward shift in the log-law region of the turbulent boundary layer. Therefore, the non-dimensional velocity profile for a rough surface can be defined as follows.

$$U^+ = \frac{1}{\kappa}\ln(y^+) + B - \Delta U^+ \tag{1}$$

where $U^+ = \frac{U}{U_\tau}$ is the non-dimensional velocity, κ is the von Karman constant equal to 0.42, $y^+ = \frac{yU_\tau}{v}$ is the non-dimensional normal distance from the boundary, B is the constant for smooth-wall log-law intercept, v is the kinematic viscosity of the fluid and $U_\tau = \sqrt{\tau_w/\rho}$ is the friction velocity where τ_w represents the shear stress at the wall and ρ represents the density. ΔU^+ represents the roughness functions depending on the roughness Reynolds number k^+, which is defined as follows:

$$k^+ = \frac{kU_\tau}{\nu} \tag{2}$$

where k is the roughness length scale. It is important to note that there is no universal roughness function. Therefore, ΔU^+ needs to be obtained experimentally for the rough condition in question. For the smooth condition, ΔU^+ is taken as zero.

In this study, barnacle-type roughness was used to represent the surface roughness around the propeller hub. Roughness functions provided in [26] were implemented in the wall function of the CFD software, Star CCM+ 14.06, 2019, to mitigate the hub vortex cavitation of a propeller. It is important to note that ΔU^+ and corresponding k values show excellent agreement with the roughness-function model of Grigson, 1992 [27], as described in Equation (3).

$$\Delta U^+ = \frac{1}{\kappa}\ln(1 + k^+) \tag{3}$$

Table 1 provides the roughness length scales and equivalent sand roughness heights of the surfaces used in this study. In Table 1, Mix and NS Mix are surface names obtained from the study of Uzun et al., 2020 [26]; h is barnacle height (mm) and k_G is the representative hydrodynamic roughness length scale (μm) that give the same roughness Reynolds numbers with corresponding roughness-function values of Grigson, 1992 [27]. It should be noted that these representative hydrodynamic roughness length scales, k_G, are not a function of measurable surface properties and may be termed as experimentally obtained equivalent roughness height. The details of the surfaces can be found in [26].

Table 1. Roughness length scales of test surfaces, adapted from [26] Uzun et al., 2020.

Test Surfaces	Surface Coverage (%)	Barnacle Height h (mm)	Representative Roughness Height k_G (μm)	Equivalent Sand Roughness Height k_S (μm)
Mix	10	5, 2.5, 1.25	94	409
NS Mix	10	5, 2.5, 1.25	136	635
Mix	20	5, 2.5, 1.25	337	1366
NS Mix	20	5, 2.5, 1.25	408	1645

3. Test Case Set-Up and Numerical Modelling

3.1. Geometry and Test Matrix

In the numerical calculations, the benchmark INSEAN E779A model propeller, four-bladed with a diameter of 0.227 m—widely used in the literature for validation purposes because of the available experimental data—was used for the investigation of hub vortex cavitation mitigation with roughness application. Table 2 summarises the test conditions explored in this study.

Table 2. Operating conditions for the INSEAN E779A propeller.

Parameter	Symbol and Unit	Value
Advance ratio	J (-)	0.71
Rotation Rate	n (rps)	36
Inflow averaged velocity	V_A (m/s)	5.8
Cavitation number	σ (-)	1.763
Vapour pressure	P_V (Pa)	2337

Here, J is the advance ratio, V_A is the averaged velocity at the propeller plane, n is the propeller rotational rate and σ_n is the cavitation number based on the propeller rotational rate. The advance ratio and cavitation number are defined as follows, respectively.

$$J = \frac{V_A}{nD} \tag{4}$$

$$\sigma_N = \frac{P_0 - P_V}{\frac{1}{2}\rho(nD)^2} \tag{5}$$

where P_0 is the static pressure, P_V is the vapour pressure, ρ is the density of the fluid, n is the propeller rotational rate and D is the propeller diameter.

The global performance characteristics (i.e., thrust (K_T), torque (K_Q) coefficients and efficiency (η_0)) of the propeller can also be calculated using the following equations.

$$K_T = \frac{T}{\rho n^2 D^4} \tag{6}$$

$$K_Q = \frac{Q}{\rho n^2 D^5} \tag{7}$$

$$\eta_0 = \frac{J}{2\pi}\frac{K_T}{K_Q} \tag{8}$$

Here, T is thrust (N) and Q is the torque of the propeller (N.m).

Figure 1 shows the roughness-application area on the hub together with the smooth propeller. Here, the black colour shows the area where the roughness is applied in this study. The four different roughness configurations given in Table 1 were applied to the black area to investigate the mitigation of hub vortex cavitation with an increase in roughness length scale.

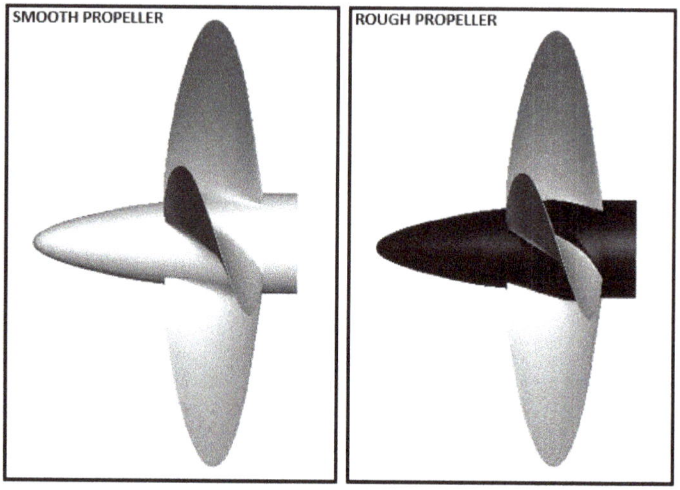

Figure 1. Representation of the roughness application area on the hub (black colour represents the area where the roughness was applied).

3.2. Numerical Modelling

3.2.1. Computational Domain and Boundary Conditions

The computational domain utilised in the numerical calculations is given in Figure 2. The total length of the domain was set to 10D. The upstream and downstream of the domain were extended to 3D and 7D, respectively, whereas the radius of the domain was set to 4D from the propeller centre. The computational domain constitutes static (blue) and rotating (green) regions, as shown in Figure 2. The rotating region was created around the propeller blades, which have a 1.1D diameter and 1.2D total length. The propeller rotational motion was specified into the rotating region, encapsulating the blades and hub. The internal interfaces enabled the transition between the rotating and static regions.

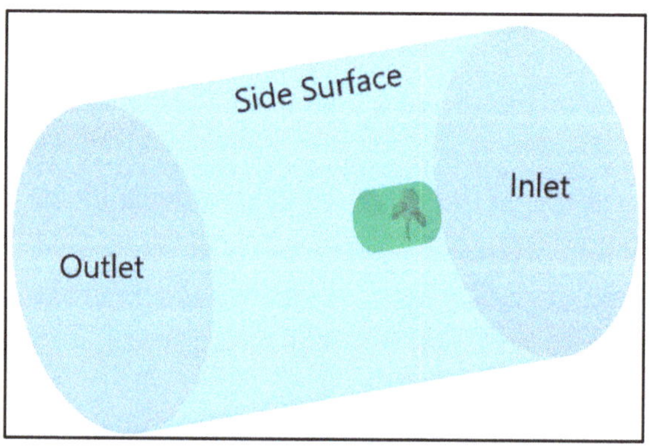

Figure 2. The computational domain used in the numerical calculations.

The inlet was defined as a velocity inlet with constant velocity given in Table 2, whereas the outlet was defined as a pressure outlet. The side surface of the cylindrical computational domain was identified as a symmetry boundary condition. The no-slip boundary condition was applied on the propeller blades, including the hub and boss cap, to satisfy the kinematic boundary condition.

3.2.2. Grid Generation

The discretisation of the computational domain is provided with the finite volume method within the facilities of the CFD solver, Star CCM+ 14.06, 2019. The trimmer mesh algorithm with hexahedral elements was used for the mesh adaptation. In this algorithm, the mesh transitions were set to 1:2 between the rotating region and further downstream. Moreover, the automated mesh tool was implemented to make the meshing process fast. The uniform grid resolution was adopted, and additional mesh refinement was applied to the blades. As the main interest of this study is hub vortex cavitation and its mitigation with roughness application, the recently introduced V-AMR (Vorticity-based Adaptive Mesh Refinement) technique by the authors [28] was not implemented in this study for the observation of tip vortex cavitation (TVC) in the propeller slipstream. This also enabled a reduction of the computational cost of the solution. As the uncertainty study was conducted with the selected propeller at the same operating condition as in our recent studies (e.g., [15,16]), the uncertainty study was not repeated. The total element count was calculated at around 10M. The grid resolution inside the computational domain is given in Figure 3.

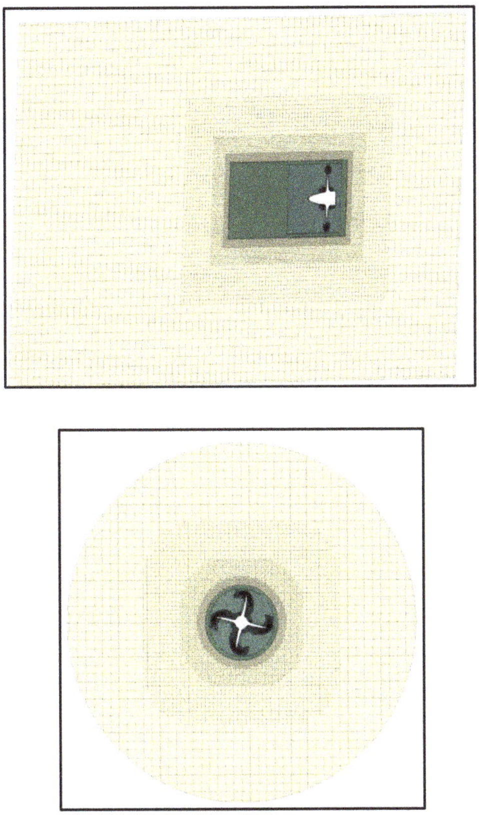

Figure 3. Grid resolution inside the computational domain.

3.2.3. Analysis Properties

The cavitating flow around the propeller was solved using the RANS method together with the k-ω SST turbulence model. The all y^+ wall treatment was used, which is a hybrid treatment combination of low y^+ for the regions where the fine mesh is present and high y^+ treatment for the coarse grids. The artificial biofouling-type roughness was imposed using the wall function of the CFD solver. In this approach, the selected y^+ needs to be higher than 30 and higher than the k^+ values to be able to reflect the roughness effects into the calculations as stated in the CFD solver. Thus, y^+ was set according to the highest roughness length scale (i.e., NSM20).

The segregated flow solver was used together with the SIMPLE algorithm to solve the continuity and momentum equations. The convection terms of the momentum equations and turbulence were discretised with the second-order scheme. Additionally, the second-order scheme was used for temporal discretisation and the time step was set to 0.5° of propeller rotational rate.

The multiphase VOF (Volume of Fraction) approach was coupled with the mass transfer cavitation model, which is Scherr–Sauer based on the reduced Rayleigh–Plesset equation, for modelling the cavitation phenomena. For the convection term of the VOF approach, High-Resolution Interface Capturing (HRIC) was used to maintain sharp interfaces between the fluid phases. The customisable cavitation parameters (i.e., nuclei density and diameter) were taken as default values in this model based on our recent investigation of its effects on the sheet, TVC and hub vortex cavitation formation [28]. Thus, the nuclei density and diameter were set to 10^{12} ($1/m^3$) and 10^{-6} (m), respectively.

The simulations were initialised in a steady manner to speed up the convergence of the numerical solution using one of the propeller rotational motion techniques, Moving Reference Frame (MRF). Following this, the simulations and propeller rotational motion technique were switched to unsteady RANS and Rigid Body Motion (RBM), respectively, and cavitation was activated. In this way, any possible stability issues were removed. In addition, when the flow field was converged, the hydrodynamic performance coefficients (i.e., thrust and torque), including total cavitation volume and velocity field, were computed using the time-averaged data corresponding to 12 propeller rotations.

4. Results

4.1. Validation of the Numerical Results in Smooth Condition

Table 3 shows the comparison of global performance characteristics (i.e., thrust and torque) of the propeller predicted using CFD with that of the experiment in a smooth condition. As shown in Table 3, the difference between the CFD and experiment is approximately 6% for thrust and torque coefficients. One of the reasons for this discrepancy can be related to the replication of the experimental set-up as the shaft in the upstream direction was eliminated in the CFD to reduce the computational cost of the solution.

Table 3. The comparison of thrust and torque coefficients between the experiment and CFD at $J = 0.71$, $\sigma = 1.763$ in the smooth condition.

Parameter	Experiment [29]	CFD
K_T	0.255	0.240
$10K_Q$	0.460	0.435

The cavitation extension predicted in the CFD calculations is also compared with the experiment at $J = 0.71$, $\sigma = 1.763$ in the smooth condition in Figure 4. In the CFD predictions, the iso-surface of volume fraction is taken as 0.1 ($\alpha_v = 0.1$). As shown in Figure 4, the sheet and hub vortex cavitation are predicted similarly both in the numerical calculations and the experiment. The sheet cavitation extension on the blades is slightly overpredicted in the CFD calculations compared to the experiment. In the experiment, the sheet cavitation rolls up into a thick and strong tip vortex cavitation. This TVC extends further downstream of the propeller. In the numerical calculations, as the V-AMR technique was not applied in this study and the hub vortex cavitation is of great interest, a similar TVC could not be predicted in the CFD calculations. Nevertheless, to show the capabilities of the V-AMR technique for the TVC modelling, the cavitation observation obtained by RANS in the authors' recent study at the same operating condition is shown in Figure 4. As can be seen in Figure 4, the TVC can be modelled in the propeller slipstream successfully using the V-AMR technique. Also, the roll-up phenomena can be clearly seen in a region where the sheet and TVC interact with each other. The less extension of TVC in the propeller slipstream is observed in the numerical calculations as compared to the experiments. This is because the excessive amount of eddy viscosity inside the vortex is produced by the RANS and this results in less extension of TVC in the numerical calculations when compared to the experiment.

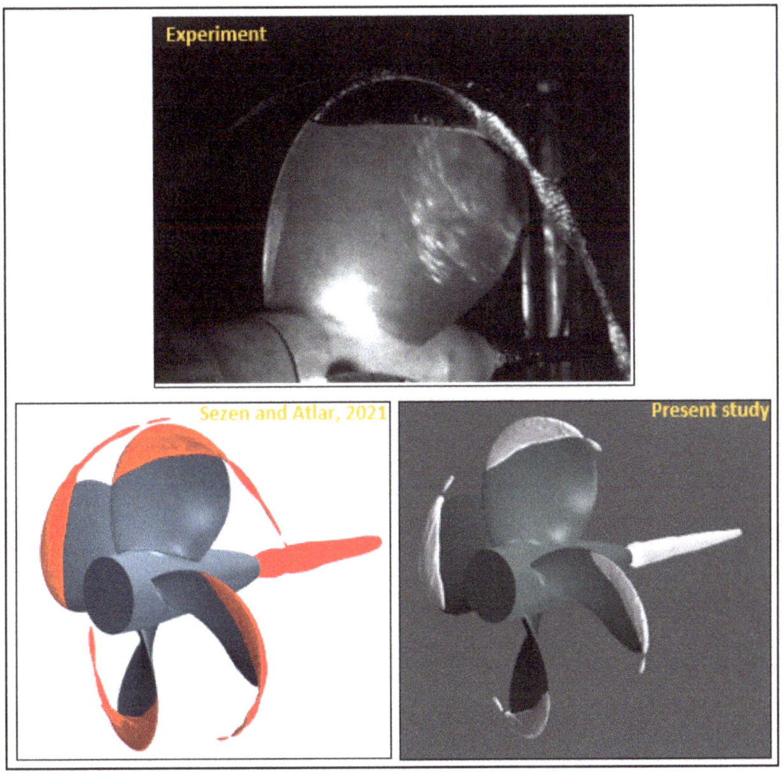

Figure 4. Comparison of cavitation extensions between CFD and experiment at $J = 0.71$, $\sigma = 1.763$ in smooth condition ($\alpha_v = 0.1$) [28].

4.2. *Influence of Roughness on Propeller Hydrodynamic Performance and Cavitation Extension*

Figures 5 and 6 show the change in thrust coefficient (i.e., K_T) and torque coefficient in the presence of roughness with the different roughness length scales, as given in Table 1. Here, the smooth condition is shown with zero roughness height. With the application of roughness, the thrust coefficient decreases with an increase in roughness length scale due to the increased drag and decreased lift, as shown in Figure 5. As expected, the thrust decrease is smaller with the application of roughness on the hub compared to that of roughness application on the blades, as explored in our previous study [16]. The maximum thrust-reduction is found at around 2% with respect to the smooth condition.

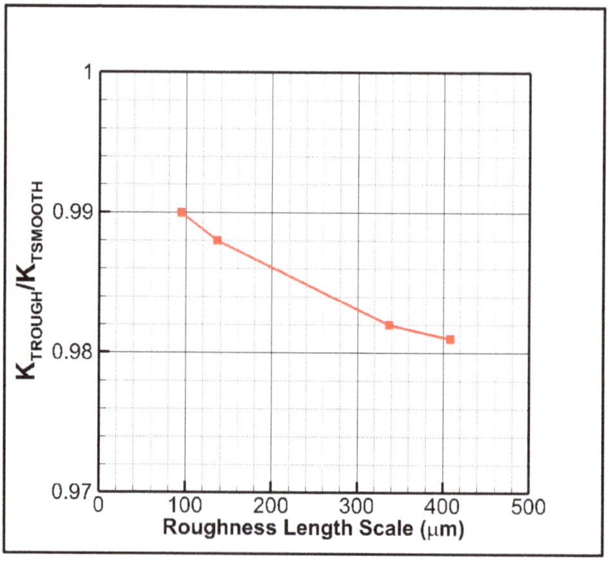

Figure 5. Change in thrust coefficient (K_T) with roughness.

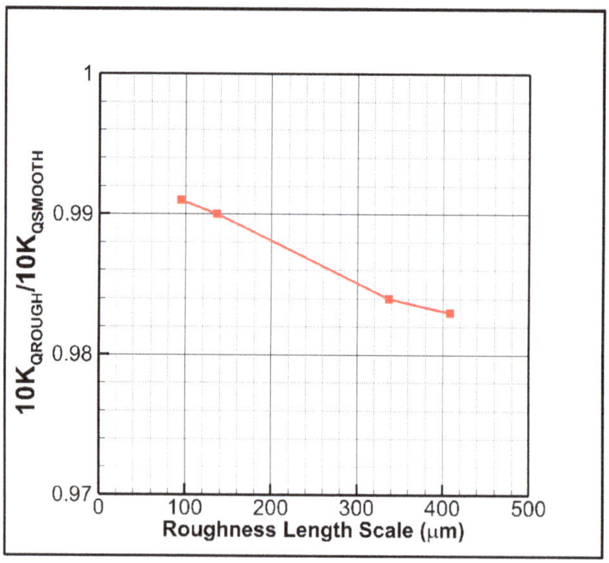

Figure 6. Change in torque coefficient $(10K_Q)$ with roughness.

The change in torque coefficient (i.e., $10K_Q$) with roughness is also shown in Figure 6. Similar to thrust coefficient, roughness applied on the hub has a degradation effect on the torque coefficient of the propeller. The maximum reduction is found to be approximately 1.5% at the maximum roughened condition.

The decreased thrust and torque coefficients with the application of roughness lead to efficiency loss for the marine propellers, as shown in Figure 7. This is the main difference between the roughness and typical PBCF applications, as the increased thrust in PBCF enables efficiency gain due to the recovery of the energy loss. Nevertheless, the efficiency loss is not high with the application of roughness on the propeller's hub and the maximum efficiency loss is found at around 0.25%, while the cost of applying roughness and PBCF is another parameter to consider when deciding which one to implement.

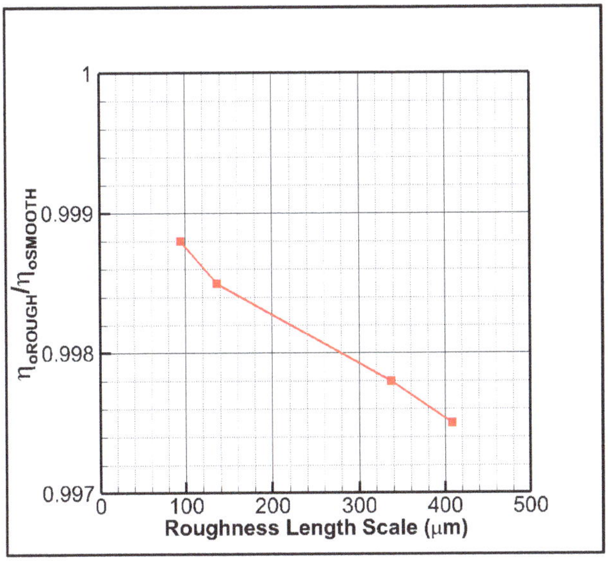

Figure 7. Efficiency (η_O) loss with roughness.

Figure 8 shows the comparison of wall shear stresses on the hub between smooth and rough conditions. As expected, the roughness applied to the hub increases the wall shear stresses.

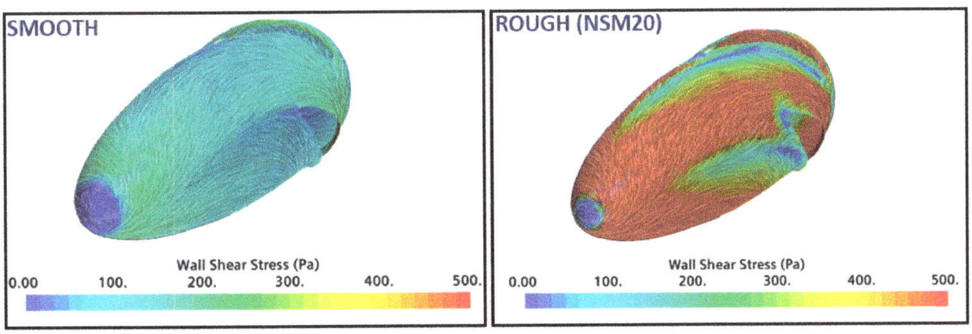

Figure 8. Comparison of wall shear stresses between smooth and rough conditions.

The detailed flow analysis is carried out in the propeller slipstream to show the influence of roughness on the hub vortex and hub vortex cavitation. Figure 9 shows the change in turbulent kinetic energy obtained directly from the turbulence model with the application of roughness. As shown in Figure 9, the turbulent kinetic energy increases considerably with the roughness due to the transformation of the vortex's circumferential momentum into turbulent kinetic energy.

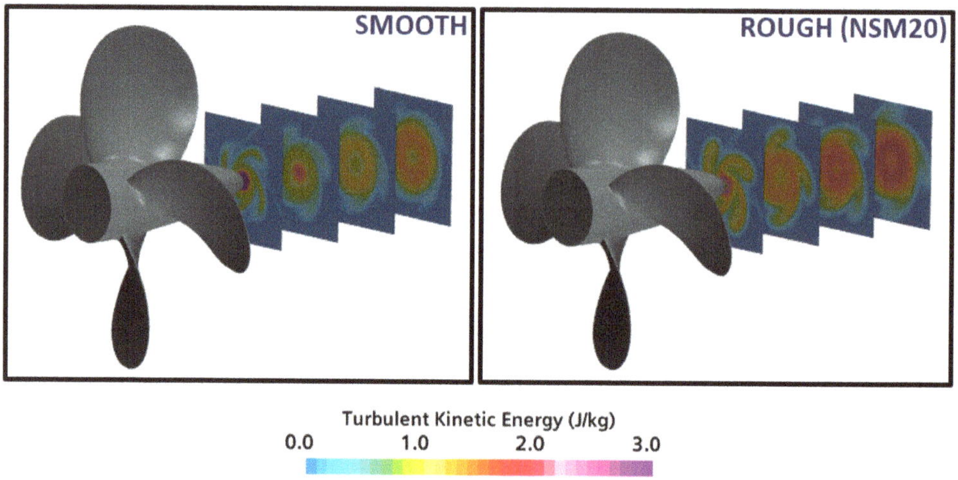

Figure 9. Change in turbulent kinetic energy with roughness.

The change in magnitude of the vortex structures with the roughness is shown at different sections in the propeller slipstream in Figure 10. Applying the roughness reduces the strength of the hub vortex with respect to the smooth condition. The reduced vortex strength enables the destabilisation process of the hub vortices and hence hub vortex disappears with the roughness application.

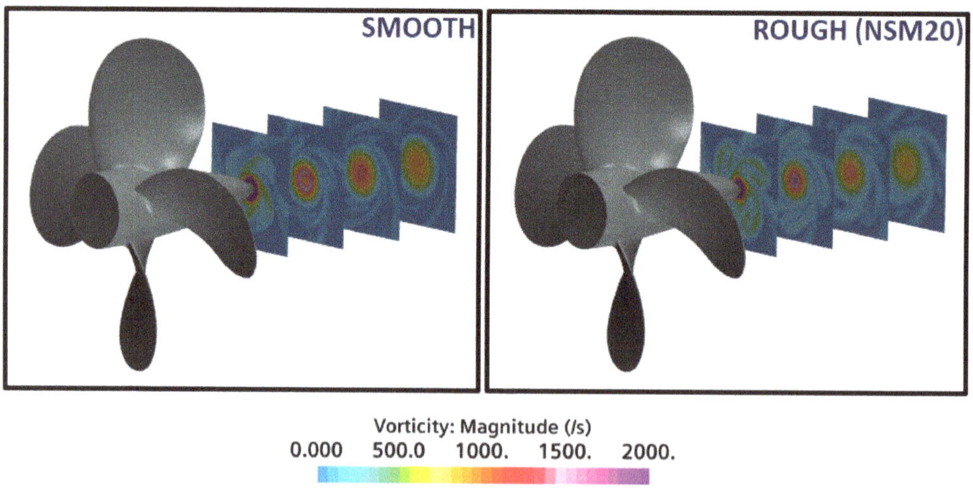

Figure 10. Change in the magnitude of the vortex structures in the propeller slipstream with roughness.

Figure 11 compares the distribution of the non-dimensional pressure coefficient ($C_p = P/0.5\rho(nD)^2$) between rough and smooth conditions. The roughness elements located around the hub interact with the hub vortices and change their velocity and pressure fields. The roughness decreases the velocity magnitudes, and hence, the pressure inside the vortex core and its surroundings increases significantly. With the application of roughness, the pressure inside the hub vortex increases, resulting in the reduction of hub vortex strength and hub vortex cavitation, as shown in Figure 12.

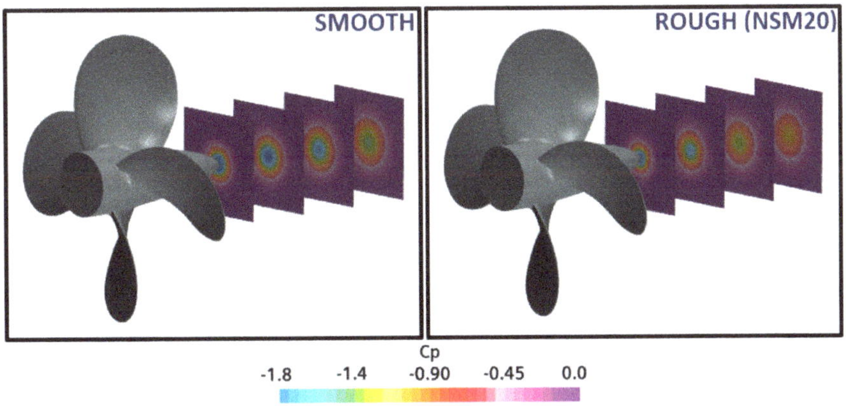

Figure 11. Change in non-dimensional pressure distribution with roughness.

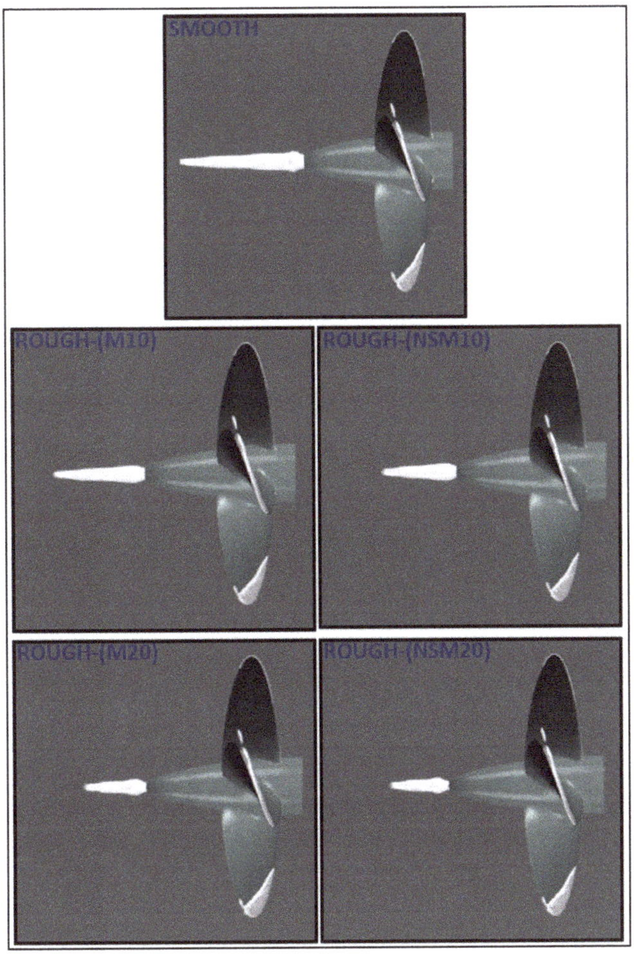

Figure 12. Mitigation of hub vortex cavitation with roughness ($\alpha_v = 0.1$).

Applying roughness leads to destabilisation of the hub vortex strength, which results in the early breakdown of the hub vortex in the propeller slipstream. The reduced strength of the hub vortex due to the increased pressure inside the vortex core results in hub vortex cavitation mitigation with roughness application, as shown in Figure 12. With an increase in roughness length scale from M10 to NSM20, the hub vortex cavitation is further reduced. The maximum hub vortex cavitation volume reduction due to the roughness is computed at around 50% with respect to the smooth condition. As the roughness is solely applied to the propeller hub and boss cap, the sheet cavitation is not affected by the roughness application.

5. Conclusions

This study presented an application of roughness onto the propeller hub as a novel concept to mitigate the hub vortex cavitation for marine propellers. In the numerical calculations, the RANS method—together with the k-w SST turbulence model—was utilised for the solution of cavitating flow around the benchmark model-scale propeller, INSEAN E779A, operating under uniform flow conditions. The Schnerr–Sauer cavitation model was used for modelling the sheet and hub vortex cavitation. A detailed flow-field analysis was also performed to understand the influence of roughness on the hub vortex, and hence, hub vortex cavitation. The crucial findings can be summarised as follows.

- The propeller's hydrodynamic characteristics (i.e., thrust and torque coefficients) and cavitation extensions showed good agreement with the experimental data and cavitation observations, with slight differences.
- Similar to the roughness application on the blades explored in our previous studies [15,16], the roughness on the propeller hub caused efficiency loss. However, the unfavourable impact of roughness applied on the hub was less than that of applying homogenously and heterogeneously distributed roughness to the blades. The maximum efficiency loss was found at 0.25% with respect to the smooth condition in the presence of roughness on the propeller hub.
- The roughness increased the turbulence kinetic energy considerably, whereas the hub vortex strength was reduced significantly due to the destabilisation effects of roughness.
- As the roughness changed the flow properties, the pressure inside the hub vortex increased. This increased pressure inside the vortex enabled hub vortex cavitation mitigation up to 50% depending on the roughness height applied to the hub with respect to the smooth condition.
- Finally, this novel concept will be further explored and it will be incorporated with the propeller URN prediction and erosion models using CFD for model- and full-scale propellers operating under non-uniform flow conditions.

Author Contributions: Conceptualization; S.S., methodology; S.S.; software, S.S.; validation, S.S.; investigation, S.S., M.A.; resources; S.S., M.A.; data curation; S.S.; writing-original draft preparation; S.S.; writing-review and editing, S.S., M.A.; visualization; S.S.; supervision, M.A.; project administration, M.A. All authors have read and agreed to the published version of the manuscript.

Funding: This research received no external funding.

Institutional Review Board Statement: Not applicable.

Informed Consent Statement: Not applicable.

Data Availability Statement: Not applicable.

Acknowledgments: The first author was sponsored by Stone Marine Propulsion Ltd. of the UK and the University of Strathclyde during his PhD study. Results were obtained using the ARCHIE-WeSt High-Performance Computer (www.archie-west.ac.uk) based at the University of Strathclyde.

Conflicts of Interest: The authors declare no conflict of interest.

References

1. *MEPC.1/Circ.833*; Guidelines for the Reduction of Underwater Noise from Commercial Shipping to Address Adverse Impacts on Marine Life. International Maritime Organization (IMO): London, UK, 2014.
2. Hildebrand, J. Sources of anthropogenic sound in the marine environment. In Proceedings of the International Policy Workshop on Sound and Marine Mammals, London, UK, 28–30 September 2004.
3. Cruz, E.; Lloyd, T.; Bosschers, J.; Lafeber, F.H.; Vinagre, P.; Vaz, G. *Study on Inventory of Existing Policy, Research and Impacts of Continuous Underwater Noise in Europe*; EMSA report EMSA/NEG/21/2020; WavEC Offshore Renewables: Lisbon, Portugal; Maritime Research Institute: Wageningen, The Netherlands, 2021.
4. Breeze, H.; Nolet, V.; Thomson, D.; Wright, A.J.; Marotte, E.; Sanders, M. Efforts to advance underwater noise management in Canada: Introduction to the Marine Pollution Bulletin Special Issue. *Mar. Pollut. Bull.* **2022**, *178*, 113596. [CrossRef] [PubMed]
5. Renilson, M.; Leaper, R.; Boisseau, O. Hydro-acoustic noise from merchant ships-impacts and practical mitigation techniques. In Proceedings of the Third International Symposium on Marine Propulsors, SMP'13, Tasmania, Australia, 5–8 May 2013.
6. Bosschers, J. Propeller Tip-Vortex Cavitation and its Broadband Noise. Ph.D. Thesis, Universtiy of Twente, Enschede, The Netherlands, 2018.
7. Carlton, J. *Marine Propellers and Propulsion*, 4th ed.; Butterworth-Heinemann: Oxford, UK, 2018.
8. Atlar, M.; Takinaci, A.C.; Korkut, E. On the Efficiency and Cavitation Performance of a Propeller with Different Boss Caps. In Proceedings of the International Symposium Honouring Tarik SABUNCU On the Occasion of His 75th Birthday, Istanbul Technical University, Istanbul, Turkey, 17 December 1998.
9. Ghassemi, H.; Mardan, A.; Ardeshir, A. Numerical Analysis of Hub Effect on Hydrodynamic Performance of Propellers with Inclusion of PBCF to Equalize the Induced Velocity. *Pol. Marit. Res.* **2012**, *19*, 17–24. [CrossRef]
10. Yakushiji, R. Mechanism of Tip Vortex Cavitation Suppression by Polymer and Water Injection. Ph.D. Thesis, Naval Architecture and Marine Engineering, University of Michigan, Ann Arbor, MI, USA, 2009.
11. Chekab, M.A.F.; Ghadimi, P.; Djeddi, S.R.; Soroushan, M. Investigation of Different Methods of Noise Reduction for Submerged Marine Propellers and Their Classification. *Am. J. Mech. Eng.* **2013**, *1*, 34–42. [CrossRef]
12. Aktas, B.; Yilmaz, N.; Atlar, M.; Sasaki, N.; Fitzsimmons, P.; Taylor, D. Suppression of Tip Vortex Cavitation Noise of Propellers using PressurePoresTM Technology. *J. Mar. Sci. Eng.* **2020**, *8*, 158. [CrossRef]
13. Asnaghi, A.; Svennberg, U.; Gustafsson, R.; Bensow, R.E. Propeller tip vortex mitigation by roughness application. *Appl. Ocean Res.* **2020**, *106*, 102449. [CrossRef]
14. Stark, C.; Shi, W. Hydroacoustic and hydrodynamic investigation of bio-inspired leading-edge tubercles on marine-ducted thrusters. *R. Soc. Open Sci.* **2021**, *8*, 210402. [CrossRef] [PubMed]
15. Sezen, S.; Uzun, D.; Turan, O.; Atlar, M. Influence of roughness on propeller performance with a view to mitigating tip vortex cavitation. *Ocean Eng.* **2021**, *239*, 109703. [CrossRef]
16. Sezen, S.; Uzun, D.; Ozyurt, R.; Turan, O.; Atlar, M. Effect of biofouling roughness on a marine propeller's performance including cavitation and underwater radiated noise (URN). *Appl. Ocean Res.* **2021**, *107*, 102491. [CrossRef]
17. Ouchi, K. Effect and Application of PBCF (Propeller Boss Cap Fins). *J. Mar. Eng. Soc. Jpn.* **1992**, *27*, 768–778. [CrossRef]
18. Ouchi, K.; Ogura, M.; Kono, Y.; Orito, H.; Shiotsu, T.; Tamashima, M.; Koizuka, H. A Research and Development of PBCF (Propeller Boss Cap Fins). *J. Soc. Nav. Arch. Jpn.* **1988**, *1988*, 66–78. [CrossRef]
19. Ouchi, K.; Tamashima, M.; Kawasaki, T.; Koizuka, H. A Research and Development of PBCF (Propeller Boss Cap Fins): 2nd Report: Study on Propeller Slipstream and Actual Ship Performance. *J. Soc. Nav. Archit. Jpn.* **1989**, *165*, 43–53. [CrossRef]
20. Kawamura, T.; Ouchi, K.; Nojiri, T. Model and full scale CFD analysis of propeller boss cap fins (PBCF). *J. Mar. Sci. Technol.* **2012**, *17*, 469–480. [CrossRef]
21. Xiong, Y.; Wang, Z.; Qi, W. Numerical study on the influence of boss cap fins on efficiency of controllable-pitch propeller. *J. Mar. Sci. Appl.* **2013**, *12*, 13–20. [CrossRef]
22. Mizzi, K.; Demirel, Y.K.; Banks, C.; Turan, O.; Kaklis, P.; Atlar, M. Design optimisation of Propeller Boss Cap Fins for enhanced propeller performance. *Appl. Ocean Res.* **2017**, *62*, 210–222. [CrossRef]
23. Sun, Y.; Wu, T.; Su, Y.; Peng, H. Numerical prediction on vibration and noise reduction effects of propeller boss cap fins on a propulsion system. *Brodogradnja* **2020**, *71*, 1–18. [CrossRef]
24. Gaggero, S.; Martinelli, M. Comparison of different propeller boss cap fins design for improved propeller performances. *Appl. Ocean Res.* **2021**, *116*, 102867. [CrossRef]
25. Siemens. *Star CCM+ 14.06, User Guide*; Siemens: Munich, Germany, 2019.
26. Uzun, D.; Ozyurt, R.; Demirel, Y.K.; Turan, O. Does the barnacle settlement pattern affect ship resistance and powering? *Appl. Ocean Res.* **2020**, *95*, 102020. [CrossRef]
27. Grigson, C. Drag Losses of New Ships Caused by Hull Finish. *J. Ship Res.* **1992**, *36*, 182–196. [CrossRef]
28. Sezen, S.; Atlar, M. An alternative Vorticity based Adaptive Mesh Refinement (V-AMR) technique for tip vortex cavitation modelling of propellers using CFD methods. *Ship Technol. Res.* **2021**, *69*, 1–21. [CrossRef]
29. Salvatore, F.; Streckwall, H.; Van Terwisga, T. Propeller Cavitation Modelling by CFD-Results from the VIRTUE 2008 Rome Workshop. In Proceedings of the First International Symposium on Marine Propulsors, SMP'09, Trondheim, Norway, 22–24 June 2009.

Journal of
Marine Science and Engineering

MDPI

Article

Longitudinal Vibration Transmission Control of Marine Propulsion Shafting with Friction Damper Integrated into the Thrust Bearing

Ganbo Zhang [1,2], Yao Zhao [1,2] and Wei Chu [1,2,*]

1 School of Naval Architecture & Ocean Engineering, Huazhong University of Science & Technology, Wuhan 430074, China
2 Hubei Key Laboratory of Naval Architecture & Ocean Engineering Hydrodynamics (HUST), Wuhan 430074, China
* Correspondence: d201677409@hust.edu.cn; Tel.: +86-027-8754-3258

Abstract: Propeller-induced longitudinal vibration resonance in marine propulsion shafting systems causes great harm to the hull structure and is the primary source of shipboard noise. Integrating a friction damper with designed parameters into thrust bearings can prevent these issues. To investigate the performance of the damper-integrated thrust bearing in longitudinal vibration transmission control, an experimental and theoretical study is carried out in a laboratory-assembled test rig, which consists of components similar to the existing marine propulsion system. We developed a prototype of a thrust bearing designed with a friction-damping generation that allows switching from two supporting states, i.e., damper-connected and damper-disconnected states. Furthermore, a nonlinear analysis method for friction dampers is proposed. By this method, the way in which the friction damper changes the dynamic characteristics of the shafting system is analyzed. Based on the test rig, the acceleration frequency response function (AFRF) of the thrust bearing with and without a friction damper is measured. By comparison, the effectiveness of the friction damper is proved. The experimental results show that the friction damper suppresses the shafting longitudinal vibration response in a broadband frequency range and also confirms the stability of the damping effect, which does not change with the shafting rotational speed or static thrust from the propeller.

Keywords: marine propulsion shafting; longitudinal vibration; friction damper; thrust bearing; test rig; parameter identification; nonlinear system

Citation: Zhang, G.; Zhao, Y.; Chu, W. Longitudinal Vibration Transmission Control of Marine Propulsion Shafting with Friction Damper Integrated into the Thrust Bearing. *J. Mar. Sci. Eng.* **2022**, *10*, 1555. https://doi.org/10.3390/jmse10101555

Academic Editors: Carlos Guedes Soares and Serge Sutulo

Received: 21 September 2022
Accepted: 14 October 2022
Published: 20 October 2022

1. Introduction

For marine vessels, the longitudinal vibration in the shafting system induced by the oscillatory propeller thrust is one of the problems of propulsion. In particular, when the excitation frequency is close or equal to the natural frequency of the shafting system, resonance will occur in the longitudinal direction and cause excessive vibration [1]. It will amplify thrust oscillations transmitted to the hull structure and potentially cause damage.

According to the objects on which the control measures are exerted, there are three methods to reduce the hull longitudinal vibration. The first is to lessen the oscillatory part of the propeller's thrust, focusing on the source of the vibration. The second method is to reduce the response of the hull under exciting forces by improving the structural design, which focuses directly on the hull structure. The last method is to manage the vibration excitation transmitted to the hull by developing or installing control devices on the shafting system, which focuses on the vibration transmission path from the propeller to the hull.

For the first method, certain approaches, such as highly skewed propeller or non-propeller propulsion systems, are used to lower the oscillatory thrust. However, these approaches are difficult to implement due to the complexity of manufacturing and the need to increase technological maturity. For the second method, since the hull structure is mainly designed to meet the requirements of structural strength and general arrangement, there are many limits in terms of structural modification. Furthermore, the vibration response of the hull is not only caused by the propeller; it is hard to achieve an overall vibration-oriented design for the hull structure. Thus, applying control measures to the shafting system to reduce longitudinal hull vibration is more acceptable in practical engineering.

The marine shafting system consists of rotational shafting and irrotational bearings. To prevent longitudinal vibration, there are two control strategies used in practice. One control strategy focuses on rotational shafting and has received much attention. Johnson et al. [2] invented an electromagnetic dynamic vibration absorber (DVA) mounted on the shafting to absorb longitudinal vibration. Baz et al. [3] developed an active pneumatic servo-controller powered by compressed air to impose pressure on the servo-piston coupled to shafting to resist longitudinal vibration. Xie et al. [4] proposed an active multi-strut assembly to attenuate vibration transmission from the shafting to the hull. The other control strategy focuses on the thrust bearing and has also been undertaken by numerous researchers. One of the typical designs proposed by Goodwin [5,6] is the hydraulic resonance changer fitted in a thrust bearing, of which the principle is anti-resonant vibration isolation. Another mechanical resonance changer was invented by William [7], but the principle is traditional vibration isolation. Other devices include magnetic thrust bearing [8,9] and thrust bearing supported by disc springs [10]. Song et al. [11] and Liu et al. [12] conducted theoretical studies on the periodic structure isolator and the anti-resonant isolator for thrust bearing, but did not involve specific devices.

In the literature, it can be observed that more researchers favor the second control strategy; thrust bearing is chosen as the primary carrier to implant the vibration control devices. The possible reasons can be summarized in two ways. On the one hand, the thrust bearing is the nearest component to the excitation source. Vibration control directly applied on thrust bearing can achieve the desired effect without interference from the intermediate parts in the transmission path from propeller to hull. Vibration control devices inserted into the thrust bearing will not interfere with the operation of the shafting system. On the other hand, the thrust bearing is a type of irrotational equipment; adding control devices to the rotational shafting will create more difficulties, for example, in its arrangement and assembly.

Damping is well known to be effective in attenuating vibration, especially in suppressing resonance peaks. However, in previous studies on the control of shafting longitudinal vibration, little attention has been paid to damping technology. One important reason is that there are no effective means to add damping to the shafting system. In this paper, the thrust bearing is also chosen as the carrier of a control device. The proposed passive solution is to introduce Coulomb damping into the thrust bearing. The Coulomb damping is a kind of non-viscous damping, and its principle of suppressing resonance is simple, dissipating vibration energy. In practice, however, there are some technical difficulties in engineering implementation. One problem is how to make a relative motion to generate Coulomb damping in the interior of the thrust bearing. Currently, the marine thrust bearings are the Kingsbury type. The leveling plates support the pivoted tilting pads, as shown in Figure 1, which are a series of upper and lower levers designed to distribute the load evenly between thrust pads. However, the leveling plates belong to the mechanical drive type and cannot damp the longitudinal vibration. To implement Coulomb damping, it is necessary to design a new support structure for thrust pads. The second problem is how to quantify the Coulomb damping to obtain a certain damping effect for shafting, or is greater Coulomb damping better? Furthermore, Coulomb damping is nonlinear; therefore, the following question arises: how can one analyze a shafting system that involves Coulomb damping? This paper examines these problems.

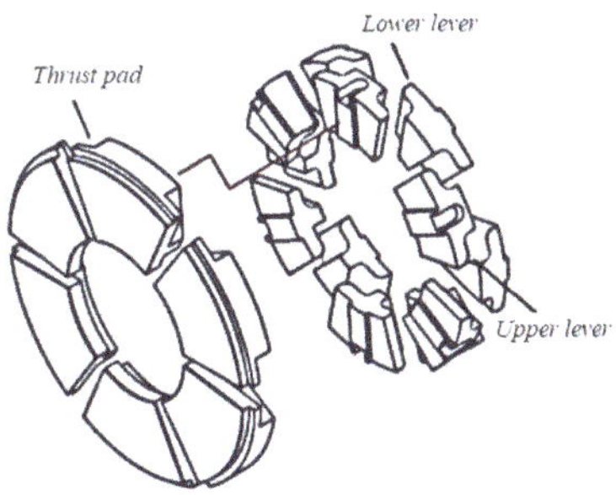

Figure 1. Support structure of the Kingsbury thrust bearing.

The friction damper was developed as a novel support structure for a thrust pad using hydraulic means [13]. It has the integrated advantages of producing Coulomb damping and distributing an uneven load between each pad and is assumed to be a potential alternative to the leveling plates. Coulomb damping has been applied in many engineering fields. In civil and architectural engineering, applications of sliding isolation, which involves inserting a friction surface between the foundation and the base mat of the structure, can be found elsewhere [14]. In automotive engineering, the dry-friction telescopic damper, as a potential substitution for traditional viscous dampers, which may be in suspension, has been developed. Simulation and experimental studies reported that the friction damper exhibited superior performance to the conventional damper [15]. One of the typical characteristics of the Coulomb damping system is discontinuity. When the external force is greater than the maximum static frictional force, the friction pair will slide relatively. Otherwise, the friction pair remains in the sticking state. As a result, the overall behavior of the Coulomb damping system is nonlinear. In view of the nonlinearity of the friction damper and the complexity of the shafting system, the research in this work is based on experiments.

The idea of applying Coulomb damping to control the longitudinal vibration of marine propulsion shafting depends on several key points to be solved in engineering practice. Firstly, a rational and acceptable design to insert the damping generating device into the thrust bearing is needed. Secondly, the level of damping that can be supplied by the damping generating device in practical operation under different conditions should be specified through a sensitive measurement. Finally, the actual effect of the longitudinal vibration control contributed by the Coulomb damping needs to be clarified.

These three key aspects of applying Coulomb damping to control the longitudinal vibration are the main focal points in this paper, which is structured as follows. Firstly, the design for integrating the friction damper into the shafting bearing is introduced, and the operating principle is described. Secondly, the frictional force produced by the friction damper is theoretically analyzed and then measured through experimental means. Thirdly, a test rig assembled to simulate a marine propulsion shafting system is set up and described with a mechanical analysis model, and some uncertain parameters are identified based on an experiment with a particle swarm optimization algorithm. Then, a nonlinear analysis method for a friction damper is proposed, and the way in which the friction damper changes the dynamic behavior of the shafting system is illustrated. Finally, the effectiveness of the friction damper employed to attenuate the transmission of longitudinal vibration in the test rig is experimentally investigated.

2. Design of the Friction Damper Integrated into the Thrust Bearing

Conceptually, the friction damping is produced by a friction surface with normal force pressing against it. As shown in Figure 2, the friction damper integrated into the shaft bearing is assembled with a number of pistons and a support ring. The support ring has a longitudinal tube shape and, around the circumference, is evenly divided into several cylindrical cavities that are interconnected through an annular interior groove to form a connected space. Each piston is assembled inside one cylindrical cavity to reciprocate in the longitudinal direction. To meet the assembling requirements of thrust bearings, the support ring adopts a symmetrical split configuration, in which both the upper and the lower parts are bolted and sealed on the interface. In addition, a radial seal is fitted for each piston for the purpose of sealing all the cavities and then forming an enclosed space; on the other hand, a friction surface is created between the pistons and the support ring. In this way, by filling the connected and enclosed space with oil from the inlet hole, a passive hydraulic system is set up inside the support ring; that is, the pressure of the hydraulic oil balances the propeller's steady thrust. The hydraulic pressure causes internal normal force at the seal outer face against the cavity wall, and consequently, with the tendency of relative motion between the pistons and the support ring, the frictional force is produced.

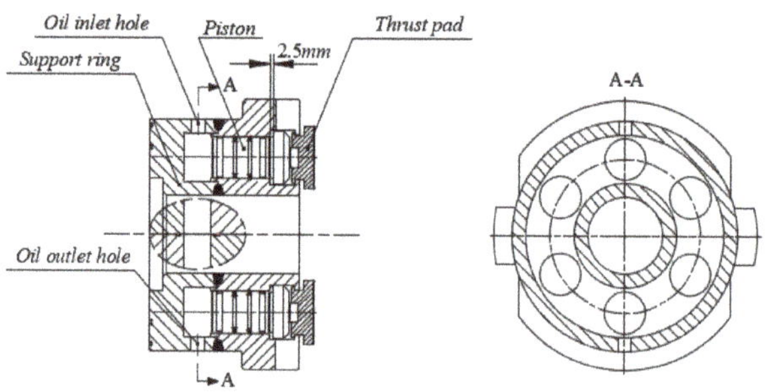

Figure 2. Friction damper integrated into the thrust bearing.

When the propeller rotates and supplies thrust, the steady part of the thrust squeezes oil in the support ring through the pistons, and correspondingly, the oil generates hydraulic pressure. Meanwhile, the oscillation part of the thrust excites the pistons to vibrate relative to the support ring. Under these conditions, the frictional force arises in the friction damper. Apparently, only when the oscillatory thrust exceeds the maximum static frictional force will the relative motion between the pistons and the support ring occur. Therefore, the exciting force transmitted to the thrust bearing is generally less than the maximum static frictional force.

From an experimental point of view, the piston of the friction damper is designed as a convex shape, and more specifically, the diameter of the piston head is larger than that of the piston rod. The piston head is positioned in the middle of the thrust pad and the support ring, with an allowable stroke of 2.5 mm. When there is no hydraulic pressure inside the support ring, for example, the oil outlet hole is open, the piston head is in contact with the support ring under the propeller thrust, and the shafting longitudinal vibration is directly transmitted through the support ring. Thus, the friction damper is short-circuited and does not work. There is, however, hydraulic pressure that balances the propeller thrust, and the piston head is separated from the support ring, while the friction damper is inserted into the shafting longitudinal vibration transmission path to play a damping role. For clarity, the state in which the piston is supported by the support ring is called the 'rigid support state', as shown in Figure 3a, while the state in which the piston is supported by hydraulic oil is called the 'hydraulic support state', as shown in Figure 3b. By switching the support states, the response of the thrust bearing with or without the friction damper can be compared; thus, the effectiveness of the friction damper can be judged.

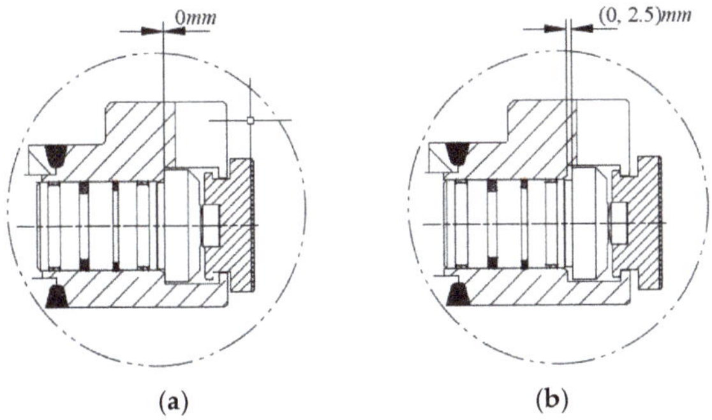

Figure 3. The state of friction damping: (**a**) rigid support state; (**b**) hydraulic support state.

3. Analysis and Test of Frictional Force

3.1. Analysis of Frictional Force

According to Coulomb's theory, the frictional force of a friction damper is

$$f = \mu F_n \tag{1}$$

where μ is the coefficient of friction, and F_n is the normal force at the piston seal outer face against the cavity.

The piston friction depends on the seal design. Taking, for instance, the sliding seal, a polytetrafluoroethylene (PTFE)-based ring pre-loaded by an O-ring, the normal force arises from two aspects. One is the contact stress due to the O-ring pre-compression being exerted on the sealing ring; the other is the pressure stress applied by the hydraulic oil being exerted on the sealing ring through the O-ring. The superposition principle is applicable for these two types of compressive stresses [16], so the normal force F_n can be expressed approximately as

$$F_n = \frac{\pi \varepsilon_o E_o + 4\mu_o(1+\mu_o)p}{4(1-\mu_o^2)} s \tag{2}$$

where ε_o, μ_o, E_o denote the O-ring pre-compression, Poisson's ratio, and elastic modulus, respectively; p is the hydraulic pressure; s is the seal outer surface area.

The coefficient of friction μ in Equation (1) is related to the sealing material, surface roughness, lubrication state, relative velocity, and so on, which does not yet have a precise mathematical formulation. Since the coefficient of friction μ is unknown, it is difficult to calculate the frictional force of the friction damper accurately. However, the conclusion can be drawn from Equation (2) that the frictional force has a positive relationship with hydraulic pressure.

3.2. Test and Determination of Frictional Force

In order to determine the actual frictional force, a test device is designed, as shown in Figure 4. The friction damper is connected to a hydraulic circuit composed of an accumulator, valve, pressure transmitter, and hand pump. The oil is pumped slowly to push the piston head to move up to a certain distance from the support ring and then the central retaining nut can be sufficiently hand-tightened. A load plate, which is used to carry a mass block, is placed on one of the pistons. Meanwhile, a dial indicator, which measures the movement, is mounted to contact the load plate. Upon installation, a number of mass blocks are placed on the load plate step by step. Once instantaneous, continuous displacement is observed by the dial indicator and the pressure gauge data and the weight of mass blocks are recorded. The single piston frictional force is equal to the difference in the weight and hydraulic pressure. By changing the initial pressure of hydraulic oil, the frictional force for various pressures can be tested.

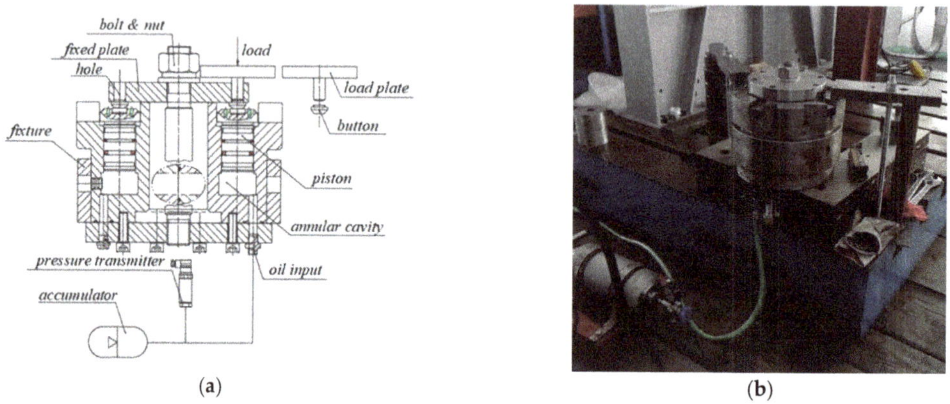

Figure 4. Test device of frictional force: (**a**) schematic diagram; (**b**) photograph.

The tests on two types of piston seals were conducted including the O-ring and the sliding seal ring. Figure 5 provides the tested frictional force. It shows that the frictional forces of the O-ring and sliding seal ring are all positively correlated with the hydraulic pressure, which is consistent with the theoretical conclusion. As the hydraulic pressure increases, the frictional force of the O-ring increases linearly, while the frictional force of the sliding seal ring increases in two stages. When the pressure is below 2 MPa, the frictional force of the sliding seal ring increases sharply. Once the pressure exceeds 2 MPa, the increase in frictional force becomes less steep. With the same hydraulic pressure, the frictional force of the sliding seal ring is much less than that of the O-ring, and with the increase in hydraulic pressure, a greater difference in frictional force between the two types of seals is obtained.

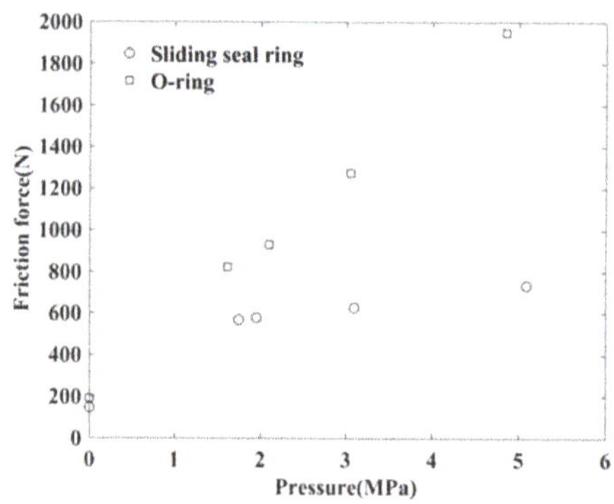

Figure 5. Test results of single piston frictional force with respect to hydraulic pressure.

Essentially, the friction damper dissipates energy under the relative motion of the piston and the support ring. Thus, the maximum static frictional force should not be so large that the relative motion cannot occur. Since the friction coefficient of the material PTFE is much lower than that of the O-ring, the sliding seal is beneficial to improve friction performance and is applied as the seal of the friction damper.

4. Nonlinear Analysis Method of Frictional Force

Because of the existence of frictional force, the friction damper is a nonlinear system. For simplicity, the nonlinear analysis method of the frictional force is explained with a two-degree-of-freedom model, as shown in Figure 6, for which the notations employed are defined as follows. $F(t)$ is the oscillatory thrust, and $f(t)$ is the induced frictional force; both $F(t)$ and $f(t)$ vary over time t. M_p denotes the total mass of forward-thrust pads and pistons, K_h denotes the longitudinal stiffness of oil in a friction damper, and M_t and K_t denote the effective mass and resultant stiffness of the supporting ring, respectively. When the friction damper is integrated into the thrust bearing, M_t and K_t denote the effective mass and resultant stiffness of the thrust bearing block, respectively. The frictional force $f(t)$ has a link with M_p and M_t. In the analysis, the loss factor of the structure has also been taken into account, which is denoted as η_t.

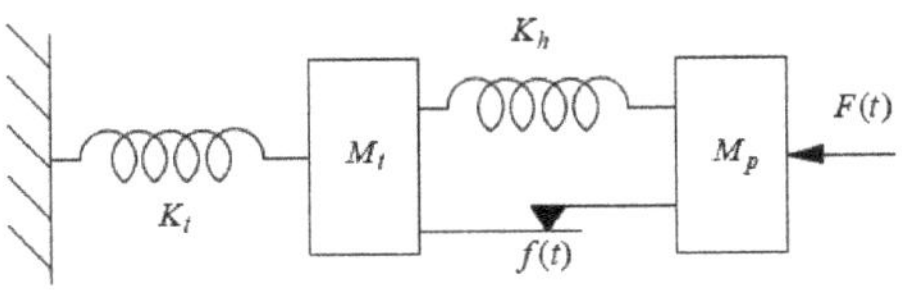

Figure 6. Two-degree-of-freedom model of friction damper.

Referring to Figure 6, with the excitation of oscillatory thrust $F(t)$, the dynamic equilibrium equation of longitudinal vibration can be written as

$$\boldsymbol{M}\ddot{\boldsymbol{x}}(t) + \boldsymbol{C}\dot{\boldsymbol{x}}(t) + \boldsymbol{K}\boldsymbol{x}(t) = \boldsymbol{L}_1 F(t) + \boldsymbol{L}_2 f(t) \tag{3}$$

where the vector $\boldsymbol{x}(t) = \{x_p(t) \quad x_t(t)\}^T$ denotes the displacement responses in the thrust pads and the supporting ring, and the vectors $\boldsymbol{L_1} = \{1 \quad 0\}^T$ *and* $\boldsymbol{L_2} = \{1 \quad -1\}^T$ denote the distribution of $F(t)$ and $f(t)$, respectively; the mass matrix $\boldsymbol{M}$, damping matrix $\boldsymbol{C}$, and stiffness matrix $\boldsymbol{K}$ in Equation (3) are defined as

$$\boldsymbol{M} = \begin{bmatrix} M_p & \\ & M_t \end{bmatrix} \tag{4}$$

$$\boldsymbol{C} = \begin{bmatrix} 0 & \\ & \frac{K_t \eta_t}{\omega} \end{bmatrix} \tag{5}$$

$$\boldsymbol{K} = \begin{bmatrix} K_h & -K_h \\ -K_h & K_h + K_t \end{bmatrix} \tag{6}$$

In Equation (3), the structural damping is transformed into equivalent viscous damping. Moreover, the frictional force $f(t)$ is moved to the right-hand side.

The motion of the friction damper consists of two different states, the sticking state and the sliding state. At any instant, the friction damper only belongs to one of the two states. The following conditions that correspond to the two states must be satisfied:

(1) In the sticking state,

$$\begin{cases} \dot{x}_p(t) = \dot{x}_t(t) \\ |f(t)| < f_{max} \end{cases} \tag{7}$$

where f_{max} is the maximum static frictional force.

(2) In the sliding state,

$$\begin{cases} \dot{x}_p(t) \neq \dot{x}_t(t) \\ f(t) = -sgn(\dot{x}_p(t) - \dot{x}_t(t)) f_{max} \end{cases} \tag{8}$$

where $sgn(\,)$ is the sign function. For simplicity, no distinction will be made between the kinetic and the maximum static frictional force.

In the sticking state, the frictional force $f(t)$ is less than the maximum static frictional force f_{max} but remains unknown. In the sliding state, however, the frictional force is constant and equal to the maximum frictional force, but with an opposite direction to the relative velocity between the thrust pads and the thrust bearings. Since Equation (3) must hold in the two states, the overall behavior of the friction damper is nonlinear. For a numerical solution, Equation (3) can be further written in the form of state space, as shown below.

$$\dot{z}(t) = Az(t) + BF(t) + Ef(t) \tag{9}$$

where $z(t) = \{\dot{x}(t) \quad x(t)\}^T$ is the state vector, and the matrixes A, B, E in Equation (9) are defined as

$$A = \begin{bmatrix} -M^{-1}C & -M^{-1}K \\ I_{2\times2} & 0_{2\times2} \end{bmatrix} \tag{10}$$

$$B = \begin{bmatrix} M^{-1}L_1 \\ 0_{2\times1} \end{bmatrix}, \tag{11}$$

$$E = \begin{bmatrix} M^{-1}L_2 \\ 0_{2\times1} \end{bmatrix} \tag{12}$$

The analytical solution of Equation (9) is

$$z(t) = e^{A(t-t_0)}z(t_0) + \int_{t_0}^{t} e^{A(t-\tau)}[BF(\tau) + Ef(\tau)]d\tau \tag{13}$$

By employing the discrete time state space formula and assuming that both $F(t)$ and $f(t)$ vary linearly within every time interval, the solution (13) can be written in the following incremental form:

$$z(k+1) = e^{A\Delta t}z(k) + B_0[BF(k) + Ef(k)] + B_1[BF(k+1) + Ef(k+1)] \tag{14}$$

where k, $k+1$ denote the kth and $(k+1)th$ time step, respectively. The coefficient matrixes are defined as $e^{A\Delta t} = \sum\limits_{i=0}^{\infty} \frac{\Delta t^i}{i!}A^i$, $B_0 = \sum\limits_{i=0}^{\infty} \frac{\Delta t^{i+1}}{i!(i+2)}A^i$, $B_1 = \sum\limits_{i=0}^{\infty} \frac{\Delta t^{i+1}}{(i+1)!}A^i - \sum\limits_{i=0}^{\infty} \frac{\Delta t^{i+1}}{i!(i+2)}A^i$, where Δt is the time interval.

It is noted that on the right-hand side of Equation (14), the frictional force $f(k+1)$ is only unknown at the kth time step. Before the state response $z(k+1)$ is evaluated, $f(k+1)$ must be determined. Unfortunately, the motion of the friction damper has two possible states at any time step. The same problem was encountered in the application of friction pendulum bearings for seismic isolation of the bridge in the reference [17], which presented an important assumption that the isolated bridge was in the sticking state at any instant, initially. Based on the same assumption, the motion of the friction damper at the $(k+1)th$ time step is in the sticking state at first, for which the velocity condition given in Equation (7) must be satisfied as

$$\dot{x}_p(k+1) - \dot{x}_t(k+1) = Dz(k+1) = 0 \tag{15}$$

where $D = \begin{bmatrix} 1 & -1 & 0 & 0 \end{bmatrix}$.

By substituting $z(k+1)$ in Equation (14) into Equation (15), the estimated frictional force at the $(k+1)th$ time step is solved as

$$\overline{f}(k+1) = -(DB_1E)^{-1}D\left\{e^{A\Delta t}z(k) + B_0[BF(k) + Ef(k)] + B_1BF(k+1)\right\} \tag{16}$$

where $\overline{f}(k+1)$ with an overbar emphasizes that the frictional force is an estimated value, obtained by assuming the friction damper to be in the sticking state, which may not be the actual state, and the sign of $\overline{f}(k+1)$ indicates the direction of frictional force produced by the friction damper.

The validity of the above assumption can be verified by the frictional force condition given in Equation (7). If $|\overline{f}(k+1)| < f_{max}$ holds, the friction damper is in the sticking state, and the actual frictional force is

$$f(k+1) = \overline{f}(k+1) \tag{17}$$

Otherwise, if $|\overline{f}(k+1)| \geq f_{max}$, the friction damper is in the sliding state, and the actual frictional force is

$$f(k+1) = sgn\left(\overline{f}(k+1)\right) f_{max} \tag{18}$$

Since the sign of $\overline{f}(k+1)$ indicates the direction of frictional force at the $(k+1)th$ time step, the replacement of the term $-sgn(\dot{x}_p(t) - \dot{x}_t(t))$ in Equation (8) by the term $sgn\left(\overline{f}(k+1)\right)$ in Equation (18) is justified. Once the actual frictional force $f(k+1)$ is determined by Equation (17) or Equation (18), the state response $z(k+1)$ at the $(k+1)th$ time step can be solved by substituting $f(k+1)$ into Equation (14).

For harmonic excitation with amplitude F and circular frequency ω, the response $z(t)$ in the time domain can be computed with the method proposed above. However, the numerical solution is the discrete time state space response $z(k)$ that corresponds to a series of discrete time steps, which is similar to the sample signals of the response in the vibration test. The discrete response $z(k)$ is processed by fast Fourier transform (FFT); then, the response amplitude $z(\omega)$ that corresponds to the excited circular frequency ω can be extracted. It should be noted that the nonlinearity introduced by the frictional force results in the occurrence of subharmonic responses, but these amplitudes are all less than $z(\omega)$. The frequency response is defined as the ratio of $z(\omega)/F$. By sweeping the excited frequency ω in an incremental frequency interval, the acceleration frequency response function (AFRF) of the friction damper can be predicted.

The computational flowchart for the nonlinear analysis method of the frictional force is given in Figure 7.

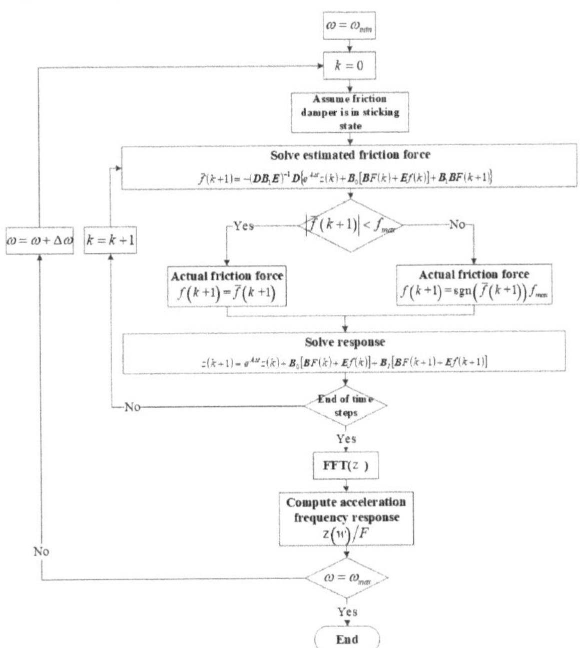

Figure 7. Computational flowchart for the nonlinear analysis method of frictional force.

5. Design and Characterization of Shafting Test Rig

5.1. Design of Shafting Test Rig

A designed test rig with a reduced scale, which is depicted in Figure 8, has been assembled to simulate the marine propulsion shafting system [18] and is taken as the benchmark to assess the effectiveness of the friction damper experimentally. The components of the test rig include a motor, flexible coupling, thrust bearing, stern shafting, intermediate journal bearing, stern journal bearing, weight block, ball spring device, air spring device, excitation rod, bracket, etc. The total length of the shafting is 5 m, and the basic diameter is 105 mm. The thrust bearing integrates the friction damper, and the flexible coupling isolates the motor side from the shafting side. The whole test rig is fixed on the base, which is welded on thick steel plates. In order to further eliminate the disturbance of motor excitation on the thrust bearing, the motor is independently installed on one base, while other test rigs are installed on another base. Furthermore, the electrical control console, hydraulic lubrication system, and air supply system are also configured to meet the use requirements.

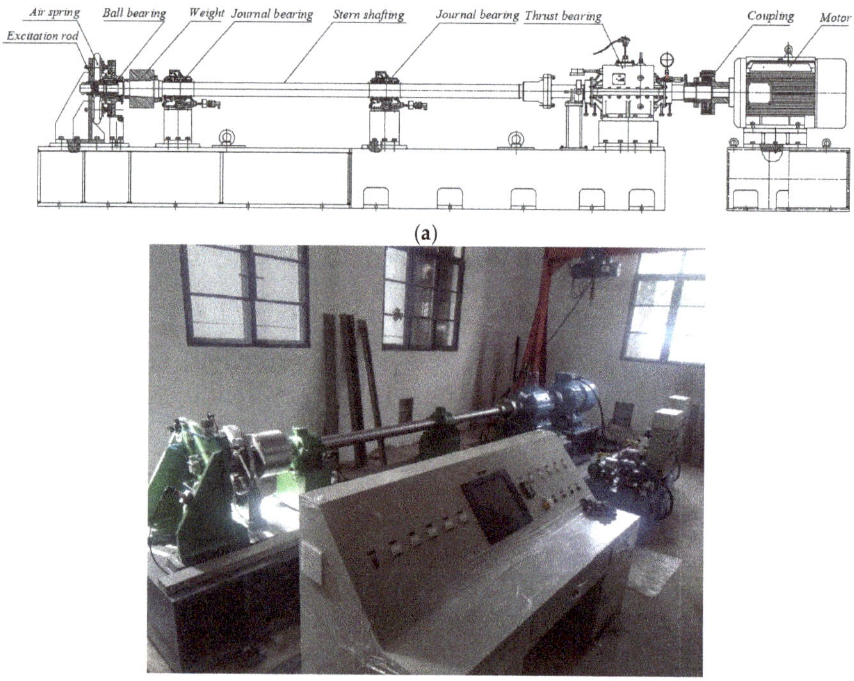

(a)

(b)

Figure 8. The designed test rig with a reduced scale in the laboratory: (**a**) schematic diagram; (**b**) photograph.

The test rig adopts the method of combining the static load and dynamic excitation at the same time to simulate the propeller propulsion. An air spring device of maximum thrust (55 kN) is fitted in the background of the test rig and is used to apply a static load onto the stern shafting to represent propeller steady thrust. This static load can be easily adjusted by changing the pressure of the air spring. An impact hammer acts as the dynamic source to simulate the excitation of the propeller. The hammer exerts dynamic force in the axial direction on an irrotational excitation rod, which is also fitted at the end of the test rig in combination with the air spring device, to generate shafting longitudinal vibration. For transmitting the static thrust and dynamic force to the shafting simultaneously, a conical ball bearing is sandwiched between the stern shafting and the air spring device. This arrangement is appropriate to solve the problem of connecting the rotational shafting with the irrotational excitation rod.

5.2. Mathematical Model of the Designed Test Rig

The test rig is modeled as a mass-spring damping discrete system, as shown in Figure 9, and the parameters of the components are listed in Table 1. This model covers the range from the air spring device to the flexible coupling and does not contain the friction damper, which is used for analyzing the dynamic characteristics of the shafting system without any control measures applied. While the air spring device imposes static thrust onto the shafting, it also constitutes the elastic boundary and is simplified as a single-degree-of-freedom system.

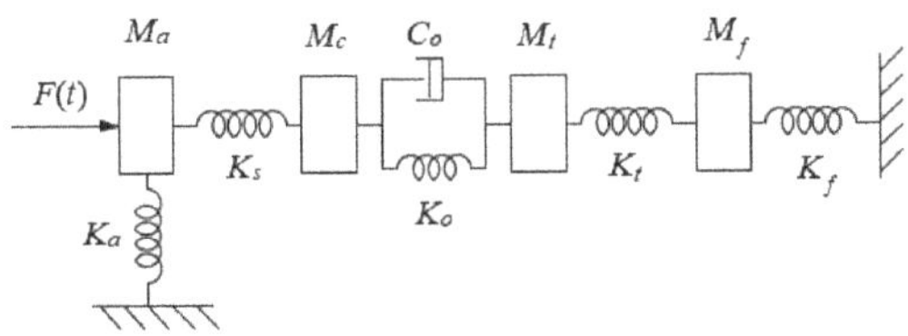

Figure 9. Mathematical model of test rig without control from friction damper.

Table 1. Parameters used in the mathematical model.

Parameter	Symbol	Value
Shafting mass	M_s	360 kg
Shafting stiffness	K_s	5.0×10^8 N/m
Total mass of air spring device and half of shafting	M_a	/
Resultant stiffness of air spring device	K_a	1.0×10^8 N/m
Loss factor of air spring device	η_a	
Total mass of flexible coupling and half of shafting	M_c	200 kg
Damping of lubricating oil film	C_o	/
Stiffness of lubricating oil film	K_o	/
Effective mass of thrust bearing	M_t	/
Resultant stiffness of thrust bearing	K_t	3.4×10^8 N/m
Loss factor of thrust bearing	η_t	/
Effective mass of foundation	M_f	/
Resultant stiffness of foundation	K_f	1.1×10^9 N/m
Loss factor of foundation	η_f	/

The AFRF of thrust bearing is chosen as the dynamic character of shafting longitudinal vibration transmission for analysis. The consideration is that the shafting longitudinal vibration is transmitted to the hull through the thrust bearing. More importantly, the AFRF of the thrust bearing is capable of being measured and calculated so as to ensure the consistency of the assessment index.

The transfer matrix method is used to formulate the AFRF of the thrust bearing. The transfer matrix equation of the test rig is expressed as

$$\begin{Bmatrix} U_f^r \\ F_f^r \end{Bmatrix} = \begin{bmatrix} T_{11}^1 & T_{12}^1 \\ T_{21}^1 & T_{22}^1 \end{bmatrix} \begin{Bmatrix} U_a^l \\ F_a^l \end{Bmatrix} \tag{19}$$

where $\boldsymbol{U}, \boldsymbol{F}$ denote the displacement and force, respectively, the superscript $\boldsymbol{l}, \boldsymbol{r}$ denote the left end and right end, respectively, the subscript $\boldsymbol{f}, \boldsymbol{a}$ denote the thrust bearing base, and air spring device, respectively; $\boldsymbol{T_{11}^1}, \boldsymbol{T_{12}^1}, \boldsymbol{T_{21}^1}, \boldsymbol{T_{22}^1}$ are frequency-dependent complex variables.

According to the rigid boundary condition $\boldsymbol{U_f^r} = \mathbf{0}$, the displacement of the air spring device is determined from Equation (19).

$$U_a^l = -\frac{T_{12}^1}{T_{11}^1} F_a^l \tag{20}$$

The transfer matrix equation of the subsystem, ranging from the air spring device to the thrust bearing, is given by

$$\begin{Bmatrix} U_t^r \\ F_t^r \end{Bmatrix} = \begin{bmatrix} T_{11}^2 & T_{12}^2 \\ T_{21}^2 & T_{22}^2 \end{bmatrix} \begin{Bmatrix} U_a^l \\ F_a^l \end{Bmatrix} \tag{21}$$

where $T_{11}^2, T_{12}^2, T_{21}^2, T_{22}^2$ are also frequency-dependent complex variables.

By substituting Equation (20) into Equation (21), the AFRF $H(\omega)$ of the thrust bearing yields

$$H(\omega) = \omega^2 \frac{U_t^r}{F_a^l} = \omega^2 \left(-\frac{T_{12}^1}{T_{11}^1} T_{11}^2 + T_{12}^2 \right) \tag{22}$$

5.3. Longitudinal Vibration Experiments and Parameter Identification

The test rig is excited in the axial direction with the impact hammer, and the impact force is superimposed onto a static thrust provided by the air spring device. Two accelerometers are attached to the thrust bearing casing for measuring the acceleration response. Another two accelerometers are diagonally attached to both sides of the thrust bearing seating. It is assumed that the frequency range of interest is between 25 Hz and 300 Hz; thus, the sampling frequency 1280 Hz is used. To minimize the response variation due to random disturbances, test results from five measurements are averaged.

The experiments are conducted in the rigid support state. When the shafting rotational speed is 60 rpm, and the applied static thrust is 30 kN, the measured AFRF of the thrust bearing is as shown in Figure 10. It is observed that the curves of AFRF measured at two points of the thrust bearing casing are almost coincidental, indicating that the casing vibrates globally. A similar situation occurs at the thrust bearing seating. Except for the amplitude, the AFRF measured at the casing and at the seating has the same variation curve. The maximum resonance peak, of which the resonant frequency is 237 Hz, is caused by the global resonance of the test rig, while the other resonance peaks are mainly governed by local resonances.

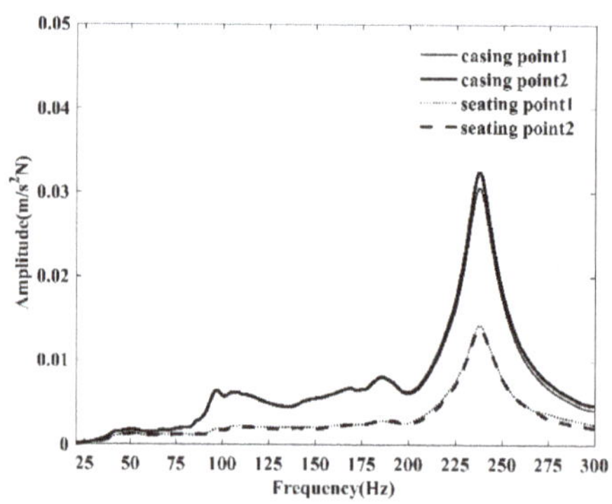

Figure 10. Measured AFRF of thrust bearing without friction damper.

For analyzing the dynamic characteristics of the designed test rig in theory, all the quantifiable parameters must be known. In Table 1, M_s, K_s can be computed accurately, and K_a, K_t, K_f can be obtained based on the results of the thrust-deformation tests. Figure 11 presents the deformations under the action of different thrusts. The thrust is loaded from 5 kN to 55 kN in increments of 5 kN, and the measurements at thrust 5 kN are provided as a reference. It is evident that the stiffnesses are the slope of the fitting lines. M_a, M_t,

M_f are not equal to the physical mass since the air spring device, thrust bearing, and foundation are assemblies. C_o, K_o are difficult to calculate accurately and can only be measured indirectly [19,20]. η_a, η_t, η_f are uncertain. These parameters need to be identified based on the experimentally measured AFRF of the thrust bearing.

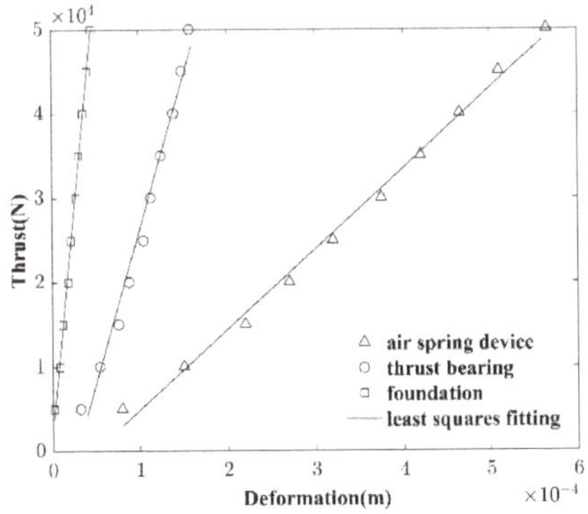

Figure 11. Deformations under the action of different thrusts.

Parameter identification is, in fact, a type of optimization for the identified parameter, and the method used, in general, is residual function optimization. Here, the residual function refers to the difference between the theoretical and experimental AFRF of the thrust bearing. The optimization problem of the shafting parameter identification is described as follows.

(1) Variables: $x = \{M_a, \eta_a, K_o, C_o, M_t, \eta_t, M_f, \eta_f\}$.

(2) Constraints: $M_a \in [0.5,\ 2] \times 215.8 + 0.5M_s$, $\eta_a \in [0.01,\ 0.35]$, $K_o \in [0.1,\ 10] \times 10^8$, $C_o \in [0.1,\ 10] \times 10^5$, $M_t \in [0.5,\ 2] \times 516.4$, $\eta_t \in [0.01,\ 0.35]$, $M_f \in [0.5,\ 5] \times 386.7$, $\eta_f \in [0.01,\ 0.35]$.

(3) Objective function: the optimization objective is to minimize the variance in the residual function, which is defined as

$$f(x) = \min \sum_{i=1}^{n} [|H(\omega)| - |H^*(\omega)|]^2 \tag{23}$$

where $H^*(\omega)$ is the experimental AFRF of the thrust bearing.

The particle swarm optimization algorithm, which is one of the intelligent evolutionary optimization technologies, is adopted to solve the above optimization problem [21]. Taking the AFRF measured at the thrust bearing casing as a representation of the thrust bearing dynamic characteristic, the parameters are identified as $M_a = 332$ kg, $\eta_a = 0.2$, $K_o = 6.3 \times 10^8$ N/m, $C_o = 1.1 \times 10^4$ Ns/m, $M_t = 284$ kg, $\eta_t = 0.15$, $M_f = 348$ kg, and $\eta_f = 0.1$. With these identified parameters, the comparison of theoretical and experimental AFRF of the thrust bearing is as shown in Figure 12. It can be observed that the calculated values from the mathematical model coincide with the experimental results very well, and it can be concluded that the model can feasibly simulate the actual longitudinal vibration of the test rig below about 300 Hz.

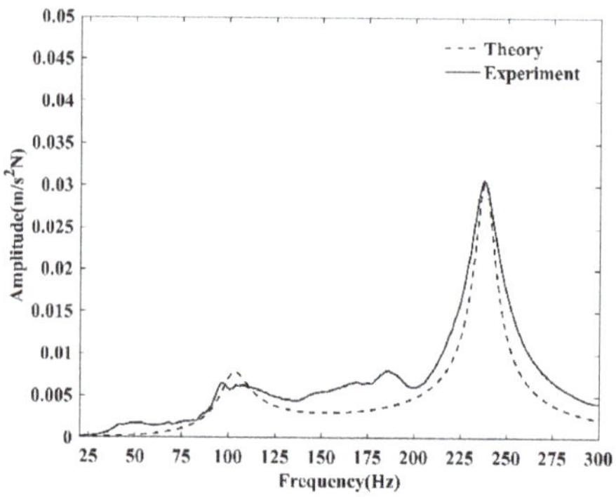

Figure 12. Comparison of theoretical and experimental AFRF of the thrust bearing.

6. Damping Characteristics of Friction Damper

The mathematical model of the test rig with the friction damper is established, as shown in Figure 13.

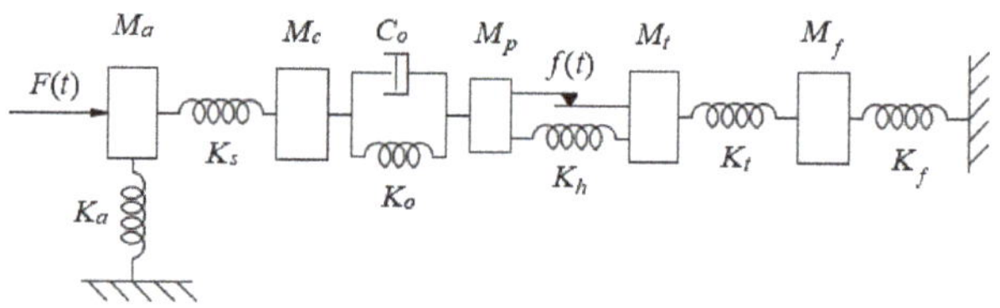

Figure 13. Mathematical model of test rig with friction damper.

In Figure 13, $\boldsymbol{M_p} = \mathbf{24}$ **kg**, $\boldsymbol{K_h} = \mathbf{1.85 \times 10^8}$ N/m, and the other dynamic parameters are all identified. The longitudinal vibration of shafting with a friction damper can be written as Equation (3), where the response vector $\boldsymbol{x}(t) = \{x_a(t) \quad x_c(t) \quad x_p(t) \quad x_t(t) \quad x_f(t)\}^T$, the distribution vectors of $\boldsymbol{F}(t)$ and $\boldsymbol{f}(t)$ are $\boldsymbol{L}_1 = \{1 \quad 0 \quad 0 \quad 0 \quad 0\}^T$, $\boldsymbol{L}_2 = \{0 \quad 0 \quad 1 \quad -1 \quad 0\}^T$, respectively, and the mass matrix $\boldsymbol{M}$, damping matrix $\boldsymbol{C}$, and stiffness matrix $\boldsymbol{K}$ are defined as

$$\boldsymbol{M} = \begin{bmatrix} M_a & & & & \\ & M_c & & & \\ & & M_p & & \\ & & & M_t & \\ & & & & M_f \end{bmatrix} \tag{24}$$

$$\boldsymbol{C} = \begin{bmatrix} \frac{K_a\eta_a}{\omega} & & & & \\ & C_o & -C_o & & \\ & -C_o & C_o & & \\ & & & \frac{K_t\eta_t}{\omega} & -\frac{K_t\eta_t}{\omega} \\ & & & -\frac{K_t\eta_t}{\omega} & \frac{K_t\eta_t}{\omega} + \frac{K_f\eta_f}{\omega} \end{bmatrix} \tag{25}$$

$$\boldsymbol{K} = \begin{bmatrix} K_a + K_s & -K_s & & & \\ -K_s & K_s + K_o & -K_o & & \\ & -K_o & K_o + K_h & -K_h & \\ & & -K_h & K_h + K_t & -K_t \\ & & & -K_t & K_t + K_f \end{bmatrix}. \tag{26}$$

By defining the friction damping ratio as $\xi = f_{max}/F$, the theoretical AFRF of the thrust bearing with a friction damper for various values of ξ is plotted in Figure 14. Compared with the AFRF of the thrust bearing without control, the use of a friction damper can effectively suppress the frequency response, especially at the frequency of the resonance peaks. An increase in resonance peaks with the increase in ξ is observed. It means that the decrease in ξ benefits the damping effect of the friction damper.

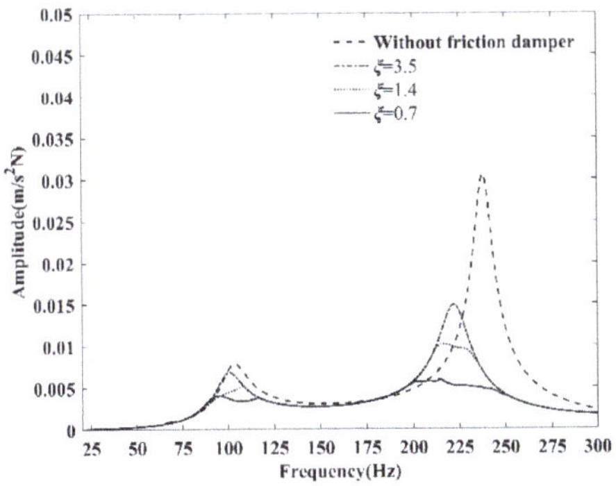

Figure 14. Theoretical AFRF of thrust bearing with friction damper for various values of ξ.

To understand the mechanism of the friction damper used for vibration control, the hysteresis loops of the friction damper for $\xi = 0.7,\ 1.4,\ 3.5$ are further presented in Figure 15. It is worth considering the case when the exciting frequency is 237 Hz, which coincides with the natural frequency of the test rig without control, and the static thrust is 30 kN, which corresponds to the maximum frictional force of 3500 N. The points in Figure 11 denote the correspondence between frictional force and relative displacement at discrete times. The area of the hysteresis loop is equal to the energy dissipated by the friction damper. As ξ increases, the area of the hysteresis loop increases, and the energy dissipation is larger.

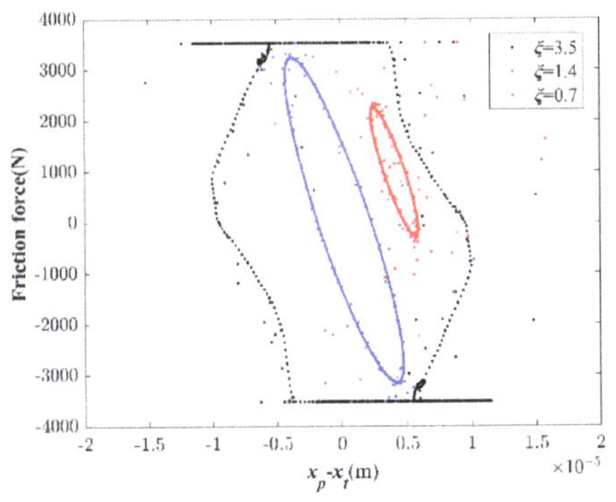

Figure 15. Hysteresis loops of friction damper for $\xi = 0.7,\ 1.4,\ 3.5$.

According to the established friction force nonlinear analysis method, we theoretically predict that the larger the friction damping ratio ξ, the larger the peak FRF, which is not conducive to Coulomb damping. Under the same excitation force, the test results show that the higher the static thrust force, the higher the corresponding peak FRF, which is consistent with the theoretical prediction. Therefore, from the engineering application point, ξ should be designed as small as possible. We suggest that ξ should be less than 1; that is, the maximum static friction should be less than the excitation force amplitude.

This paper uses the particle swarm optimization algorithm to identify the unknown dynamic parameters of the test rig based on the experimental FRF (Section 5.3). Therefore, consistency is the combined result of all the model's dynamic parameters, and each parameter's variations have different influences on FRF. We use the univariate analysis method to change the identification values of the above dynamic parameters one by one, such as increasing them by 10%. The comparison is shown in the Figure 16. It can be observed that, with the accurate identification of the dynamic parameters, changing only one of the dynamic parameter values will not cause a significant deviation between the theoretical FRF and the experimental FRF, which indicates that the established shafting model has good robustness and provides a more accurate model for analyzing the damping effect of friction dampers.

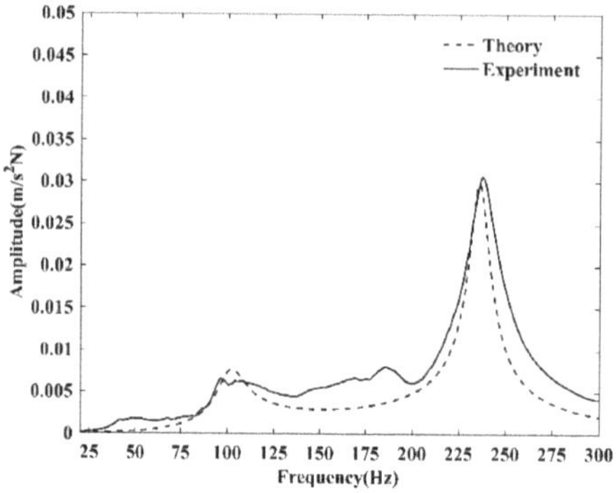

(**a**)

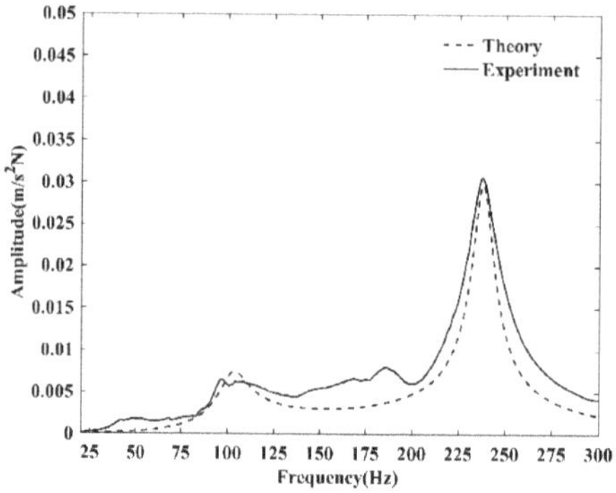

(**b**)

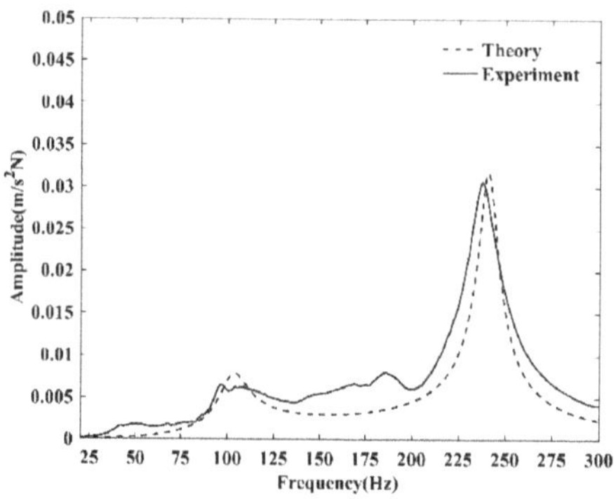

(**c**)

Figure 16. *Cont.*

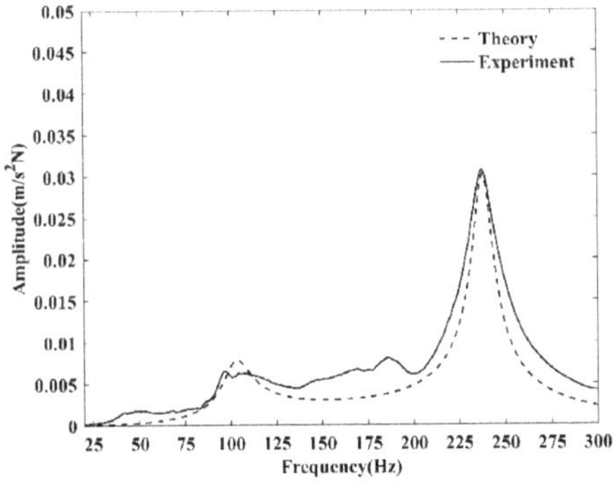

(**d**)

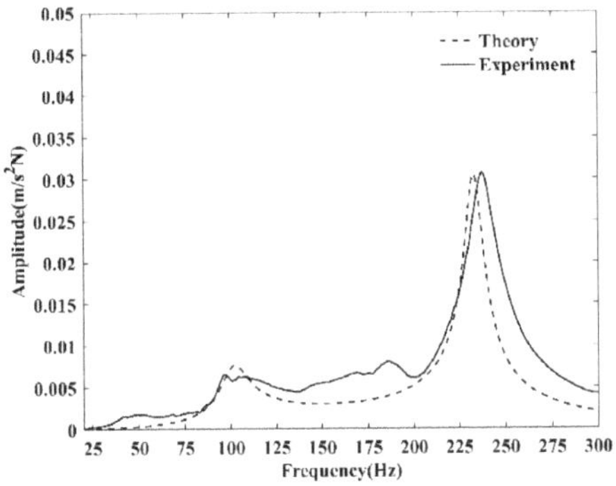

(**e**)

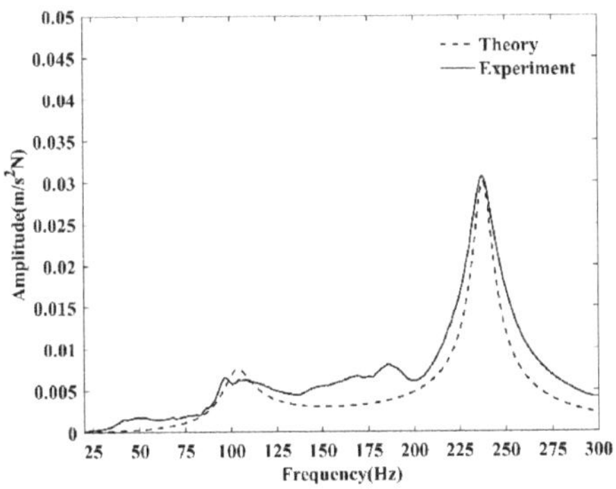

(**f**)

Figure 16. *Cont.*

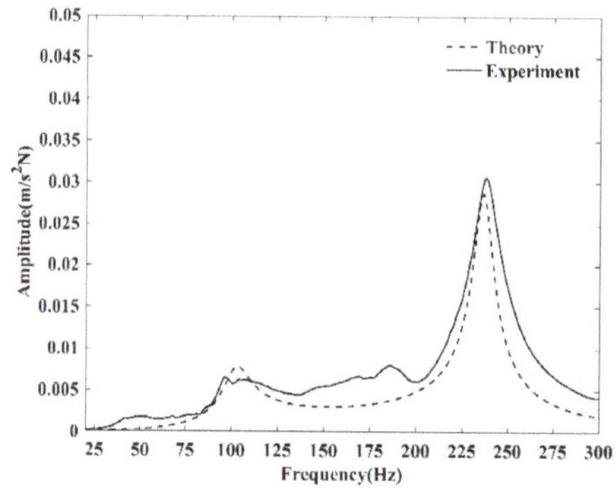

(g)

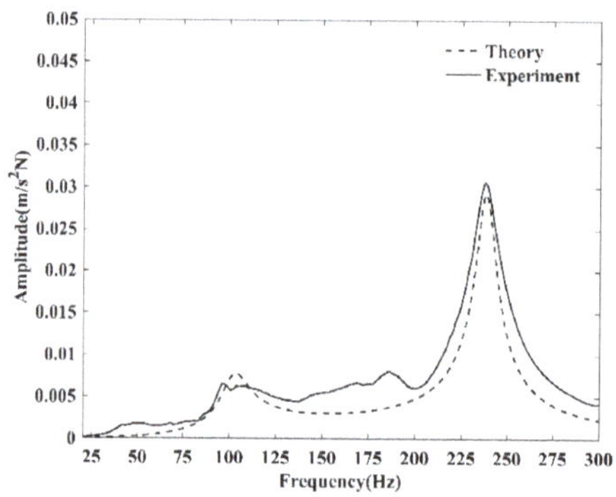

(h)

Figure 16. Effects of different dynamic parameters: (**a**) M_a; (**b**) η_a; (**c**) K_o; (**d**) C_o; (**e**) M_t; (**f**) η_t; (**g**) M_f; (**h**) η_f.

7. Experimental Investigation of Friction Damper

7.1. Damping Effect on Longitudinal Vibration

Following the same test procedure, the AFRF of the thrust bearing with a friction damper was measured. By considering the case that the shafting rotational speed is at 60 rpm and the applied static thrust is 30 kN, the comparison of experimental AFRF of the thrust bearing with and without a friction damper is plotted in Figure 17. It is observed that the response of the thrust bearing is effectively suppressed by the friction damper in a broadband frequency range, which shows the effectiveness of the friction damper in reducing longitudinal vibration transmission in the shafting system. In particular, the attenuation of resonance peaks is very significant, with a maximum reduction of about 75.5%. The straightforward explanation is that the friction damper increases the damping of the thrust bearing, and the friction damping dissipates energy.

Furthermore, the AFRF of the thrust bearing with a friction damper is also predicted by employing the proposed nonlinear analysis method, which is indicated by the dotted line in Figure 17. In the calculation, the value of excitation force $F(t)$ is the same as the sample signals of the hammer impact force in the experiments. As can be observed, the predicted AFRF is close to the measured experimental result in general, except that there is some difference in the maximum resonance peak. This difference may be introduced by other nonlinear factors, such as the assembly clearance that exists in the air spring device and thrust bearing. Generally, the proposed nonlinear analysis method can accurately predict the damping effect of the friction damper and design.

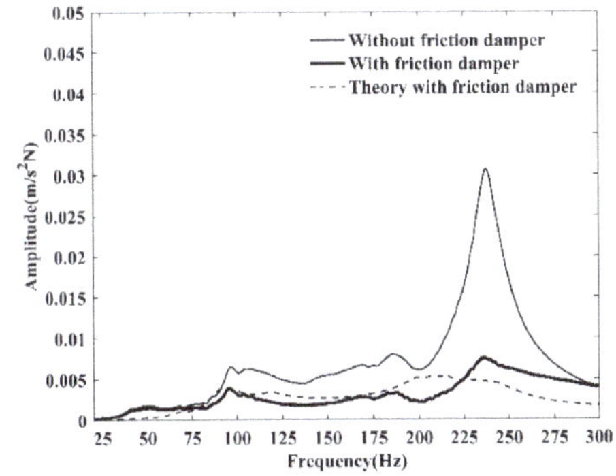

Figure 17. Comparison of experimental AFRF of thrust bearing with and without friction damper.

7.2. Stability of Damping Effect

In order to ensure the stability of the damping effect of the friction damper, a series of tests are carried out for various static thrusts and rotational speeds. The effects of these two variables are studied by means of a single factor test.

Firstly, the influence of rotational speed is investigated in the condition of the same static thrust. The range of rotational speed covers from 60 rpm to 160 rpm, of which the increment 20 rpm is used. Figure 18 shows the comparisons of the AFRF of the thrust bearing with and without a friction damper at different rotational speeds. By comparison, when the static thrust is constant, the resonance peaks of the thrust bearing with a friction damper are always significantly less than those without a friction damper for various rotational speeds. The experimental results indicate that the rotational speed does not affect the damping performance of the friction damper, and the longitudinal vibration characteristics of the shafting system are unrelated to the rotational speed.

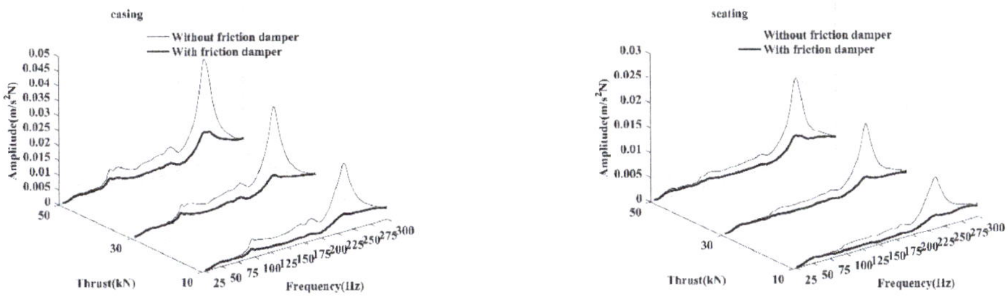

(**a**)

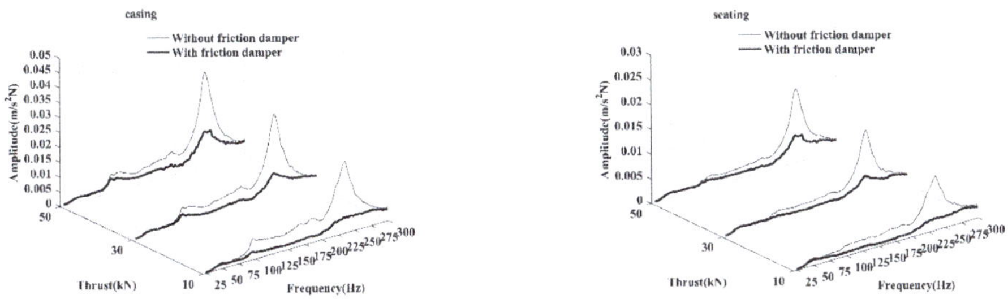

(**b**)

Figure 18. Comparisons of the AFRF of thrust bearing with and without friction damper at different rotational speeds: (**a**) rotational speed at 60 rpm; (**b**) rotational speed at 160 rpm.

The influence of static thrust is investigated under the condition of the same rotational speed. The range of static thrust covers from 10 kN to 50 kN, of which the increment 20 kN is used. Figure 19 presents a comparison of AFRF of the thrust bearing under different static thrusts. Compared with the measurements without control, the AFRFs of thrust bearings with friction dampers under various thrusts are all greatly suppressed in the broadband frequency range, and the attenuation of the maximum resonance peak is particularly significant. It means that the variation in static thrust also has little influence on the performance of the friction damper.

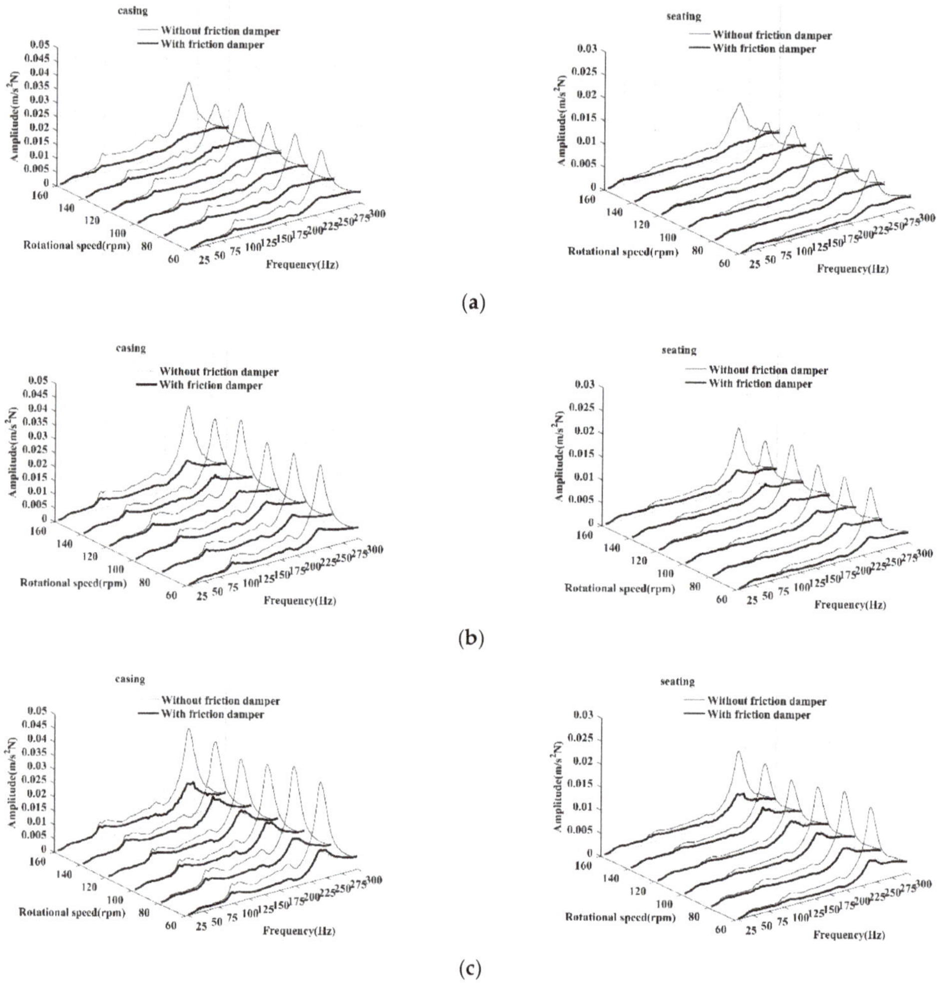

Figure 19. Comparisons of AFRF of thrust bearing under different static thrusts: (**a**) static thrust is 10 kN; (**b**) static thrust is 30 kN; (**c**) static thrust is 50 kN.

One observation is that the resonance peak of AFRF of the thrust bearing with a friction damper increases with the static thrust. Such a variation in resonance is related to the increase in frictional force, which results in an increase in the friction damping ratio. As predicted in theory, the resonance peak increases with the friction damping ratio. The experiments verify the theoretical conclusion.

8. Conclusions

In this work, a friction damper integrated into the thrust bearing was presented to tackle the problem of shafting longitudinal vibration. The choice of using friction damping as a vibration control measure in marine applications is non-conventional; this work conducts an exploratory study in this field.

A test rig was assembled to simulate a marine propulsion shafting system. A prototype of a novel thrust bearing, which provides two operation states for the friction damper device, was also developed. Compared with the traditional thrust bearing, of which the support structure includes upper and lower levers, the invented thrust bearing used hydraulic means. Based on the test rig, the AFRF of the thrust bearing with and without a friction damper was tested. By taking the measurements carried out without control as the benchmark, the effectiveness of the friction damper was assessed. In theory, a nonlinear analysis method for friction dampers was proposed.

The experimental and theoretical results show that the shafting longitudinal vibration response that is normally transmitted through thrust bearings is effectively suppressed by the friction damper in a broadband frequency range. For the test rig, the maximum resonance peak reduction reaches 75.5%, and the damping effect of the friction damper is considerably stable and does not change much with rotational speed and static thrust. The friction damping ratio is the critical parameter that affects the damping effect; a decrease in the friction damping ratio can improve the damping effect. In engineering, the friction damper should be designed according to propeller oscillatory thrust to avoid a redundant piston seal.

Author Contributions: Conceptualization, G.Z. and Y.Z.; methodology, G.Z.; software, G.Z.; validation, W.C., Y.Z.; formal analysis, G.Z.; investigation, G.Z.; resources, Y.Z.; data curation, W.C.; writing—original draft preparation, G.Z.; writing—review and editing, Y.Z.; visualization, W.C.; supervision, Y.Z.; project administration, Y.Z.; funding acquisition, G.Z. All authors have read and agreed to the published version of the manuscript.

Funding: This work was funded by the National Natural Science Foundation of China under Grant No. 51479078 and No. 52101354. Huazhong University of Science and Technology funds the APC.

Institutional Review Board Statement: Not applicable.

Informed Consent Statement: Not applicable.

Data Availability Statement: Not applicable.

Acknowledgments: The authors would like to acknowledge the academic guidance of Chen Xuedong, who helped with the experiments.

Conflicts of Interest: The authors declare no conflict of interest.

References

1. Zhang, G.; Zhao, Y.; Li, T.; Zhu, X. Propeller excitation of longitudinal vibration characteristics of marine propulsion shafting system. *Shock. Vib.* **2014**, *2014*, 1–19. [CrossRef]
2. Johnson, B.G.; Hockney, R.; Eisenhaure, D.; Fenn, R. System and Method for Damping Narrow Band Axial Vibrations of a Rotating Device. U.S. Patent 5291975A, 27 October 1992.
3. Baz, A.; Gilheany, J.; Steimel, P. Active vibration control of propeller shafts. *J. Sound Vib.* **1990**, *3*, 361–372. [CrossRef]
4. Xie, X.L.; Yang, D.Q.; Wu, D. Theoretical analysis on vibration transmission control in a shaft-hull system excited by propeller forces via an active multi-strut assembly. *Ocean. Eng.* **2021**, *221*, 108511. [CrossRef]
5. Goodwin, A.J.H. The design of a resonance changer to overcome excessive axial vibration of propeller shafting. *Trans. Inst. Mar. Eng.* **1960**, *72*, 37–63.
6. Goodwin, A.J.H. Vibration Isolators. U.S. Patent 3202388, 24 August 1965.
7. William, P.W. Thrust Bearing Resonance Changer. U.S. Patent 3937534, 10 February 1976.
8. Lewis, D.W.; Allaire, P.E. Active magnetic control of oscillatory axial shaft vibrations in ship shaft transmission systems part 1: System natural frequencies and Laboratory scale model. *Tribol. Trans.* **1989**, *32*, 170–178. [CrossRef]
9. Lewis, D.W.; Allaire, P.E. Active magnetic control of oscillatory axial shaft vibrations in ship shaft transmission systems part 2: Control Analysis and Response of Experimental System. *Tribol. Trans.* **1989**, *32*, 179–188. [CrossRef]

10. Guo, Q.; Li, Q.; Chen, W.; Yu, Q.; Wang, L.; Li, D. Vibration Reduction of Thrust Bearing with Disc-Spring. Chinese Patent CN102927137A, 13 February 2013.
11. Song, Y.; Wen, J.; Yu, D.; Liu, Y.; Wen, X. Reduction of vibration and noise radiation of an underwater vehicle due to propeller forces using periodically layered isolators. *J. Sound Vib.* **2014**, *333*, 3031–3043. [CrossRef]
12. Liu, N.; Li, C.; Yin, C.; Dong, X.; Hua, H. Application of a dynamic anti-resonant vibration isolator to minimize the vibration transmission in underwater vehicles. *J. Vib. Control.* **2018**, *24*, 3819–3829. [CrossRef]
13. Zhao, Y.; Zhang, G.; Chu, W.; Chang, L. A hydraulic Support Structure of Marine Sliding Thrust Bearing. Chinese Patent CN202011485606.0, 2013.
14. Naeim, F.; Kelly, J.M. *Design of Seismic Isolated Structures: From Theory to Practice*; Wiley: New York, NY, USA, 1999.
15. Guglielmino, E.; Edge, K. A controlled friction damper for vehicle applications. *Control. Eng. Pract.* **2004**, *12*, 431–443. [CrossRef]
16. Müller, H.K.; Nau, B.S. *Fluid Sealing Technology*; Dekker: New York, NY, USA, 1998.
17. Wang, W.; Liu, B.; Zhang, Y.; Shao, X.; Allaire, P.E. Theoretical and experimental study on the static and dynamic characteristics of tilting-pad thrust bearing. *Tribol. Trans.* **2018**, *123*, 26–36. [CrossRef]
18. Zeng, F.M.; Wu, J.M.; Pang, Z.Y. *Principle of Marine Power Plant*; National Defense Industry Press: Beijing, China, 2009.
19. Wang, Y.P.; Chung, L.L.; Liao, W.H. Seismic response analysis of bridges isolated with friction pendulum bearings. *Earthq. Eng. Struct. Dyn.* **1998**, *27*, 1069–1093. [CrossRef]
20. Wang, L.; Fu, Y.; Pei, S.; Xu, H. Theoretical and Experimental Study on the Axial Oil Film Stiffness of Tilting Pad Thrust Bearings. *Tribol. Trans.* **2017**, *60*, 419–427. [CrossRef]
21. Pan, F.; Li, W.X.; Gao, Q. *Particle Swarm Optimizer and Multi-Object Optimization*; Beijing Institute of Technology Press: Beijing, China, 2013.

Journal of
Marine Science and Engineering

MDPI

Article

A New Theoretical Dynamic Analysis of Ship Rolling Motion Considering Navigational Parameters, Loading Conditions and Sea State Conditions

José M. Pérez-Canosa, José A. Orosa *, María Isabel Lamas Galdo and Juan José Cartelle Barros

Department of Navigation Science and Marine Engineering, University of A Coruña, Paseo de Ronda, 51, 15011 A Coruña, Spain

* Correspondence: jose.antonio.orosa@udc.es; Tel.: +34-981-167-000 (ext. 4320)

Abstract: Despite the IMO's efforts and the large quantity of research carried out over the years concerning the sudden loss of stability in fishing vessels, and even the damage done to merchant fleets due to cargo shifting, accidents with very relevant consequences continue to occur. This paper can be considered as a continuation of the recent research of authors which was carried out with ships in static conditions, with pure beam seas and without resistance. The aim of the present research is to provide a reference for ships' operators to improve the ship's behavior and seakeeping, to alter the ship's loading conditions or the navigational parameters (heading and speed), and even be aware of the time available to carry out these modifications before reaching dangerous situations. For this, all sea state conditions were mathematically modelled for, including the ship's rolling motion both in static and in realistic and dynamic conditions, with the waves influencing the vessel by coming from any direction. Relevant results of easy comprehension for ships' operators are shown in each of the models, which were validated with a representative real case study.

Keywords: rolling motion; angle of roll; waves; dynamic conditions; capsizing; vessels; operability

Citation: Pérez-Canosa, J.M.; Orosa, J.A.; Galdo, M.I.L.; Barros, J.J.C. A New Theoretical Dynamic Analysis of Ship Rolling Motion Considering Navigational Parameters, Loading Conditions and Sea State Conditions. *J. Mar. Sci. Eng.* **2022**, *10*, 1646. https://doi.org/10.3390/jmse10111646

Academic Editor: Kostas Belibassakis

Received: 29 September 2022
Accepted: 19 October 2022
Published: 3 November 2022

1. Introduction

Fishing vessels are the most common ship type by number in the world, with 25,607. This represents 21.3% of the total world fleet, most of them being small- and medium-sized ships [1]. In the case of Spain, its fishing fleet is one of the most important in relation to the number of vessels belonging to European Union countries, corresponding to 22% or 335,000 GT [2]. In 2020, Spanish authorities had 8937 registered fishing vessels, the small- and medium-size fishing vessels being the most common [3]. Furthermore, the daily work on board this type of vessel is considered to be some of the most dangerous, with the mortality per year very high in comparison with other sectors [4–6]. According to the last report available from EMSA (European Maritime Safety Agency) about marine casualties and incidents [7], from 2014–2020 a general decrease is visible in the average occurrence indicators of all type of ships, except for fishing vessels. A loss of stability, with the result of capsizing or large heeling, characterizes the type of accidents in which more deaths are registered than any other, in spite of many vessels being compliant with IMO (International Maritime Organization) stability regulations. However, the lack of stability is not only an issue related to fishing vessels, but also to merchant vessels and passenger ships in spite of the efforts of the IMO to improve stability requirements [8].

The IMO current stability criteria might not consider the dynamic, heterogeneous and inconstant waves conditions that ships can encounter during the sea passage or during fishing tasks. As a consequence, since the start of the IMO and due to serious incidents occurring, specific research has attempted to understand and reduce the number of incidents related with stability. However, there are still produced accidents related to large amplitude motions, resulting in many cases in capsizing and the loss of lives. Among these undesired accidents, those which have most attracted the attention of researchers and institutions such as the IMO are the causes that lead to ships capsizing. For this reason, some research was carried out to improve the safety of, specifically, the fishing fleet. Mantari et al. (2011) [9] studied the intact stability of ships, taking into account the actions of fishing gear loads and beam waves, and the wind influence, via a comparison of inclining and righting arms lever points of view for different operational conditions, including the influence of hull parameters and even the selection of fishing vessel machinery. Tello et al. (2011) [4], through new operability criteria, proposed limiting values based on seakeeping calculations of several Portuguese fishing vessels in some sea states conditions, establishing certain hull shapes which optimize and fulfil the proposed criteria from an early ship's design point of view. Mata-Santullano et al. (2014) [5] studied the intact stability and operability characteristics of some Galician fishing vessels which were lost, and considered the stability curves in certain loading conditions, defined from KG and GZ values which allow us to fulfill the IMO intact stability criteria, but only in two sea state conditions. These results were confronted with decommissioned vessels, which ended their service life in a regular way and which were operated in similar conditions that sank ones. Míguez, et al. (2018) [10] carried out an investigation to ascertain and obtain simplified models for roll response for medium-sized stern trawlers in regular beam and longitudinal waves and in static conditions. Alvite et al. (2020) [2] proposed a new criterion for small Galician vessels (length less than 24 m) based on metacentric height (GM), dynamic stability and the value of the critical wave height according to specific Galician fishing grounds conditions. Santiago-Caaamaño et al. (2018) [11] consider that most accidents in fishing fleets are caused by, among others, an incorrect operation of the ship and sailing in very bad weather conditions, which can lead to stability problems and, in consequence, sudden capsizing. For that reason, they proposed a guidance system which was as simple as possible based on naval architecture software and the Fast Fourier Transform which, without interaction from the crew, can predict automatically and in real time the ship's initial stability based on natural roll frequency, transverse mass moment of inertia and displacement, although the proposed methodology was only tested on head waves. In the same sense, Santiago-Caamaño et al. (2019) [12] proposed another system to monitor in real time the stability-integrating advanced signal processing, which was tested in beam seas but without considering the dynamic influence, especially the rough weather conditions. Although, obviously they are very interesting proposals, the question remains is if the small fishing vessel's master would know how to interpret the presented values of GM, and most important, how they should proceed to decrease the risk level immediately with navigational parameters. Furthermore, some of the previously discussed research referred only to specific sea state conditions and static conditions, without considering the navigational parameters, which can be modified in real time by the duty officers.

Concerning merchant fleets, which also suffer stability and cargo shifting with the consequent claims, following Woo et al. (2021) [13] the evaluation of ship's stability only considering GM is not sufficient, because ships have to fulfill eight intact stability parameters. With that objective, a methodology was proposed with empirical formulas to calculate all these parameters once it has first calculated GM, in such a way that the duty officer can evaluate if a vessel is complying IMO requirements. Although a ship may fulfill the IMO requirement in static conditions, it will likely not do so in dynamic situations sailing in waves. For this reason, it is important to keep in mind that most of the operational life of ships takes place in the presence of waves, and occasionally quite severe conditions are faced. However, in some cases, fatal accidents have also occurred without facing extreme conditions, specified in the fishing fleet; in addition to the claims by damages, the shifting of cargo on board merchant ships can also produce the capsizing. This was demonstrated in the interesting research of Spandonidis et al. (2016) [14], where systematic numerical research of granular material and vessel motion in regular and pure beam seas of different amplitudes was described.

Of the six degrees of freedom of a ship, the rolling motion is the most important, followed by the large amplitude motion and finally other dynamic effects which can produce the sudden loss of a ship [10]. For this reason, the aim of this paper is to analyze the ship's behavior in rolling motion, considering the navigational parameters involved in possible causes of capsizing, which are not treated by the current 2008 IS Code (IMO). The sudden loss of vessels is determined by the large roll angle at which downflooding occurs or when a synchronism situation is reached. Then, in order to reduce accidents improving the ship's behavior, it is necessary that operators have the previous knowledge and tools about how they should proceed when the rolling motion starts reaching unacceptable levels for safety. Amongst the ship navigational parameters that a duty officer can alter immediately are the ship's heading and speed. To do so, an officer needs to know how much time they have to alter these parameters, i.e., the specified time (t) elapsed from the rolling motion start at a certain loading condition until reaching a capsizing or downflooding situations. Therefore, in addition to the safe time frame available, the present work discusses also the headings and speed the ship's operators need to adopt in order to conduct the navigation safely. These results are not only valid for fishing fleets, but also for others such as general cargo vessels, container vessels or RORO vessels. In these vessels, the avoidance of large roll angles is essential in order to minimize the possibility of cargo shifting, and to reduce the claims at the destination port and even diminish the loss of stability due to the lack of sufficient stability [13].

It is important to note that the sudden losses of ships are concentrated in small fishing vessels [15], specifically in some cases where the crew training in stability matters can be considered subpar. Therefore, in these ships the stability matters are arranged by the masters according to their own experience, i.e., without conducting a deep theoretical study of stability parameters [11,16]. This aspect was also mentioned in the last available report of the EMSA about analysis on marine casualties and incidents involving fishing vessels [17] where it is considered that not correctly assessing of the vessel's limitations (bad weather conditions) had contributed to listing/capsizing accidents. For this reason, it is essential to provide these crews with the tools and knowledge, and to make it as easy as possible to act accordingly, altering speed and heading in order to deal with unforeseen wave conditions. However, capsizing is a dynamic phenomenon, from a stability and environmental conditions point of view. Thus, if the operator's stability training is subpar, understanding these motions while a ship is sailing in bad weather conditions can be a little difficult for them. This is due to lack of knowledge of dynamic stability and the large number of parameters involved. With the objective of minimizing the probability of accident, specifically large-amplitude roll angles and capsizing, in this paper a new methodological proposal is proposed to be used by ship operators in order to improve the seakeeping, considering parameters very easy to use and obtain. The first parameter for controlling the stability is the ship's loading conditions, expressed as natural roll period (Td), which can be easily measured by any sailor in calm conditions. The second parameter is the waves' period (Tw), measured on board and taken as reference to determined points along the ship's length and, finally, the encounter period (Te), which can be calculated using a very simple equation.

Operability is a very important factor for the ship's master in order to establish vessel safety. As limited literature analyses maneuverability in waves, this paper intents to deal the maneuverability and seakeeping jointly, despite the fact that in the literature they were typically treated separately. Furthermore, the present research can be considered as a continuation of previous research carried out by authors [18] about the study of ship motion around the longitudinal axis, i.e., the rolling motion. In that case, the ship's rolling motion sailing between pure beam waves was studied, both without resistance and in static conditions. The aim of the present paper is to carry out research about the ship's behavior from a point of view closer to reality, studying the ship's performance sailing in any type of waves, with resistance and the ship being hit from any influence angle. Thus, the optimal heading and speed to be set by the ship's operator are being studied implicitly in order to keep the sea passage as safe and comfortable as possible. Furthermore, the specified time (t) elapsed from the rolling motion start until a capsizing or downflooding situations occurs is being studied in order to give to masters an idea about the available time to modify or alter certain parameters. For this mission, different mathematical models were applied to rolling motion, which is independently studied in the following sections.

2. Materials and Methods

As the ship's behavior leading to capsizing in irregular waves is similar to that in regular waves [19], in the first subsection in which it is presented the mathematical models of regular waves corresponded to any sea state conditions registered. Secondly, the mathematical model which represents the ship's rolling motion sailing in pure beam waves and with resistance was studied. Finally, considering the two previous sections together, the mathematical model that allows for determining the ship's behavior of sailing, while receiving the waves from any constant direction and considering the ship's navigational parameters which can be modified by the master, is studied i.e., the speed and the heading.

2.1. Waves

As was already established in previous research, it is difficult to accurately predict the ship motion in irregular seas [18]. Thus, as usual, the waves with a trochoidal profile, which correspond to tender regular sea (after the wind ceases), are studied here from a mathematical point of view. In the field of waves' study, three parameters have always been considered: wavelength (Lw), translation velocity (Vw) and period (Tw). With that objective, empirical observations of any sea state condition have allowed the building a scale of 21 types of waves relating to these three parameters. Considering the values of the period (Tw) and the registered empirical observations, the following several polynomial regression equations were obtained (wavelength in m; velocity in knots and velocity in $m{\cdot}s^{-1}$), as a function of wave period.

$$\mathrm{Lw(m)} = 1.5838{\cdot}\mathrm{Tw}^2 - 0.5558{\cdot}\mathrm{Tw} + 1.4737 \quad (1)$$

$$\mathrm{Vw(knots)} = -0.0006{\cdot}\mathrm{Tw}^2 + 1.5196{\cdot}\mathrm{Tw} + 0.1165 \quad (2)$$

$$\mathrm{Vw(m{\cdot}s^{-1})} = 0.0009{\cdot}\mathrm{Tw}^2 + 0.7632{\cdot}\mathrm{Tw} + 0.3669 \quad (3)$$

In Figure 1 the high precision for the determination factor of Equations (1)–(3) can be observed.

However, one of the easiest and most practical parameters for sailors to use is the wave amplitude (Aw), i.e., half of the height (Hw), so it was necessary to relate it with the three previous parameters (wavelength, period and velocity). However, as the wave height (Hw) is affected by, among others, the fetch and duration of the wind swell generator, there are no formulas which allow the linking together of all these parameters. Nevertheless, it is possible to link the maximum wave slope (θ_{MW}) with the wavelength (Lw) and the height (Hw), as is shown in the following equation.

$$\sin(\theta_{MW}) = \pi{\cdot}\frac{\mathrm{Hw}}{\mathrm{Lw}} \quad (4)$$

The empirical observation of waves around the world has allowed for registering the wave characteristics [20], concluding that θ_{MW}, regardless of smooth or very rough sea, is almost constant and equal to 0.1047 (6°). Therefore, with the help of Equation (4), and wavelength calculated according to Equation (1), Hw and Aw can be obtained.

Table 1 includes all data calculated according to Equations (1)–(4) for the full spectrum of regular waves considered in the present research.

In Appendix A (Table A1) is included all symbols and definitions used.

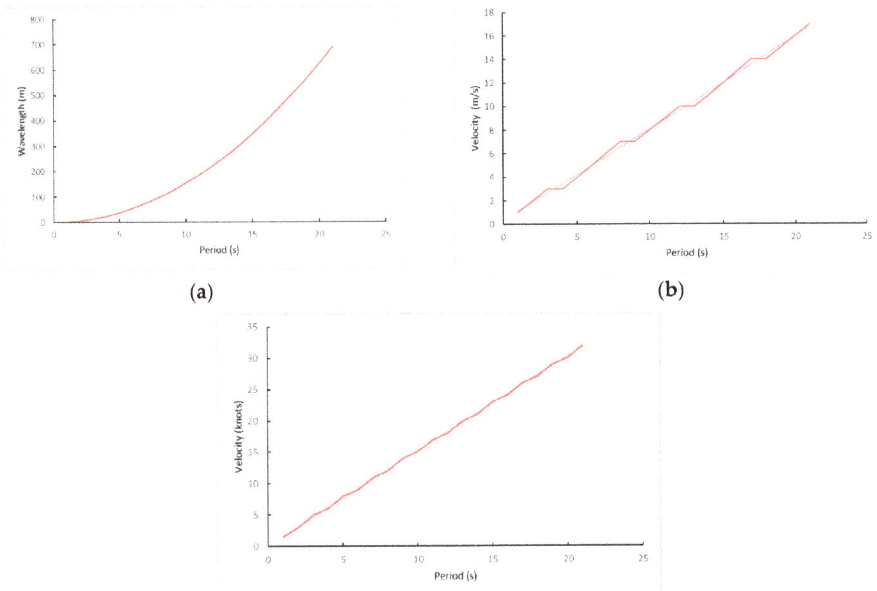

(**a**) (**b**) (**c**)

Figure 1. (**a**) Wavelength as a function of period. Precision 100%. (**b**) Velocity in m·s^{-1} as a function of period. Precision 99.67%. (**c**) Velocity in knots as a function of period. Precision 99.93%.

Table 1. Data of studied waves.

Tw (s)	Lw (m)	Vw (m·s^{-1})	Vw (knots)	Hw (m)	Aw (m)
1	2.5017	1.131	1.6355	0.08	0.04
2	6.6973	1.8969	3.1533	0.22	0.11
3	14.0605	2.6646	4.6699	0.47	0.23
4	24.5913	3.4341	6.1853	0.82	0.41
5	38.2897	4.2054	7.6995	1.28	0.64
6	55.1557	4.9785	9.2125	1.84	0.92
7	75.1893	5.7534	10.7243	2.50	1.25
8	98.3905	6.5301	12.2349	3.28	1.64
9	124.7593	7.3086	13.7443	4.15	2.08
10	154.2957	8.0889	15.2525	5.14	2.57
11	186.9997	8.871	16.7595	6.23	3.11
12	222.8713	9.6549	18.2653	7.42	3.71
13	261.9105	10.4406	19.7699	8.72	4.36
14	304.1173	11.2281	21.2733	10.13	5.06
15	349.4917	12.0174	22.7755	11.64	5.82
16	398.0337	12.8085	24.2765	13.25	6.63
17	449.7433	13.6014	25.7763	14.98	7.49
18	504.6205	14.3961	27.2749	16.80	8.40
19	562.6653	15.1926	28.7723	18.74	9.37
20	623.8777	15.9909	30.2685	20.78	10.39
21	688.2577	16.791	31.7635	22.92	11.46

2.2. Rolling Motion in Static Conditions and under the Influence of Beam Waves with Resistance

In reality, the free rolling motion produced by an external and punctual force is affected by the resistance of water and air, which cause the roll angle amplitude to be damped until the ship is at rest again. In the case of air resistance, its value is very low, therefore, its practically null effect on roll damping means that air resistance can be disregarded. The same does not happen with water resistance, so the inertia moment, the moment of water resistance and the ship's righting torque are involved in the equation of the ship's rolling motion sailing in beam waves with resistance. In the ship's righting torque (Ma) the angle of wave's slope, θ_w, is introduced so in the initial stability, i.e., for low rolling angles, it can be established:

$$\text{Ma} = \text{D}\cdot\text{GM}\cdot(\theta - \theta_W) \tag{5}$$

where D is the ship's displacement; GM is the initial transverse metacentric height; and $(\theta - \theta_W)$ is the angle of relative rolling between the normal to the wave's slope and the diametrical plane. In consequence, the equation of the ship's rolling motion sailing between transversal waves and with resistance is expressed as follows:

$$\text{Ig}\cdot\frac{d^2\theta}{dt^2} + A_R\cdot\frac{d(\theta - \theta_W)}{dt} + \text{D}\cdot\text{GM} \times (\theta - \theta_W) = 0 \tag{6}$$

where Ig is the inertia moment of the ship's mass about a longitudinal axis through its center of gravity and A_R is a damping coefficient.

Considering that in regular seas it is common that the angle of wave' slope is low compared to the absolute rolling angle between the perpendicular to the surface of the water in calm conditions and the ship's center line, Equation (6) can be simplified as follows:

$$\frac{d(\theta - \theta_W)}{dt} \approx \frac{d\theta}{dt} \tag{7}$$

$$\text{Ig}\cdot\frac{d^2\theta}{dt^2} + A_R\cdot\frac{d\theta}{dt} + \text{D}\cdot\text{GM}\cdot(\theta - \theta_W) = 0 \tag{8}$$

Now, several transformations on Equation (8) are performed with the aim of obtaining the relationship between rolling periods. So, in the first stage Equation (8) is divided by the inertia moment Ig:

$$\frac{d^2\theta}{dt^2} + \frac{A_R}{\text{Ig}}\cdot\frac{d\theta}{dt} + \frac{\text{D}\cdot\text{GM}}{\text{Ig}}\cdot(\theta - \theta_W) = 0 \tag{9}$$

Knowing that the inertia moment is:

$$\text{Ig} = \frac{D}{g}\cdot k^2 \tag{10}$$

where g is the gravity acceleration and k is the turning radius of the ship's mass with respect to the longitudinal axis passing through its center of gravity.

Operating for convenience:

$$2\lambda = \frac{A_R}{\text{Ig}} = \frac{A_R\cdot g}{D\cdot k^2} \tag{11}$$

Replacing Equations (10) and (11) in Equation (9):

$$\frac{d^2\theta}{dt^2} + 2\lambda\cdot\frac{d\theta}{dt} + \frac{\text{D}\cdot\text{GM}}{\frac{D}{g}\cdot k^2}\cdot(\theta - \theta_W) = 0 \tag{12}$$

$$\frac{d^2\theta}{dt^2} + 2\lambda\cdot\frac{d\theta}{dt} + \frac{g\cdot\text{GM}}{k^2}\cdot(\theta - \theta_W) = 0 \tag{13}$$

Knowing that natural roll frequency is;

$$\omega^2 = \frac{g \cdot GM}{k^2} \tag{14}$$

Equation (14) is now introduced in Equation (13):

$$\frac{d^2\theta}{dt^2} + 2\lambda \cdot \frac{d\theta}{dt} + \omega^2 \cdot (\theta - \theta_W) = 0 \tag{15}$$

$$\frac{d^2\theta}{dt^2} + 2\lambda \cdot \frac{d\theta}{dt} + \omega^2 \cdot \theta - \omega^2 \cdot \theta_W = 0 \tag{16}$$

In Equation (16), the first three terms match with the general equation of the rolling motion sailing in calm water conditions with resistance, the fourth term ($-\omega^2 \cdot \theta$) being the disturbing (induced) force of the waves. When studying the rolling motion of a ship sailing in waves without resistance, the same disturbing force of the wave's slope is obtained, being the resulting term;

$$-\omega^2 \cdot \theta_{MW} \cdot \sin \omega_W \cdot t \tag{17}$$

Replacing Equation (17) in Equation (16):

$$\frac{d^2\theta}{dt^2} + 2\lambda \cdot \frac{d\theta}{dt} + \omega^2 \cdot \theta - \omega^2 \cdot \theta_{MW} \cdot \sin \omega_W \cdot t = 0 \tag{18}$$

If it is taken as a particular solution of Equation (18),

$$\theta = E_K \cdot \sin \omega_W \cdot t + F_K \cdot \cos \omega_W \cdot t \tag{19}$$

And Equation (19) is derived twice:

$$\frac{d\theta}{dt} = E_K \cdot \omega_W \cdot \cos \omega_W \cdot t - F_K \cdot \omega_W \cdot \sin \omega_W \cdot t \tag{20}$$

$$\frac{d^2\theta}{dt^2} = -E_K \cdot \omega_W^2 \cdot \sin \omega_W \cdot t - F_K \cdot \omega_W^2 \cdot \cos \omega_W \cdot t \tag{21}$$

Replacing the particular solution and its derivatives in Equation (18):

$$\begin{gathered}(-E_K \cdot \omega_W^2 \cdot \sin \omega_W \cdot t - F_K \cdot \omega_W^2 \cdot \cos \omega_W \cdot t) + 2\lambda(E_K \cdot \omega_W \cdot \cos \omega_W \cdot t - F_K \cdot \omega_W \cdot \sin \omega_W \cdot t) + \\ +\omega^2 \cdot (E_K \cdot \sin \omega_W \cdot t + F_K \cdot \cos \omega_W \cdot t) - \omega^2 \cdot \theta_{MW} \cdot \sin \omega_W \cdot t = 0\end{gathered} \tag{22}$$

Taking out common factor $\sin \omega_W \cdot t$ and $\cos \omega_W \cdot t$ of Equation (22):

$$\sin \omega_W \cdot t(-E_K \cdot \omega_W^2 - 2\lambda \cdot F_K \cdot \omega_W + E_K \cdot \omega^2 - \omega^2 \cdot \theta_{MW}) + \cos \omega_W \cdot t(-F_K \cdot \omega_W^2 + 2\lambda \cdot E_K \cdot \omega_W + F_K \cdot \omega^2) = 0 \tag{23}$$

Equation (23) must satisfy any value of t, and for that mission it is necessary that

$$(-E_K \cdot \omega_W^2 - 2\lambda \cdot F_K \cdot \omega_W + E_K \cdot \omega^2 - \omega^2 \cdot \theta_{MW}) = 0 \tag{24}$$

$$(-F_K \cdot \omega_W^2 + 2\lambda \cdot E_K \cdot \omega_W + F_K \cdot \omega^2) = 0 \tag{25}$$

Clearing F_K in Equation (25):

$$F_K = \frac{2\lambda \cdot E_K \cdot \omega_W}{\omega_W^2 - \omega^2} \tag{26}$$

In order to find E_K, $\omega^2 \cdot \theta_{MW}$ is cleared in Equation (24) and the value of F_K found is introduced;

$$-E_K \cdot \omega_W^2 - 2\lambda \cdot F_K \cdot \omega_W + E_K \cdot \omega^2 = \omega^2 \cdot \theta_{MW} \tag{27}$$

$$-E_K \cdot \omega_W^2 - 2\lambda \cdot \frac{2\lambda \cdot E_K \cdot \omega_W}{\omega_W^2 - \omega^2} \cdot \omega_W + E_K \cdot \omega^2 = \omega^2 \cdot \theta_{MW} \tag{28}$$

$$-E_K \cdot \omega_W^2 - \frac{4\lambda^2 \cdot E_K \cdot \omega_W^2}{\omega_W^2 - \omega^2} + E_K \cdot \omega^2 = \omega^2 \cdot \theta_{MW} \tag{29}$$

Taking out common factor E_K

$$E_K \cdot \left[\omega^2 - \omega_W^2 + \frac{4\lambda^2 \cdot \omega_W^2}{\omega^2 - \omega_W^2} \right] = \omega^2 \cdot \theta_{MW} \tag{30}$$

$$E_K \cdot \left[\frac{\left(\omega^2 - \omega_W^2\right)^2 + 4\lambda^2 \cdot \omega_W^2}{\omega^2 - \omega_W^2} \right] = \omega^2 \cdot \theta_{MW} \tag{31}$$

Considering that:

$$\tan \beta = -\frac{F_K}{E_K} \tag{32}$$

This Equation (32), according to Equation (26), can be also expressed as follows:

$$\tan \beta = -\frac{2\lambda \cdot E_K \cdot \omega_W}{E_K \cdot (\omega_W^2 - \omega^2)} \tag{33}$$

$$\tan \beta = \frac{2\lambda \cdot \omega_W}{\omega^2 - \omega_W^2} \tag{34}$$

Developing the trigonometric relationship:

$$1 + \tan^2 \beta = \frac{1}{\cos^2 \beta} \tag{35}$$

$$1 + \frac{4\lambda^2 \cdot \omega_W^2}{\left(\omega^2 - \omega_W^2\right)^2} = \frac{1}{\cos^2 \beta} \tag{36}$$

$$\frac{\left(\omega^2 - \omega_W^2\right)^2 + 4\lambda^2 \cdot \omega_W^2}{\omega^2 - \omega_W^2} = \frac{\omega^2 - \omega_W^2}{\cos^2 \beta} \tag{37}$$

The first term of Equation (37) is equal to the first parenthesis of the first term of Equation (31), therefore:

$$E_K \cdot \left[\frac{\omega^2 - \omega_W^2}{\cos^2 \beta} \right] = \omega^2 \cdot \theta_{MW} \tag{38}$$

Clearing E_K is obtained:

$$E_K = \frac{\omega^2 \cdot \theta_{MW} \cdot \cos^2 \beta}{\omega^2 - \omega_W^2} \tag{39}$$

Equation (19) is divided by E_K, which is the particular solution of the differential equation of the ship's rolling motion between waves, in which the resistance of water Equation (18) is considered, and considering Equation (32):

$$\frac{\theta}{E_K} = \sin \omega_W \cdot t - \tan \beta \cdot \cos \omega_W \cdot t \tag{40}$$

$$\frac{\theta}{E_K} = \sin \omega_W \cdot t - \frac{\sin \beta}{\cos \beta} \cdot \cos \omega_W \cdot t \tag{41}$$

$$\frac{\theta}{E_K} = \frac{\sin \omega_W \cdot t \cdot \cos \beta - \sin \beta \cdot \cos \omega_W \cdot t}{\cos \beta} \tag{42}$$

$$\frac{\theta}{E_K} = \frac{\sin(\omega_W \cdot t - \beta)}{\cos \beta} \tag{43}$$

Then, the absolute rolling angle is equal to:

$$\theta = E_K \cdot \frac{\sin(w_W \cdot t - \beta)}{\cos \beta} \tag{44}$$

If in Equation (44) E_K is replaced by its value of Equation (39):

$$\theta = \frac{\omega^2 \cdot \theta_{MW} \cdot \cos^2 \beta}{\omega^2 - \omega_W^2} \cdot \frac{\sin(\omega_W \cdot t - \beta)}{\cos \beta} \tag{45}$$

$$\theta = \frac{\theta_{MW} \cdot \cos \beta \cdot \sin(\omega_W \cdot t - \beta)}{1 - \frac{\omega_W^2}{\omega^2}} \tag{46}$$

Being;

$$\frac{\omega_W^2}{\omega^2} = \frac{(2\pi)^2/Tw^2}{(2\pi)^2/Td^2} = \frac{Td^2}{Tw^2} \tag{47}$$

$$\theta = \frac{\theta_{MW} \cdot \cos \beta \cdot \sin(\omega_W \cdot t - \beta)}{1 - \frac{Td^2}{Tw^2}} \tag{48}$$

Equation (48) is similar to the equation of rolling motion induced by transversal waves, but without the resistance studied in the previous research of authors [18], although it is now included by the β angle, which depends on damping factor λ. Thus, the rolling angle of Equation (48) is the angle of rolling motion induced by waves' disturbances.

According to Equation (48) and with the equation of rolling motion in calm water conditions with resistance [21], the equation of a ship's rolling motion sailing in waves and with resistance can be expressed as follows:

$$\theta = \theta_M \cdot e^{-\lambda \cdot t} \cdot \left[\cos \omega \cdot t + \frac{\lambda}{\omega} \cdot \sin \omega \cdot t\right] + \frac{\theta_{MW} \cdot \cos \beta \cdot \sin(\omega_W \cdot t - \beta)}{1 - \frac{Td^2}{Tw^2}} \tag{49}$$

The first term represents the damped rolling, where λ is the factor damping and θ_M is the initial maximum angle of roll.

If the damping factor is equaled by a factor sub-1 by time unit:

$$\lambda = \frac{\lambda_1}{Td} \tag{50}$$

In addition, the ships' and the waves' frequencies are replaced by their corresponding periods (Td and Tw), Equation (49) is:

$$\theta = \theta_M \cdot e^{\frac{-\lambda_1 \cdot t}{Td}} \cdot \left[\cos \frac{2\pi}{Td} \cdot t + \frac{\lambda_1}{2\pi} \cdot \sin \frac{2\pi}{Td} \cdot t\right] + \frac{\theta_{MW} \cdot \cos \beta \cdot \sin\left[\frac{2\pi}{Tw} \cdot t - \beta\right]}{1 - \frac{Td^2}{Tw^2}} \tag{51}$$

Moreover, considering that:

$$\tan \beta = \frac{2\lambda_1}{Td} \cdot \frac{\omega_W}{\omega^2 - \omega_W^2} \tag{52}$$

Numerator and denominator of the second term of Equation (52) is divided by ω^2:

$$\tan \beta = \frac{2\lambda_1}{Td} \cdot \frac{\frac{\omega_W}{\omega^2}}{1 - \frac{\omega_W^2}{\omega^2}} \tag{53}$$

Equation (53) can be expressed, taking into account the corresponding periods (Td and Tw):

$$\tan\beta = \frac{2\lambda_1}{\text{Td}} \cdot \frac{\frac{2\pi}{\text{Tw}} \cdot \frac{\text{Td}^2}{4\pi^2}}{1 - \left[\frac{4\pi^2}{\text{Tw}^2} \cdot \frac{\text{Td}^2}{4\pi^2}\right]} \tag{54}$$

$$\tan\beta = \frac{\lambda_1}{\pi} \cdot \frac{\frac{\text{Td}}{\text{Tw}}}{1 - \frac{\text{Td}^2}{\text{Tw}^2}} \tag{55}$$

2.3. *Rolling Motion in Underway Conditions and Under the Influence of Waves of Any Constant Direction*

In this section, the ship rolling motion for when waves are hitting it from any constant value of α angle is analyzed, formed between the ship's center line (heading), which can be modified by the duty officer, and the waves' direction.

According to basic plane trigonometry, the encounter velocity (Ve) can be expressed as follows:

$$\text{Ve} = \text{Vw} + \text{Vs} \cdot \cos\alpha \tag{56}$$

where Vw is the waves' velocity and Vs is the ship's velocity. Furthermore, considering that the relationship between the waves' length (Lw), the waves' period (Tw) and the waves' velocity (Vw) is:

$$\text{Tw} = \frac{\text{Lw}}{\text{Vw}} \tag{57}$$

It can be expressed in the following relationship:

$$\text{Te} = \frac{\text{Lw}}{\text{Ve}} \tag{58}$$

Replacing in Equation (58) the encounter velocity (Ve) given in Equation (56):

$$\text{Te} = \frac{\text{Lw}}{\text{Vw} + \text{Vs} \cdot \cos\alpha} \tag{59}$$

The term $\text{Vs} \cdot \cos\alpha$ is positive when α is in the first or fourth quadrant (when the swell comes from port and starboard bow), and negative when is in the second and third quadrant (when the swell comes from port and starboard quarter), considering the line forward-aft as a positive axis. According to Equation (59), when the ship receives the waves from the bow, the encounter period (Te) is lower than the wave period (Tw); when it receives them from the quarter, the encounter period is higher than the wave period.

The wave's slope depends on its length (Lw) and height (Hw). Decomposing the incoming wave in a longitudinal and a transversal wave, the length of the last one (L_{BEAM}) is:

$$L_{ABEAM} = \text{Lw} \cdot \sin\alpha \tag{60}$$

As a consequence, the transversal component of the wave's slope is affected by sin α and by the maximum wave's slope ($\theta_{MW\ ROLLING}$), which is of interest in the study of induced rolling motion:

$$\theta_{MW\ ROLLING} = \theta_{MW} \cdot \sin\alpha \tag{61}$$

Therefore, Equation (51) of rolling motion sailing in waves with resistance can be generalized by introducing the encounter period (Te) and the maximum wave's slope of the transversal component, which can be expressed as follows:

$$\theta = \theta_M \cdot e^{\frac{-\lambda_1 \cdot t}{\text{Td}}} \cdot \left[\cos\frac{2\pi}{\text{Td}} \cdot t + \frac{\lambda_1}{2\pi} \cdot \sin\frac{2\pi}{\text{Td}} \cdot t\right] + \frac{\theta_{MW} \cdot \sin\alpha \cdot \cos\beta \cdot \sin\left[\frac{2\pi}{\text{Te}} \cdot t - \beta\right]}{1 - \frac{\text{Td}^2}{\text{Te}^2}} \tag{62}$$

Considering that, according to Equation (55);

$$\tan\beta = \frac{\lambda_1}{\pi}\cdot\frac{\frac{Td}{Te}}{1-\frac{Td^2}{Te^2}} \tag{63}$$

The obtained Equation (62) represents the real rolling motion of a ship in waves, where the two navigational parameters are included. This can be adjusted by operators at any time, as can the heading and ship's speed. The heading is included in the second term as sin α, and the ship's speed is indirectly included in the Te variable, while at the same time it also has influence the heading (α). Furthermore, that the rolling angle is a function and changes with the time (t) has been shown, and so in the Results section a deep analysis applied to the case studies is carried out.

3. Results

This section provides a study of a ship's behavior, expressed as a function of the rolling angle reached in the linear time domain. In the first subsection, the rolling motion in static conditions was analyzed under the influence of beam and regular waves with resistance. In a way, this approach is considered to be interesting due to it being the worst situation found when the ship was at zero speed. Following the same approach, afterwards the rolling motion was studied from a more realistic point of view, i.e., when the ship's movement is underway at any ship's speed and heading, and under the influence of all registered waves coming from any direction.

In order to validate the obtained results in the previous section checking that they are valid to any sea state condition, simulations were conducted with the 21 swell conditions, which were previously modelled in Equations (1)–(4).

For the objective proposed in this paper, the mathematical models, the appearance of figures and the general conclusions obtained from them are the most important aspects of this paper. For the case study used for validations, the usual parameters found in some of the fishing vessels of the Spanish fleet were selected. These have lost stability in the last 20 years, resulting in the loss of many lives. Although the exact loading condition in the moment when an accident happens is unknown, Bonilla de la Corte (1994) [22] considers that usual Td for fishing vessels is between 8 s and 14 s. Furthermore, in Bonilla de la Corte (1994) [22] we can find that Td can be estimated as follows:

$$Td = 0.77\cdot\frac{B}{GM} \tag{64}$$

where B is the beam and GM is the metacentric height.

Moreover, the classification society DNV (2016) [23] considers that the relationship between GM and B is the following:

$$GM = 0.07\cdot B \tag{65}$$

In addition, according to DNV (2016) [23], knowing the beam (B) and the metacentric height (GM), the Td can also be calculated by the following equation:

$$Td = \frac{2\cdot 0.39\cdot B}{\sqrt{GM}} \tag{66}$$

Then, included in Table 2 are the most relevant data of several capsized fishing vessels, and the calculated Td according to previous equations.

Obviously, Equations (64)–(66) only allow for an estimated idea of Td and GM. This is because the exact value of both parameters depends on the specific ship's loading condition, expressed as natural roll period (Td). For that reason, from the proceedings literature it can be extracted Td values are substantially different from those calculated as per the mentioned equations. Miguez et al. (2016, 2017, 2018) [24–26] consider that, in a trawler fishing vessel with a beam of 8.0 m, the Td assigned was between 11.1 s and 10.3 s, while the calculated Td according to equations would be 8.2 s. Oliva-Remola (2017) [27] considers that in a trawler fishing vessel with a beam of 11.50 m, the assigned Td was 10.6 s, and the calculated Td was 9.9 s.

Table 2. Some of the Spanish fishing vessels capsized with the result of loss of lives reason for capsizing.

Name	Year	Crewmembers Dead	Beam	Calculated Td
Villa de Pitanxo	2022	21	9.70 m	9.0 s
Siempre Gueto	2021	2	6.0 m	7.1 s
Sin Querer Dos	2018	4	6.1 m	7.2 s
Cordero	2008	5	7.2 m	7.8 s
Nuevo Pepita Aurora	2007	8	5.8 m	7.0 s
Siempre Casina	2005	8	5.3 m	6.7 s
O Bahía	2004	10	5.2 m	6.6 s
Nuevo Pilín	2004	5	5.0 m	6.5 s

Considering the above, and with the objective proposed in this research of being as generic as possible, a natural roll period (Td) of 9.0 s can be considered representative enough of the type of fishing vessel fleet which has sustained problems in terms of the stability of sailing on waves. This value is in line with the Td values used in previous works [12], where the models were evaluated with 7.8 s < Td < 12.2 s for a fishing trawler with a length of 34.5 m and a beam of 8.00 m.

Regarding the initial maximum angle of roll, θ_M, 10° was considered in our simulations, understanding that, in bad weather conditions, it can be a roll angle easily reached by medium-sized fishing vessels. In terms of the roll-damping factor, λ, following recommendations of the literature review [28,29], a median coefficient of 0.015 was selected. As the roll-damping factor depends on each rolling angle reached, and they are not constant, the results of proposed models can be considered as simplified and a first approach to be validated in scale model tests or in computer programs. There, the influence of different underwater shapes (bilge keels or not), block coefficients and appendices as different types of roll stabilizations systems (active fins) could be discussed.

3.1. *Rolling Motion in Static Conditions under Beam—Regular Wave's Influence with Resistance*

In this subsection, the ship's performance during rolling motion is calculated from a mathematical point of view, both for when the ship is in static conditions and receiving the waves from beam, although with resistance.

Analyzing the first term of Equation (49) and the variation with the time (t) of the damped rolling, it is concluded that when the ship is with zero speed and under pure beam seas influence, it would end up disappearing due to its decreasing geometric ($e^{-\lambda \cdot t}$) quality. Theoretically, the ship would end up rolling according to the waves' period (Tw) instead of its natural roll period (Td), i.e., according to the second term of the equation, which corresponds to the induced rolling (forced), reaching a synchronism condition. However, due to the lack of regularity in the waves and the tendency of a ship to roll according to its natural period, in the reality it is usual that synchronism is not reached. Considering this premise, the maximum rolling angle is produced with the following periodicity:

$$t = \left(\frac{(2 \cdot n + 1)}{4}\right) \cdot Tw + \frac{Tw}{2\pi} \cdot \beta \quad (67)$$

This interesting conclusion expressed in Equation (67) can be corroborated in Figure 2, where it was represented by the rolling motion under the influence of the 21 types of waves

(Tw). For clarifying purposes, the representation was divided into two graphs, i.e., from Tw = 1 s to Tw = 10 s (Figure 2a), and from Tw = 11 s to Tw = 21 s (Figure 2b). Furthermore, it is observed that, with Tw = 8.99 s, it theoretically produced the synchronism phenomenon (Td = Tw), with the appearance of asymptotic curves. However, another relevant result is that the amplitude of rolling for Tw = 8 s and Tw = 10 s is not the same, even though both are 1 s apart from the Tw = 9 s.

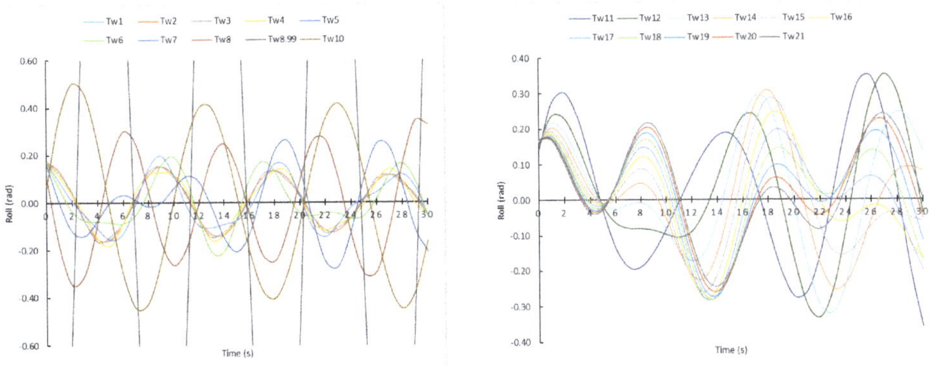

(**a**) (**b**)

Figure 2. Rolling motion induced by pure beam seas with resistance and with the ship in static condition. (**a**) Tw from 1 s to 10 s. (**b**) Tw from 11 s to 21 s.

Moreover, although the relative maximum rolling angle is reached according to our results expressed in Equation (67), it is also noted that the maximum amplitude does not always have the same value, and it varies as time increases. The reason for this is that the amplitude of roll depends on cos β in such a way that, as β increases, the roll angles decrease, being consistent with the water resistance.

Another relevant observation is that the absolute maximum angle is not always reached in the first moments, as it can be deduced from, for example, Tw = 11 s and Tw = 12 s. That is to say, maybe that master would have enough time to act and then, improve the ship's behavior, for example, modifying the Td, i.e., ballasting, deballasting or transferring the ballast vertically.

3.2. *Rolling Motion in Underway Conditions and under the Influence of Waves of Any Constant Direction*

In this subsection the ship's behavior in rolling motions closest to the reality experienced by a ship during the sea passage is studied, i.e., for a given ship's speed and heading. The last component, together with the waves' influence direction, defines the angle α, i.e., the angle formed between the ship's heading and the waves influencing direction.

Once analyzed in deep the Equation (62), it is observed that the first term is related with the damped factor rolling, and then that it will decrease geometrically ($e^{\frac{-\lambda_1 \cdot t}{Td}}$) as time (t) goes on. In consequence, over time, the roll motion will be determined by the second term of Equation (62), i.e., following to encounter period (Te) and the angle α of influence of the waves. According to this reasoning, and considering the ship's speed and heading as constant, the maximum roll angle is reached when:

$$t = \left(\frac{(2 \cdot n + 1)}{4}\right) \cdot Te + \frac{Te}{2\pi} \cdot \beta \quad (68)$$

In order to validate our conclusion mentioned in Equation (68), and according to Equation (62), it is necessary to establish the ship's speed (Vs) and the α angle. The Vs can be considered between 0–12 knots, and the α angle between 000° and 180°, regardless of whether the waves are received from port or starboard side. This is because the only difference is the rolling side, not the amplitude or the moments (t) when a certain roll angle is reached. As it is not practical to represent here all the graphs for all these combinations, in this paper the simulations carried out are shown, with a selected speed of 8 knots and an α angle of 45°, i.e., for receiving the waves by the bow. In Figure 3, the influence of all sea state conditions (21 types) is depicted in a ship with the parameters commented above. In this case, unlike those established in the previous subsection, the influence of waves is represented through Te instead of Tw. As it is shown, in all sea state conditions our Results expressed in Equation (68) are satisfied.

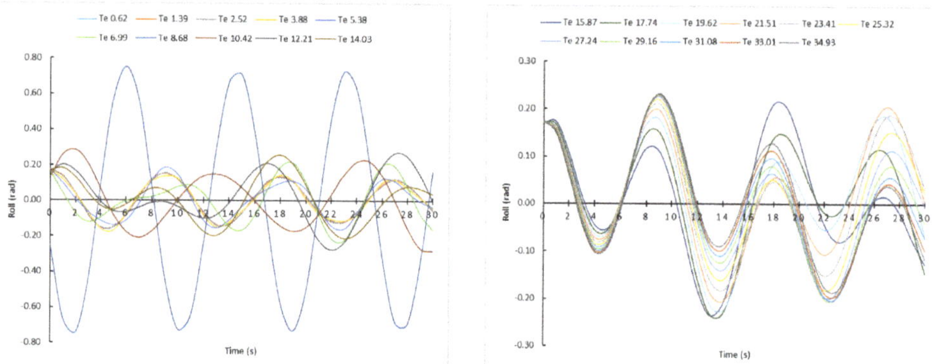

(**a**) (**b**)

Figure 3. Rolling motion sailing at 8.0 knots and induced by waves with resistance with an α angle of 45°. (**a**) Te from 0.62 s to 14.3 s. (**b**) Tw from 15.87 s to 34.93 s.

As it can be observed from Figure 3a, with a Te = 8.68 s, corresponding to Tw = 7 s, the ship would reach a roll angle of about 0.75 rad (43° approx.). However, no asymptotic curve corresponding to the synchronism phenomenon is observed, although theoretically it should be registered. For this reason, it is noted that, in real situations, Te must be analyzed in depth, far more so than Tw. This consideration is considered important because, although waves' characteristics found at sea cannot be modified, the ship's master can influence Te, altering the ship's speed and/or the heading.

In order to check more in detail what the worst Te, and in consequence the Tw, are to face the rolling motion, in Figure 4 the rolling motion was represented for Te = 8.85 s (Tw = 7.1 s); Te = 8.94 s (Tw = 7.15) and Te = 9.02 s (Tw = 7.2 s). For that, it was necessary to appeal to Equations (1)–(3) of waves, which were previously modelled. In Figure 4, it can be observed how the curves tend to be asymptotic, being a dangerous situation for the ship's safety, even though all the ship's parameters were in keeping with the optimal range. Therefore, with the results presented here, it is demonstrated that it is necessary to carry out a more detailed analysis of each sea state condition in order to not compromise the ship's stability. Then, if ships masters pay attention only to the categorization of the sea state according to the wave period (Tw) typical from the weather forecast, for example, they can think that the ship is sailing in a more or less safe situation, when this is not the case.

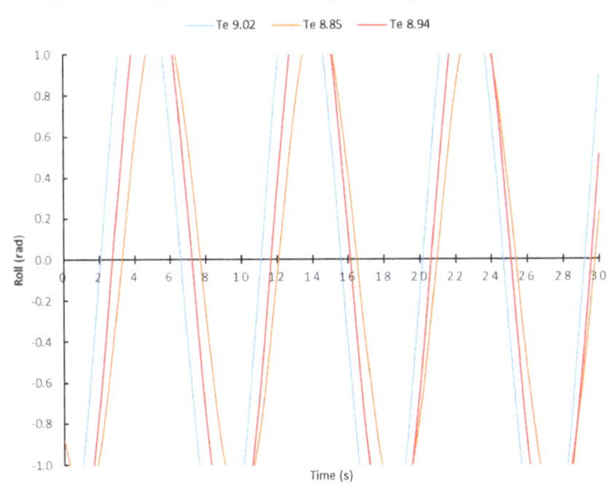

Figure 4. Rolling motion sailing at 8.0 knots and induced by waves with resistance (Tw = 7.1 s; Tw = 7.15 s and Tw = 7.2 s) with an α angle of 45°.

4. Discussion

In our simulations, 10° was considered to be the initial maximum angle of roll. If we consider that initially the ship is in the upright position it is relevant to note that, in simulations with the ship in a static position and between pure beam seas, as it can be observed in Figure 5, for Tw = 8 and 10 s the amplitude of the maximum rolling angle is lower than when the initial maximum roll is 10°. In the rest of the wave conditions (Tw), the amplitude of the rolling angle is lower with the ship initially in an upright position. Therefore, depending on which position is taken as the origin of time (t), the obtained results can be surprising, because as it is shown, it would be better that initially the ship has a maximum roll angle of 10° that it would be upright position, something that goes against good seamanship.

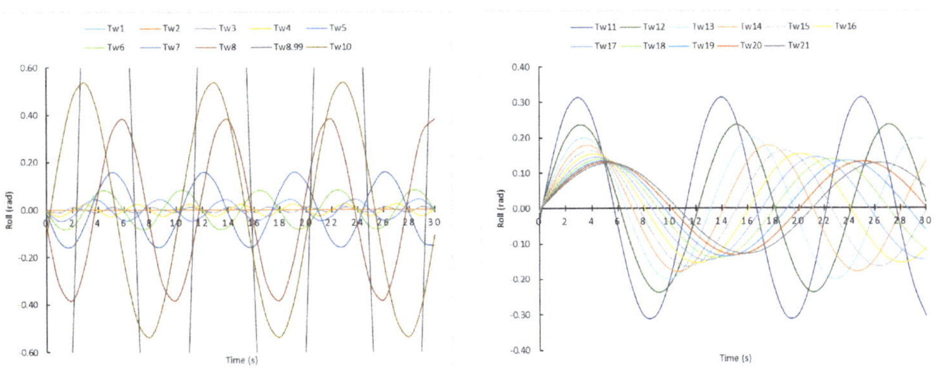

(**a**) (**b**)

Figure 5. Rolling motion induced by pure beam seas with resistance and with the ship in static and upright positions (Td = 9 s). (**a**) Tw from 1 s to 10 s. (**b**) Tw from 11 s to 21 s.

In dynamic conditions, with respect to the parameters simulated in our research referred to real conditions, i.e., Vs = 8.0 knots, α = 45°, θ_{MAX} = 10° and Td = 9 s, if Td is altered from 9 s to 8 s and 10 s, i.e., the closest loading conditions which can be reached most easily during sea passage altering weights, relevant conclusions are obtained too.

As it can be observed in Figure 6, with Td = 8 s, the amplitude of maximum roll angle is greatly reduced almost in half for Te = 8.68 s (Tw = 7 s), i.e., for the same sea conditions that with Td = 9 s. It is evident that if the ship's operators are alert well in advance, the fact of reducing 1 s the Td, can suppose avoid very dangerous situations or even the ship's capsizing. In the other Te conditions, the amplitude of angle rolling is almost the same.

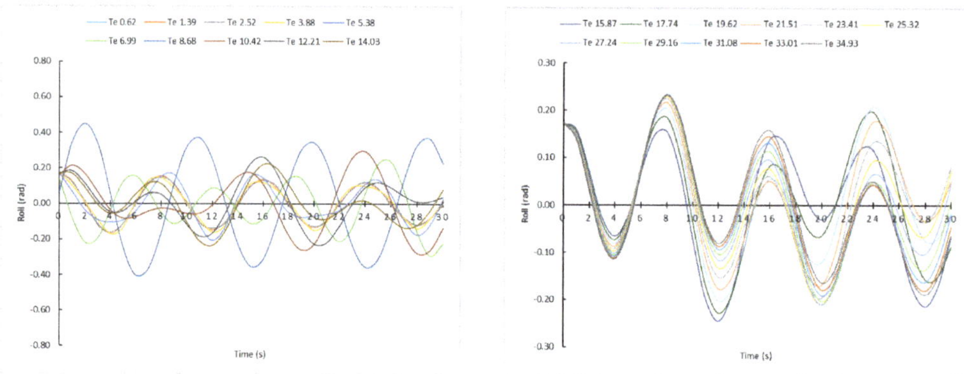

(a) (b)

Figure 6. Rolling motion sailing at 8.0 knots and induced by waves with resistance with an α angle of 45°. Td = 8 s. θ_{MAX} = 10°. (**a**) Te from Te 0.62 s to 14.03 s. (**b**) Te from Te 15.87 s to 34.93 s.

However, if simulations are carried out with Td = 10 s, the most important variation is the Te (Tw) condition in which the maximum amplitude of roll is reached, in this case Te = 10.42 (Tw = 8 s). For other waves' conditions, as it can be noted in Figure 7, the amplitude of maximum roll angle hardly changes, except obviously the moment (t) when these relative maximum roll angles are reached, as was concluded previously in the Results section.

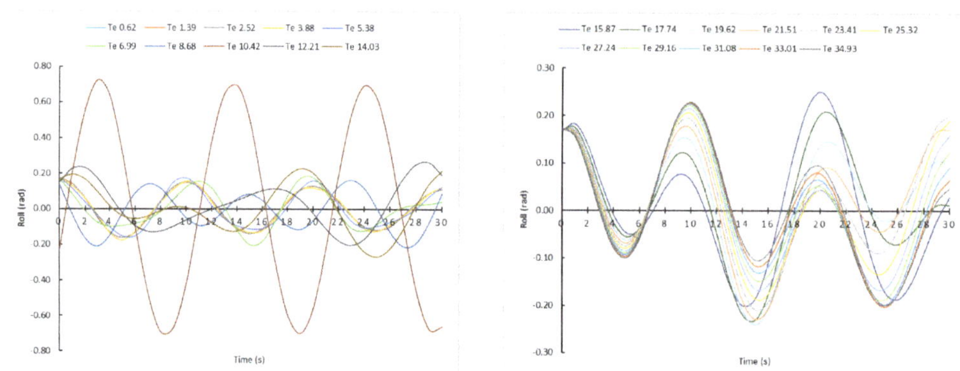

(a) (b)

Figure 7. Rolling motion sailing at 8.0 knots and induced by waves with resistance with an α angle of 45°. Td = 10 s. θ_{MAX} = 10°. (**a**) Te from Te 0.62 s to 14.03 s. (**b**) Te from Te 15.87 s to 34.93 s.

As is shown in Figure 8, if the ship's operator decides to change the ship's heading so that an α angle of 120° is reached, the ship's behavior and the maximum rolling angles are substantially improved. There are hardly any significant variations between the different sea state conditions.

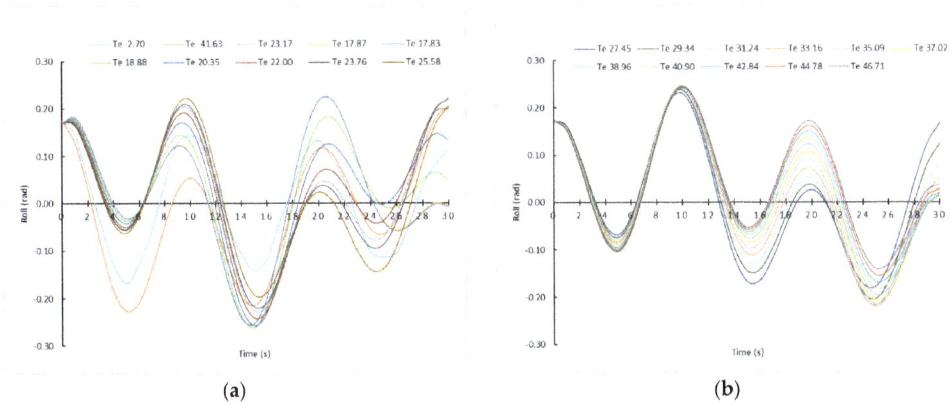

Figure 8. Rolling motion sailing at 8.0 knots and induced by waves with resistance with an α angle of 120°. Td = 9 s. θ_{MAX} = 10°. (**a**) Te from Te −2.70 s to 25.58 s. (**b**) Te from Te 27.45 s to 46.71 s.

Once again, it is clear the importance for ship's operators not only to control the ship's loading conditions to improve the ship's behavior sailing in waves, but also to control the waves' parameters and the navigation parameters as the ship's speed and heading, which are included in Te variable. Furthermore, sailing in waves conditions, it is shown again that it is more important to verify Te than Tw, with the consequence that maybe Tw lower implies a worse ship's behavior.

It is worth mentioning that, sailing in longitudinal waves (α =000° or 180°), theoretically no rolling motion is registered, so it is an option that the master should be in mind. However, in bad weather conditions receiving the waves exactly by the bow or astern are not recommended for safety of seakeeping conditions due to different factors as slamming, propeller's emergence and others.

Finally, in Figure 9 a schematic diagram of waves with period of 12 s (Tw), which corresponds with a height of 7.42 m and a ship with a loading condition of Td = 9 s. Over the wave profile, it is shown the maximum rolling angle and the moment (t), when they are reached, and in different situations as are the follows:

- In static conditions, under a beam wave's influence without resistance (blue colour).
- In static conditions, under the abeam waves influence with resistance (green colour).
- In dynamic conditions, sailing with a ship's speed of 8 knots and an α angle of 45° (orange colour).
- In dynamic conditions, sailing with a ship's speed of 4 knots and an α angle of 120° (yellow colour).

In this Figure 9, it is graphically shown that, in dynamic conditions, in some cases, the fact of altering substantially the combination between heading and ship's speed there hardly causes differences in the maximum rolling angle reached, but the time to reach this rolling angle changes considerably. Furthermore, it is also relevant that the maximum rolling angle in static conditions and realistic conditions (with resistance), under the influence of pure beam seas, is reached at 27 s, i.e., quite some time after starting the rolling motion, unlike what happens in dynamic conditions and in theoretical conditions (pure beam seas without resistance).

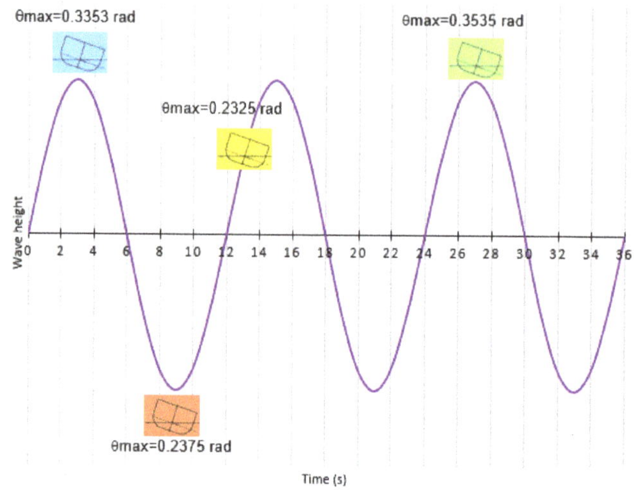

Figure 9. Maximum rolling angles and the time (t) when they are reached under the influence of waves Tw = 12 s in different supposed conditions (Td = 9 s).

5. Conclusions

In this research, the ship's behavior was studied and modeled in rolling motion at zero-speed conditions with beam seas, as well as in underway conditions receiving the waves from any direction. Once the corresponding models had been obtained, it was concluded that as time goes on, in static conditions, the ship will roll according to waves' period (Tw) instead of its natural roll period (Td), and that in dynamic conditions, the ship will roll according to the encounter period (Te) instead of Td or Tw. As in the Te value, the heading and ship's speed and included, and it is essential that ship operators have under control the influence of Te in the ship's behavior. What is more, these results permit vessel operators to know how they must proceed in order to change heading and/or ship's speed to improve the ship's behavior, from a scientific point of view and without leaving these changes at discretion of master experience. This is required because the watch officer can be a novice operator without the necessary experience to face adverse conditions. Furthermore, given that the particular sea state conditions are imposed during the sea passage, it of the utmost importance for ships' operators to know the navigational parameters (heading and speed) or the loading conditions to be set, modified or avoided. Thus, using the obtained models several simulations were carried out in order to obtain relevant results. For a given loading condition (Td), the available time (t) before reaching the maximum rolling angle that ships' operators have in order to act and thus change the ship's behavior was obtained. This is a direct function, in static conditions, of waves' period (Tw), expressed as $t = \left(\frac{(2\cdot n+1)}{4}\right)\cdot Tw + \frac{Tw}{2\pi}\cdot\beta$ and in dynamic conditions, of the encounter period (Te), shown as $t = \left(\frac{(2\cdot n+1)}{4}\right)\cdot Te + \frac{Te}{2\pi}\cdot\beta$. Knowing the available time (t) before reaching the relative maximum angle is essential in order to alter the navigational parameters (heading and speed) or even the loading condition, ballasting, deballasting or transferring ballast. It must be taken into account that reaching high rolling angles compromises the safety of the ship, crewmembers and cargo, and can even produce capsizing or reaching a downflooding angle.

The action of change of heading and ship's speed is a usual procedure followed by ship masters in order to improve the ship's behavior when sailing in rough seas. However, these changes are left in many cases to the discretion of master experience or based on empirical facts experienced in real time on which the master adjusts according to ship's performance, i.e., without considering an objective and scientific analysis. Through the case studies included in the Discussions section, it was demonstrated, from a scientific point of view, that through modifying the loading condition or altering the heading, the ship's behavior improves and the maximum relative rolling angle reached is reduced very considerably. However, many of the obtained results are not within what is normally expected in the ship's behavior by a sailor. These results show the importance for ships operators of of knowing the available time to act accordingly in stable conditions to improve the ship's behavior, both to potentially avoid a very dangerous situation, or what is more usual and faster during sea passage, that is, alter the heading and/or the speed.

Finally, future research can be guided towards validating these theoretical results in a scale model test in a seakeeping basin or perform simulations by means of a computer program and in different vessel configurations.

Author Contributions: Conceptualization, J.M.P.-C. and J.A.O.; methodology, J.M.P.-C., J.A.O., J.J.C.B. and M.I.L.G.; validation, J.M.P.-C., J.A.O., J.J.C.B. and M.I.L.G.; formal analysis, J.M.P.-C., J.A.O., J.J.C.B. and M.I.L.G.; investigation, J.M.P.-C., J.A.O., J.J.C.B. and M.I.L.G.; data curation, J.M.P.-C., J.A.O., J.J.C.B. and M.I.L.G.; writing—original draft preparation, J.M.P.-C. and J.A.O.; writing—review and editing, J.M.P.-C., J.A.O., J.J.C.B. and M.I.L.G. All authors have read and agreed to the published version of the manuscript.

Funding: This research received no external funding.

Institutional Review Board Statement: Not applicable.

Informed Consent Statement: Not applicable.

Data Availability Statement: Not applicable.

Conflicts of Interest: The authors declare no conflict of interest.

Appendix A

Table A1. Symbols and Its Definitions.

Symbol	Definition
Lw	Wavelength
Tw	Wave period
Vw	Wave translation velocity
θ_{MW}	Maximum wave slope
Hw	Wave height
Aw	Wave amplitude
Ma	Ship's righting torque
θ_W	Wave slope
θ	Initial roll angle
D	Ship's displacement
GM	Transverse metacentric height
Ig	Inertia moment of the ship's mass
A_R	Damping coefficient
2λ	Relation between A_R and Ig
g	Gravity acceleration
k	Turning radius of the ship's mass
ω	Ship's natural roll frequency
ω_W	Wave frequency
E_K	Constant, particular solution

Table A1. *Cont.*

Symbol	Definition
F_K	Constant, particular solution
β	Relation (-) between F_K and E_K
λ	Rolling damping factor
λ_1	$Td \cdot \lambda$
θ_M	Maximum initial rolling angle
Ve	Encounter velocity
Vs	Ship's velocity
α	Angle formed between heading and wave influence direction
Te	Encounter period
L_{ABEAM}	Length of transversal wave component
$\theta_{MW\ ROLLING}$	Transversal maximum wave's slope
B	Ship's beam

References

1. EQUASIS. The 2020 World Merchant Fleet Statistics from Equasis. Available online: http://www.equasis.org/EquasisWeb/public/HomePage (accessed on 2 September 2022).
2. Alvite-Castro, J.; Orosa, J.A.; Vergara, D.; Costa, A.M.; Bouzón, R. A New Design Criterion to Improve the Instact Stability of Galician Small Fishing Vessels. *J. Mar. Sci. Eng.* **2020**, *8*, 499. [CrossRef]
3. Secretariat-General for Fisheries; Ministry of Agriculture, Fisheries and Food; Spain Government. 2021 Annual Report on the Activity of the Spanish Fishing Fleet. Available online: https://oceans-and-fisheries.ec.europa.eu/system/files/2021-09/2020-fleet-capacity-report-action-plan-spain_en.pdf (accessed on 1 September 2022).
4. Tello, M.; Ribeiro e Silva, S.; Guedes Soares, C. Seakeeping performance of fishing vessels in irregular waves. *Ocean Eng.* **2011**, *38*, 763–773. [CrossRef]
5. Mata-Álvarez-Santullano, F.; Souto-Iglesias, A. Stability, safety and operability of small fishing vessels. *Ocean Eng.* **2014**, *179*, 81–91. [CrossRef]
6. Lapa, K.N. Estimation of Stability for a Fishing Vessel and some Considerations. *IOP Conf. Ser. Mater. Sci. Eng.* **2018**, *400*, 82014. [CrossRef]
7. European Maritime Safety Agency. Annual Overview of Marine Casualties and Incidents 2021. Available online: https://www.emsa.europa.eu/publications/reports/item/4266-annual-overview-of-marine-casualties-and-incidents-2020.html (accessed on 11 October 2022).
8. Woo, D.; Im, N.-K. A Methodology for Simply Evaluating the Safety of a Passenger Shp Stability Using the Index for the Intact Stability Appraisal Module. *Sensors* **2022**, *22*, 1938. [CrossRef] [PubMed]
9. Mantari, J.L.; Ribeiro e Silva, S.; Guedes Soares, C. Intact stability of fishing vessels under combined action of fishing gear, beam waves and wind. *Ocean Eng.* **2011**, *38*, 1989–1999. [CrossRef]
10. Míguez González, M.; Bulian, G. Influence of Ship Dynamics Modelling on the Prediction of Fishing Vessels Roll Response in Beam and Longitudinal Waves. *Ocean Eng.* **2018**, *148*, 312–330. [CrossRef]
11. Santiago-Caamaño, L.; Míguez-González, M.; Díaz Casás, V. On the Feasibility of a Real Time Stability Assessment for Fishing Vessels. *Ocean Eng.* **2018**, *159*, 76–87. [CrossRef]
12. Santiago-Caamaño, L.; Galeazzi, R.; Nielsen, U.D.; Míguez-González, M.; Díaz Casás, V. Real-time Detection of Transverse Stability Changes in Fishing Vessels. *Ocean Eng.* **2019**, *189*, 106369. [CrossRef]
13. Woo, D.; Choe, H.; Im, N.-K. Analysis of the Relationship between GM and IMO Intact Stability Parameters to Propose Simple Evaluation Methodology. *J. Mar. Sci. Eng.* **2021**, *9*, 735. [CrossRef]
14. Spandonidis, C.C.; Spyrou, K.J. Coupled vessel-dry-granular-cargo roll dynamics in regular beam seas. *Ocean Eng.* **2016**, *120*, 238–245. [CrossRef]
15. Maceiras, C.; Pérez-Canosa, J.M.; Vergara, D.; Orosa, J.A. Detailed Identification of Classificatory Variables in Ship Accidents: A Spanish Case Study. *J. Mar. Sci. Eng.* **2021**, *9*, 192. [CrossRef]
16. Míguez-González, M.; Caamaño-Sobrino, P.; Tedín-Álvarez, R.; Díaz-Casás, V.; Martínez-López, A.; López-Peña, F. Fishing vessel stability assessment system. *Ocean Eng.* **2012**, *41*, 67–78. [CrossRef]

17. European Maritime Safety Agency. Safety Analysis of Data Reported in EMCIP. Analysis on Marine Casualties and Incidents Involving Fishing Vessels. 2018. Available online: https://www.emsa.europa.eu/we-do/safety/accident-investigation/items.html?cid=141&id=3406 (accessed on 11 October 2022).
18. Pérez-Canosa, J.M.; Orosa, J.A.; Fraguela, F.; López-Varela, P. Proposal of Optimal Operation in Ship Rolling Motion Considering Sea State Conditions. *J. Mar. Sci. Eng.* **2022**, *10*, 669. [CrossRef]
19. Gourlay, T.; Lilienthal, T. Dynamic Stability of Ships in Waves. In Proceedings of the Pacific 2002 International Maritime Conferencia, Sydney, Australia, 29–31 January 2022.
20. Medina, M. *The Sea and the Weather. Nautical Meteorology*, 3rd ed.; Juventud: Barcelona, Spain, 2007.
21. Olivella-Puig, J. *Ship's Theory: Trochoidal Wave, Movements and Forces*, 2nd ed.; Universitat Politécnica de Catalunya: Barcelona, Spain, 2011.
22. Bonilla de la Corte, A. *Ship's Theory*, 4th ed.; Librería San José: Vigo, Spain, 1994.
23. DNV. Rules for Classification of Ships. Part 3. Chapter 1. Hull Structural Design—Ships with Length 100 Metres and Above. 2016. Available online: http://rules.dnvgl.com/docs/pdf/dnv/rulesship/2016-01/ts301 (accessed on 30 August 2022).
24. Míguez González, M.; Díaz Casás, V.; Santiago Caamaño, L. Real-Time Stability Assessment in Mid-Sized Fishing Vessels. In Proceedings of the 15th International Ship Stability Workshop, Stockholm, Sweden, 13–15 June 2016.
25. Míguez González, M.; Bulian, G.; Santiago Caamaño, L.; Díaz Casás, V. Towards real-time identification of initial stability from ship roll motion analysis. In Proceedings of the 16 International Ship Stability Workshop, Belgrade, Serbia, 5–7 June 2017.
26. Míguez González, M.; Santiago Caamaño, L.; Díaz Casás, V. On the applicability of real time stability monitoring for increasing the safety of fishing vessels. In Proceedings of the 13th International Conference on the Stability of Ships and Ocean Vehicles, Kobe, Japan, 16–21 September 2018.
27. Oliva-Remola, A.; Pérez-Rojas, L. Contributions on Roll Damping Coefficient for Fishing Vessels. In Proceedings of the 16th International Ship Stability Workshop, Belgrade, Serbia, 5–7 June 2017.
28. Marín-López, J.R.; Pinto-Loor, J.C. Coeficiente de amortiguamiento en balance del modelo de un buque pesquero con velocidad de avance. Global Partnerships for Development and Engineering Education. In Proceedings of the 15th LACCEI International Multi-Conference for Engineering, Education and Technology, Boca Ratón, FL, USA, 19–21 July 2017. [CrossRef]
29. Valle-Cabezas, J. Estudio teórico experimental de las no linealidades del amortiguamiento en el movimiento de balance de buques. Ph.D. Thesis, Universidad Politécnica de Madrid, Madrid, Spain, 1998.

Journal of
Marine Science and Engineering

MDPI

Article

Solitary Wave Interaction with a Floating Pontoon Based on Boussinesq Model and CFD-Based Simulations

Sarat Chandra Mohapatra *, Hafizul Islam, Thiago S. Hallak and C. Guedes Soares

Centre for Marine Technology and Ocean Engineering (CENTEC), Instituto Superior Técnico, Universidade de Lisboa, Av. Rovisco Pais, 1049-001 Lisboa, Portugal
* Correspondence: sarat.mohapatra@centec.tecnico.ulisboa.pt

Abstract: A mathematical model of solitary wave interaction with a pontoon-type rigid floating structure over a flat bottom is formulated based on Boussinesq-type equations under weakly nonlinear dispersive waves. Based on the higher-order Boussinesq equations, the solitary wave equation is derived, and a semi-analytical solution is obtained using the perturbation technique. On the other hand, brief descriptions of the application of wave2Foam and OceanWave3D on the aforementioned problem are presented. The analytical solitary wave profiles in the outer region are compared with Computational Fluid Dynamics (CFD) and OceanWave 3D model simulations in different cases. The comparison shows a good level of agreement between analytical, wave2Foam, and OceanWave3D. In addition, based on the wave2Foam and coupled OceanWave3D model, the horizontal, vertical wave forces, and the pressure distributions around the pontoon are analysed. Further, the effect of the Ursell number, pontoon length, and water depth on the solitary wave profiles are analysed based on the analytical solution. The paper validates each of the three models and performs intercomparison among them to assess their fidelity and computational burden.

Keywords: floating structure; solitary wave; semi-analytical solution; Boussinesq equations; CFD; OceanWave3D

Citation: Mohapatra, S.C.; Islam, H.; Hallak, T.S.; Soares, C.G. Solitary Wave Interaction with a Floating Pontoon Based on Boussinesq Model and CFD-Based Simulations. *J. Mar. Sci. Eng.* **2022**, *10*, 1251. https://doi.org/10.3390/jmse10091251

Academic Editor: Decheng Wan

Received: 22 July 2022
Accepted: 1 September 2022
Published: 5 September 2022

1. Introduction

In coastal and ocean engineering applications, nonlinear and shallow-water wave characteristics become important for coastal engineers in order to understand the influence of shallow water waves on nonlinear forms. One frequently used nonlinear water wave model is the solitary-wave model, in order to simplify some boundary value problems based on Boussinesq-type equations or nonlinear kind of equations associated without structures to analyse the characteristics and behaviour of solitary wave profiles by comparing various numerical wave models over flat or variable bottom topography (see [1–4]).

Due to the world's energy requirement for future activities, the global community has driven towards design techniques for the study of wave energy devices for energy recovery from renewable sources like waves. Apart from the analysis of various environmental and design conditions, nonlinear wave load analysis on floating structures performed based on various analytical methods is of importance for achieving a better understanding of nonlinear waves with structures. Therefore, various Boussinesq models have been proposed for analysing the mutual interaction of nonlinear waves and floating structures in different water depths applications to marine activities and wave energy converters (see [5–8]).

Another interesting aspect of the nonlinear or solitary wave is the interaction with floating or submerged structures based on analytical, numerical, and experimental studies. Some of the relevant studies are discussed. An integrated analytical–numerical model was presented in [9] to simulate solitary wave interaction with a fixed and partially immersed body in two dimensions, and experiments were conducted to examine the effects of wave amplitude, structural submergence, and structural width on the transmitted and reflected waves. The interactions between solitary waves and structures of single and twin rectangular cylinders were studied to analyse the hydrodynamic forces with different incoming wave amplitudes based on a finite element method (FEM) in two dimensions (see [10]). The FEM-based code has also been used in coastal engineering studies to investigate the solitary wave impact on other types of structures (see [11,12]). The shoaling of broad-crested internal solitary waves using immersed boundary method was studied in [13], and they concluded that different solitary wave breaking mechanisms are related to the internal Iribarren number.

On the other hand, several researchers have simulated the floating structures based on a numerical CFD model in the interaction of solitary waves using OpenFOAM solver. For instance, the impact of solitary waves on a simplified bridge deck model without girder in [14], whereas another bridge model with girders, both experimentally and with CFD simulations, was studied in [15]. Another influential experimental investigation using a 1:10th scale model of periodic wave impact on a bridge was conducted in [16]. Additionally, the analysis of the impact of the solitary wave on straight bridge decks was performed in larger-scale (1:5) experiments in [17,18]. A modification of the solitary wave generation method proposed by [19] was proposed by [20]. Lather, in ref. [21], confirmed that the method proposed by [19] works better than Goring's method (see [20]). However, their soliton generated in the experimental facility was off from the target by 10%. The solitary wave interaction with a vegetated platform breakwater in the framework of the 2D numerical model using a macroscopic approach based on OlaFlow solver was studied in [22]. The dispersive characteristics of solitary waves while propagating over a sloping area were studied in [23]. Although numerous previous studies have been performed for solitary waves, each of the studies had a different target, and the analysis of the results was performed based on the target. The study of the solitary wave interaction with a breakwater and the subsequent results remain to be investigated in detail.

In addition to analytical studies, a floating fixed box based on a CFD model was simulated and compared the results against analytical and experimental model results in different cases (see [24]). A Boussinesq model associated with wave diffraction by a floating cylinder and results were compared with CFD model simulations in [7,8]. On the other hand, very recently, a comparison of solitary waves over flat- and variable-bottom topographies was performed between Boussinesq model and CFD simulations in [4]. However, studies focusing on the interaction of solitary waves with a floating pontoon-type structure based on Boussinesq formulation are very few. The study of solitary wave interaction considering the viscous effect and their effect on load predictions are also not common.

Therefore, in the present paper, solitary wave interaction with a pontoon-type floating structure over the flat bottom is presented based on semi-analytical, CFD and coupled (OceanWave 3D) model simulations. The paper is organized as follows: Section 2 presents the mathematical model formulation based on Boussinesq equations, and a solitary wave solution in the outer region is obtained. Section 3 discusses the coupled model set up between CFD and OceanWave 3D, and their importance. Section 4 presents a comparison of the free-surface solitary wave profiles among analytical, CFD, and OceanWave 3D model simulations. Section 5 analyses the wave forces, pressure distributions around the floating pontoon, wave energy analysis and further, the effect of Ursell number, pontoon length, and water depth on the solitary wave profiles are presented from the analytical solution. Finally, some significant conclusions are discussed in Section 6.

2. Mathematical Model Based on Boussinesq Equations

The mathematical model of the aforemention model is in a three-dimensional Cartesian coordinate system (x^*, y^*, z^*), with x^*-y^* being the horizontal plane and z^* pointing downward. The fluid occupies the region $-h^* < z^* < 0$, $-\infty < x^* < \infty$ and a pontoon of length $2l^*$ is floating at $z^* = 0$ with draft d^* and is balanced by the neutral buoyance over water depth at $z = -h^*$, where '*' indicates a dimensional variable. Hence, the fluid region is divided into two regions: an interior region that corresponds to the area below the pontoon and an outer region (incident wave region) (see Figure 1). The velocity potential and the free surface elevation in the outer region are denoted by Φ^* and ζ^*, respectively. Further, the fluid is assumed to be inviscid, incompressible, and the motion is irrotational and simple harmonic in time with angular frequency ω. In the case of long waves, two important non-dimensional parameters characterize the problem: the nonlinearity parameter $\varepsilon = a_0/h_0$ and the dispersion parameter $\mu = h_0/\lambda$, where h is the length representative of the depth scaled by h_0, λ is the typical wavelength, and a_0 is the representative wave amplitude. The expressions for the non-dimensional quantities associated with the above variables are deferred here as it is quite common sense in the ocean engineering field.

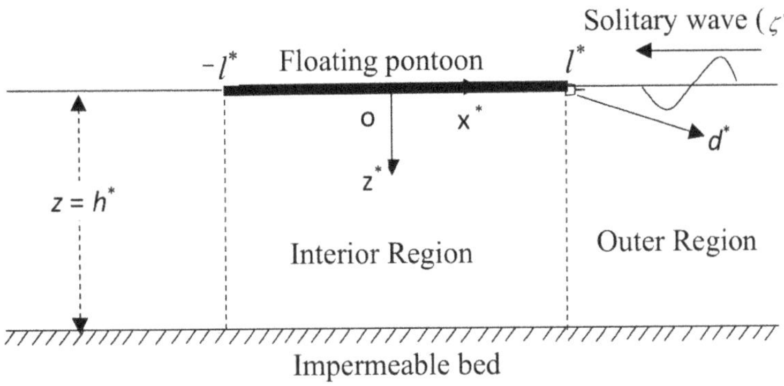

Figure 1. Schematic diagram: solitary wave interaction with a floating pontoon.

In this mathematical model, the right-hand referencing frame has its x-axis pointing in the positive (right) direction, the z-axis is pointing downward negative direction, and the y-axis is expanding laterally. The coordinate origin is located at the water level of the undisturbed fluid region. The non-dimensionalised boundary-value problem can be formulated as the governing equation and associated boundary conditions. These equations are listed as follows:

$$\mu^2\nabla^2\Phi + \Phi_{zz} = 0 \text{ for } -h \leq z \leq \varepsilon\zeta, \tag{1}$$

where ∇^2 is the two-dimensional Laplacian operator in the horizontal plane.

The non-dimensional kinematic conditions in the outer region (free surface) region is given by

$$\Phi_z = \mu^2(\zeta_t + \varepsilon\nabla\Phi\nabla\zeta) \text{ at } z = \varepsilon\zeta. \tag{2}$$

The non-dimensional dynamic condition is read as:

$$\Phi_t + g\zeta + \frac{1}{2}\varepsilon|\nabla\Phi|^2 + \frac{U_r}{2}\Phi_z^2 = 0 \text{ at } z = \varepsilon\zeta,\ x \in (l, \infty), \tag{3}$$

where $U_r = \varepsilon/\mu^2$ is known as the Ursell number (as in [7]). The boundary condition on the pontoon cover yields

$$\Phi_z = 0 \text{ at } x \in (-l, l), z = d, \tag{4}$$

where the subscripts 'z' and 't' are denote to the partial derivatives with respect to z and t, respectively.

Next, the bottom boundary condition at $z = -h$ is given by

$$\Phi_z = 0 \ \text{ at } \ z = -h. \tag{5}$$

Now, integrating Equation (1) w. r. t. z and using boundary conditions (2) and (4), the following is obtained:

$$\zeta_t = -\nabla.\left[\int_{-h}^{\varepsilon\zeta} \nabla.\Phi dz\right] \text{ for } x \in (l, \infty). \tag{6}$$

The next subsections will present the governing equation and boundary conditions in the interior region and solitary wave solutions in the outer region.

2.1. Governing Equations of the Interior Region and Matching Conditions

For the interior region beneath the pontoon, as shown in Figure 1, the velocity potential $\Phi^I(x, z; t)$ satisfies the Laplace equation

$$\left(\frac{\partial^2}{\partial x^2} + \frac{\partial^2}{\partial z^2}\right)\Phi^I = 0 \text{ on } -h \le z \le -d. \tag{7}$$

The kinematic boundary conditions underneath the pontoon can be applied as follows:

$$\frac{\partial \Phi^I}{\partial z} = 0 \text{ at } z = -d. \tag{8}$$

The matching conditions applied to the velocities and velocity potentials at the interface $x = l$ between the outer and interior fluid domains, as:

$$\frac{\partial \Phi^O}{\partial x} = 0 \text{ for } -d \le z \le \varepsilon\zeta. \tag{9}$$

where Equation (9) holds because the wave cannot overtop the pontoon.

$$\Phi^O = \Phi^I \text{ for } -h \le z \le -d, \tag{10}$$

$$\frac{\partial \Phi^O}{\partial x} = \frac{\partial \Phi^I}{\partial x} \text{ for } -h \le z \le -d. \tag{11}$$

The next subsection derives the solitary waveform in the outer region based on second-order approximation and using a depth-integrated approach.

2.2. Velocity Potential in Interior Region

Based on the method of separation of variables, the analytical solution of the velocity potentials in the interior region is expressed by solving the Laplace Equation (7) along with the boundary conditions (4) and (8) can be expressed as:

$$\Phi^I(x, z; t) = \sum_{n=0}^{\infty} \{A_n \cosh(\beta_n x) + B_n \sinh(\beta_n x)\} \cos\beta_n(z+h), \tag{12}$$

where $\beta_n = n\pi/(h-d)$ and $\cos\beta_n(z+h)$ are the eigenfunctions and are orthogonal over the interval $-h \le z \le -d$.

2.3. Solitary Waveform ζ in Outer Region

The velocity potential $\Phi(x, y, z; t)$ is expanded in powers of μ^2 as follows:

$$\Phi(x, y, z; t) = \sum_{n=0}^{\infty} \mu^{2n} \frac{(z+h)^{2n}}{(2n)!} \nabla_n^2 \varphi_n(x, y; t) \text{ for } x \in (l, \infty), \tag{13}$$

where $\Phi(x, y, z; t)$ satisfies Equations (1) and (5). The higher-order Boussinesq equations are obtained by substituting Equation (13) into Equations (2) and (3) and on simplification and taking into account terms up to the order $\varepsilon^2, \varepsilon\mu^2, \mu^4$, as follows:

$$
\begin{aligned}
\nabla^2\varphi - \varphi_{tt} = \; & \tfrac{\varepsilon}{2}(\nabla\varphi)_t^2 + \varepsilon(\nabla\varphi)(\nabla\varphi_t) + \varepsilon\varphi_t\nabla^2\varphi + \tfrac{\mu^2}{6}\nabla^4\varphi - \tfrac{\mu^2}{2}\nabla^2\varphi_{tt} + \tfrac{\varepsilon^2}{2}(\nabla\varphi)(\nabla(\nabla\varphi)^2) \\
& + \tfrac{\varepsilon^2}{2}\nabla^2\varphi(\nabla\varphi) + \tfrac{\varepsilon\mu^2}{2}\left\{(\nabla\varphi)^2 - \varphi_t\nabla^2\varphi_t - (\nabla\varphi)(\nabla\nabla^2\varphi)\right\}_t - \tfrac{\varepsilon\mu^2}{2}(\nabla\nabla^2\varphi)(\nabla\varphi_t) \\
& - \tfrac{\varepsilon\mu^2}{2}(\nabla\nabla^2\varphi_t)(\nabla\varphi) - \tfrac{\varepsilon\mu^2}{2}(\nabla^2\varphi)(\nabla^2\varphi_t) + \tfrac{\varepsilon\mu^2}{2}\varphi_t\nabla^4\varphi + o(\varepsilon^2, \varepsilon\mu^2, \mu^4), \text{ for } x \in (l, \infty),
\end{aligned} \tag{14}
$$

$$
\begin{aligned}
\zeta = \; & \varphi_t + \tfrac{\mu^2}{2}\nabla^2\varphi_t + \nabla^2.\varphi_t + \tfrac{\varepsilon}{2}(\nabla.\varphi)^2 - \tfrac{\varepsilon\mu^2}{2}\left\{(\nabla.\varphi)^2 - (\nabla.\varphi)(\nabla\nabla^2.\varphi) - \varphi_t\nabla^2\varphi_t\right\} \\
& + o(\varepsilon^2, \varepsilon\mu^2, \mu^4), \text{ for } x \in (l, \infty)
\end{aligned} \tag{15}
$$

Now, the solitary waveform (ζ) and velocity potential φ are the solutions of the above Equations to the lateral boundary at $x = l$ ($l < x < \infty$ and left going) can be expressed as:

$$
\zeta(x;t) = \frac{3}{4}\left\{\left(\frac{4}{3} - \varepsilon\right)\text{sech}^2\alpha(x - ct - l) + \varepsilon\text{sech}^4\alpha(x - ct - l)\right\}, \tag{16}
$$

$$
\varphi(x;t) = \frac{\text{tanh}\alpha(x - ct - l)}{\alpha c} + \varepsilon\text{tanh}\alpha(x - ct - l)\left[\frac{1}{3}\text{tanh}^2\alpha(x - ct - l) - \frac{1}{4}\right], \tag{17}
$$

where $c = (1/2)(2 + \varepsilon)$ is the wave celerity and $\alpha = (\sqrt{3}/16)(8 - 5\varepsilon)\sqrt{U_r}$.

3. Description of Numerical Models

In this section, the brief descriptions of the numerical models based on CFD and OceanWave3D along with their model set up for the solitary wave interaction with a fixed floating pontoon are presented in the subsequent sections.

3.1. Brief Description of the CFD Model

To run the CFD simulations, an open-source CFD toolkit, OpenFOAM, was used. OpenFOAM is an open-source library that numerically solves a wide range of problems in fluid dynamics from laminar to turbulent flows, with single and multi-phases. The solver has been elaborately described by [25]. The governing equations for the solver are the momentum or the Navier–Stokes equation for incompressible two-phase flow (18) and the continuity Equation (19). In vector form, the Navier–Stokes and continuity equations are given by:

$$
\rho\left(\frac{\partial v}{\partial t} + v.\nabla v\right) = -\nabla p + \mu\nabla^2 v + \rho g, \tag{18}
$$

$$
\nabla.v = 0, \tag{19}
$$

where v is the velocity, p is the pressure, μ is the dynamic viscosity, and g is the acceleration due to gravity.

The volume of fluid (VOF) method is used to model the fluid as one continuum of mixed properties. This method determines the fraction of each fluid that exists in each cell, thus tracking the free surface elevation. The volume fraction is obtained by:

$$
\frac{\partial\alpha}{\partial t} + \nabla.(\alpha v) = 0 \tag{20}
$$

where α is the volume fraction of water in the cell, varying from 0 to 1, with 0 representing a cell full of air and 1 representing a cell full of water. The Finite Volume Method (FVM) is used to discretize the governing equations. Pressure–velocity coupling is obtained through the PIMPLE algorithm. The PIMPLE algorithm can be used to ensure multiple iterations between the flow parameters (outer correctors) to produce a stable and reliable solution without unwanted dissipation. An adaptive time-stepping approach was used to ensure a constant Courant–Friedrichs–Lewy (CFL) number during the simulation. All interpolation schemes in OpenFOAM, used for flux calculation in advection, operate with the MULES (Multidimensional Universal Limiter for Explicit Solution)-based solver. For convection (divergence) and diffusion terms, the Gauss linear Upwind and Gauss linear schemes are used.

To numerically reproduce the benchmark cases, simulations were run using OpenFOAM version 2.4.0 (https://openfoam.org/version/2-4-0/), together with waves2Foam in [26] module for wave generation. For solitary wave generation, the solitary wave model from [27] was used within waves2Foam. All benchmark simulations were run in two dimensions, to follow the proposed case setup. Since the Reynolds number for the simulated cases is relatively low, the flow was assumed to be laminar, and therefore no turbulence model was considered. Several wave gauges were placed along the simulation domain to probe the wave elevation.

CFD Setup

The study involves a rectangular pontoon encountering a solitary wave in a two-dimensional (2D) wave flume. The case setup consists of a 150-m-long 2D wave flume with a constant still water depth of 1.0 m. The rectangular pontoon has a width of 4 m, a draft of 0.4 m and 0.2 m above water height. Two wave gauges are placed in the domain at positions $x = -31.0$ m and $x = 26.5$ m, assuming the box is located at the centre ($x = 0$) of the domain (as shown in Figure 2).

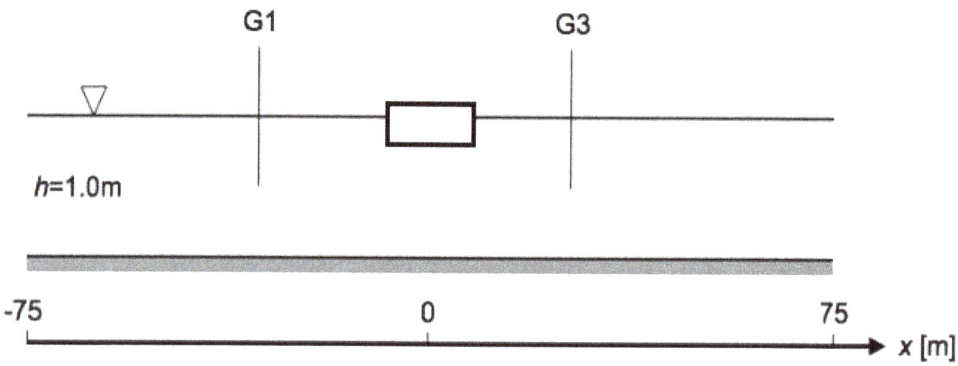

Figure 2. Schematic diagram of the case setup of the solitary wave scattering by a pontoon (not to scale).

To replicate the study numerically, simulations were run using OpenFOAM version 2.4.0, together with the waves2Foam module and OceanWave3D for wave generation. For solitary wave generation, initially, the wave model described in ref. [27] was used from waves2Foam. The simulations were run using 2D static mesh, to follow the proposed case setup. The flow was assumed to be laminar, and no turbulence model was used, considering the relatively low Reynolds number involved. Several wave gauges were placed along the simulation domain to probe the wave elevation.

The mesh domain for the case was generated following the case setup. For the first case, using the waves2Foam (see [27]) solitary wave model, the blockMeshDict was used to generate the simulation domain, and a single layer of refinement was applied using toposetDict to ensure proper propagation of the incoming wave. Finally, snappyHexMesh was used to integrate the rectangular pontoon into the simulation domain. The total resolution of the mesh was 3.1 million, with a spacing of 0.005 m in horizontal and vertical directions, in the wave propagation area. The resolution represents 200 cells per meter height and width of the pontoon. In terms of the soliton, a minimum of 20 cells per wave height was maintained. Several wave gauges were placed along the wave propagation area in the simulation domain to observe the wave propagation. A general mesh assembly used for the simulations is shown in Figure 3. A sufficiently large simulation domain was generated to ensure that on reflection is observed from the outlet of the domain.

The simulation was terminated after a steady free surface profile was observed after the interaction, which happened much earlier before the propagating wave could reach the outlet of the domain. A mesh dependency study for wave structure interaction was previously reported in [28], and thus was not repeated here.

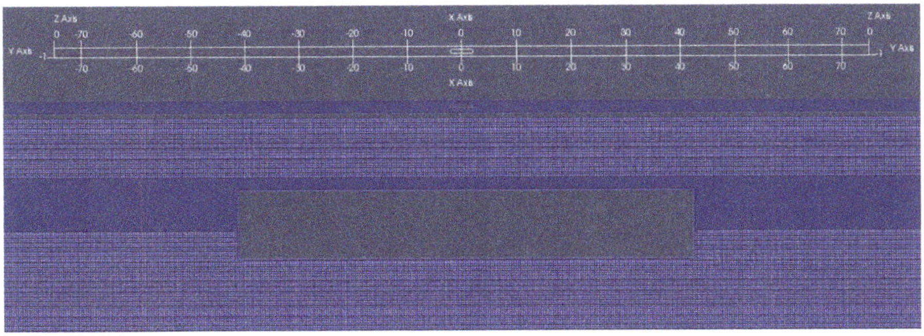

Figure 3. General mesh assembly for the study of solitary wave scattering by a pontoon—OpenFOAM mode.

3.2. *OceanWave3D Model Description*

Just like the Boussinesq-type equations, the Fully Nonlinear Potential Flow (FNPF) formulation used in the OceanWave3D (OW3D) is based on the incompressible NS equations with further assumptions of zero vorticity and zero viscosity. The equations of balance are solved for each point of the domain (3D), considering all the remaining linear and nonlinear terms of the obtained equation. Thus, the formulation handles more complex problems than Boussinesq models, but is less complex than CFD. The FNPF formulation is therefore very suitable for nonlinear gravity waves propagation, including shoaling, reflection and dispersion of waves, as long as the waves do not break or acquire vorticity. It is also considerably less time-consuming than CFD techniques.

Once the flow assumes a potential $\varphi(x, z; t)$, such that $V = (u, w) = (\nabla\varphi,\ \partial_z\varphi)$, where x is the horizontal position and $\nabla = (\partial_x, \partial_y)$ is the horizontal gradient operator, the kinematic and dynamic and kinematic boundary conditions at the free surface read:

$$\partial_t\zeta = -\nabla\zeta.\nabla\overline{\varphi} + \overline{w}(1 + \nabla\zeta.\nabla\zeta), \tag{21}$$

$$\partial_t\overline{\varphi} = -g\zeta - \frac{1}{2}\left\{\nabla\overline{\varphi}.\nabla\overline{\varphi} - \overline{w}^2(1 + \nabla\zeta.\nabla\zeta)\right\}, \tag{22}$$

where ζ and g are the same as defined in Section 3.1. The boundary conditions at the free surface and sea bottom read:

$$\varphi = \overline{\varphi} \ \text{ for } z = \zeta, \tag{23}$$

$$\partial_z\varphi = -\nabla h.\nabla\varphi \ \text{ at } \ z = -h, \tag{24}$$

and the Hamilton equation reads as:

$$\nabla^2\varphi + \partial_{zz}\varphi = 0 \text{ for } -h < z < \zeta. \tag{25}$$

For further details on the numerical scheme used to solve this nonlinear problem, the reader is referred to [29].

A coupled OW3D and OpenFOAM module is used to propagate the solitons over a fixed depth and simulate the interaction between a soliton and a fixed pontoon for varying soliton amplitudes. However, the solitons must be generated before being propagated. For that, the MATLAB® code presented in [30] was considered. It is important to mention that the FNPF one soliton propagating stably over fixed depth may be defined by: (1) water depth and soliton's amplitude or, (2) water depth and soliton celerity's. In non-dimensional values, a soliton may be simply defined by a single parameter, which may be the non-dimensional soliton amplitude (a/h) or the Froude-depth number $(F_n = c/\sqrt{gh})$.

OceanWave3D Setup

As for OceanWave3D, the coupling of the OW3D domain to that of the OpenFOAM happens at the inlet and the outlet of the OpenFOAM mesh domain. The inlet velocity boundary of the mesh domain is set to the *waveVelocity* received from the OW3D domain. The outlet boundary is also set to *waveVelocity*, which transfers the outward wave back to the OW3D domain. As before, the simulation solver is *waveFoam*, which solves static mesh with waves. The mesh was generated with *blockMeshDict* and *snappyHexMesh*, without any refinement. The total resolution used in the mesh was around 1.3 million, with grid size in the horizontal and vertical direction being 0.005 m and 0.005 m, respectively. The average time step size was 0.007 (s). The general mesh assembly used in the simulation is shown in Figure 4. Unlike the previous case, here, instead of placing the pontoon at $x = 0$, the pontoon was placed at $x = 100$. This was done to accommodate the OW3D domain, which starts at $x = 0$. The OW3D domain was larger by 50 m on each side (x-axis) to ensure the entry of a stable soliton at the inlet and no reflection from the outlet after solitary wave leaves the OpenFOAM domain after interaction.

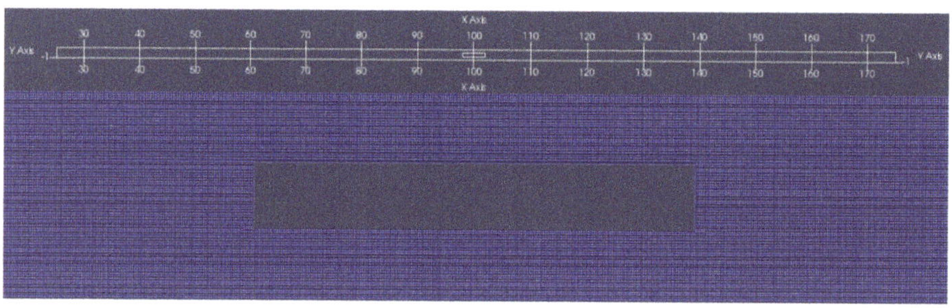

Figure 4. General mesh assembly for solitary wave scattering by a pontoon-coupled model with OW3D run.

4. Comparison of Wave Elevation from Different Models

Figure 5 shows comparisons of the solitary waveform ζ(m) in the outer region against CFD model simulations for different nonlinear quantities (a) $\varepsilon = 0.1$, (b) $\varepsilon = 0.3$, and (c) $\varepsilon = 0.5$ and (d) their comparison with an external numerical model (see [31]) versus time t (s). From Figure 5a,b, it can be observed that the peak amplitude of the solitary wave is in good agreement, and the solitary wave profile trend is similar to the CFD model results. However, the peak amplitude for $\varepsilon = 0.5$ CFD is higher than that of analytical results, but the trend is similar. The discrepancy in the results between the analytical and CFD simulations may be due to the consideration of second-order approximation in analytical and full nonlinear in the CFD model. Figure 5d shows that the analytical solitary wave profile is very close to the CFD and EFD model simulations. Further, in Figure 5c, the peak amplitude of the solitary wave from the analytical model is 0.5 m, while the CFD model is 0.56 m, which is 6% larger. However, the agreement of solitary wave profiles between analytical and CFD is almost 95%.

Figure 6 compares the results of a solitary wave from analytical and OceanWave3D model simulations in the outer region for different degrees of nonlinearity (a) $\varepsilon = 0.1$, (b) $\varepsilon = 0.3$, and (c) $\varepsilon = 0.5$ versus time t (s). In this case, it can be seen that the peak amplitudes between analytical and OceanWave3D results are in good agreement, and their profiles also show a similar trend. However, in Figure 6c there is a small discrepancy in their trends, maybe for a similar reason to that described in Figure 5.

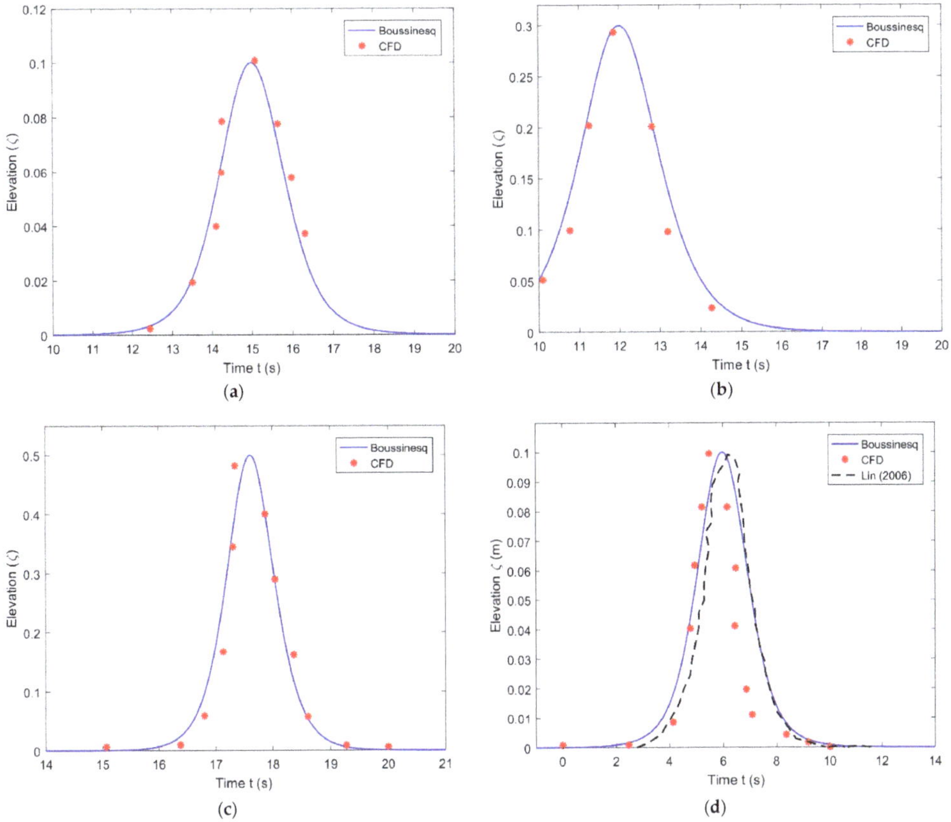

Figure 5. Comparisons of the solitary waveform in the outer region against CFD model simulations for different nonlinear quantities with (**a**) Comparison for $\varepsilon = 0.1$, (**b**) Comparison for $\varepsilon = 0.3$, (**c**) Comparison for $\varepsilon = 0.5$, and (**d**) Comparison: analytical, CFD, and external numerical model [31] (see [31]) versus time t (s).

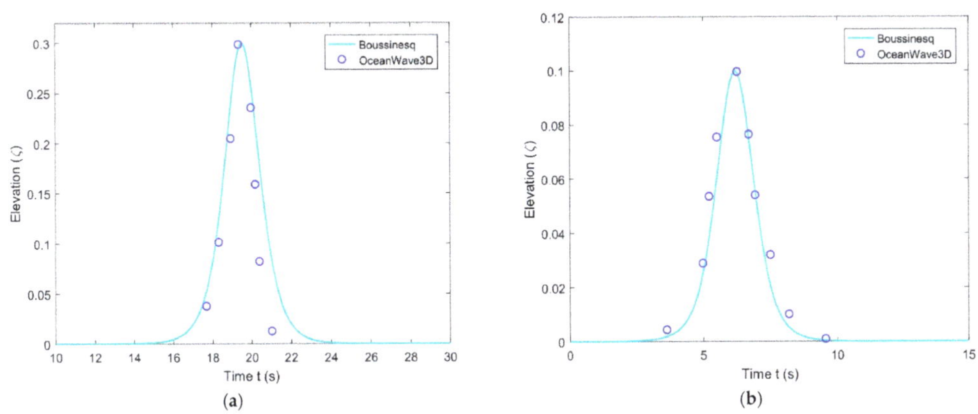

Figure 6. *Cont.*

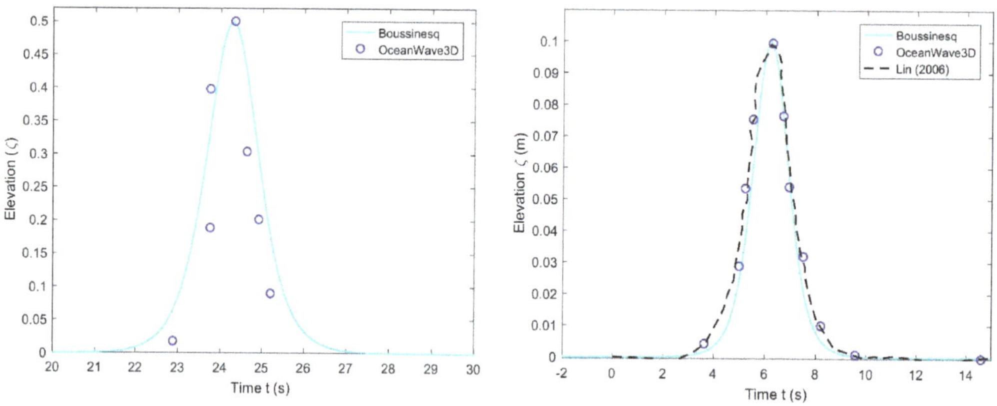

(c) (d)

Figure 6. Comparison of the solitary waveform between analytical and OceanWave3D model simulations with (**a**) Comparison for $\varepsilon = 0.1$, (**b**) Comparison for $\varepsilon = 0.3$, (**c**) Comparison for $\varepsilon = 0.5$ versus time t (s) and (**d**) comparison between analytical, CFD and external numerical model [31].

In Figure 6a–c, it is shown that the agreement of peak amplitude of the solitary waves between analytical and OceanWave3D is about 97%. On the other hand, Figure 6d demonstrates that the solitary wave profile between analytical, OceanWave3D, and Numerical model (NM) from other calculations (see [31]) are in good agreement.

5. Analysis of Results from CFD, OceanWave3D, and Analytical Model Simulations

In this section, the effect on the wave forces and pressure distributions around the pontoon for different $\varepsilon = a/h$ is analysed based on the CFD and OceanWave3D model simulations. Further, the effect of the Ursell number, pontoon length, and water depth on the solitary wave profiles are analysed based on the analytical solution. In the CFD study, six simulations are performed in total, with three different solitary wave amplitudes, using two different wave generation formulations. For initial validation of the results, simulations were performed for the non-dimensional solitary wave amplitude, $a/h = 0.1(0.1$ m) and compared with the results of [31]. Later, two more cases are simulated using the amplitude of 0.3 m $(a/h = 0.3)$ and 0.5 m $(a/h = 0.5)$.

The intention is to assess the effectiveness of a pontoon type wave breaker against solitary waves of different steepness. It may be mentioned that no experimental data are available for the case; however, several numerical studies have been performed previously with VOF-RANS (see [31]), σ-transformed NS (see [31]) and layer-averaged RANS (see [32]).

5.1. Solitary Waveform Generated Using the Chappelear Model (Waves2Foam)

The Chappelear model (see [30]) is a second-order solitary wave generation model that is used to generate and propagate the solitary wave in the domain. Initially, simulation is performed with amplitude 0.1 m so that the results can be compared with available data. During the simulation, wave elevations are measured at different positions of the domain. Measurements from all the gauges are shown in Figure 7. The gauge numbers indicate the location of the gauge in the domain. The results show that after interacting with the fixed pontoon, a portion of the solitary wave advances, while a reflected wave propagates backwards. The forwarding soliton propagates with roughly 0.06 m amplitude after the interaction. Meanwhile, the reflected wave propagates backwards with a crest and a trough of roughly 0.04 m.

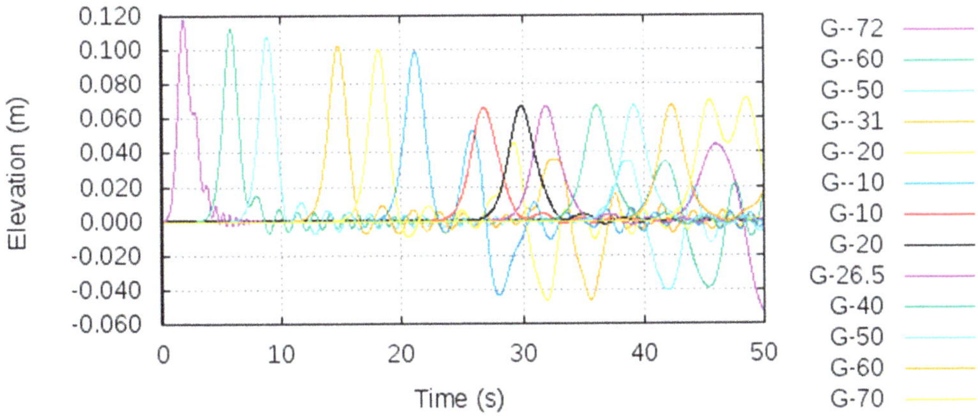

Figure 7. Free surface elevations measured by different wave gauges along the simulation domain (see [27], $a/h = 0.1$).

Figure 8 shows a relative comparison among the previous numerical results (see [31]) and the present simulation results for wave gauges G1 and G3. The figure shows simulation results for $a/h = 0.1$, 0.3, and 0.5, whereas Lin (see [31]) only presented results for $a/h = 0.1$. The timeline for the numerical study from Lin (see [31]) also differs from the simulation timeline of OpenFOAM, so the time values (x-axis) were shifted to bring the curves closer. A difference of 10 s was observed for soliton propagation comparing to Lin [26]. The difference most probably come from the wave initiation time and position, which might have varied for the two solutions. Generally, in RANS simulations, a time delay is applied on wave generation and propagation to ensre better simulation stability.

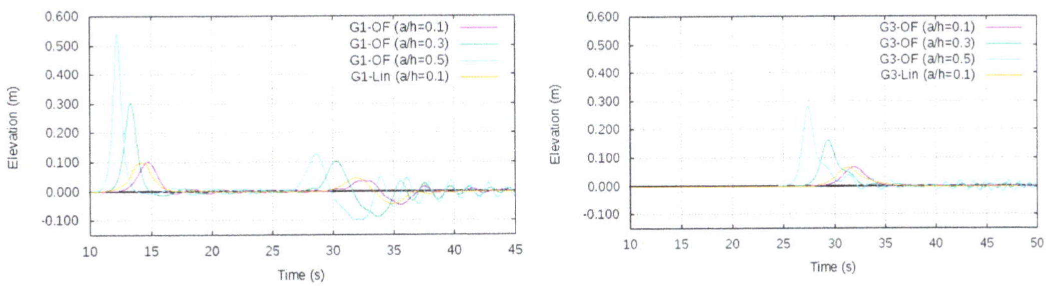

Figure 8. Relative comparison among the wave gauge results from the CFD simulations (OpenFOAM) for gauge G1 (**left**) and G3 (**right**) at $a/h = 0.1$, 0.3, and 0.5, and the results from [31] at $a/h = 0.1$.

From the simulations, wave elevation at gauge G1 and G3 are measured. The figure shows that the wave model generates the soliton with the right wave amplitude (gauge G1, at −31 m) of steepness 0.1. The figure shows good agreement among the results for steepness 0.1, for both the propagating soliton and the reflected waves. As for the amplitude of the propagating wave after interaction with the pontoon (gauge G3, at 26.5 m), the simulation results show slight over-prediction compared to previous numerical predictions. Nevertheless, the presented simulation results in Figure 5 show a good agreement with previous numerical data.

The results from the wave gauges (Figure 7) show that the soliton shows a dispersive tail during propagation and as a result, the free surface shows slight oscillation. This dispersive tail reduced the soliton energy slightly, causing its amplitude to reduce, over the period of propagation. The dispersive tail is not observed in the numerical results of [31], which might be due to the simplified assumptions for wave modelling. Experimental studies performed with solitary waves also show dispersive tail in the channel (as in [30]), which confirms that the observed scenario in the simulation is not unphysical.

The horizontal and vertical forces encountered in the pontoon are also measured during the wave encounters, and the results are shown in Figures 9 and 10. The forces are important for understanding the mooring force required to hold the pontoon in place or for studying the feasibility of installing a wave energy converter for dual purposes. However, the pontoon studied in this paper is fixed, and no mooring lines are used in the simulations. While inclusion of mooring lines would surely affect the forces and moments, the static results still provide a reference regarding the required mooring line properties.

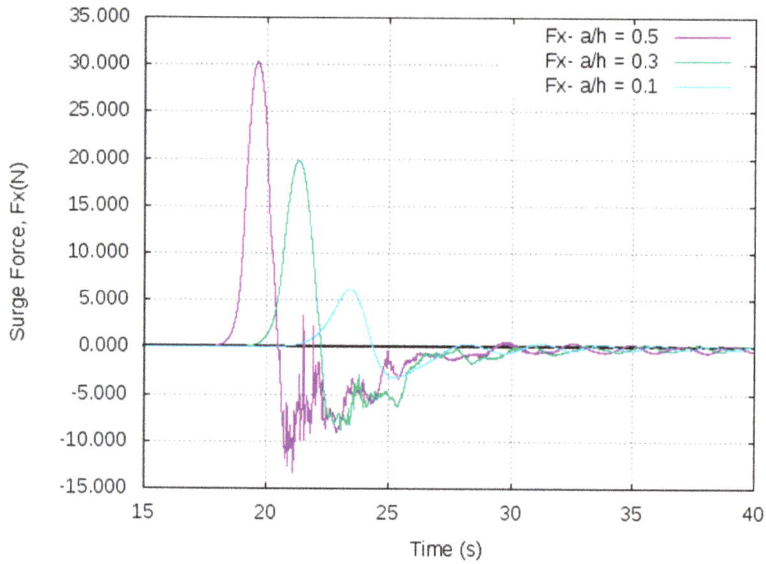

Figure 9. Surge or longitudinal force experienced by the pontoon while encountering solitary wave (for $a/h = 0.1$, 0.3, and 0.5).

The high-frequency oscillations observed in force come from the overtopping of water on the pontoon. After the solitary wave interacts with the fixed pontoon, for the $a/h = 0.3$ and 0.5 cases, wave overtopping occurs. As such, the force–time history is the summation of forces on top of the pontoon and below it. The oscillating part comes from the integrated force on top of the pontoon that comes from the overtopped water.

As expected, the force results show a gradual increase in surge and heave forces with increasing amplitude of the soliton. The heave force encountered by the pontoon is substantially larger compared to the encountered surge force. For the high amplitude cases, the pontoon encounters almost the same level of positive and negative heave force while encountering the solitary wave. The time stamp on the force history also provides a relative idea regarding the propagation velocity of the solitary waves. As expected, high amplitude waves propagate faster compared to low amplitude waves. The propagation velocity is also affected by the vertical steepness of the propagation area (see [4]). The pitch moment is also of high interest for soliton–pontoon interaction, since it is generated from localized pressure on certain section of the pontoon, which in turn creates a high overturning moment. In fact, refs. [11,33,34] demonstrated that the pitch moment in such cases is high enough to over-stress the structural components and connections. Thus, the pitch moment predicted from the simulations is also shown in Figure 11. The results show that the pontoon experiences a sharp rise in pitch moment after a certain period of interaction.

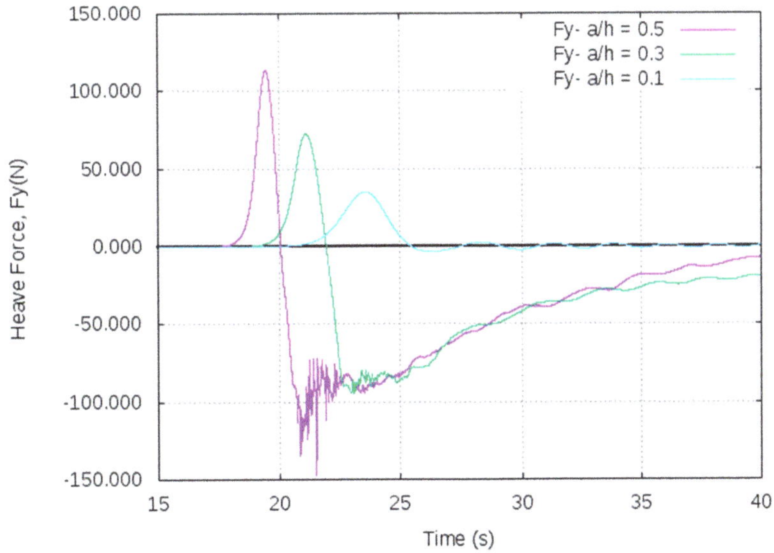

Figure 10. Heave or vertical force experienced by the pontoon while encountering solitary waves (for $a/h = 0.1$, 0.3, and 0.5).

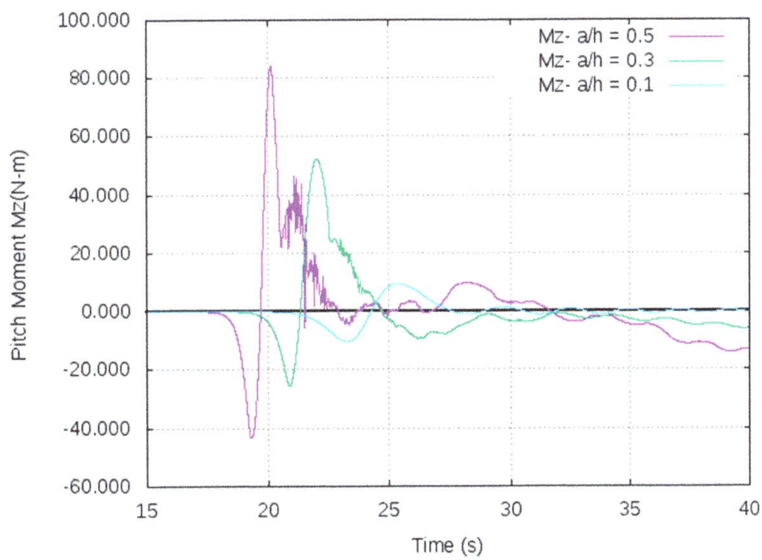

Figure 11. Pitch moment experienced by the pontoon while encountering solitary waves (for a/h = 0.1, 0.3, and 0.5).

In Figure 12, pressure distribution in the flow field close to the pontoon is shown. The figure better illustrates the wave interaction with the static pontoon. Pressure distribution in the fluid domain is shown for the cases with amplitude 0.1 m, 0.3 m, and 0.5 m in Figure 12a–c. Figure 12 shows the solitary wave breaking part and shows the interaction of the wave with the fixed pontoon. While Figure 12 shows the sequence of interaction, the other two cases show simulation time. Overtopping of the wave is observed for amplitudes

0.3 m and 0.5 m. For all three cases, small eddies or vortices are observed at the sharp ends of the pontoons. All three images use a different pressure scale, since the aim is to present the wave interaction and not to measure the encountered pressure.

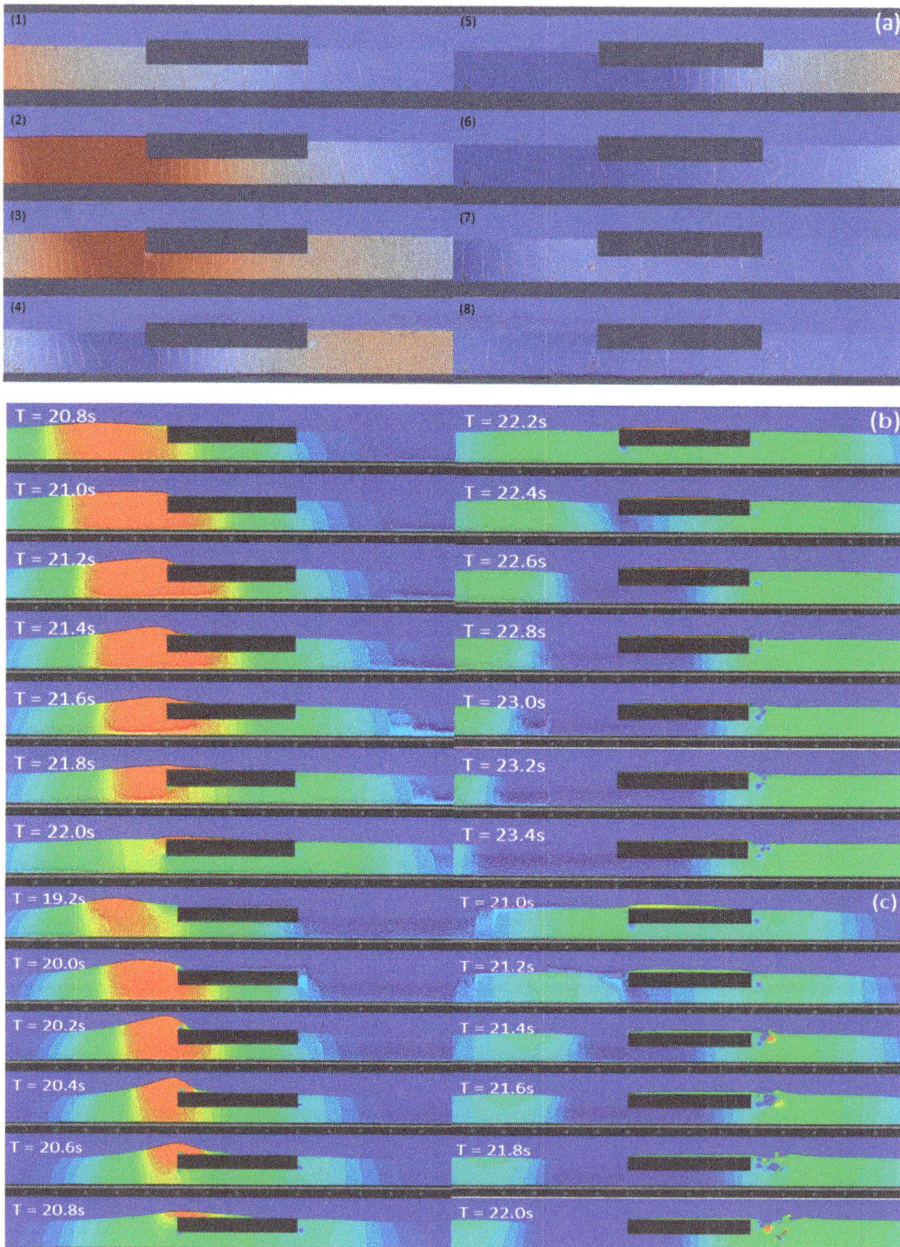

Figure 12. Pressure distribution around the pontoon during the wave encounter in the fluid domain, while the solitary wave interacts with the fixed pontoon; (**a**) $a/h = 0.1$, (**b**) $a/h = 0.3$, and (**c**) $a/h = 0.5$.

5.2. Solitons Generated Using the OceanWave3D Model

Next, simulations are performed using a coupled model between OpenFOAM and OceanWave3D (OW3D), a Fully Nonlinear Potential Flow (FNPF) code by [29]. The solitons are generated by inputting the exact wave flux of an FNPF soliton solution obtained from the conformal mapping technique in [30]. Initially, the solitons are simulated on OW3D separately to check for the stability of the solution. After checking the stability of the soliton, the input parameters are used in the coupled model. In theory, solitons propagating over fixed depth are stable up to the limit of around $a/h = 0.76$.

As before, initially, simulation is performed with the amplitude of 0.1 m ($a/h = 0.1$) and compared with the available numerical data, as shown in Figure 5. Compared to Figure 5, the results show improvement in measured wave elevations. The presented CFD results, coupled with the OW3D model, show better agreement with previous numerical results compared to the Chappelear wave model (see [27]).

The wave elevations measured by all the wave gauges (for $a/h = 0.1,\ 0.3,$ and 0.5) are shown in Figure 12. The respective numbers after the G here indicate the position of the gauge in the domain. G-31 and G-26.5 represent G1 and G3 in the original setup. They are represented as G-31 and G-26.5, instead of being referred to by their actual position in the domain, to avoid confusion.

When compared with Figure 7 (for $a/h = 0.1$), Figure 13 shows more stable wave propagation and negligible dispersive tail after the soliton. This is most probably linked to the viscosity effect, which is not well captured in the coupled model, hence the deviation from the experimental observation. Furthermore, due to the positioning of the wave probes, the simulation was also able to capture the peak of the solitary wave before breaking, as shown by gauge 97. However, OW3D shows a loss of wave amplitude by 30%, in contrast to the second-order Chappelear model (see [27]), which showed a 35% loss. This can be related to the stable propagation of the solitary wave without dispersive tail, which causes loss of energy.

As for $a/h = 0.3$ and $a/h = 0.5$, similar to the previous case, due to the positioning of the wave probes, the probe placed at position 97 captures the peak of the soliton before breaking. Furthermore, the Figures suggest, for higher amplitude cases, the soliton generated by OW3D also shows a dispersive tail. Nevertheless, the generated solitons using OW3D are more stable and does not lose amplitude before interacting with the pontoon. According to the figures after the interaction, the soliton propagates with an amplitude of 0.068 m. Whereas, the reflected wave propagates backwards with a crest and trough of roughly 0.04 m amplitude.

As before, a comparison is shown in Figure 14 among the CFD results for G1 and G3 wave gauges $a/h = 0.1,\ 0.3,$ and 0.5, and the results from [31] at $a/h = 0.1$. The comparison among $a/h = 0.1$ results shows good agreement. Like the previous case, the time for the Lin results (see [31]) is also adjusted here for overlapping. In the case of OW3D results, the $a/h = 0.1$ case shows wave propagation in between the 0.3 and 0.5 cases. This is because the domain sizes for the $a/h = 0.3$ and 0.5 cases are different from the 0.1 cases. The domain configuration had to be changed to create a stable solitary wave with the right amplitude. Thus, they show different times of interaction.

Next, a relative comparison among the surge and heave forces experienced for different wave amplitudes is shown in Figures 15 and 16. For all the cases, the surge force results show a slight underprediction, compared to the Chappelear model ([27], non-coupled solution), in predicting the positive force, while they slightly overpredict the negative force. As for the heave force, the results show minor overprediction of positive forces compared to the Chappelear model (see [27]), whereas negative forces show a notable underprediction. This difference might be partly due to the difference in mesh resolution. However, the major difference should be a result of the difference between the two models.

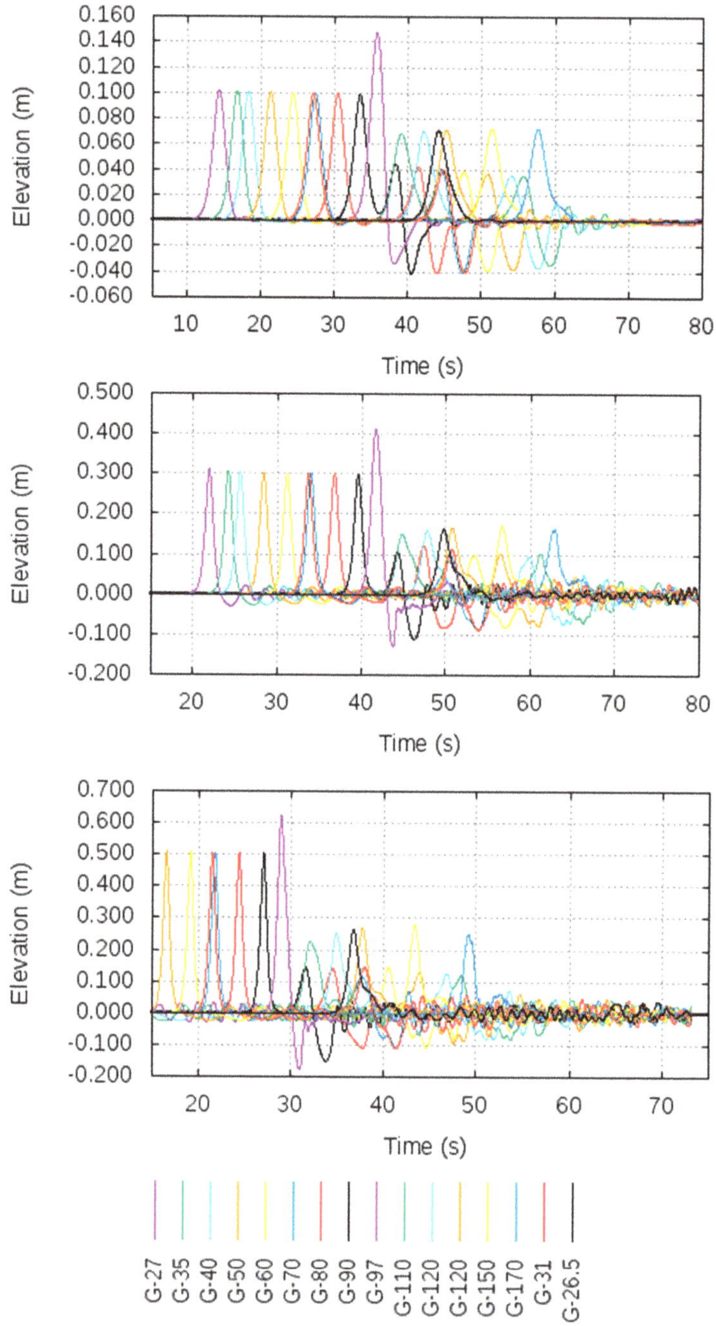

Figure 13. Free surface elevations measured by different wave gauges along the simulation domain $a/h = 0.1,\ 0.3,$ and 0.5(from top to bottom, respectively; the legend is shown at the bottom) versus time t (s).

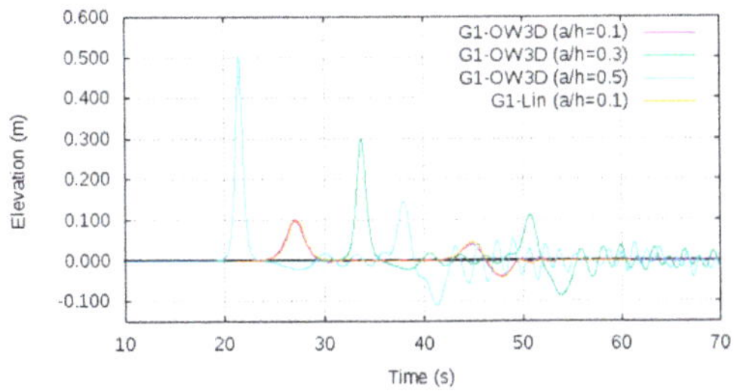

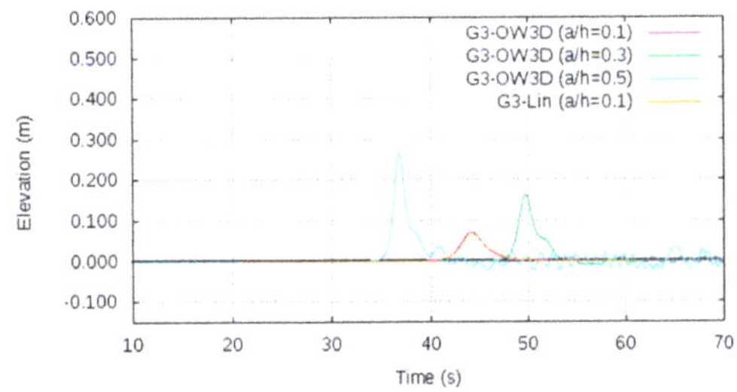

Figure 14. Relative comparison among the wave gauge results from the CFD simulations (coupled OF and OW3D model) for gauge G1 (**left**) and G3 (**right**) at $a/h = 0.1, 0.3,$ and 0.5, and the results from [31] at a/h = 0.1.

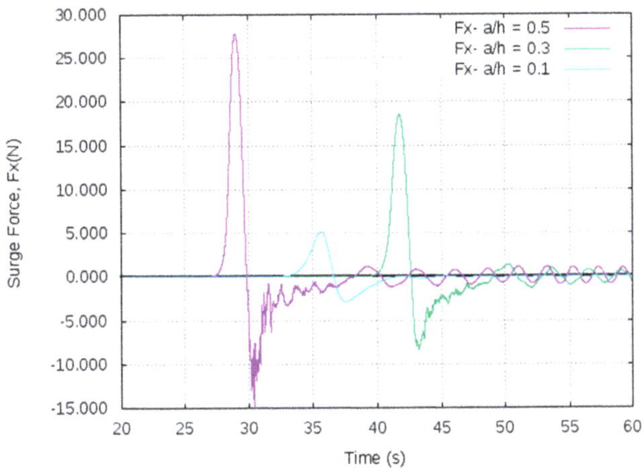

Figure 15. Surge or longitudinal force experienced by the pontoon while encountering solitary wave (OW3D, $a/h = 0.1, 0.3,$ and 0.5).

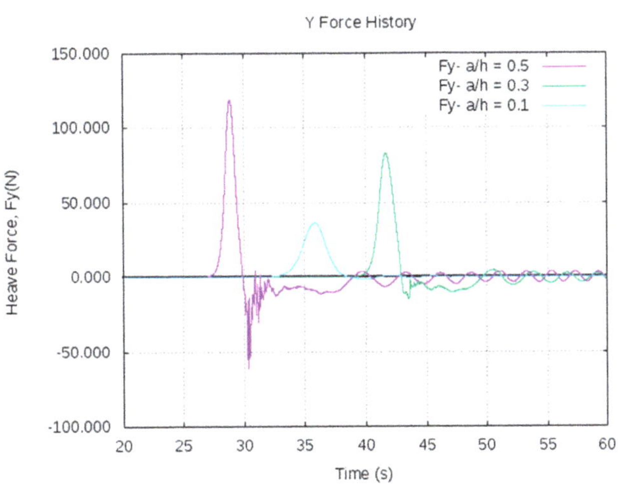

Figure 16. Heave or vertical force experienced by the pontoon while encountering solitary wave (OW3D, $a/h = 0.1, 0.3,$ and 0.5).

Finally, pressure distribution in the flow field close to the pontoon is shown for all three cases in Figure 17. Pressure distribution is shown for the cases with amplitude 0.1 m, 0.3 m, and 0.5 m in Figure 17a–c, respectively. As before, overtopping of a wave is observed for amplitudes of 0.3 m and 0.5 m. However, for the coupled simulation, the small eddies or vortices are less visible below the pontoon. The images here use a different pressure scale, as well, since the same scale comparison will not be sufficient to show the slight gradients in pressure during the encounter.

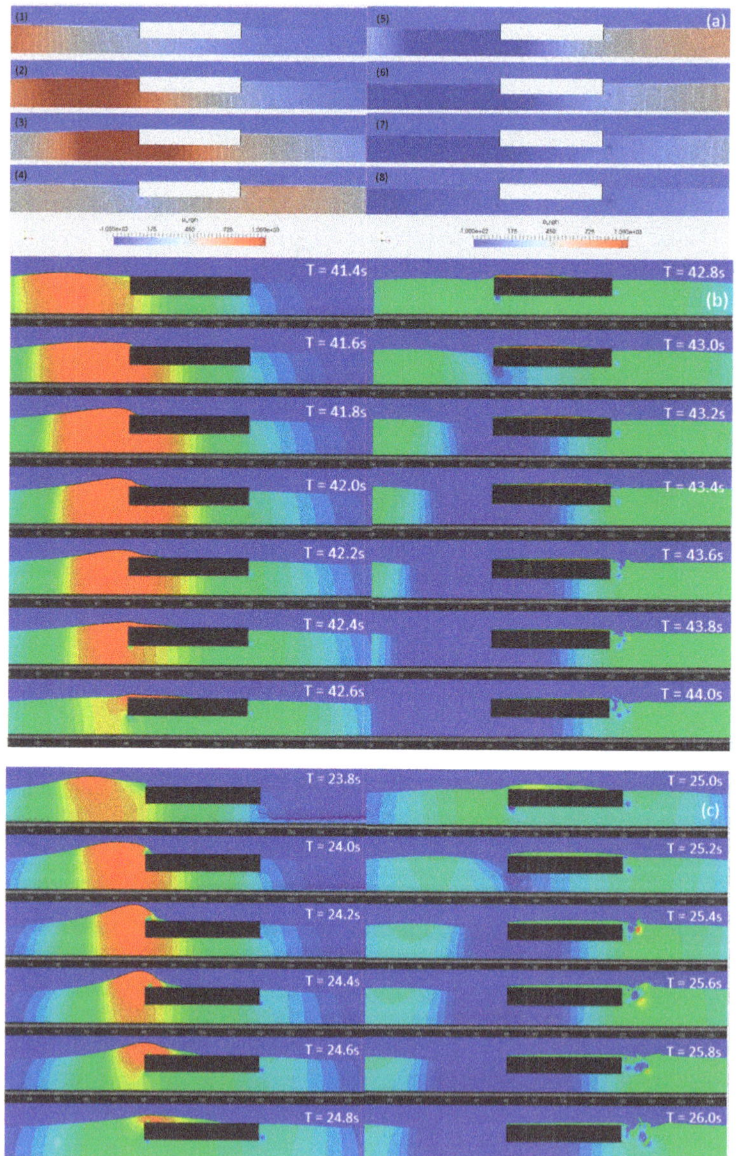

Figure 17. Pressure distribution (**a**) around the pontoon during the wave encounter (1–8 shows successive time steps) (**a**) OW3D, $a/h = 0.1$, in the fluid domain, while the solitary wave with the fixed pontoon (**b**) OW3D, $a/h = 0.3$, and (**c**) OW3D, $a/h = 0.5$.

Overall, it may be concluded that both the Chappelear (see [27]) and coupled OW3D models show similar predictions, and both might be used for studying floating body interaction with a solitary wave. However, the solitary wave generated using OW3D shows better stability compared to the Chappelear model present in waves2Foam. Furthermore, OW3D coupled simulations showed significantly shorter run-time (20 h) compared to the Chappelear model in waves2Foam (100 h). Considering the computational cost and relative accuracy, the OW3D coupled model seems more practical for such studies. The numerical computations associated with analytical expressions were performed in a desktop machine with Intel® core i7-4790 CPU with a 3.60 GHz processor and 32 GB of RAM. On average, each case took roughly 10–15 min to finish.

5.3. Behavior of Solitary Waves Based on Analytical Solution

In this subsection, the effect of different parameters, such as the Ursell number, plate length, and water depth on the behaviour of the solitary wave is analysed based on the analytical solution. To start, Figure 18 shows the effect of solitary wave profiles in the outer region for different Ursell numbers versus water plane length x (m). It can be clearly seen that, as the Ursell number increases, the solitary wave profile ζ compresses, having the same peak for each Ursell number.

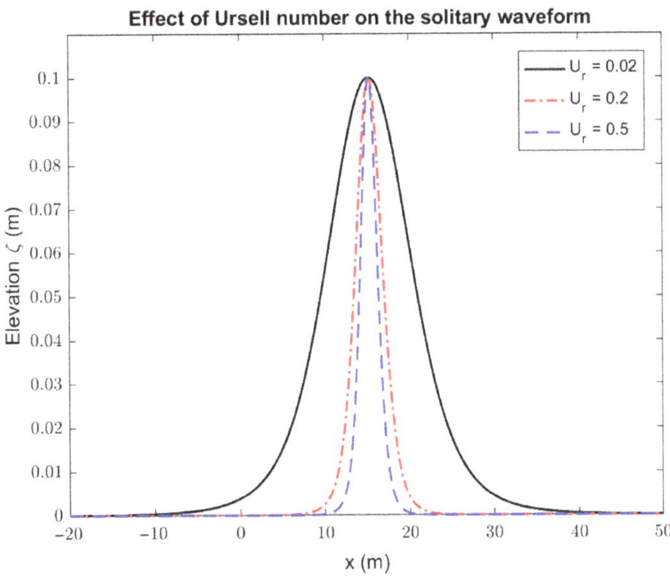

Figure 18. Effect of Ursell number on the solitary wave profiles in the outer region versus water plane length x (m).

In Figure 19, the effect of solitary wave profiles in the outer region for different pontoon lengths versus water plane length x (m) is plotted. It can be observed that, as the length of the pontoon increases, the solitary wave shifts towards the right, while maintaining similar wave amplitude throughout the water plane length. A certain shifting to right in the solitary wave profiles is observed which is due to the constructive/destructive interference of the incident and reflected waves at the edge of the structure.

Figure 20 shows the variation in solitary wave profiles for different water depths versus x (m). It can be seen that the effect of water depth on the solitary wave profiles is insignificant. However, it caused a small effect at the bottom of the solitary waves, but not at the peak.

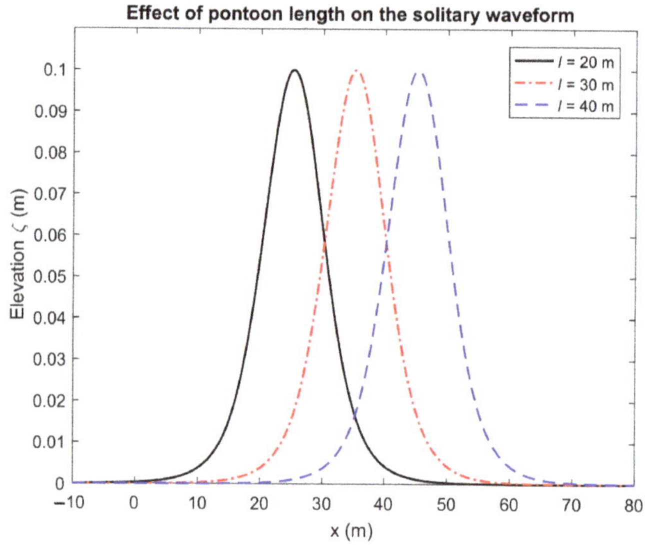

Figure 19. Effect of pontoon length on the solitary wave profiles in the outer region versus water plane length x (m).

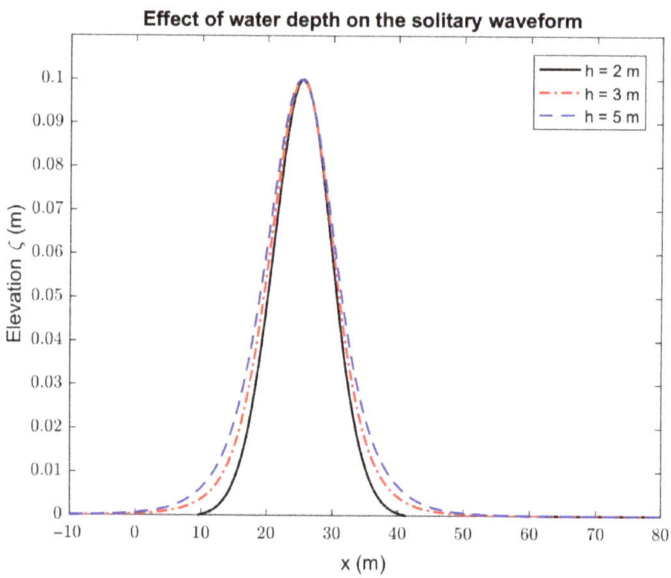

Figure 20. Effect of water depth on the solitary wave profiles in the outer region versus water plane length x (m).

5.4. Wave Energy Analysis

Finally, energy analysis is performed on the OpenFOAM CFD results to calculate the loss of energy of the solitary wave after interaction with the pontoon. The energy calculation is performed based on the incident and transmitted solitary wave amplitudes, and the energy formulation for exact solitary solutions. However, the energy of the reflected wave is not calculated. To perform the calculation, soliton amplitudes are recorded from the wave gauges used in the simulation domain, as shown in Figures 7 and 21. For the propagating wave after the interaction, the stable average amplitude is considered. As can be seen from the figure in the case of a/h = 0.5, the soliton breaks after propagating forward.

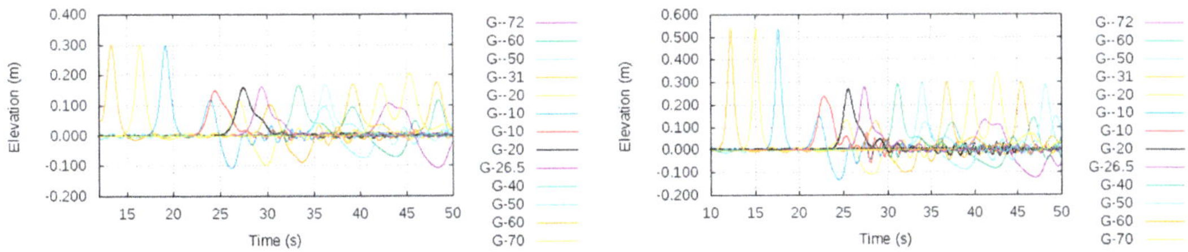

Figure 21. Free surface elevations measured by different wave gauges along the simulation domain for a/h = 0.3 and 0.5 versus time t (s) using OpenFOAM.

The energy analysis results are shown in Table 1. The table shows that the higher the initial amplitude, the higher the loss of solitary energy. In addition, energy is being reflected, absorbed, and dissipated. The dissipation of energy also increases with initial amplitude, which can be observed in Figures 7 and 21, which show longer dispersive tails for higher-amplitude cases.

Table 1. Energy analysis results for the solitary wave interacting with a fixed pontoon.

Amplitude (a/h)	Input Energy (J)	Transmitted Energy (J)	Loss of Energy after Interaction	Loss of Amplitude after Interaction
0.1	5.10×10^{-2}	2.76×10^{-2}	46%	33%
0.3	2.82×10^{-1}	1.23×10^{-1}	57%	41%
0.5	6.87×10^{-1}	2.71×10^{-1}	60%	45%

6. Conclusions

This paper presents a new mathematical model, which is the Boussinesq formulation and solitary wave solution. Moreover, the application of the technique to the considered problem is new, and the comparison with two different independent numerical model simulations and its analysis are also new. Furthermore, the coupled numerical model between CFD and OceanWave 3D is another new contribution of the present study. The solitary waves are generated using 2nd-order solitary wave theory and a coupled FNPF model. The analytical solitary wave is compared with CFD and OceanWave3D (FNPF) model simulations for three-different soliton amplitudes. The solitary wave solution in the outer region associated with the aforesaid model is derived and comparison results with two different higher-fidelity numerical models are presented. From the present study, it can be concluded that:

1. The compared results of the analytical model showed that the solitary waveform in the outer region is very close to CFD, OceanWave3D, and previous numerical results from the literature.
2. From both the CFD and OceanWave3D, the analysis of wave forces indicated that with the increase in the amplitude of solitons, the surge and heave force gradually increase. The quantitative information is important to ensure station keeping of such structures. Further, the heave force encountered by the pontoon is substantially larger than the encountered surge force, which confirms the higher potential for application of the OWC type devices. The pressure distribution around the pontoon indicates that vortices are observed at the sharp ends of the pontoon, which might induce vibration and erosion of structure.
3. The higher the initial soliton amplitude, the higher the loss of energy during interaction with the pontoon, and the higher the dissipation of energy into the dispersive tail.

4. The Ursell number has a significant effect on the solitary wave profiles than those of water depth and pontoon length.

Finally, in terms of fidelity, the RANS solution with the 2nd-order solitary wave model (OpenFOAM) shows the greatest accuracy, followed by the coupled FNPF (OceanWave3D) model. Finally, the Boussinesq formulation-based analytical solution shows good accuracy in capturing the propagating solitary wave. In terms of computational burden, the OpenFOAM model is the most expensive, followed by the coupled model (OceanWave3D), and finally the analytical model. Thus, analytical solutions may be a reliable way of obtaining initial predictions for solitary wave propagation in coastal areas, whereas coupled models may be more suitable for the study of wave–structure interaction with reliable accuracy. Therefore, the present analytical formulation will be helpful to generalize the problem of solitary wave interaction with flexible floating structure over the Boussinesq approach and comparison with different model simulations. Furthermore, future studies will attempt to expand the methodology presented herein to other types of flexible structures connected with mooring lines and it is important to investigate the hydroelastic analysis including the structural properties (flexibility and compressive force) as well as the uplift wave loads, which will not be subjected to the aforementioned limitations.

Author Contributions: The concept of the considered problem is developed by S.C.M. and C.G.S. The CFD and coupled analysis is performed by H.I. and OceanWave3D simulation is performed by T.S.H. The writing of the original draft manuscript is done by S.C.M., H.I., T.S.H. and C.G.S. All authors have read and agreed to the published version of the manuscript.

Funding: This work was performed within the Project MIDWEST, Multi-fidelity Decision making tools for Wave Energy Systems, which is co-funded by European Union's Horizon 2020 research and innovation programme under the framework of OCEANERA-NET (http://oceaneranet.eu (accessed on 1 January 2021)) and by the Portuguese Foundation for Science and Technology (Fundação para a Ciência e Tecnologia—FCT) under contract (OCEANERA/0006/2014). The first author has been contracted as a Researcher by the Portuguese Foundation for Science and Technology (Fundação para a Ciência e Tecnologia—FCT), through Scientific Employment Stimulus, Individual support under Contract No. CEECIND/04879/2017. The third author has a PhD Scholarship by the Portuguese Foundation for Science and Technology (Fundação para a Ciência e Tecnologia—FCT), under the contract No. SFRH/BD/145602/2019. This work contributes to the Strategic Research Plan of the Centre for Marine Technology and Ocean Engineering (CENTEC), which is financed by the Portuguese Foundation for Science and Technology (Fundação para a Ciência e Tecnologia—FCT) under contract UIDB/UIDP/00134/2020.

Institutional Review Board Statement: Not applicable.

Informed Consent Statement: Not applicable.

Data Availability Statement: Not applicable.

Conflicts of Interest: The authors declare no conflict of interest.

References

1. Mohapatra, S.C.; Guedes Soares, C. Comparing solutions of the coupled Boussinesq equations in shallow water. In *Maritime Technology and Engineering*; Guedes Soares, C., Santos, T.A., Eds.; Taylor & Francis Group: London, UK, 2015; pp. 947–954, ISBN 978-1-138-02727-5.
2. Mohapatra, S.C.; Fonseca, R.B.; Guedes Soares, C. A comparison between analytical and numerical simulations of solutions of the coupled Boussinesq equations. In *Maritime Technology and Engineering*; Guedes Soares, C., Santos, T.A., Eds.; Taylor & Francis Group: London, UK, 2016; pp. 1175–1180. ISBN 978-1-138-03000-8.
3. Mohapatra, S.C.; Fonseca, R.B.; Guedes Soares, C. Comparison of analytical and numerical simulations of long nonlinear internal waves in shallow water. *J. Coast. Res.* **2018**, *34*, 928–938. [CrossRef]
4. Hallak, T.S.; Islam, H.; Mohapatra, S.C.; Guedes Soares, C. Comparing numerical and analytical solutions of solitary water waves over finite and variable depth. In Proceedings of the International Conference on Offshore Mechanics and Arctic Engineering, Cancún, Mexico, 5–6 April 2021; ASME: New York, NY, USA, 2021. Paper No: OMAE2021-62642, V006T06A062.
5. Chang, C.H. Interaction of a Solitary Wave with Vertical Fully/Partially Submerged Circular Cylinders with/without a Hollow Zone. *J. Mar. Sci. Eng.* **2020**, *8*, 1022. [CrossRef]

6. Wang, Q.; Fang, Y.; Liu, H. An experimental study of run-up and loads on a vertical truncated cylinder in a solitary wave. *Ocean Eng.* **2021**, *219*, 108346. [CrossRef]
7. Mohapatra, S.C.; Islam, H.; Guedes Soares, C. Boussinesq Model and CFD Simulations of Non-Linear Wave Diffraction by a Floating Vertical Cylinder. *J. Mar. Sci. Eng.* **2020**, *8*, 575. [CrossRef]
8. Mohapatra, S.C.; Islam, H.; Guedes Soares, C. Wave diffraction by a floating fixed truncated vertical cylinder based on Boussinesq equations. In *Advances in Renewable Energies Offshore*; Guedes Soares, C., Ed.; Taylor & Francis Group: London, UK, 2019; pp. 281–289, ISBN 978-1-138-58535-5.
9. Lu, X.; Wang, K.H. Modeling a solitary wave interaction with a fixed floating body using an integrated analytical–numerical approach. *Ocean Eng.* **2015**, *109*, 691–704. [CrossRef]
10. Sun, J.I.; Wang, C.Z.; Wu, G.X.; Khoo, B.C. Fully nonlinear simulations of interactions between solitary waves and structures based on the finite element method. *Ocean Eng.* **2015**, *108*, 202–215. [CrossRef]
11. Xiang, T.; Istrati, D.; Yim, S.C.; Buckle, I.G.; Lomonaco, P. Tsunami Loads on a Representative Coastal Bridge Deck: Experimental Study and Validation of Design Equations. *J. Waterw. Port Coast. Ocean Eng.* **2020**, *146*, 04020022. [CrossRef]
12. Hasanpour, A.; Istrati, D. Extreme Storm Wave Impact on Elevated Coastal Buildings. In Proceedings of the 3rd International Conference on Natural Hazards & Infrastructure (ICONHIC2022), Athens, Greece, 5–7 July 2022.
13. Zhu, H.; Wang, L.L.; Avital, E.J.; Tang, H.W.; Williams, J.J.R. Numerical Simulation of Shoaling Broad-Crested Internal Solitary Waves. *J. Hydraul. Eng.* **2017**, *143*, 04017006. [CrossRef]
14. Seiffert, B.; Hayatdavoodi, M.; Ertekin, R.C. Experiments and computations of solitary-wave forces on a coastal-bridge deck. Part I: Flat plate. *Coast. Eng.* **2014**, *88*, 194–209. [CrossRef]
15. Hayatdavoodi, M.; Seiffert, B.; Ertekin, R.C. Experiments and computations of solitary-wave forces on a coastal-bridge deck. Part II: Deck with girders. *Coast. Eng.* **2014**, *88*, 210–228. [CrossRef]
16. Cuomo, G.; Shimosako, K.I.; Takahashi, S. Wave-in-deck loads on coastal bridges and the role of air. *Coast. Eng.* **2009**, *56*, 793–809. [CrossRef]
17. Istrati, D.; Buckle, I.G.; Lomonaco, P.; Yim, S.; Itani, A. Tsunami induced forces in bridges: Large-scale experiment and the role of air-entrapment. *Coast. Eng. Proc.* **2017**, *35*, 1–14. [CrossRef]
18. Martinelli, L.; Lamberti, A.; Gaeta, M.G.; Tirindelli, M.; Alderson, J.; Schimmels, S. Wave loads on exposed jetties: Description of large scale experiments and preliminary results. *Coast. Eng. Proc.* **2011**, *32*, 1–13. [CrossRef]
19. Wu, N.-J.; Tsay, T.-K.; Chen, Y.-Y. Generation of stable solitary waves by a piston-type wave maker. *Wave Motion* **2014**, *51*, 240–255. [CrossRef]
20. Goring, D.G. Tsunamis—The Propagation of Long Waves onto a Shelf. In *W. M. Keck Laboratory of Hydraulics and Water Resources*; California Institute of Technology: Pasadena, CA, USA, 1978.
21. Wu, N.J.; Hsiao, S.C.; Chen, H.H.; Yang, R.Y. The study on solitary waves generated by a piston-type wave maker. *Ocean Eng.* **2016**, *117*, 114–129. [CrossRef]
22. Wang, Y.; Yin, Z.; Liu, Y. Numerical study of solitary wave interaction with a vegetated platform. *Ocean Eng.* **2019**, *192*, 106561. [CrossRef]
23. Martins, K.; Bonneton, P.; Michallet, H. Dispersive characteristics of nonlinear waves propagating and breaking over a mildly sloping laboratory beach. *Coast. Eng.* **2021**, *167*, 103917. [CrossRef]
24. Gadelho, J.F.M.; Mohapatra, S.C.; Guedes Soares, C. CFD analysis of a fixed floating box-type structure under regular waves. In *Development in Maritime Transportation and Harvesting of Sea Resources*; Guedes Soares, C., Teixeira, Â.P., Eds.; Taylor & Francis Group: London, UK, 2017; pp. 513–520.
25. Jasak, H. OpenFOAM: Open source CFD in research and industry. *Int. J. Nav. Archit. Ocean Eng.* **2009**, *1*, 89–94.
26. Jacobsen, N.G.; Fuhrman, D.R.; Fredsoe, J. A wave generation toolbox for the open-source CFD library: OpenFOAM. *Int. J. Numer. Methods Fluids* **2011**, *70*, 1073–1088. [CrossRef]
27. Chappelear, J.E. Shallow-water waves. *J. Geophys. Res.* **1962**, *67*, 4693–4704. [CrossRef]
28. Islam, H.; Mohapatra, S.C.; Gadelho, J.; Guedes Soares, C. OpenFOAM analysis of the wave radiation by a box-type floating structure. *Ocean Eng.* **2019**, *193*, 106532. [CrossRef]
29. Engsig-Karup, A.P.; Bingham, H.B.; Lindberg, O. An efficient flexible-order model for 3D nonlinear water waves. *J. Comput. Phys.* **2009**, *228*, 2100–2118. [CrossRef]
30. Dutykh, D.; Clamond, D. Efficient computation of steady solitary gravity waves. *Wave Motion* **2014**, *51*, 86–99. [CrossRef]
31. Lin, P. A multiple-layer σ-coordinate model for simulation of wave–structure interaction. *Comput. Fluids* **2006**, *35*, 147–167. [CrossRef]
32. Rijnsdorp, D.P.; Zijlema, M. Simulating waves and their interactions with a restrained ship using a non-hydrostatic wave-flow model. *Coast. Eng.* **2016**, *114*, 119–136. [CrossRef]
33. Sarfaraz, M.; Pak, A. SPH numerical simulation of tsunami wave forces impinged on bridge superstructures. *Coast. Eng.* **2017**, *121*, 145–157. [CrossRef]
34. Istrati, D.; Buckle, I.; Lomonaco, P.; Yim, S. Deciphering the tsunami wave impact and associated connection forces in open-girder coastal bridges. *J. Mar. Sci. Eng.* **2018**, *6*, 148. [CrossRef]

Journal of
Marine Science and Engineering

MDPI

Article

Hydrodynamic Behaviour of a Floating Polygonal Platform Centrally Placed within a Polygonal Ring Structure under Wave Action

Jeong Cheol Park * and Chien Ming Wang

School of Civil Engineering, University of Queensland, St Lucia, QLD 4072, Australia
* Correspondence: jeongcheol.park@uq.net.au

Abstract: In this paper, a semi-analytical method has been developed for the hydrodynamic analysis of a floating polygonal platform that is centrally placed within a floating polygonal ring structure under wave action. In view to understand the wave interactions inside the ring structure, the formulation considers two cases: when the platform and ring structure oscillate individually, and when they oscillate together under wave action. The polygonal shapes of the floating structures can be created from a parametric equation involving the cosine-type radial perturbation. The formulation and computer code are verified by comparing the results with those obtained from the commercial software ANSYS AQWA. When floating ring structures are used, trapped waves are created in the inner water basin resulting in resonance. The interactions among the trapped waves, inner floating platforms and outer ring structures are investigated by performing parametric studies. By changing the dimensions of the platform and ring structure such as the drafts, the radii of platforms and polygonal shapes, their effects on major hydrodynamic quantities may be understood.

Keywords: 3D hydrodynamic analysis; resonance; floating polygonal structures; wave energy harvesting; cosine-type radial perturbation; eigenfunction expansion method

Citation: Park, J.C.; Wang, C.M. Hydrodynamic Behaviour of a Floating Polygonal Platform Centrally Placed within a Polygonal Ring Structure under Wave Action. *J. Mar. Sci. Eng.* **2022**, *10*, 1430. https://doi.org/10.3390/jmse10101430

Academic Editors: Carlos Guedes Soares and Serge Sutulo

Received: 2 September 2022
Accepted: 20 September 2022
Published: 4 October 2022

1. Introduction

Floating platforms have been used for offshore oil and gas rigging, offshore renewable energy farms, aquaculture farms, floating hotels, floating parks, floating houses and floating entertainment/leisure facilities [1–3]. Floating breakwaters create a sheltered sea space that allows safe operation and maintenance of floating solar farms [4], fish farming [5], ship harbouring [6–8], etc. By placing a floating platform within a floating ring breakwater, one may have a practical solution for operating the aforementioned activities in an open sea. For example, Figure 1 shows a conceptual design of a mega offshore floating fish farm surrounded by a hexagonal floating breakwater in an open ocean. The internal floating hexagonal platform houses the control centre, power production and storage facility, fish processing plant, offices, workers' quarters, etc. Alternatively, the floating ring structure may be designed to trap waves to create a high wave energy environment with the view to harvest wave energy using the piston-like internal floating cylinder, i.e., Wave Energy Converter (WEC) device [9,10].

Garrett [8] determined the wave motion inside a thin-walled bottomless harbour using an analytical method. Mavrakos [11] and Mavrakos [12] extended Garrett [8] study for thick-walled floating bottomless circular cylinders and solved the diffraction and radiation problems, respectively. Later, Mavrakos and Chatjigeorgiou [13] tackled the second-order waves for the same problem in order to improve the inaccuracy of the linear wave potential theory due to the trapped waves in the inner water basin resulting in the highly amplified resonant waves. For two concentric floating circular cylinders, Mavrakos [14] and Mavrakos [15] obtained the wave exciting forces and hydrodynamic coefficients. Mavrakos, et al. [16] addressed tightly moored two concentric floating circular cylinders under first and second-order waves. Konispoliatis, Mazarakos and Mavrakos [10] presented analytical solutions for an array of Oscillating Water Column (OWC) devices. Each OWC device consists of concentric circular cylinders.

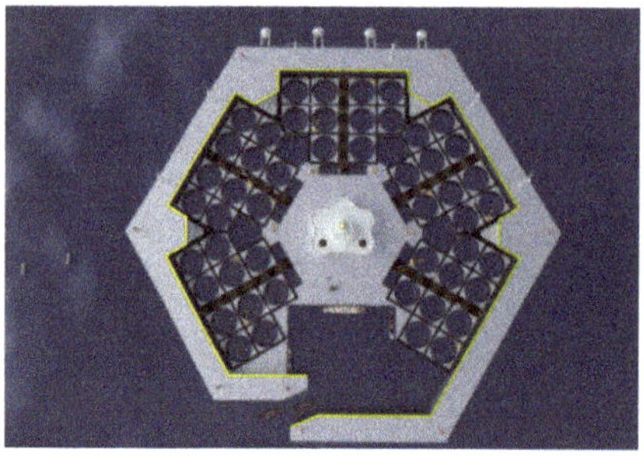

(a)

(b)

Figure 1. HEXAGON: a mega offshore floating fish farm: (**a**) plan view; (**b**) isometric view.

However, the existing formulations are mostly based on floating circular cylinders as it is analytically tractable in the cylindrical coordinate system. In practice, non-circular models are used for marine structure applications [17–19]. Recently, Park and Wang [20] and Park and Wang [21] respectively investigated the hydrodynamic behaviours of floating polygonal platforms and floating polygonal ring structures by solving the diffraction and radiation problems. The shape of the polygonal platform or ring structure was created by using the cosine-type radial perturbation [22]. It has been shown that these hydrodynamic problems can be solved analytically by using the Eigenfunction Expansion Method. In continuing this line of study that uses the analytical method for hydrodynamic analysis of floating structures, this paper further investigates the hydrodynamic behaviour of a floating polygonal platform that is centrally placed within a floating polygonal ring structure. In order to understand the wave interactions inside the ring structure, the study will investigate the cases where the floating platform and ring structure oscillate individually as well as when they oscillate together under wave action. A parametric study involving different drafts, widths of the floating polygonal platforms and polygonal shapes will be performed to understand the hydrodynamic actions inside the floating polygonal ring structure that exhibits resonance phenomena at specific wave frequencies.

The contents of the paper are laid out as follows: Section 2 defines the problem at hand and Section 3 presents the governing equation and boundary conditions for the problem. Section 4 solves the diffracted and radiated potentials by using the semi-analytical approach. Sections 5–7 deal with the determination of the wave exciting forces, hydrodynamic radiation forces and motion responses of the floating platform and ring structure, respectively. Section 8 demonstrates the verification of the present semi-analytical approach by comparing the results with those obtained from the commercial software ANSYS AQWA. Section 9 furnishes the hydrodynamic results for parametric studies. Finally, concluding remarks are given in Section 10.

2. Problem Definition

Consider a floating rigid regular polygonal platform that is centrally placed within a floating rigid regular polygonal ring structure as shown in Figure 2. They are allowed to oscillate together or individually but are assumed to be kept in place as the current and drift force are not considered in this study. The considered water depth is h and the incident wave, having a period T and amplitude A, impacts the floating platform and ring structure at an oblique angle β. The freeboard is assumed to be sufficiently high to prevent wave overtopping. The drafts of the platform and ring structure are d_1 and d_2, respectively.

The cylindrical coordinates (r, θ, z) are adopted with the origin at the centre of the regular polygonal platform.

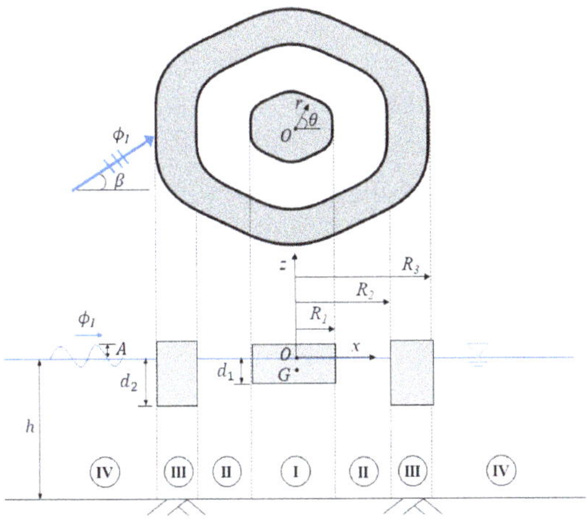

Figure 2. Floating polygonal platform and ring breakwater.

The plan shape of the polygonal platform or polygonal ring structure is generated by using a radius function defined by the cosine-type radial perturbation given by [22]

$$R_l(\theta) = R_{0_l}\{1 + \varepsilon_l \cos n_{p_l}(\theta - \theta_{0_l})\}, \quad l = 1,2,3 \tag{1}$$

where R_{0_l}, ε_l, n_{p_l} and θ_{0_l} are parameters to be chosen by the analyst. This radius function can be used to construct all kinds of regular polygonal shapes. For example, polygonal shapes such as an equilateral triangular, square, pentagon and hexagon can be straightforwardly created by choosing the appropriate values for the dimensionless parameters ε_l, n_{p_l} and θ_{0_l}, which are summarised in Table 1. The size and shape of the polygonal platform and ring structure are predominantly controlled by R_l $(l = 1, 2, 3)$ where $l = 1$ represents the platform boundary whilst $l = 2$ and 3 represent the inner and outer boundaries of the ring structure, respectively. In addition, one can freely orientate the polygonal shapes by changing θ_{0_l}.

Table 1. Regular polygonal platform and ring shapes created from the cosine-type radial perturbation S_{0_q} denotes the plan area of the polygonal platform for $q = 1$ and the polygonal ring structure for $q = 2$. The values in the bracket are in turn associated with R_1, R_2 and R_3.

Polygonal Shapes	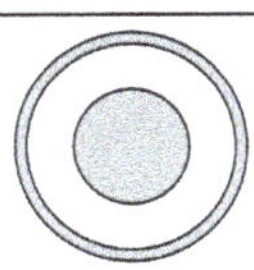	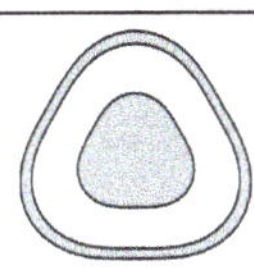	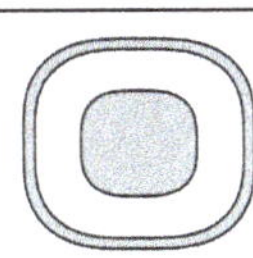	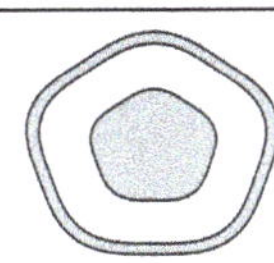	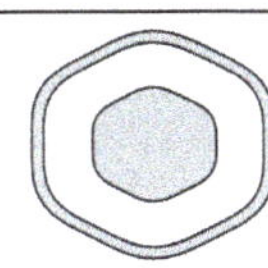
	Circle	**Triangle**	**Square**	**Pentagon**	**Hexagon**
S_{0_1}	A_{0_1}	$1.005 \times A_{0_1}$	$1.002 \times A_{0_1}$	$1.001 \times A_{0_1}$	$1.000 \times A_{0_1}$
S_{0_2}	A_{0_2}	$1.005 \times A_{0_2}$	$1.002 \times A_{0_2}$	$1.001 \times A_{0_2}$	$1.000 \times A_{0_2}$
ε_l	[0, 0, 0]	[0.1, 0.1, 0.1]	[0.06, 0.06, 0.06]	[0.04, 0.04, 0.04]	[0.03, 0.03, 0.03]
n_{p_l}	[0, 0, 0]	[3, 3, 3]	[4, 4, 4]	[5, 5, 5]	[6, 6, 6]
θ_{0_l}	–	$\left[\frac{\pi}{2}, \frac{\pi}{2}, \frac{\pi}{2}\right]$	$\left[\frac{\pi}{4}, \frac{\pi}{4}, \frac{\pi}{4}\right]$	$\left[\frac{\pi}{2}, \frac{\pi}{2}, \frac{\pi}{2}\right]$	$\left[\frac{\pi}{6}, \frac{\pi}{6}, \frac{\pi}{6}\right]$

In this study, the following hydrodynamic properties of floating polygonal platform and ring structure are to be determined: (i) diffracted and radiated potentials, (ii) wave exciting forces, (iii) added mass, (iv) radiation damping, (v) RAOs (Response Amplitude Operators), (vi) wave field.

3. Governing Equation and Boundary Conditions

The hydrodynamic analysis will be performed in the frequency domain. The fluid domain is divided into 4 regions as shown in Figure 2. Region I is the sea space underneath the platform structure, Region II is the sea space between the platform and ring structure, Region III is the sea space underneath the ring structure and Regions IV is the sea space outside the ring structure. The fluid is assumed to be incompressible, inviscid and irrotational, and hence, the linear potential theory may be applied. Accordingly, the fluid motion is governed by the following Laplace equation

$$\frac{1}{r}\frac{\partial}{\partial r}\left(r\frac{\partial\phi}{\partial r}\right)+\frac{1}{r^2}\frac{\partial^2\phi}{\partial\theta^2}+\frac{\partial^2\phi}{\partial z^2}=0 \tag{2}$$

where ϕ is the velocity potential given by

$$\phi=\phi_I+\phi_D+\phi_R \tag{3}$$

in which ϕ_I is the incident potential, ϕ_D the diffracted potential and ϕ_R the radiated potential. The radiated potential may be expressed as the sum of 6 radiation modes corresponding to the six degrees of freedom as

$$\phi_R=\sum_{j=1}^{6}\left(-i\omega\xi_j\phi_R^{(j)}\right) \tag{4}$$

where i is the imaginary unit, ω the wave angular frequency, ξ_j the motion amplitude of the ring structure for the j-th radiation mode and $\phi_R^{(j)}$ the normalised radiated potential for the j-th radiation mode.

The velocity potential must satisfy the following boundary conditions

$$\left.\frac{\partial\phi}{\partial z}\right|_{z=0}=\left.\frac{\omega^2}{g}\phi\right|_{z=0} \quad \text{on the free surface} \tag{5}$$

$$\left.\frac{\partial\phi}{\partial z}\right|_{z=-h}=0 \quad \text{on the seabed} \tag{6}$$

$$\lim_{r\to\infty}\left(\frac{\partial\phi_{D,R}}{\partial r}-ik\phi_{D,R}\right)=0 \quad \text{at infinity} \tag{7}$$

$$\nabla\phi_D\cdot\mathbf{n}_{s_l}=-\nabla\phi_I\cdot\mathbf{n}_{s_l} \quad \text{at wetted surface for diffraction problem} \tag{8}$$

$$\nabla\phi_R^{(j)}\cdot\mathbf{n}_{s_l}=\mathbf{n_j}\cdot\mathbf{n}_{s_l}\quad(j=1,2,\cdots,6) \quad \text{at wetted surface for radiation problem} \tag{9}$$

where k is the wave number, g the gravitational acceleration and $\mathbf{n}_j$ the generalised motion normal for 6 DOFs (degrees of freedom), i.e., $\mathbf{n}_1=\mathbf{n}_x$, $\mathbf{n}_2=\mathbf{n}_y$, $\mathbf{n}_3=\mathbf{n}_z$, $\mathbf{n}_4=-(z-z_G)\mathbf{n}_y+(y-y_G)\mathbf{n}_z$, $\mathbf{n}_5=(z-z_G)\mathbf{n}_x-(x-x_G)\mathbf{n}_z$ and $\mathbf{n}_6=-(y-y_G)\mathbf{n}_x+(x-x_G)\mathbf{n}_y$, where (x_G,y_G,z_G) are the coordinates of the floating structure's centre of gravity and $\mathbf{n}_{s_l}$ the unit normal vector to the polygonal body surface pointing out of the floating body. In order to consider the polygonal geometries, the surface function $S_l(r,\theta)=r-R_l(\theta)$, $(l=1,2,3)$ is introduced and its derivative with respect to θ is defined as $\frac{\partial S_l}{\partial \theta} = -\frac{\partial R_l(\theta)}{\partial \theta}$ [23]. By using the surface function, the unit normal vector to the wetted body surface pointing to the fluid $\mathbf{n}_{s_l}$ is given by [23]

$$\mathbf{n}_{s_l} = \frac{\nabla S_l(r,\theta)}{\left|\mathbf{n}_{s_l}\right|} = \frac{\frac{\partial S_l}{\partial r}\vec{\mathbf{r}} + \frac{1}{r}\frac{\partial S_l}{\partial \theta}\vec{\mathbf{\theta}} + \frac{\partial S_l}{\partial z}\vec{\mathbf{z}}}{\left|\mathbf{n}_{s_l}\right|} = \frac{1}{\sqrt{1+\left(\frac{1}{r}\frac{\partial S_l}{\partial \theta}\right)^2}}\left(\vec{\mathbf{r}} + \frac{1}{r}\frac{\partial S_l}{\partial \theta}\vec{\mathbf{\theta}} + 0\vec{\mathbf{z}}\right) \tag{10}$$

where ∇ denotes the del operator in a cylindrical coordinate system for obtaining the gradient of a vector. Thus, the normal velocity on the floating polygonal platform and ring structure wetted surface can be in general calculated by using the divergence operator in the cylindrical coordinate system as

$$\nabla\phi\cdot\mathbf{n}_{s_l} = \frac{1}{\sqrt{1+\left(\frac{1}{r}\frac{\partial S_l}{\partial \theta}\right)^2}}\left(\frac{\partial \phi}{\partial r} + \frac{1}{r^2}\frac{\partial S_l}{\partial \theta}\frac{\partial \phi}{\partial \theta}\right) \tag{11}$$

4. Solutions for Diffracted and Radiated Potentials

The assumed solutions for the diffraction and radiation problems can be expressed as a unified form as given by

$$\phi_1^{(j,p)} = \phi_{p_1}^{(j,p)} + \sum_{m=-\infty}^{\infty}\left\{A_{m0}^{(j,p)}\left(\frac{r}{b_1}\right)^{|m|} + \sum_{n=1}^{\infty} A_{mn}^{(j,p)}\frac{I_m(p_n r)}{I_m(p_n b_1)}\cos p_n(z+h)\right\}e^{im\theta} \tag{12}$$

$$\phi_2^{(j,p)} = \sum_{m=-\infty}^{\infty}\left[\left\{B_{m0}^{(j,p)} J_m(kr) + C_{m0}^{(j,p)}\frac{H_m(kr)}{H_m(ka_1)}\right\}\frac{Z_0(z)}{Z_0(0)} + \sum_{n=1}^{\infty}\left\{B_{mn}^{(j,p)}\frac{I_m(k_n r)}{I_m(k_n b_2)} + C_{mn}^{(j,p)}\frac{K_m(k_n r)}{K_m(k_n a_1)}\right\}\frac{Z_n(z)}{Z_n(0)}\right]e^{im\theta} \tag{13}$$

$$\begin{aligned}\phi_3^{(j,p)} = \phi_{p3}^{(j,p)} + \; & D_{00}^{(j,p)}\ln\frac{r}{a_2} + E_{00}^{(j,p)}\ln\frac{b_3}{r} + \sum_{\substack{m=-\infty \\ m\neq 0}}^{\infty}\left\{D_{m0}^{(j,p)}\left(\frac{r}{b_3}\right)^{+|m|} + E_{m0}^{(j,p)}\left(\frac{r}{a_2}\right)^{-|m|}\right\}e^{im\theta} \\ & + \sum_{m=-\infty}^{\infty}\sum_{n=1}^{\infty}\left\{D_{mn}^{(j,p)}\frac{I_m(q_n r)}{I_m(q_n b_3)} + E_{mn}^{(j,p)}\frac{K_m(q_n r)}{K_m(q_n a_2)}\right\}\cos q_n(z+h)e^{im\theta}\end{aligned} \tag{14}$$

$$\phi_4^{(j,p)} = \sum_{m=-\infty}^{\infty}\left\{F_{m0}^{(j,p)}\frac{H_m(kr)}{H_m(ka_3)}\frac{Z_0(z)}{Z_0(0)} + \sum_{n=1}^{\infty}F_{mn}^{(j,p)}\frac{K_m(k_n r)}{K_m(k_n a_3)}\frac{Z_n(z)}{Z_n(0)}\right\}e^{im\theta} \tag{15}$$

where the superscript j denotes the diffraction when $j = 0$, otherwise the radiation mode for 6 DOFs ($j = 1$ for surge, $j = 2$ for sway, $j = 3$ for heave, $j = 4$ for roll, $j = 5$ for pitch and $j = 6$ for yaw) and the superscript p denotes the oscillating body, i.e., $p = 1$ for the central platform, $p = 2$ for the ring structure, $p = 3$ for the two bodies oscillating monolithically or individually and $p = 0$ for zero-motion to address the diffraction problem. Hence, in order to obtain the solution for the radiation problem, the two cases for $p = 1$ and 2 must be added. $\phi_{p_1}^{(j,p)}$ and $\phi_{p_3}^{(j,p)}$ are respectively the particular solutions for Regions I and III and the vertical eigenfunction Z_n for both Regions II and IV is given by

$$Z_0(z) = \frac{\cosh k(z+h)}{\sqrt{N_0}}, \quad N_0 = \frac{1}{2}\left[1 + \frac{\sinh 2kh}{2kh}\right], \; (n=0) \tag{16}$$

$$Z_n(z) = \frac{\cos k_n(z+h)}{\sqrt{N_n}}, \quad N_n = \frac{1}{2}\left[1 + \frac{\sin 2k_n h}{2k_n h}\right], \; (n = 1,2,\cdots,\infty) \tag{17}$$

$A_{mn}^{(j,p)}$, $B_{mn}^{(j,p)}$, $C_{mn}^{(j,p)}$ and $D_{mn}^{(j,p)}$ are the unknown complex coefficients to be determined; J_m is the Bessel function of the first kind of order m, I_m and K_m are respectively the modified Bessel function of the first and the second kinds of order m; a_l and b_l are respectively the shortest and the longest distance from the origin to the structure surface along the radial direction at $r = R_l(\theta)$ $(l = 1,\ 2,\ 3)$. The wavenumber k, and the vertical eigenvalue k_n for Regions II and IV are given by

$$\frac{\omega^2}{g} - k\tanh k = 0 \tag{18}$$

$$k_0 = -ik, \quad \frac{\omega^2}{g} + k_n \tan k_n = 0 \ \ (n = 1, 2, \infty) \tag{19}$$

and vertical eigenvalues p_n and q_n for Regions I and III are respectively given by

$$p_n = \frac{\pi n}{h - d_1} \ \ (n = 0, 1, 2, \infty) \tag{20}$$

$$q_n = \frac{\pi n}{h - d_2} \ \ (n = 0, 1, 2, \infty) \tag{21}$$

The particular solutions $\phi_{p_1}^{(j,p)}$ and $\phi_{p_3}^{(j,p)}$ are given by

$$\begin{aligned} \phi_{p_1}^{(j,p)}(r,\theta,z) &= -\delta_{0j}\delta_{0p}\phi_I \\ &+ \tfrac{4(z+h)^2-(1+\delta_{3j})r^2}{8(h-d_1)}\left\{\delta_{3j} + \delta_{4j}\left(r\sin\theta - y_{G_p}\right) - \delta_{5j}\left(r\cos\theta - x_{G_p}\right)\right\}\delta_{1p} + \tfrac{\left(\delta_{4j}y_{G_p} - \delta_{5j}x_{G_p}\right)r^2}{8(h-d_1)} \end{aligned} \tag{22}$$

$$\begin{aligned} \phi_{p_3}^{(j,p)}(r,\theta,z) &= -\delta_{0j}\delta_{0p}\phi_I \\ &+ \tfrac{4(z+h)^2-(1+\delta_{3j})r^2}{8(h-d_2)}\left\{\delta_{3j} + \delta_{4j}\left(r\sin\theta - y_{G_p}\right) - \delta_{5j}\left(r\cos\theta - x_{G_p}\right)\right\}\delta_{2p} + \tfrac{\left(\delta_{4j}y_{G_p} - \delta_{5j}x_{G_p}\right)r^2}{8(h-d_2)} \end{aligned} \tag{23}$$

where δ_{ij} is the Kronecker delta (1 if $i = j$, 0 if $i \neq j$) and (x_{G_p}, y_{G_p}) the horizontal coordinates of the centre of gravity of the oscillating body p. The incident velocity potential in the cylindrical coordinate system is given by

$$\phi_I(r,\theta,z) = -\frac{igA}{\omega}\sum_{m=-\infty}^{\infty} J_m(kr)e^{im(\theta+\frac{\pi}{2}-\beta)}\frac{\cosh k(z+h)}{\cosh kh} \tag{24}$$

The matching conditions for the pressure and velocity continuities are

$$\nabla\phi_2^{(j,p)}\cdot\mathbf{n_{s_1}} = \begin{bmatrix} \{-\delta_{0p}\nabla\phi_I + (\delta_{1p}+\delta_{3p})\mathbf{n}_j\}\cdot\mathbf{n_{s_1}} \\ \nabla\phi_1^{(j,p)}\cdot\mathbf{n_{s_1}} \end{bmatrix} \quad \text{at } r = R_1 \text{ and } \begin{cases} -d_1 \le z \le 0 \\ -h \le z \le -d_1 \end{cases} \tag{25}$$

$$\phi_{1,p} = \phi_{2,p} \quad \text{at } r = R_1 \text{ and } -h \le z \le -d_1 \tag{26}$$

$$\nabla\phi_2^{(j,p)}\cdot\mathbf{n_{s_2}} = \begin{bmatrix} \{-\delta_{0p}\nabla\phi_I + (\delta_{2p}+\delta_{3p})\mathbf{n}_j\}\cdot\mathbf{n_{s_2}} \\ \nabla\phi_3^{(j,p)}\cdot\mathbf{n_{s_2}} \end{bmatrix} \quad \text{at } r = R_2 \text{ and } \begin{cases} -d_2 \le z \le 0 \\ -h \le z \le -d_2 \end{cases} \tag{27}$$

$$\phi_{3,p} = \phi_{2,p} \quad \text{at } r = R_2 \text{ and } -h \le z \le -d_2 \tag{28}$$

$$\nabla\phi_4^{(j,p)}\cdot\mathbf{n_{s_3}} = \begin{bmatrix} \{-\delta_{0p}\nabla\phi_I + (\delta_{2p}+\delta_{3p})\mathbf{n}_j\}\cdot\mathbf{n_{s_3}} \\ \nabla\phi_3^{(j,p)}\cdot\mathbf{n_{s_3}} \end{bmatrix} \quad \text{at } r = R_3 \text{ and } \begin{cases} -d_2 \le z \le 0 \\ -h \le z \le -d_2 \end{cases} \tag{29}$$

$$\phi_3^{(j,p)} = \phi_4^{(j,p)} \quad \text{at } r = R_3 \text{ and } -h \le z \le -d_2 \tag{30}$$

where R_l $(l = 1,\ 2,\ 3)$ denotes the radius function as defined in Equation (1) and $\mathbf{n_j}\cdot\mathbf{n_{s_l}}$ $(l = 1,\ 2,\ 3)$ for 6 DOFs is provided in Appendix A.

We consider the horizontal coordinates of the centre of gravity to coincide with the origin, i.e., $(x_{G_1}, y_{G_1}) = (x_{G_2}, y_{G_2}) = (0,\ 0)$; however, the vertical coordinate of the centre of gravity may not be zero, $z_{G_1} \neq 0$ and $z_{G_2} \neq 0$. The assumed velocity potentials given in Equations (12) and (15) are substituted into the matching conditions given in Equations (25)–(30). This furnishes

$$\begin{aligned} &\sum_{m=-\infty}^{\infty}\sum_{n=0}^{\infty}\left[B_{mn}^{(j,p)}\left\{R_1^2\tfrac{\mathcal{I}'_{mn}(k_nR_1)}{\mathcal{I}_{mn}(k_nb_2)} + im\tfrac{\mathcal{I}_{mn}(k_nR_1)}{\mathcal{I}_{mn}(k_nb_2)}S_{1,\theta}\right\} + C_{mn}^{(j,p)}\left\{R_1^2\tfrac{\mathcal{K}'_{mn}(k_nR_1)}{\mathcal{K}_{mn}(k_na_1)} + im\tfrac{\mathcal{K}_{mn}(k_nR_1)}{\mathcal{K}_{mn}(k_na_1)}S_{1,\theta}\right\}\right]\tfrac{Z_n(z)}{Z_n(0)}e^{im\theta} \\ &= (\delta_{0,p} + \delta_{1,p} + \delta_{3,p})\mathcal{H}_1^{(j)} \\ &\qquad (r = R_1, -d_1 \le z \le 0,\ 0 \le \theta \le 2\pi) \end{aligned} \tag{31}$$

$$\begin{aligned}
&\sum_{m=-\infty}^{\infty}\sum_{n=0}^{\infty}\Big[B_{mn}^{(j,p)}\Big\{R_1^2\tfrac{\mathcal{I}'_{mn}(k_nR_1)}{\mathcal{I}_{mn}(k_nb_2)}+im\tfrac{\mathcal{I}_{mn}(k_nR_1)}{\mathcal{I}_{mn}(k_nb_2)}S_{1,\theta}\Big\}+C_{mn}^{(j,p)}\Big\{R_1^2\tfrac{\mathcal{K}'_{mn}(k_nR_1)}{\mathcal{K}_{mn}(k_na_1)}S_{1,\theta}+im\tfrac{\mathcal{K}_{mn}(k_nR_1)}{\mathcal{K}_{mn}(k_na_1)}S_{1,\theta}\Big\}\Big]\tfrac{Z_n(z)}{Z_n(0)}e^{im\theta}\\
&-\sum_{m=-\infty}^{\infty}\sum_{n=0}^{\infty}A_{mn}^{(j,p)}\Big\{R_1^2\tfrac{\mathcal{I}'_{mn}(p_nR_1)}{\mathcal{I}_{mn}(p_nb_1)}+im\tfrac{\mathcal{I}_{mn}(p_nR_1)}{\mathcal{I}_{mn}(p_nb_1)}S_{1,\theta}\Big\}\cos p_n(z+h)e^{im\theta}\\
&=(\delta_{0,p}+\delta_{1,p}+\delta_{3,p})\tilde{\mathcal{P}}_1^{(j)}\\
&\qquad(r=R_1,-h\le z\le -d_1,\ 0\le\theta\le 2\pi)
\end{aligned}\tag{32}$$

$$\begin{aligned}
&\sum_{m=-\infty}^{\infty}\sum_{n=0}^{\infty}\Big\{B_{mn}^{(j,p)}\frac{\mathcal{I}_{mn}(k_nR_1)}{\mathcal{I}_{mn}(k_nb_2)}+C_{mn}^{(j,p)}\frac{\mathcal{K}_{mn}(k_nR_1)}{\mathcal{K}_{mn}(k_na_1)}\Big\}\frac{Z_n(z)}{Z_n(0)}e^{im\theta}-\sum_{m=-\infty}^{\infty}\sum_{n=0}^{\infty}A_{mn}^{(j,p)}\frac{\mathcal{I}_{mn}(p_nR_1)}{\mathcal{I}_{mn}(p_nb_1)}\cos p_n(z+h)e^{im\theta}\\
&=(\delta_{0,p}+\delta_{1,p}+\delta_{3,p})\mathcal{P}_1^{(j)}\\
&\qquad(r=R_1,-h\le z\le -d_1,\ 0\le\theta\le 2\pi)
\end{aligned}\tag{33}$$

$$\begin{aligned}
&\sum_{m=-\infty}^{\infty}\sum_{n=0}^{\infty}\Big[B_{mn}^{(j,p)}\Big\{R_2^2\tfrac{\mathcal{I}'_{mn}(k_nR_2)}{\mathcal{I}_{mn}(k_nb_2)}+im\tfrac{\mathcal{I}_{mn}(k_nR_2)}{\mathcal{I}_{mn}(k_nb_2)}S_{2,\theta}\Big\}+C_{mn}^{(j,p)}\Big\{R_2^2\tfrac{\mathcal{K}'_{mn}(k_nR_2)}{\mathcal{K}_{mn}(k_na_1)}+im\tfrac{\mathcal{K}_{mn}(k_nR_2)}{\mathcal{K}_{mn}(k_na_1)}S_{1,\theta}\Big\}\Big]\tfrac{Z_n(z)}{Z_n(0)}e^{im\theta}\\
&=(\delta_{0,p}+\delta_{2,p}+\delta_{3,p})\mathcal{H}_2^{(j)}\\
&\qquad(r=R_2,-d_2\le z\le 0,\ 0\le\theta\le 2\pi)
\end{aligned}\tag{34}$$

$$\begin{aligned}
&\sum_{m=-\infty}^{\infty}\sum_{n=0}^{\infty}\Big[B_{mn}^{(j,p)}\Big\{R_2^2\tfrac{\mathcal{I}'_{mn}(k_nR_2)}{\mathcal{I}_{mn}(k_nb_2)}+im\tfrac{\mathcal{I}_{mn}(k_nR_2)}{\mathcal{I}_{mn}(k_nb_2)}S_{2,\theta}\Big\}+C_{mn}^{(j,p)}\Big\{R_2^2\tfrac{\mathcal{K}'_{mn}(k_nR_2)}{\mathcal{K}_{mn}(k_na_1)}+im\tfrac{\mathcal{K}_{mn}(k_nR_2)}{\mathcal{K}_{mn}(k_na_1)}S_{1,\theta}\Big\}\Big]\tfrac{Z_n(z)}{Z_n(0)}e^{im\theta}\\
&-\sum_{m=-\infty}^{\infty}\sum_{n=0}^{\infty}\Big[D_{mn}^{(j,p)}\Big\{R_2^2\tfrac{\mathcal{I}'_{mn}(q_nR_2)}{\mathcal{I}_{mn}(q_nb_3)}+im\tfrac{\mathcal{I}_{mn}(q_nR_2)}{\mathcal{I}_{mn}(q_nb_3)}S_{2,\theta}\Big\}\\
&+E_{mn}^{(j,p)}\Big\{R_2^2\tfrac{\mathcal{K}'_{mn}(q_nR_2)}{\mathcal{K}_{mn}(q_na_2)}+im\tfrac{\mathcal{K}_{mn}(q_nR_2)}{\mathcal{K}_{mn}(q_na_2)}S_{2,\theta}\Big\}\Big]+\cos q_n(z+h)e^{im\theta}=(\delta_{0,p}+\delta_{2,p}+\delta_{3,p})\tilde{\mathcal{Q}}_2^{(j)}\\
&\qquad(r=R_2,-h\le z\le -d_2,\ 0\le\theta\le 2\pi)
\end{aligned}\tag{35}$$

$$\begin{aligned}
&\sum_{m=-\infty}^{\infty}\sum_{n=0}^{\infty}\Big\{B_{mn}^{(j,p)}\frac{\mathcal{I}_{mn}(k_nR_2)}{\mathcal{I}_{mn}(k_nb_2)}+C_{mn}^{(j,p)}\frac{\mathcal{K}_{mn}(k_nR_2)}{\mathcal{K}_{mn}(k_na_1)}\Big\}\frac{Z_n(z)}{Z_n(0)}e^{im\theta}\\
&-\sum_{m=-\infty}^{\infty}\sum_{n=0}^{\infty}\Big\{D_{mn}^{(j,p)}\frac{\mathcal{I}_{mn}(q_nR_2)}{\mathcal{I}_{mn}(q_nb_3)}+E_{mn}^{(j,p)}\frac{\mathcal{K}_{mn}(q_nR_2)}{\mathcal{K}_{mn}(q_na_2)}\Big\}\cos q_n(z+h)e^{im\theta}=(\delta_{0,p}+\delta_{2,p}+\delta_{3,p})\mathcal{Q}_2^{(j)}\\
&\qquad(r=R_2,-h\le z\le -d_2,\ 0\le\theta\le 2\pi)
\end{aligned}\tag{36}$$

$$\begin{aligned}
&\sum_{m=-\infty}^{\infty}\sum_{n=0}^{\infty}F_{mn}^{(j,p)}\Big\{R_3^2\frac{\mathcal{K}'_{mn}(k_nR_3)}{\mathcal{K}_{mn}(k_na_3)}+im\frac{\mathcal{K}_{mn}(k_nR_3)}{\mathcal{K}_{mn}(k_na_3)}S_{3,\theta}\Big\}\frac{Z_n(z)}{Z_n(0)}e^{im\theta}=(\delta_{0,p}+\delta_{2,p}+\delta_{3,p})\mathcal{H}_3^{(j)}\\
&\qquad(r=R_3,-d_2\le z\le 0,\ 0\le\theta\le 2\pi)
\end{aligned}\tag{37}$$

$$\begin{aligned}
&\sum_{m=-\infty}^{\infty}\Big[\sum_{n=0}^{\infty}F_{mn}^{(j,p)}\Big\{R_3^2\tfrac{\mathcal{K}'_{mn}(k_nR_3)}{\mathcal{K}_{mn}(k_na_3)}+im\tfrac{\mathcal{K}_{mn}(k_nR_3)}{\mathcal{K}_{mn}(k_na_3)}S_{3,\theta}\Big\}\tfrac{Z_n(z)}{Z_n(0)}e^{im\theta}\Big]\\
&-\sum_{m=-\infty}^{\infty}\sum_{n=0}^{\infty}\Big[D_{mn}^{(j,p)}\Big\{R_3^2\tfrac{\mathcal{I}'_{mn}(q_nR_3)}{\mathcal{I}_{mn}(q_nb_3)}+im\tfrac{\mathcal{I}_{mn}(q_nR_3)}{\mathcal{I}_{mn}(q_nb_3)}S_{3,\theta}\Big\}\\
&+E_{mn}^{(j,p)}\Big\{R_3^2\tfrac{\mathcal{K}'_{mn}(q_nR_3)}{\mathcal{K}_{mn}(q_na_2)}+im\tfrac{\mathcal{K}_{mn}(q_nR_3)}{\mathcal{K}_{mn}(q_na_2)}S_{3,\theta}\Big\}\Big]\cos q_n(z+h)e^{im\theta}=(\delta_{0,p}+\delta_{2,p}+\delta_{3,p})\tilde{\mathcal{Q}}_3^{(j)}\\
&\qquad(r=R_3,-h\le z\le -d_2,\ 0\le\theta\le 2\pi)
\end{aligned}\tag{38}$$

$$\begin{aligned}
&\sum_{m=-\infty}^{\infty}\sum_{n=0}^{\infty}F_{mn}^{(j,p)}\frac{\mathcal{K}_{mn}(k_nR_3)}{\mathcal{K}_{mn}(k_na_3)}\frac{Z_n(z)}{Z_n(0)}e^{im\theta}-\sum_{m=-\infty}^{\infty}\sum_{n=0}^{\infty}\Big\{D_{mn}^{(j,p)}\frac{\mathcal{I}_{mn}(q_nR_3)}{\mathcal{I}_{mn}(q_nb_3)}+E_{mn}^{(j,p)}\frac{\mathcal{K}_{mn}(q_nR_3)}{\mathcal{K}_{mn}(q_na_2)}\Big\}\cos q_n(z+h)e^{im\theta}\\
&=(\delta_{0,p}+\delta_{2,p}+\delta_{3,p})\mathcal{Q}_3^{(j)}\\
&\qquad(r=R_3,-h\le z\le -d_2,\ 0\le\theta\le 2\pi)
\end{aligned}\tag{39}$$

where $S_{l,\theta}$ denotes $\left.\frac{\partial S_l}{\partial\theta}\right|_{r=R_l(\theta)}$ $(l=1,\ 2,\ 3)$, $\mathcal{K}'_{mn}$ and $\mathcal{I}'_{mn}$ are respectively the derivatives of $\mathcal{K}_{mn}$ and $\mathcal{I}_{mn}$ with respect to r. $\mathcal{I}_{mn}(k_nr)$, $\mathcal{I}_{mn}(p_nr)$, $\mathcal{I}_{mn}(q_nr)$, $\mathcal{K}_{mn}(q_nr)$ and $\mathcal{K}_{mn}(k_nr)$ are given by

$$\mathcal{I}_{mn}(k_nr)=\begin{cases}J_m(kr) & \text{for } n=0\\ \frac{I_m(k_nr)}{I_m(k_nb_2)} & \text{for } n>0\end{cases}\tag{40}$$

$$\mathcal{I}_{mn}(p_nr)=\begin{cases}\left(\frac{r}{b_1}\right)^{|m|} & \text{for } n=0\\ \frac{I_m(p_nr)}{I_m(p_nb_1)} & \text{for } n>0\end{cases}\tag{41}$$

$$\mathcal{I}_{mn}(q_nr)=\begin{cases}\ln\frac{r}{a_2} & \text{for } n=0 \text{ and } m=0\\ \left(\frac{r}{b_3}\right)^{|m|} & \text{for } n=0 \text{ and } m\neq 0\\ \frac{I_m(q_nr)}{I_m(q_nb_2)} & \text{for } n>0\end{cases}\tag{42}$$

$$\mathcal{K}_{mn}(q_nr)=\begin{cases}\ln\frac{b_3}{r} & \text{for } n=0 \text{ and } m=0\\ \left(\frac{r}{a_2}\right)^{-|m|} & \text{for } n=0 \text{ and } m\neq 0\\ \frac{K_m(q_nr)}{K_m(q_nb_2)} & \text{for } n>0\end{cases}\tag{43}$$

$$\mathcal{K}_{mn}(k_n r) = \begin{cases} \frac{H_m(kr)}{H_m(ka_3)} & \text{for } n = 0 \\ \frac{K_m(k_n r)}{K_m(k_n a_3)} & \text{for } n > 0 \end{cases} \tag{44}$$

$\mathcal{H}_l^{(j)}\mathcal{P}_l^{(j)}, \mathcal{Q}_l^{(j)}, \tilde{\mathcal{P}}_l^{(j)}$ and $\tilde{\mathcal{Q}}_l^{(j)}$ $(j = 0, 1, \cdots, 6; l = 1, 2, 3)$ are provided in Appendix B.

In the foregoing formulation, numerical integration should be used for establishing the associated simultaneous equations because the radius functions are substituted into the arguments of radial terms such as the Bessel function, modified Bessel function and radial polynomial functions. Liu, et al. [24] replaced such radial terms with the Fourier expansion series so that the integration could be performed analytically. Recently, Park and Wang [20] suggested an improved method that enables the computational time to be significantly reduced by using minimal numbers of Fourier coefficient sets. By using Park and Wang [20] improved method, 13 sets of Fourier coefficients such as $a_{m,n,q}^{(1)}$, $b_{m,n,q}^{(1),(2)}$, $c_{m,n,q}^{(1),(2)}$, $d_{m,n,q}^{(2),(3)}$, $e_{m,n,q}^{(2),(3)}$, $f_{m,q}^{(3)}$ and $g_{m,q}^{(1,2,3)}$ are introduced for solving diffraction and radiation problems. The applied Fourier expansions are as given in Table 2 and their derivatives with respect to r can be directly obtained by using the formerly obtained Fourier coefficients (see Table 3). The Fourier coefficients can be calculated by the functional orthogonality, i.e., by multiplying $e^{-iq\theta}$ and integrating over $[0,\ 2\pi]$.

Table 2. Fourier expansions for the functions used in velocity potential.

Velocity Potential	Condition	Functions Used		r	Fourier Expansions
ϕ_1	$n = 0$	Incoming waves	$r^{\|m\|}$	R_1	$\sum_{q=-\infty}^{\infty} a_{m,n,q}^{(1)} e^{iq\theta}$
	$n > 0$		$I_m(p_n r)$		
ϕ_2	$n = 0$	Incoming waves	$J_m(kr)$	$R_{1,2}$	$\sum_{q=-\infty}^{\infty} b_{m,n,q}^{(1,2)} e^{iq\theta}$
	$n > 0$		$I_m(k_n r)$		
	$n = 0$	Outgoing waves	$H_m(kr)$		$\sum_{q=-\infty}^{\infty} c_{m,n,q}^{(1,2)} e^{iq\theta}$
	$n > 0$		$K_m(k_n r)$		
ϕ_3	$n = 0$, $m = 0$	Incoming waves	$\ln \frac{r}{a_2}$	$R_{2,3}$	$\sum_{q=-\infty}^{\infty} d_{m,n,q}^{(2,3)} e^{iq\theta}$
	$n = 0$, $m \neq 0$		$r^{\|m\|}$		
	$n > 0$		$I_m(q_n r)$		
	$n = 0$, $m = 0$	Outgoing waves	$\ln \frac{b_3}{r}$		$\sum_{q=-\infty}^{\infty} e_{m,n,q}^{(2,3)} e^{iq\theta}$
	$n = 0$, $m \neq 0$		$r^{-\|m\|}$		
	$n > 0$		$K_m(q_n r)$		
ϕ_4	$n = 0$	Outgoing waves	$H_m(k_n r)$	R_1	$\sum_{q=-\infty}^{\infty} f_{m,n,q}^{(1)} e^{iq\theta}$
	$n > 0$		$K_m(k_n r)$		
ϕ_I	N/A	Incident waves	$J_m(kr)$	$R_{1,2,3}$	$\sum_{q=-\infty}^{\infty} g_{m,n,q}^{(1,2,3)} e^{iq\theta}$

Table 3. Fourier expansions for the derivative of functions used in velocity potential.

Derivative of Velocity Potential	Condition	Functions Used		r	Fourier Expansions
$\frac{\partial\phi_1}{\partial r}$	$n=0$	Incoming waves	$\lvert m\rvert r^{\lvert m\rvert-1}$	R_1	$\frac{\lvert m\rvert}{r}\sum\limits_{q=-\infty}^{\infty} a^{(1)}_{m,0,q}\mathrm{e}^{iq\theta}$
	$n>0$		$I'_m(p_n r)$		$\frac{p_n}{2}\sum\limits_{q=-\infty}^{\infty}\left(a^{(1)}_{m-1,n,q}+a^{(1)}_{m+1,n,q}\right)\mathrm{e}^{iq\theta}$
$\frac{\partial\phi_2}{\partial r}$	$n=0$	Incoming waves	$J'_m(kr)$	$R_{1,2}$	$\frac{k}{2}\sum\limits_{q=-\infty}^{\infty}\left(b^{(1,2)}_{m-1,0,q}-b^{(1,2)}_{m+1,0,q}\right)\mathrm{e}^{iq\theta}$
	$n>0$		$I'_m(k_n r)$		$\frac{k_n}{2}\sum\limits_{q=-\infty}^{\infty}\left(b^{(1,2)}_{m-1,n,q}+b^{(1,2)}_{m+1,n,q}\right)\mathrm{e}^{iq\theta}$
	$n=0$	Outgoing waves	$H'_m(kr)$		$\frac{k}{2}\sum\limits_{q=-\infty}^{\infty}\left(c^{(1,2)}_{m-1,0,q}-c^{(1,2)}_{m+1,0,q}\right)\mathrm{e}^{iq\theta}$
	$n>0$		$K'_m(k_n r)$		$-\frac{k_n}{2}\sum\limits_{q=-\infty}^{\infty}\left(c^{(1,2)}_{m-1,n,q}+c^{(1,2)}_{m+1,n,q}\right)\mathrm{e}^{iq\theta}$
$\frac{\partial\phi_3}{\partial r}$	$n=0$, $m=0$	Incoming waves	$\ln\frac{r}{a_2}$	$R_{2,3}$	$\sum\limits_{q=-\infty}^{\infty} d^{(2,3)}_{0,0,q}\mathrm{e}^{iq\theta}$
	$n=0$, $m\neq 0$		$r^{\lvert m\rvert}$		$\frac{\lvert m\rvert}{r}\sum\limits_{q=-\infty}^{\infty} d^{(2,3)}_{m,0,q}\mathrm{e}^{iq\theta}$
	$n>0$		$I_m(q_n r)$		$\frac{q_n}{2}\sum\limits_{q=-\infty}^{\infty}\left(d^{(2,3)}_{m-1,n,q}+d^{(2,3)}_{m+1,n,q}\right)\mathrm{e}^{iq\theta}$
	$n=0$, $m=0$	Outgoing waves	$\ln\frac{b_3}{r}$		$-\sum\limits_{q=-\infty}^{\infty} e^{(2,3)}_{0,0,q}\mathrm{e}^{iq\theta}$
	$n=0$, $m\neq 0$		$r^{-\lvert m\rvert}$		$-\frac{\lvert m\rvert}{r}\sum\limits_{q=-\infty}^{\infty} e^{(2,3)}_{m,0,q}\mathrm{e}^{iq\theta}$
	$n>0$		$K_m(q_n r)$		$-\frac{q_n}{2}\sum\limits_{q=-\infty}^{\infty}\left(e^{(2,3)}_{m-1,n,q}+e^{(2,3)}_{m+1,n,q}\right)\mathrm{e}^{iq\theta}$
$\frac{\partial\phi_4}{\partial r}$	$n=0$	Outgoing waves	$H'_m(k_n r)$	R_3	$\frac{k}{2}\sum\limits_{q=-\infty}^{\infty}\left(f^{(3)}_{m-1,0,q}-f^{(3)}_{m+1,0,q}\right)\mathrm{e}^{iq\theta}$
	$n>0$		$K'_m(k_n r)$		$-\frac{k_n}{2}\sum\limits_{q=-\infty}^{\infty}\left(f^{(3)}_{m-1,n,q}+f^{(3)}_{m+1,n,q}\right)\mathrm{e}^{iq\theta}$
$\frac{\partial\phi_I}{\partial r}$	N/A	Incident waves	$J'_m(kr)$	$R_{1,2,3}$	$\frac{k}{2}\sum\limits_{q=-\infty}^{\infty}\left(g^{(1,2,3)}_{m-1,0,q}-g^{(1,2,3)}_{m+1,0,q}\right)\mathrm{e}^{iq\theta}$

Also, the radius function and the derivative of the surface function with respect to θ can be obtained by using $a_{1,0,q}$, $d_{1,0,q}$ and $e_{1,0,q}$.

$$r\big|_{r=R_1(\theta)} = \sum_{n_r=-\infty}^{\infty} a^{(1)}_{1,0,n_r}\mathrm{e}^{in_r\theta} \tag{45}$$

$$r\big|_{r=R_{2,3}(\theta)} = \sum_{n_r=-\infty}^{\infty} d^{(2,3)}_{1,0,n_r}\mathrm{e}^{in_r\theta} \tag{46}$$

$$\frac{1}{r}\bigg|_{r=R_{2,3}(\theta)} = \sum_{n_r=-\infty}^{\infty} e^{(2,3)}_{1,0,n_r}\mathrm{e}^{in_r\theta} \tag{47}$$

$$\frac{\partial S_1}{\partial\theta}\bigg|_{S_1=0} = -\sum_{n_r=-\infty}^{\infty}(in_r)a^{(1)}_{1,0,n_r}\mathrm{e}^{in_r\theta} \tag{48}$$

$$\frac{\partial S_{2,3}}{\partial\theta}\bigg|_{S_{2,3}=0} = -\sum_{n_r=-\infty}^{\infty}(in_r)d^{(2,3)}_{1,0,n_r}\mathrm{e}^{in_r\theta} \tag{49}$$

Equations (31)–(39) can be thus expressed by using the Fourier coefficients obtained in Tables 2 and 3 as

$$\sum_{m=-\infty}^{\infty}\sum_{n_r=-\infty}^{\infty}\sum_{q=-\infty}^{\infty} e^{i(m+n_r+q)\theta}\sum_{n=0}^{\infty}\left[B_{mn}^{(j,p)}\tilde{b}'^{(1)}_{m,n,q}+C_{mn}^{(j,p)}\tilde{c}'^{(1)}_{m,n,q}\right]\frac{Z_n(z)}{Z_n(0)}=(\delta_{0,p}+\delta_{1,p}+\delta_{3,p})\mathcal{H}_1^{(j)} \quad (r=R_1,-d_1\le z\le 0,\ 0\le\theta\le 2\pi) \tag{50}$$

$$\begin{aligned}&\sum_{m=-\infty}^{\infty}\sum_{n_r=-\infty}^{\infty}\sum_{q=-\infty}^{\infty} e^{i(m+n_r+q)\theta}\sum_{n=0}^{\infty}\left[B_{mn}^{(j,p)}\tilde{b}'^{(1)}_{m,n,q}+C_{mn}^{(j,p)}\tilde{c}'^{(1)}_{m,n,q}\right]\frac{Z_n(z)}{Z_n(0)}-\sum_{m=-\infty}^{\infty}\sum_{nq=-\infty}^{\infty}\sum_{q=-\infty}^{\infty} e^{i(m+n_r+q)\theta}\sum_{n=0}^{\infty}A_{mn}^{(j,p)}\tilde{a}'^{(1)}_{m,n,p}\\&=(\delta_{0,p}+\delta_{1,p}+\delta_{3,p})\tilde{\mathcal{P}}_1^{(j)}\\&\qquad (r=R_1,-h\le z\le -d_1,\ 0\le\theta\le 2\pi)\end{aligned} \tag{51}$$

$$\begin{aligned}&\sum_{m=-\infty}^{\infty}\sum_{q=-\infty}^{\infty} e^{i(m+q)\theta}\sum_{n=0}^{\infty}\left\{B_{mn}^{(j,p)}\tilde{b}'^{(1)}_{m,n,p}+C_{mn}^{(j,p)}\tilde{c}^{(1)}_{m,n,p}\right\}\frac{Z_n(z)}{Z_n(0)}-\sum_{m=-\infty}^{\infty}\sum_{q=-\infty}^{\infty} e^{i(m+q)\theta}\sum_{n=0}^{\infty}A_{mn}^{(j,p)}\tilde{a}'^{(1)}_{m,n,p}\cos p_n(z+h)\\&=(\delta_{0,p}+\delta_{1,p}+\delta_{3,p})\mathcal{P}_1^{(j)}\\&\qquad (r=R_1,-h\le z\le -d_1,\ 0\le\theta\le 2\pi)\end{aligned} \tag{52}$$

$$\sum_{m=-\infty}^{\infty}\sum_{n_r=-\infty}^{\infty}\sum_{q=-\infty}^{\infty} e^{i(m+n_r+q)\theta}\sum_{n=0}^{\infty}\left[B_{mn}^{(j,p)}\tilde{b}'^{(2)}_{m,n,q}+C_{mn}^{(j,p)}\tilde{c}'^{(2)}_{m,n,q}\right]\frac{Z_n(z)}{Z_n(0)}=(\delta_{0,p}+\delta_{2,p}+\delta_{3,p})\mathcal{H}_2^{(j)} \quad (r=R_2,-d_2\le z\le 0,\ 0\le\theta\le 2\pi) \tag{53}$$

$$\begin{aligned}&\sum_{m=-\infty}^{\infty}\sum_{n_r=-\infty}^{\infty}\sum_{q=-\infty}^{\infty} e^{i(m+n_r+q)\theta}\sum_{n=0}^{\infty}\left[B_{mn}^{(j,p)}\tilde{b}'^{(2)}_{m,n,q}+C_{mn}^{(j,p)}\tilde{c}'^{(2)}_{m,n,q}\right]\frac{Z_n(z)}{Z_n(0)}\\&-\sum_{m=-\infty}^{\infty}\sum_{n_r=-\infty}^{\infty}\sum_{q=-\infty}^{\infty} e^{i(m+n_r+q)\theta}\sum_{n=0}^{\infty}\left[D_{mn}^{(j,p)}\tilde{d}'^{(2)}_{m,n,q}+E_{mn}^{(j,p)}\tilde{e}'^{(2)}_{m,n,q}\right]\cos q_n(z+h)\\&=(\delta_{0,p}+\delta_{2,p}+\delta_{3,p})\tilde{Q}_2^{(j)}\\&\qquad (r=R_2,-h\le z\le -d_2,\ 0\le\theta\le 2\pi)\end{aligned} \tag{54}$$

$$\begin{aligned}&\sum_{m=-\infty}^{\infty}\sum_{q=-\infty}^{\infty} e^{i(m+q)\theta}\sum_{n=0}^{\infty}\left[B_{mn}^{(j,p)}\tilde{b}^{(2)}_{m,n,q}+C_{mn}^{(j,p)}\tilde{c}^{(2)}_{m,n,q}\right]\frac{Z_n(z)}{Z_n(0)}\\&-\sum_{m=-\infty}^{\infty}\sum_{q=-\infty}^{\infty} e^{i(m+q)\theta}\sum_{n=0}^{\infty}\left[D_{mn}^{(j,p)}\tilde{d}^{(2)}_{m,n,q}+E_{mn}^{(j,p)}\tilde{e}^{(2)}_{m,n,q}\right]\cos q_n(z+h)=(\delta_{0,p}+\delta_{2,p}+\delta_{3,p})Q_2^{(j)}\\&\qquad (r=R_2,-h\le z\le -d_2,\ 0\le\theta\le 2\pi)\end{aligned} \tag{55}$$

$$\sum_{m=-\infty}^{\infty}\sum_{n_r=-\infty}^{\infty}\sum_{q=-\infty}^{\infty} e^{i(m+n_r+q)\theta}\sum_{n=0}^{\infty}F_{mn}^{(j,p)}\tilde{f}'^{(3)}_{m,n,q}\frac{Z_n(z)}{Z_n(0)}=(\delta_{0,p}+\delta_{2,p}+\delta_{3,p})\mathcal{H}_3^{(j)} \quad (r=R_3,-d_2\le z\le 0,\ 0\le\theta\le 2\pi) \tag{56}$$

$$\begin{aligned}&\sum_{m=-\infty}^{\infty}\sum_{n_r=-\infty}^{\infty}\sum_{q=-\infty}^{\infty} e^{i(m+n_r+q)\theta}\sum_{n=0}^{\infty}F_{mn}^{(j,p)}\tilde{f}'^{(3)}_{m,n,q}\frac{Z_n(z)}{Z_n(0)}\\&-\sum_{m=-\infty}^{\infty}\sum_{n_r=-\infty}^{\infty}\sum_{q=-\infty}^{\infty} e^{i(m+n_r+q)\theta}\sum_{n=0}^{\infty}\left[D_{mn}^{(j,p)}\tilde{d}'^{(3)}_{m,n,q}+E_{mn}^{(j,p)}\tilde{e}'^{(3)}_{m,n,q}\right]\cos q_n(z+h)\\&=(\delta_{0,p}+\delta_{2,p}+\delta_{3,p})\tilde{Q}_3^{(j)}\\&\qquad (r=R_3,-h\le z\le -d_2,\ 0\le\theta\le 2\pi)\end{aligned} \tag{57}$$

$$\begin{aligned}&\sum_{m=-\infty}^{\infty}\sum_{q=-\infty}^{\infty} e^{i(m+q)\theta}\sum_{n=0}^{\infty}F_{mn}^{(j,p)}\tilde{f}^{(3)}_{m,n,q}\frac{Z_n(z)}{Z_n(0)}-\sum_{m=-\infty}^{\infty}\sum_{q=-\infty}^{\infty} e^{i(m+q)\theta}\sum_{n=0}^{\infty}\left[D_{mn}^{(j,p)}\tilde{d}^{(3)}_{m,n,q}+E_{mn}^{(j,p)}\tilde{e}^{(3)}_{m,n,q}\right]\cos q_n(z+h)\\&=(\delta_{0,p}+\delta_{2,p}+\delta_{3,p})Q_3^{(j)}\\&\qquad (r=R_3,-h\le z\le -d_2,\ 0\le\theta\le 2\pi)\end{aligned} \tag{58}$$

where the reduced forms of $\tilde{a}^{(1)}_{m,n,q}$, $\tilde{b}^{(1),(2)}_{m,n,q}$, $\tilde{c}^{(1),(2)}_{m,n,q}$, $\tilde{d}^{(2),(3)}_{m,n,q}$, $\tilde{e}^{(2),(3)}_{m,n,q}$, $\tilde{f}^{(3)}_{m,n,q}$ and their normal derivatives are given in Appendix C.

By multiplying the corresponding vertical eigenfunctions $\frac{1}{h}Z_n(z)$ in Equations (50), (51), (53), (54), (56) and (57), $\frac{1}{h-d_1}\cos p_n(z+h)$ in Equation (52), $\frac{1}{h-d_2}\cos q_n(z+h)$ in Equations (55) and (58) and the angular eigenfunction $e^{-im\theta}$ in Equations (50)–(58) and integrating the equations for the associated integral intervals, one can combine Equation (50) with (51), Equation (53) with (54) and Equation (56) with (57). By truncating the series terms at $m=M$, $n=N$, $n_r=N_r$ and $q=N_q$, $9(2M+1)(N+1)$ equations for the monolithic motion or $18(2M+1)(N+1)$ equations for the individual motion, and the same number of unknowns are given for the diffraction problem and each radiation mode. Consequently, the unknown complex coefficients ($A_{mn}^{(j,p)}$, $B_{mn}^{(j,p)}$, $C_{mn}^{(j,p)}$, $D_{mn}^{(j,p)}$, $E_{mn}^{(j,p)}$, $F_{mn}^{(j,p)}$) can be solved by linear algebra.

5. Determination of Wave Exciting Force

From Bernoulli's equation, the fluid pressure p is given by

$$p = \rho i \omega \phi \tag{59}$$

where ρ is the water density.

By integrating over the wetted areas, the wave exciting force and rotational moment for 6 DOFs are obtained by

$$F_{w_j}^{(q)} = \rho i \omega \int\limits_{S_{w_q}} (\phi_I + \phi_D) \cdot \mathbf{n}_j \mathrm{d}S_{w_q} \tag{60}$$

where S_{w_q} denotes the wetted surface of the floating body q on which the wave exciting forces are acting.

By substituting the incident velocity potential and diffracted potentials with obtained unknown complex coefficients into Equation (60), the wave exciting forces in 3 DOFs (i.e., surge, heave and pitch) acting on the floating polygonal platform ($q = 1$) and ring structure ($q = 2$) are given below:

$$\begin{aligned}
F_{w_1}^{(1)} &= \rho i \omega \int_0^{2\pi} \int_{-d_1}^{0} (\phi_I + \phi_{D_2})\big|_{r=R_1} R_1 \mathbf{n_x} \cdot (-\mathbf{n_{s_1}}) \mathrm{d}z \mathrm{d}\theta \\
&= \frac{\rho g A}{\cosh kh} \sum_{m=-\infty}^{\infty} \mathrm{e}^{im(\frac{\pi}{2}-\beta)} \int_0^{2\pi} J_m(kR_1) \mathrm{e}^{im\theta} R_1 \mathbf{n_1} \cdot (-\mathbf{n_{s_1}})\, \mathrm{d}\theta \int_{-d_1}^{0} \cosh k(z+h) \mathrm{d}z \\
&+ \rho i \omega \sum_{m=-\infty}^{\infty} \sum_{n=0}^{\infty} \int_0^{2\pi} \{B_{mn} \mathcal{I}_{mn}(k_n R_1) + C_{mn} \mathcal{K}_{mn}(k_n R_2)\} \mathrm{e}^{im\theta} R_1 \mathbf{n_1} \cdot (-\mathbf{n}_{s_1}) \mathrm{d}\theta \int_{-d_1}^{0} \frac{Z_n(z)}{Z_n(0)} \mathrm{d}z
\end{aligned} \tag{61}$$

$$\begin{aligned}
F_{w_3}^{(1)} &= \rho i \omega \int_0^{2\pi} \int_{R_1}^{R_2} (\phi_I + \phi_{D_2})\big|_{z=-d_1} \mathbf{n_z} \cdot (\mathbf{n_z}) r \mathrm{d}r \mathrm{d}\theta \\
&= \rho i \omega \sum_{m=-\infty}^{\infty} \sum_{n=0}^{\infty} \cos p_n(h-d_1) \int_0^{2\pi} \int_0^{R_1} A_{mn} \mathcal{I}_{mn}(p_n r) \mathrm{e}^{im\theta} r \mathrm{d}r \mathrm{d}\theta
\end{aligned} \tag{62}$$

$$\begin{aligned}
F_{w_5}^{(1)} &= \rho i \omega \int_0^{2\pi} \Big[\int_{-d_1}^{0} (z - z_{G_1}) (\phi_I + \phi_{D_2})\big|_{r=R_1} R_1 \mathbf{n}_x \cdot (-\mathbf{n_{s_1}}) \mathrm{d}z - \int_0^{R_1} (\phi_I + \phi_{D_1})\big|_{z=-d_1} (r\cos\theta) \mathbf{n_z} \cdot (\mathbf{n}_z) r \mathrm{d}r \Big] \mathrm{d}\theta \\
&= \frac{\rho g A}{\cosh kh} \sum_{m=-\infty}^{\infty} \mathrm{e}^{im(\frac{\pi}{2}-\beta)} \Big[\int_0^{2\pi} J_m(kR_1) R_1 \mathbf{n_1} \cdot (-\mathbf{n}_{s_1}) \mathrm{e}^{im\theta} \mathrm{d}\theta \int_{-d_1}^{0} (z - z_{G_1}) \cosh k(z+h) \mathrm{d}z \\
&- \cosh k(h-d_1) \int_0^{2\pi} \int_0^{R_1} J_m(kr) \mathrm{e}^{im\theta} r^2 \cos\theta \mathrm{d}r \mathrm{d}\theta \Big] \\
&+ \rho i \omega \sum_{m=-\infty}^{\infty} \sum_{n=0}^{\infty} \Big[\int_0^{2\pi} \{B_{mn} \mathcal{I}_{mn}(k_n R_1) + C_{mn} \mathcal{K}_{mn}(k_n R_1)\} \mathrm{e}^{im\theta} R_1 \mathbf{n_1} \cdot (-\mathbf{n}_{s_1}) \mathrm{d}\theta \int_{-d_1}^{0} (z - z_{G_1}) \frac{Z_n(z)}{Z_n(0)} \mathrm{d}z \\
&- \cos p_n(h-d_1) \int_0^{2\pi} \int_0^{R_1} A_{mn} \mathcal{I}_{mn}(p_n r) \mathrm{e}^{im\theta} r^2 \cos\theta \mathrm{d}r \mathrm{d}\theta \Big]
\end{aligned} \tag{63}$$

$$\begin{aligned}
F_{w_1}^{(2)} &= \rho i \omega \int_0^{2\pi} \int_{-d_2}^{0} \Big\{ (\phi_I + \phi_{D_2})\big|_{r=R_2} R_2 \mathbf{n_x} \cdot \mathbf{n_{s_2}} + (\phi_I + \phi_{D_4})\big|_{r=R_3} R_3 \mathbf{n_x} \cdot (-\mathbf{n_{s_3}}) \Big\} \mathrm{d}z \mathrm{d}\theta \\
&= \frac{\rho g A}{\cosh kh} \sum_{m=-\infty}^{\infty} \mathrm{e}^{im(\frac{\pi}{2}-\beta)} \int_0^{2\pi} \{J_m(kR_2) R_2 \mathbf{n_1} \cdot \mathbf{n_{s_2}} + J_m(kR_3) R_3 \mathbf{n_1} \cdot (-\mathbf{n_{s_3}})\} \mathrm{e}^{im\theta}\, \mathrm{d}\theta \int_{-d_2}^{0} \cosh k(z+h) \mathrm{d}z \\
&+ \rho i \omega \sum_{m=-\infty}^{\infty} \sum_{n=0}^{\infty} \int_0^{2\pi} [\{B_{mn} \mathcal{I}_{mn}(k_n R_2) + C_{mn} \mathcal{K}_{mn}(k_n R_2)\} R_2 \mathbf{n_1} \cdot \mathbf{n}_{s_2} + F_{mn} \mathcal{K}_{mn}(k_n R_3) R_3 \mathbf{n_1} \cdot (-\mathbf{n}_{s_3})] \mathrm{e}^{im\theta} \mathrm{d}\theta \int_{-d_2}^{0} \frac{Z_n(z)}{Z_n(0)} \mathrm{d}z
\end{aligned} \tag{64}$$

$$\begin{aligned}
F_{w_3}^{(2)} &= \rho i \omega \int_0^{2\pi} \int_{R_2}^{R_3} (\phi_I + \phi_{D_3})\big|_{z=-d_2} \mathbf{n_z} \cdot (\mathbf{n_z}) r \mathrm{d}r \mathrm{d}\theta \\
&= \rho i \omega \sum_{m=-\infty}^{\infty} \sum_{n=0}^{\infty} \cos p_n(h-d_2) \int_0^{2\pi} \int_{R_2}^{R_3} \{D_{mn} \mathcal{I}_{mn}(q_n r) + E_{mn} \mathcal{K}_{mn}(q_n r)\} \mathrm{e}^{im\theta} r \mathrm{d}r \mathrm{d}\theta
\end{aligned} \tag{65}$$

$$
\begin{aligned}
F_{w_5}^{(2)} &= \rho i\omega \int_0^{2\pi}\Big[\int_{-d_2}^{0}(z-z_{G_2})\Big\{(\phi_I+\phi_{D_2})\big|_{r=R_2}R_2\mathbf{n}_x\cdot\mathbf{n}_{\mathbf{s_2}} + (\phi_I+\phi_{D_4})\big|_{r=R_3}R_3\mathbf{n}_x\cdot(-\mathbf{n}_{\mathbf{s_3}})\Big\}\mathrm{d}z \\
&- \int_{R_2}^{R_3}(\phi_I+\phi_{D_3})\big|_{z=-d_2}(r\cos\theta)\mathbf{n_z}\cdot(\mathbf{n}_z)r\mathrm{d}r\Big]\mathrm{d}\theta \\
&= \frac{\rho g A}{\cosh kh}\sum_{m=-\infty}^{\infty}\mathrm{e}^{im(\frac{\pi}{2}-\beta)}\Big[\int_0^{2\pi}\{J_m(kR_2)R_2\mathbf{n_1}\cdot\mathbf{n}_{s_2}+J_m(kR_3)R_3\mathbf{n_1}\cdot\mathbf{n}_{s_3}\}\mathrm{e}^{im\theta}\mathrm{d}\theta\int_{-d_2}^{0}(z \\
&-z_{G_2})\cosh k(z+h)\mathrm{d}z - \cosh k(h-d_2)\int_0^{2\pi}\int_{R_2}^{R_3}J_m(kr)\mathrm{e}^{im\theta}r^2\cos\theta\mathrm{d}r\mathrm{d}\theta\Big] \\
&+\rho i\omega\sum_{m=-\infty}^{\infty}\sum_{n=0}^{\infty}\int_0^{2\pi}[\{B_{mn}\mathcal{I}(k_nR_2)+C_{mn}\mathcal{K}_{mn}(k_nR_2)\}R_2\mathbf{n_1}\cdot\mathbf{n}_{s_2}+F_{mn}\mathcal{K}_{mn}(k_nR_3)R_3\mathbf{n_1} \\
&\cdot(-\mathbf{n}_{s_3})]\mathrm{e}^{im\theta}\mathrm{d}\theta\int_{-d_2}^{0}(z-z_{G_2})\frac{Z_n(z)}{Z_n(0)}\mathrm{d}z \\
&-\rho i\omega\sum_{m=-\infty}^{\infty}\sum_{n=0}^{\infty}\cos p_n(h-d_2)\int_0^{2\pi}\int_{R_2}^{R_3}\{D_{mn}\mathcal{I}_{mn}(q_nr)+E_{mn}\mathcal{K}_{mn}(q_nr)\}\mathrm{e}^{im\theta}r^2\cos\theta\mathrm{d}r\mathrm{d}\theta
\end{aligned} \tag{66}
$$

where the unknown complex coefficients for the diffraction $A_{mn}^{(0,0)}$, $B_{mn}^{(0,0)}$, $C_{mn}^{(0,0)}$, $D_{mn}^{(0,0)}$, $E_{mn}^{(0,0)}$ and $F_{mn}^{(0,0)}$ are expressed as A_{mn}, B_{mn}, C_{mn}, D_{mn}, E_{mn} and F_{mn} for brevity. z_{G_1} and z_{G_2} are the z-coordinates of the centre of gravity for the floating polygonal platform and ring structure, respectively. The wave exciting forces for other DOFs such as sway, roll and yaw can be similarly obtained by using the unit normal vector $(-\mathbf{n_s}+\mathbf{n}_z)$ and the generalised motion normal $\mathbf{n}_j$. The divergence $\mathbf{n}_j\cdot\mathbf{n}_s$ are given in Appendix A.

6. Determination of Radiation Forces

The radiation force is obtained by

$$
F_{ij}^{(q,p)} = i\omega\rho\int\limits_{S_{w_q}}\Big\{-i\omega\xi_j^{(p)}\phi_R^{(j,p)}\Big\}\,\mathbf{n}_j\mathrm{d}S_{w_q} = \xi_j\Big(\omega^2\mu_{ij}^{(q,p)}+i\omega\lambda_{ij}^{(q,p)}\Big) \tag{67}
$$

where $\mu_{ij}^{(q,p)}$ and $\lambda_{ij}^{(q,p)}$ are the added mass and the radiation damping of the floating body q by the oscillating body p for the i-th mode of force and the j-th mode of motion, which are respectively defined as

$$
\mu_{ij}^{(q,p)} = \mathrm{Re}\Big(\rho f_{ij}^{(q,p)}\Big) \tag{68}
$$

$$
\lambda_{ij}^{(q,p)} = \mathrm{Im}\Big(\rho\omega f_{ij}^{(q,p)}\Big) \tag{69}
$$

$\mathrm{Re}(\cdot)$ denotes the real part and $\mathrm{Im}(\cdot)$ the imaginary part. $f_{ij}^{(q,p)}$ is the integral form for i-th mode of force and j-th mode of motion associated with the floating body q by the oscillating body p. By substituting the normalised radiated velocity potentials with the obtained unknown complex coefficients into Equation (67), $f_{ij}^{(q,p)}$ ($p=1,2$ and $q=1,2$) for 3 DOFs are given by

$$
\begin{aligned}
f_{11}^{(1,p)} &= \int_0^{2\pi}\int_{-d_1}^{0}\phi_{R_2}^{(1,p)}\Big|_{r=R_1}\mathbf{n}_x\cdot(-\mathbf{n}_{\mathbf{s_1}})R_1\mathrm{d}z\mathrm{d}\theta \\
&= \sum_{m=-\infty}^{\infty}\sum_{n=0}^{\infty}\int_0^{2\pi}\Big\{\Big\{B_{mn}^{(1,p)}\mathcal{I}_{mn}(k_nR_1)+C_{mn}^{(1,p)}\mathcal{K}_{mn}(k_nR_1)\Big\}R_1\mathbf{n}_x\cdot(-\mathbf{n}_{\mathbf{s_1}})\Big\}\mathrm{e}^{im\theta}\mathrm{d}\theta\int_{-d_1}^{0}\frac{Z_n(z)}{Z_n(0)}\mathrm{d}z
\end{aligned} \tag{70}
$$

$$
\begin{aligned}
f_{51}^{(1,p)} &= \int_0^{2\pi}\Big\{\int_{-d_1}^{0}(z-z_{G_1})\phi_{R_2}^{(1,p)}\Big|_{r=R_1}\mathbf{n}_x\cdot(-\mathbf{n}_{\mathbf{s_1}})R_1\mathrm{d}z - \int_0^{R_1}\phi_{R_1}^{(1,p)}\Big|_{z=-d_1}(r\cos\theta)\mathbf{n_z}\cdot(\mathbf{n}_z)r\mathrm{d}r\Big\}\mathrm{d}\theta \\
&= \sum_{m=-\infty}^{\infty}\sum_{n=0}^{\infty}\int_0^{2\pi}\Big\{B_{mn}^{(1,p)}\mathcal{I}_{mn}(k_nR_1)+C_{mn}^{(1,p)}\mathcal{K}_{mn}(k_nR_1)\Big\}R_1\mathbf{n}_x\cdot(-\mathbf{n}_{\mathbf{s_1}})\mathrm{e}^{im\theta}\mathrm{d}\theta\int_{-d_1}^{0}(z-z_{G_1})\frac{Z_n(z)}{Z_n(0)}\mathrm{d}z \\
&- \sum_{m=-\infty}^{\infty}\sum_{n=0}^{\infty}\cos p_n(h-d_1)\int_0^{2\pi}\int_0^{R_1}A_{mn}^{(1,p)}\mathcal{I}_{mn}(p_nr)\mathrm{e}^{im\theta}r^2\cos\theta\mathrm{d}r\mathrm{d}\theta
\end{aligned} \tag{71}
$$

$$
\begin{aligned}
f_{33}^{(1,p)} &= \int_0^{2\pi}\int_0^{R_1}\phi_{R_2}^{(3,p)}\Big|_{z=-d_1}\mathbf{n_z}\cdot(\mathbf{n}_z)r\mathrm{d}r\mathrm{d}\theta \\
&= \delta_{1p}\int_0^{2\pi}\int_0^{R_1}\Big\{\frac{(h-d_1)}{2}-\frac{r^2}{4(h-d_1)}\Big\}r\mathrm{d}r\mathrm{d}\theta \\
&+ \sum_{m=-\infty}^{\infty}\sum_{n=0}^{\infty}\cos p_n(h-d_1)\int_0^{2\pi}\int_0^{R_1}A_{mn}^{(3,p)}\mathcal{I}_{mn}(p_nr)\mathrm{e}^{im\theta}r\mathrm{d}r\mathrm{d}\theta
\end{aligned} \tag{72}
$$

$$
\begin{aligned}
f_{55}^{(1,p)} &= \int_0^{2\pi}\left[\int_{-d_1}^{0}(z-z_{G_1})\,\phi_{R_2}^{(5,p)}\Big|_{r=R_1}\mathbf{n}_x\cdot(-\mathbf{n}_{s_1})R_1\mathrm{d}z-\int_0^{R_1}\phi_{R_1}^{(5,p)}\Big|_{z=-d_1}(r\cos\theta)\mathbf{n}_{\mathbf{z}}\cdot(\mathbf{n}_z)r\mathrm{d}r\right]\mathrm{d}\theta\\
&= \sum_{m=-\infty}^{\infty}\sum_{n=0}^{\infty}\int_0^{2\pi}\left\{A_{mn}^{(5,p)}\mathcal{I}_{mn}(p_nR_1)\mathrm{e}^{im\theta}R_1\mathbf{n}_x\cdot(-\mathbf{n}_{s_1})\right\}\mathrm{d}\theta\int_{-d_1}^{0}(z-z_{G_1})\frac{Z_n(z)}{Z_n(0)}\mathrm{d}z\\
&-\left\{\delta_{1p}\int_0^{2\pi}\int_0^{R_1}\frac{-4(h-d_1)^2+r^2}{8(h-d_1)}r\cos\theta r^2\cos\theta\mathrm{d}r\mathrm{d}\theta\right.\\
&\left.+\sum_{m=-\infty}^{\infty}\sum_{n=0}^{\infty}\cos p_n(h-d_1)\int_0^{2\pi}\int_0^{R_1}A_{mn}^{(5,p)}\mathcal{I}_{mn}(p_nr)\mathrm{e}^{im\theta}r^2\cos\theta\mathrm{d}r\mathrm{d}\theta\right\}
\end{aligned}
\tag{73}
$$

$$
\begin{aligned}
f_{15}^{(1,p)} &= \int_0^{2\pi}\int_{-d_1}^{0}\phi_{R_2}^{(5,p)}\Big|_{r=R_1}\mathbf{n}_x\cdot(-\mathbf{n}_{s_1})R_1\mathrm{d}z\mathrm{d}\theta\\
&= \sum_{m=-\infty}^{\infty}\sum_{n=0}^{\infty}\int_0^{2\pi}\left\{B_{mn}^{(5,p)}\mathcal{I}_{mn}(k_nR_1)+C_{mn}^{(5,p)}\mathcal{K}_{mn}(k_nR_1)\right\}R_1\mathbf{n}_x\cdot(-\mathbf{n}_{s_1})\mathrm{e}^{im\theta}\mathrm{d}\theta\int_{-d_1}^{0}\frac{Z_n(z)}{Z_0(z)}\mathrm{d}z
\end{aligned}
\tag{74}
$$

$$
\begin{aligned}
f_{11}^{(2,p)} &= \int_0^{2\pi}\int_{-d_2}^{0}\left\{\phi_{R_2}^{(1,p)}\Big|_{r=R_2}\mathbf{n}_x\cdot(\mathbf{n}_{s_2})R_2+\phi_{R_4}^{(1,p)}\Big|_{r=R_3}\mathbf{n}_x\cdot(-\mathbf{n}_{s_3})R_3\right\}\mathrm{d}z\mathrm{d}\theta\\
&= \sum_{m=-\infty}^{\infty}\sum_{n=0}^{\infty}\int_0^{2\pi}\left[\left\{B_{mn}^{(1,p)}\mathcal{I}_{mn}(k_nR_2)+C_{mn}^{(1,p)}\mathcal{K}_{mn}(k_nR_2)\right\}R_2\mathbf{n}_x\cdot(\mathbf{n}_{s_2})+F_{mn}^{(1,p)}\mathcal{K}_{mn}(R_3)R_3\mathbf{n}_x\right.\\
&\left.\cdot(-\mathbf{n}_{s_3})\right]\mathrm{e}^{im\theta}\mathrm{d}\theta\int_{-d_2}^{0}\frac{Z_n(z)}{Z_n(0)}\mathrm{d}z
\end{aligned}
\tag{75}
$$

$$
\begin{aligned}
f_{51}^{(2,p)} &= \int_0^{2\pi}\left\{\int_{-d_2}^{0}(z-z_{G_2})\left\{\phi_{R_2}^{(1,p)}\Big|_{r=R_2}\mathbf{n}_x\cdot(\mathbf{n}_{s_2})R_2+\phi_{R_4}^{(1,p)}\Big|_{r=R_3}\mathbf{n}_x\cdot(-\mathbf{n}_{s_3})R_3\right\}\mathrm{d}z\right.\\
&\left.-\int_{R_2}^{R_3}\phi_{R_3}^{(1,p)}\Big|_{z=-d_2}(r\cos\theta)\mathbf{n}_{\mathbf{z}}\cdot(\mathbf{n}_z)r\mathrm{d}r\right\}\mathrm{d}\theta\\
&= \sum_{m=-\infty}^{\infty}\sum_{n=0}^{\infty}\int_0^{2\pi}\left\{\left\{B_{mn}^{(1,p)}\mathcal{I}_{mn}(k_nR_2)+C_{mn}^{(1,p)}\mathcal{K}_{mn}(k_nR_2)\right\}R_2\mathbf{n}_x\cdot(\mathbf{n}_{s_2})+F_{mn}^{(1,p)}\mathcal{K}_{mn}(k_nR_3)R_3\mathbf{n}_x\right.\\
&\left.\cdot(-\mathbf{n}_{s_3})\right\}\mathrm{e}^{im\theta}\mathrm{d}\theta\int_{-d}^{0}(z-z_{G_2})\frac{Z_n(z)}{Z_n(0)}\mathrm{d}z\\
&-\sum_{m=-\infty}^{\infty}\sum_{n=0}^{\infty}\cos p_n(h-d_2)\int_0^{2\pi}\int_{R_2}^{R_3}\left\{D_{mn}^{(1,p)}\mathcal{I}_{mn}(q_nr)+E_{mn}^{(1,p)}\mathcal{K}_{mn}(q_nr)\right\}\mathrm{e}^{im\theta}r^2\cos\theta\mathrm{d}r\mathrm{d}\theta
\end{aligned}
\tag{76}
$$

$$
\begin{aligned}
f_{33}^{(2,p)} &= \int_0^{2\pi}\int_{R_2}^{R_3}\phi_{R_3}^{(3,p)}\Big|_{z=-d_2}\mathbf{n}_{\mathbf{z}}\cdot(\mathbf{n}_z)r\mathrm{d}r\mathrm{d}\theta\\
&= \delta_{2p}\int_0^{2\pi}\int_{R_2}^{R_3}\left\{\frac{(h-d_2)}{2}-\frac{r^2}{4(h-d_2)}\right\}r\mathrm{d}r\mathrm{d}\theta\\
&+\sum_{m=-\infty}^{\infty}\sum_{n=0}^{\infty}\cos p_n(h-d_2)\int_0^{2\pi}\int_{R_2}^{R_3}\left\{D_{mn}^{(3,p)}\mathcal{I}_{mn}(q_nr)+E_{mn}^{(3,p)}\mathcal{K}_{mn}(q_nr)\right\}\mathrm{e}^{im\theta}r\mathrm{d}r\mathrm{d}\theta
\end{aligned}
\tag{77}
$$

$$
\begin{aligned}
f_{55}^{(2,p)} &= \int_0^{2\pi}\left[\int_{-d_2}^{0}(z-z_{G_2})\left\{\phi_{R_2}^{(5,p)}\Big|_{r=R_2}\mathbf{n}_x\cdot(\mathbf{n}_{s_2})R_2+\phi_{R_4}^{(5,p)}\Big|_{r=R_3}\mathbf{n}_x\cdot(-\mathbf{n}_{s_3})R_3\right\}\mathrm{d}z-\int_{R_2}^{R_3}\phi_{R_3}^{(5,p)}\Big|_{z=-d_2}(r\cos\theta)\mathbf{n}_{\mathbf{z}}\cdot(\mathbf{n}_z)r\mathrm{d}r\right]\mathrm{d}\theta\\
&= \sum_{m=-\infty}^{\infty}\sum_{n=0}^{\infty}\int_0^{2\pi}\left[\left\{B_{mn}^{(5,p)}\mathcal{I}_{mn}(k_nR_2)+C_{mn}^{(5,p)}\mathcal{K}_{mn}(k_nR_2)\right\}R_2\mathbf{n}_x\cdot(\mathbf{n}_{s_2})+F_{mn}^{(5,p)}\mathcal{K}_{mn}(k_nR_3)R_3\mathbf{n}_x\cdot(-\mathbf{n}_{s_3})\right]\mathrm{e}^{im\theta}\mathrm{d}\theta\int_{-d_2}^{0}(z-z_{G_2})\frac{Z_n(z)}{Z_n(0)}\mathrm{d}z\\
&-\left\{\delta_{2p}\int_0^{2\pi}\int_{R_2}^{R_3}\frac{-4(h-d_2)^2+r^2}{8(h-d_2)}r\cos\theta r^2\cos\theta\mathrm{d}r\mathrm{d}\theta\right.\\
&\left.+\sum_{m=-\infty}^{\infty}\sum_{n=0}^{\infty}\cos p_n(h-d_2)\int_0^{2\pi}\int_{R_2}^{R_3}\left\{D_{mn}^{(5,p)}\mathcal{I}_{mn}(q_nr)+E_{mn}^{(5,p)}\mathcal{K}_{mn}(q_nr)\right\}\mathrm{e}^{im\theta}r^2\cos\theta\mathrm{d}r\mathrm{d}\theta\right\}
\end{aligned}
\tag{78}
$$

$$
\begin{aligned}
f_{15}^{(2,p)} &= \int_0^{2\pi}\int_{-d_2}^{0}\left\{\phi_{R_2}^{(5,p)}\Big|_{r=R_2}\mathbf{n}_x\cdot(\mathbf{n}_{s_2})R_1+\phi_{R_4}^{(5,p)}\Big|_{r=R_3}\mathbf{n}_x\cdot(-\mathbf{n}_{s_3})R_3\right\}\mathrm{d}z\mathrm{d}\theta\\
&= \sum_{m=-\infty}^{\infty}\sum_{n=0}^{\infty}\int_0^{2\pi}\left\{\left\{B_{mn}^{(5,p)}\mathcal{I}_{mn}(k_nR_2)+C_{mn}^{(5,p)}\mathcal{K}_{mn}(k_nR_2)\right\}R_2\mathbf{n}_x\cdot(\mathbf{n}_{s_2})+F_{mn}^{(5,p)}\mathcal{K}_{mn}(k_nR_3)R_3\mathbf{n}_x\right.\\
&\left.\cdot(-\mathbf{n}_{s_3})\right\}\mathrm{e}^{im\theta}\mathrm{d}\theta\int_{-d_2}^{0}\frac{Z_n(z)}{Z_0(z)}\mathrm{d}z
\end{aligned}
\tag{79}
$$

Note that $f_{15}^{(q,p)}$ and $f_{51}^{(p,q)}$ are the same when the geometry of the floating body is symmetrical about the x-axis.

7. Motion Responses of Floating Ring Structure

Consider the horizontal coordinates of the centre of gravity of the floating regular polygonal ring structure coincide with the origin, i.e.,$(x_{G_1},y_{G_1})=(x_{G_2},y_{G_2})=(0,0)$ but $z_{G_1}\neq 0$ and $z_{G_2}\neq 0$. As regular polygons are geometrically symmetrical about certain axes, the products of inertia of the floating regular polygonal platform/structure will become zero regardless of their orientation. If the motion of the floating body is relatively small, the following equations of motion in 3 DOFs (i.e., surge, heave and pitch) are satisfied.

$$
\begin{aligned}
&\left[-\omega^2\left(m^{(1)}+\mu_{11}^{(1,1)}\right)-i\omega\lambda_{11}^{(1,1)}\right]\left\{\xi_1^{(1)}\right\}+\left[-\omega^2\mu_{15}^{(1,1)}-i\omega\lambda_{15}^{(1,1)}\right]\left\{\xi_5^{(1)}\right\}+\left[-\omega^2\mu_{11}^{(1,2)}-i\omega\lambda_{11}^{(1,2)}\right]\left\{\xi_1^{(2)}\right\}\\
&+\left[-\omega^2\mu_{15}^{(1,2)}-i\omega\lambda_{15}^{(1,2)}\right]\left\{\xi_5^{(2)}\right\}=\left\{F_1^{(1)}\right\}
\end{aligned}
\tag{80}
$$

$$
\left[-\omega^2\left(m^{(1)}+\mu_{33}^{(1,1)}\right)-i\omega\lambda_{33}^{(1,1)}+\rho gA_w^{(1)}\right]\left\{\xi_3^{(1)}\right\}+\left[-\omega^2\mu_{33}^{(1,2)}-i\omega\lambda_{33}^{(1,2)}\right]\left\{\xi_3^{(2)}\right\}=\left\{F_3^{(1)}\right\}
\tag{81}
$$

$$
\begin{aligned}
&\left[-\omega^2\mu_{51}^{(1,1)}-i\omega\lambda_{51}^{(1,1)}\right]\left\{\xi_1^{(1)}\right\}+\left[-\omega^2\left(I_{22}^{(1)}+\mu_{55}^{(1,1)}\right)-i\omega\lambda_{55}^{(1,1)}+m^{(1)}g\overline{GM}_L^{(1)}\right]\left\{\xi_5^{(1)}\right\}+\left[-\omega^2\mu_{51}^{(1,2)}-i\omega\lambda_{51}^{(1,2)}\right]\left\{\xi_1^{(2)}\right\}\\
&+\left[-\omega^2\mu_{55}^{(1,2)}-i\omega\lambda_{55}^{(1,2)}\right]\left\{\xi_5^{(2)}\right\}=\left\{F_5^{(1)}\right\}
\end{aligned}
\tag{82}
$$

$$\left[-\omega^2\left(m^{(2)}+\mu_{11}^{(2,2)}\right)-i\omega\lambda_{11}^{(2,2)}\right]\left\{\bar{\xi}_1^{(2)}\right\}+\left[-\omega^2\mu_{15}^{(2,2)}-i\omega\lambda_{15}^{(2,2)}\right]\left\{\bar{\xi}_5^{(2)}\right\}+\left[-\omega^2\mu_{11}^{(2,1)}-i\omega\lambda_{11}^{(2,1)}\right]\left\{\bar{\xi}_1^{(1)}\right\} \\ +\left[-\omega^2\mu_{15}^{(2,1)}-i\omega\lambda_{15}^{(2,1)}\right]\left\{\bar{\xi}_5^{(1)}\right\}=\left\{F_1^{(2)}\right\} \tag{83}$$

$$\left[-\omega^2\left(m^{(2)}+\mu_{33}^{(2,2)}\right)-i\omega\lambda_{33}^{(2,2)}+\rho g A_w^{(2)}\right]\left\{\bar{\xi}_3^{(2)}\right\}+\left[-\omega^2\mu_{33}^{(2,1)}-i\omega\lambda_{33}^{(2,1)}\right]\left\{\bar{\xi}_3^{(1)}\right\}=\left\{F_3^{(2)}\right\} \tag{84}$$

$$\left[-\omega^2\mu_{51}^{(2,2)}-i\omega\lambda_{51}^{(2,2)}\right]\left\{\bar{\xi}_1^{(2)}\right\}+\left[-\omega^2\left(I_{22}^{(2)}+\mu_{55}^{(2,2)}\right)-i\omega\lambda_{55}^{(2,2)}+m^{(2)}g\overline{GM}_L^{(2)}\right]\left\{\bar{\xi}_5^{(2)}\right\}+\left[-\omega^2\mu_{51}^{(2,1)}-i\omega\lambda_{51}^{(2,1)}\right]\left\{\bar{\xi}_1^{(1)}\right\} \\ +\left[-\omega^2\mu_{55}^{(2,1)}-i\omega\lambda_{55}^{(2,1)}\right]\left\{\bar{\xi}_5^{(1)}\right\}=\left\{F_5^{(2)}\right\} \tag{85}$$

where $m^{(p)}$ is the mass of the oscillating body p, $\bar{\xi}_i^{(p)}$ the unknown displacement of the oscillating body p at i-th radiation mode, $F_i^{(q)}$ the wave exciting force acting on the floating body q, $I_{ii}^{(p)}$ $(i=1,2,3)$ the mass moment of inertia of the oscillating body p about i-axis, $A_w^{(p)}$ the waterplane area of the oscillating body p, $V_w^{(p)}$ the wetted volume of the oscillating body p, $\overline{GM}_T^{(p)}\left(=\frac{A_{w11}^{(p)}}{V_w^{(p)}}+z_B^{(p)}-z_G^{(p)}\right)$ the transverse metacentric height of the oscillating body p, $\overline{GM}_L^{(p)}\left(=\frac{A_{w22}^{(p)}}{V_w^{(p)}}+z_B^{(p)}-z_G^{(p)}\right)$ the longitudinal metacentric height of the oscillating body p and $A_{w_{ii}}^{(p)}$ $(i=1,2)$ is the second moment over the waterplane area of the oscillating body p about i-axis.

By linear algebra, the motion responses $\bar{\xi}_i^{(p)}$ $(i=1,2,\cdots,6)$ are solved by using Equations (80)–(85). By substituting the solutions into Equation (4), one can obtain the radiated potential.

8. Verification of Semi-Analytical Approach and Computer Code

In order to verify the semi-analytical approach and the computer code, we compare the hydrodynamic results (i.e., added mass, radiation damping, wave exciting forces, RAO and wave field) with those obtained from the commercial software ANSYS AQWA based on the Boundary Element Method.

For the verification exercise, we consider a floating hexagonal platform that is placed within a floating hexagonal ring structure whose geometries are defined by the radius function as given in Equation (1) with $R_{0_1}=50$ m, $R_{0_2}=90$ m, $R_{0_3}=100$ m, $\varepsilon_{1,2,3}=0.03$, $n_{p_{1,2,3}}=6$ and $\theta_{0_{1,2,3}}=\frac{\pi}{6}$. The drafts d_1 for the platform and d_2 for the ring structure are equally 10 m. Figure 3 shows the structural shape. It is assumed that the centre of gravity coincides with the origin and the water depth is 50 m. The incident wave is along the x-axis (i.e., $\beta=0°$).

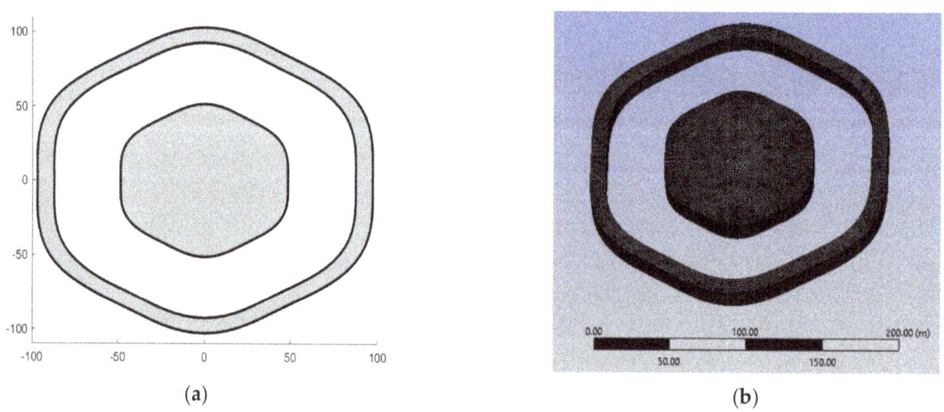

(a) (b)

Figure 3. Hexagonal ring structure for verification study: (**a**) plan view; (**b**) 3D mesh model (AQWA).

For the 3D model in AQWA, the maximum mesh size is 2.8 m which leads to a total of 11,459 diffracting elements to be used. It should be noted that a fine mesh model is required to obtain a reasonably converged inner free water elevation. The regular wave angular frequency domain for AQWA is divided into 20 frequencies for the interval [0.3866 rad/s, 1.2528 rad/s], which is equivalent to $kh = [1,\ 8]$.

Figure 4 presents the added mass and radiation damping obtained from the present semi-analytical method and AQWA. The numbers of truncated terms for the present semi-analytical method were taken as $M = 30$, $N = 5$, $N_r = 12$ and $N_q = 24$ for parametric studies. It is unnecessary to perform the integration for all the m-series terms, but series index m can be restricted to

$$\begin{gathered} m = \pm 1,\ \pm(n_p \pm 1) \quad \text{for surge or sway} \\ m = 0,\ \pm n_p \quad \text{for heave} \\ m = \pm 1,\ \pm(n_p \pm 1),\ \pm(2n_p \pm 1) \quad \text{for roll or pitch} \\ m = 0, \pm 2, \pm n_p,\ \pm(n_p \pm 2), \pm(2n_p \pm 1),\ \pm 2n_p \quad \text{for yaw} \end{gathered} \tag{86}$$

where n_p is the parameter of the radius function as given in Equation (1). Also, it should be noted that N_r and N_q can be taken as multiple numbers of n_p. For instance, $N_r = 12$ and $N_q = 24$ are respectively 2 and 4 times $n_p = 6$ for a hexagonal shape. This rule can be equally applied for calculating the wave exciting forces. Hence, such a selective calculation can speed up the hydrodynamic analysis in composing the global system matrix as well as obtaining the wave exciting forces and radiation forces.

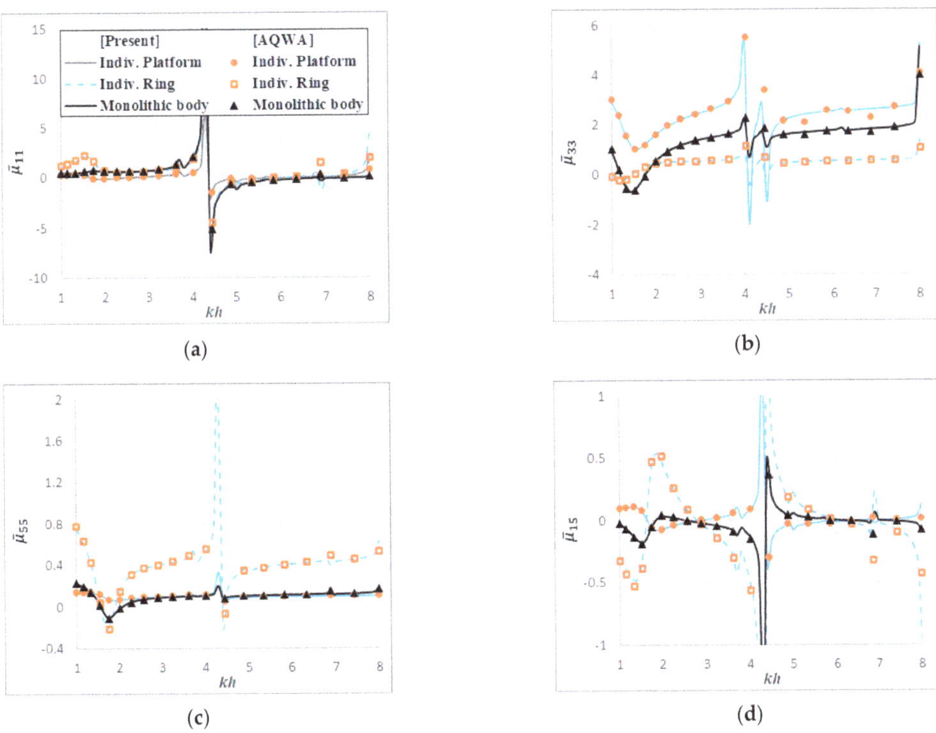

Figure 4. *Cont.*

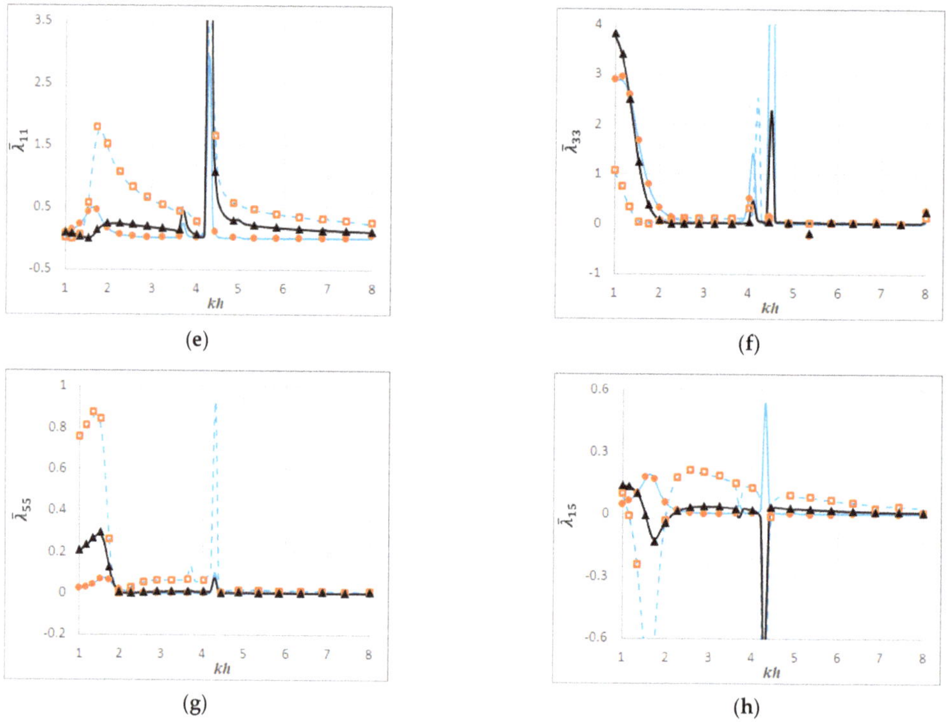

Figure 4. Comparison of normalised added mass and radiation damping obtained from AQWA and present approach: (**a**) added mass for surge; (**b**) added mass for heave; (**c**) added mass for pitch; (**d**) added mass for surge-pitch; (**e**) radiation damping for surge; (**f**) radiation damping for heave; (**g**) radiation damping for pitch; (**h**) radiation damping for surge-pitch.

The added mass was normalised by

$$\overline{\mu}_{11}^{(q,p)} = \frac{\mu_{11}^{(q,p)}}{\rho S_{0_q} d_q};\ \overline{\mu}_{33}^{(q,p)} = \frac{\mu_{33}^{(q,p)}}{\rho S_{0_q} d_q};\ \overline{\mu}_{55}^{(q,p)} = \frac{\mu_{55}^{(q,p)}}{\rho S_{0_q}^2 d_q};\ \overline{\mu}_{15}^{(q,p)} = \frac{\mu_{15}^{(q,p)}}{\rho S_{0_q}^{1.5} d_q} \tag{87}$$

while the radiation damping was normalised by

$$\overline{\lambda}_{11}^{(q,p)} = \frac{\lambda_{11}^{(q,p)}}{\rho\omega S_{0_q} d_q};\ \overline{\lambda}_{33}^{(q,p)} = \frac{\lambda_{33}^{(q,p)}}{\rho\omega S_{0_q} d_q};\ \overline{\lambda}_{55}^{(q,p)} = \frac{\lambda_{55}^{(q,p)}}{\rho\omega S_{0_q}^2 d_q};\ \overline{\lambda}_{15}^{(q,p)} = \frac{\lambda_{15}^{(q,p)}}{\rho\omega S_{0_q}^{1.5} d_q} \tag{88}$$

where S_{0_q} is the cross-sectional area of the polygonal platform for $q = 1$ and that of the polygonal ring structure for $q = 2$. The present semi-analytical results and AQWA results are in excellent agreement; thereby confirming the validity, convergence and accuracy of the semi-analytical approach. It can be seen that the magnitudes of crests near resonance frequencies are sensitively varying with minor discrepancies between AQWA and present results (see Figure 4b,d).

The wave exciting forces and RAOs were calculated for $\beta = 0°$. The horizontal wave force, vertical wave force and rotational moment were normalised as follows:

$$\overline{F_x}^{(q)} = \frac{F_x^{(q)}}{\rho g A S_{0_q}};\ \overline{F_z}^{(q)} = \frac{F_z^{(q)}}{\rho g A S_{0_q}};\ \overline{M_y}^{(q)} = \frac{M_y^{(q)}}{\rho g A S_{0_q} d_q}$$

The RAO (Response Amplitude Operator) is defined as the motion response normalised by the incident wave amplitude A. Figure 5 compares the wave exciting forces and the RAOs obtained from the present semi-analytical method and those computed from ANSYS AQWA. It can be seen that the results are in close agreement; thereby verifying the present semi-analytical formulation and method of solution.

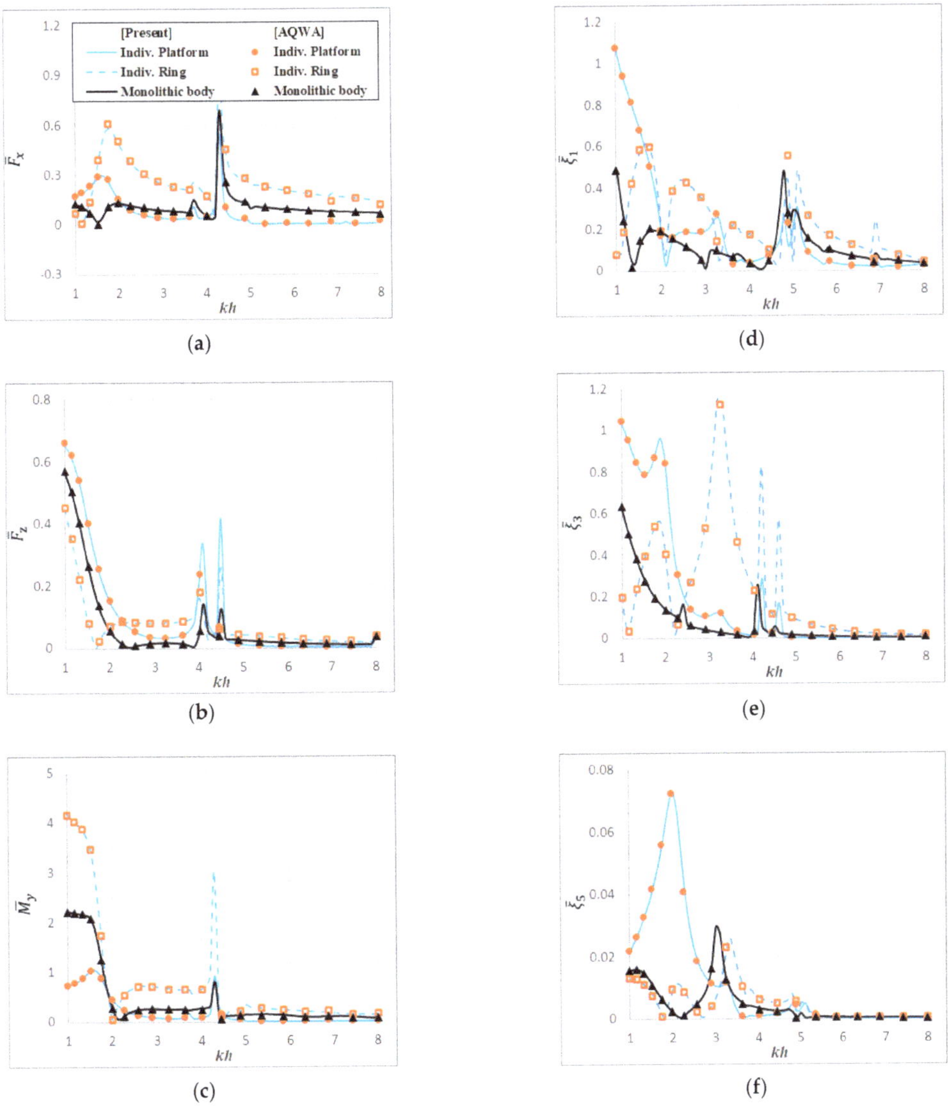

Figure 5. Comparison of wave exciting forces and RAOs obtained from AQWA and present approach: (**a**) surge force; (**b**) heave force; (**c**) pitch moment; (**d**) RAO for surge; (**e**) RAO for heave; (**f**) RAO for pitch.

The wave field and profiles for the inner water basin of the floating hexagonal platform and ring structure were calculated as shown in Figure 6 (for T = 10 s). The floating bodies are oscillating together or individually. Wave profiles are drawn along the x-axis at y = 0 m and $y = 50$ m, which are normalised by the incident wave amplitude for the wave period 10 s. The wave profiles obtained from the present approach and AQWA are well matched. Note that the wave profiles belonging to AQWA were extracted from AQWA FLOW by writing a computer code to process the raw data at selected points.

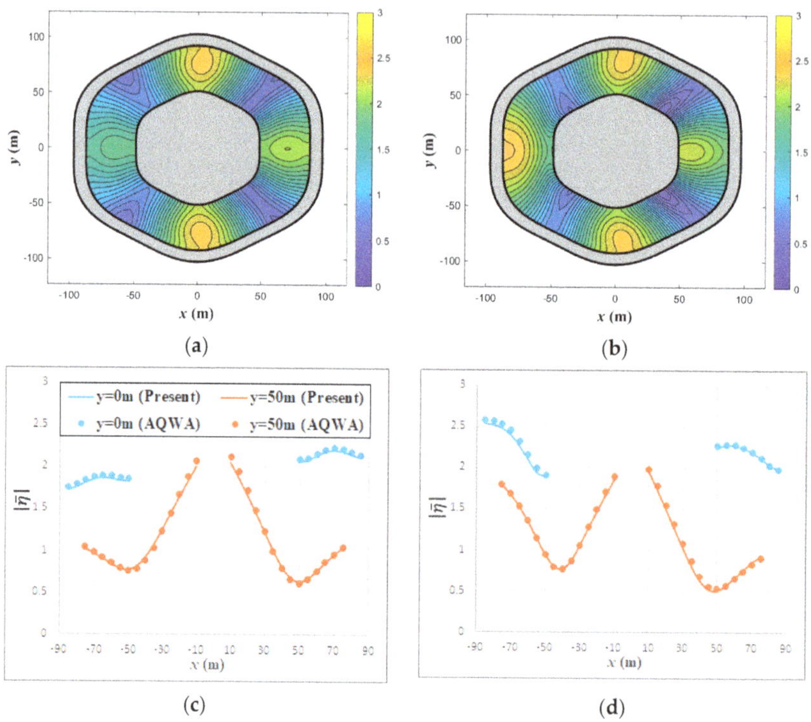

Figure 6. Comparison of normalised wave elevations and profiles for together or individually oscillating floating hexagonal platform and ring structure obtained from AQWA and present approach (for $T = 10$ s): (**a**) wave field for monolithic motion; (**b**) wave field for individual motion; (**c**) wave profile for monolithic motion; (**d**) wave profile for individual motion.

9. Results and Discussion

Further hydrodynamic analyses have been carried out to investigate various effects due to drafts, the radii of platforms and polygonal shapes. Floating hexagonal platform and ring structure that oscillate individually are considered. The drafts for two individual floating bodies are respectively taken as: (1) $d_1/h = 0.1$, $d_2/h = 0.1$, (2) $d_1/h = 0.1$, $d_2/h = 0.2$, (3) $d_1/h = 0.2$, $d_2/h = 0.1$ and (4) $d_1/h = 0.2$, $d_2/h = 0.2$. The wave exciting forces and RAOs are given in Figure 7 while the added mass and radiation damping are presented in Figure 8. Note that "P" and "R" used in legends stand for "Platform" and "Ring", respectively. Overall, the ring structure shows greater dominance in surge and pitch forces while the platform shows dominance in heave forces as shown in Figure 7 as the ring structure primarily prevents wave propagation. For longer waves (say $kh < 2$), as more waves are transmitted into the inner water basin, they become trapped; thereby, creating a high energetic environment within the ring structure. Accordingly, this aggravates the hydrodynamic interaction on the platform. However, the situations are reversed when considering the RAOs for heave and pitch (see Figure 7e,f). Similar phenomena are observed for added mass and radiation damping in the case of heave (see Figure 8b,f). In general, when the ratio of the draft-to-water depth of the ring structure (i.e., d_2/h) is the same, the hydrodynamic results such as the wave exciting forces, RAOs, added mass and radiation damping associated with the ring structure are very close to each other regardless of the ratio of the draft-to-water depth of the platform (i.e., d_1/h). Likewise, when d_1/h of the platform is the same, the hydrodynamic results associated with the platform are similarly irrelevant to d_2/h.

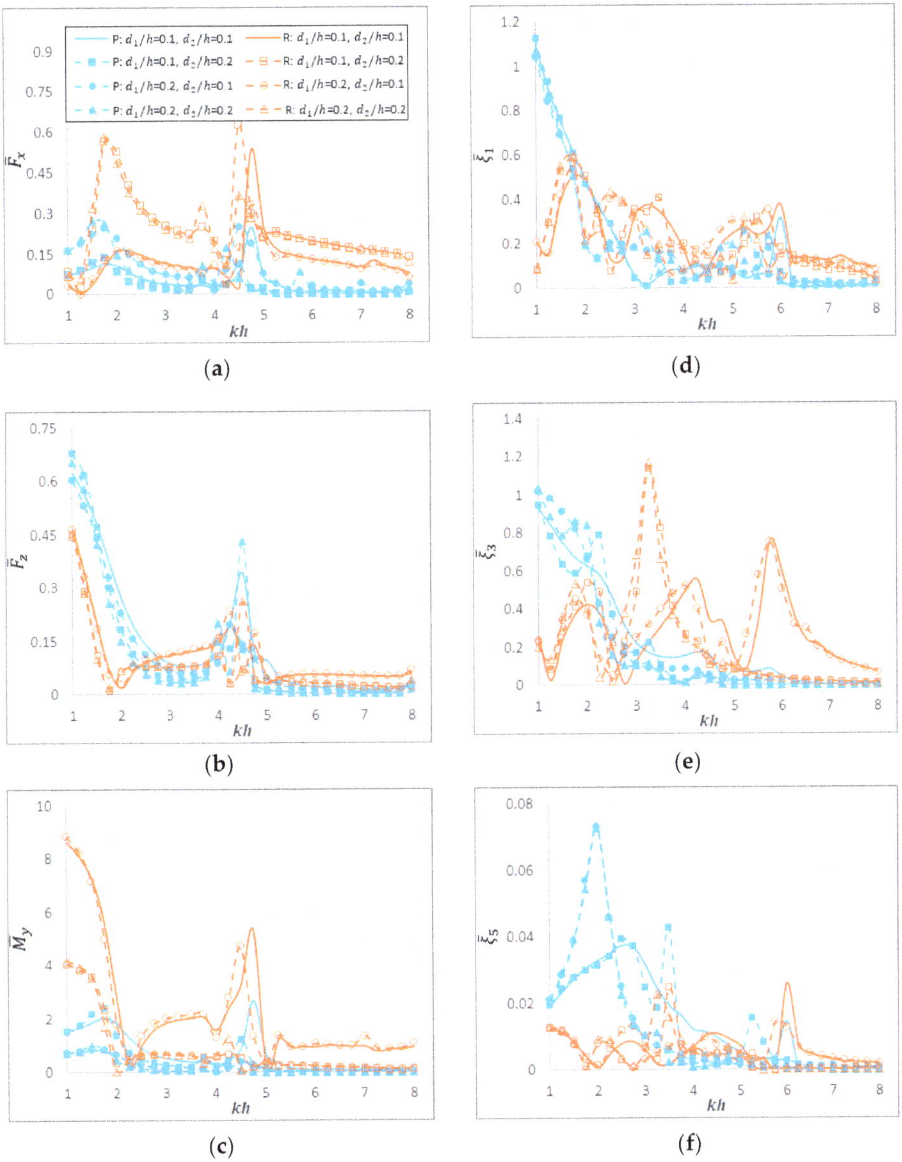

Figure 7. Normalised wave exciting forces and RAOs for floating hexagonal platforms and ring structures with various draft-to-water depth ratios: (**a**) surge force; (**b**) heave force; (**c**) pitch moment; (**d**) RAO for surge; (**e**) RAO for heave; (**f**) RAO for pitch.

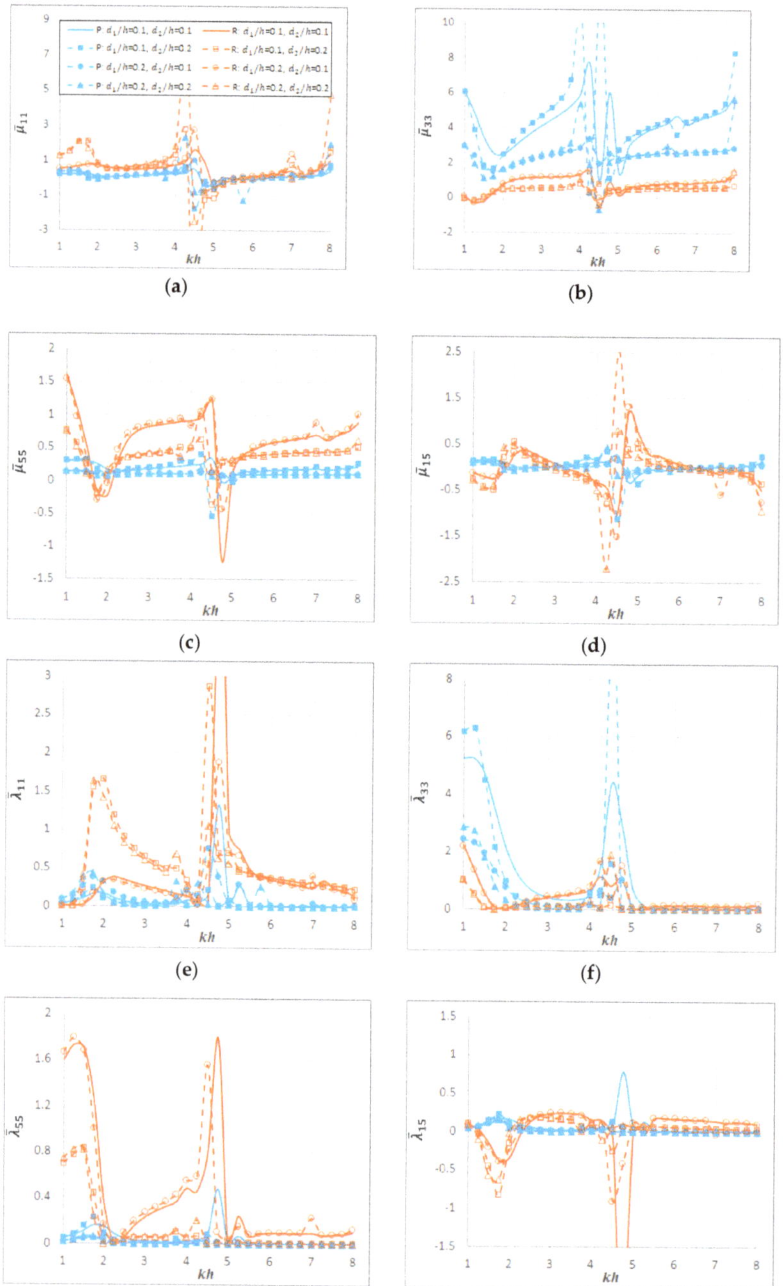

Figure 8. Normalised added mass and radiation damping for floating hexagonal platforms and ring structures with various draft-to-water depth ratios: (**a**) added mass for surge; (**b**) added mass for heave; (**c**) added mass for pitch; (**d**) added mass for surge-pitch; (**e**) radiation damping for surge; (**f**) radiation damping for heave; (**g**) radiation damping for pitch; (**h**) radiation damping for surge-pitch.

In Figure 9, the wave exciting forces and RAOs for floating circular platform and ring structure are presented for various radii of the circular platform, i.e., $R_1/h = 0.5, 1,\ 1.5$. It can be seen that the wave exciting forces increase with increasing R_1/h. On the other hand, the RAOs for the platform decrease with increasing R_1/h, whereas the RAOs for the ring structure do not show a distinctive relationship. This implies that a large working platform placed within a floating ring breakwater is relatively stable in terms of motions when compared with a small platform. On the other hand, when a small oscillating cylindrical platform is deployed within a floating ring structure, relatively large kinetic energy can be obtained.

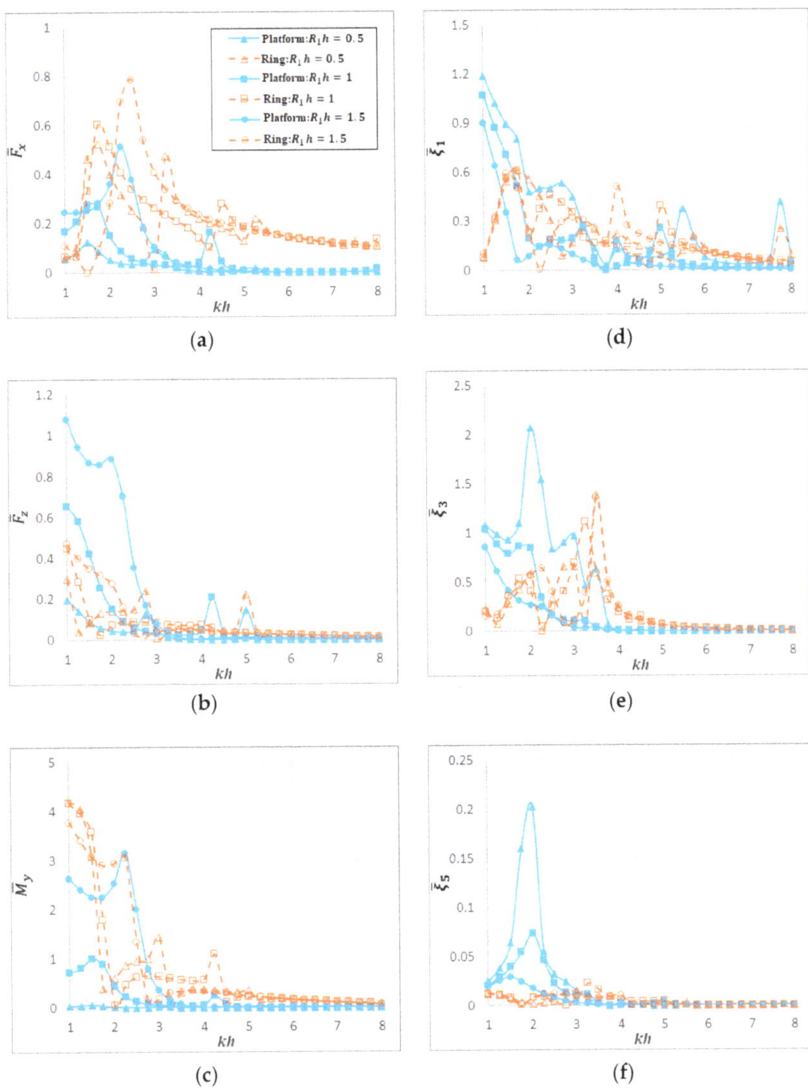

Figure 9. Normalised wave exciting forces and RAOs for floating circular platform and circular ring structure with $R_1/h = 0.5, 1,\ 1.5$: (**a**) surge force; (**b**) heave force; (**c**) pitch moment; (**d**) RAO for surge; (**e**) RAO for heave; (**f**) RAO for pitch.

Next, the free water surface elevation in the inner water basin will be investigated for the various geometries of floating polygonal platform and ring structure. Added mass, radiation damping and RAOs do not represent significant differences for similar sizes of plan shapes as when their geometries are created by using the radius function as given in Equation (1), the plan shapes for every kind of polygons are almost the same as each other. In addition, as far as the wave exciting forces are concerned, the horizontal forces such as surge or sway forces are distinctively changed when polygonal shapes are appropriately used with their orientations. This phenomenon was already reported by Park and Wang [20]. Hence, in this parametric study, the free water surface elevation in the inner water basin will be the focus.

The floating circular platform and circular ring structure are initially considered to investigate the inner wave fields. The draft-to-water depth ratio for the circular floating bodies $d_1/h = 0.1, 0.2$ and $d_2/h = 0.1, 0.2$ are combined into 4 cases. The inner wave fields at significant resonance frequencies are presented in Figure 10 by assuming the water depth $h = 50$ m. The maximum value of the scale bar is set to 4; implying wave fields larger than 4 are included in the maximum value (yellow colour). It can be learned that when larger d_2 and smaller d_1 are considered, there will be more waves trapped in the inner water basin, resulting in higher free water surface elevation (see Figure 10b) and vice versa. Thus, Figure 10b may be used for harvesting more wave energy, whereas Figure 10c is used for creating a calm patch of water space.

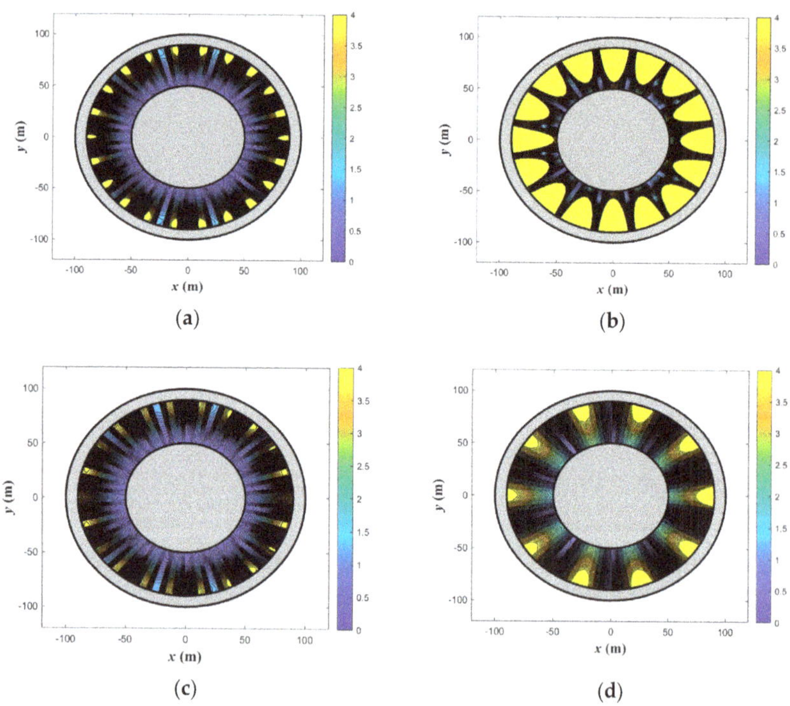

Figure 10. Normalised inner wave fields for individually oscillating floating circular platforms and ring structures at significant resonance frequencies: (**a**) $d_1/h = 0.1, d_2/h = 0.1$ $(kh = 7.5)$; (**b**) $d_1/h = 0.1, d_2/h = 0.2$ $(kh = 5.5)$; (**c**) $d_1/h = 0.2, d_2/h = 0.1$ $(kh = 7.5)$; (**d**) $d_1/h = 0.2, d_2/h = 0.2$ $(kh = 3.75)$.

In order to investigate the shape effect on the inner wave fields, several combinations of different polygonal shapes such as a square, hexagon and circle are considered for

$d_1/h = d_2/h = 0.2$ by assuming the water depth $h = 50$ m. Normalised inner wave fields for various floating polygonal platforms and ring structures at significant resonance frequencies are presented in Figure 11. By referring to Figure 10d together with Figure 11, it can be found that the inner wave fields at the significant resonance frequency for circular shapes tend to spread waves to multiple directions (see Figure 10d). For the square shapes, the inner wave fields are mild when compared with other cases (see Figure 11a) and amplified waves appeared to be in one direction (see Figure 11c,e). For the hexagonal shapes, waves are more amplified than the circular shapes and their propagations are in multiple directions. This implies that the square shapes are beneficial in terms of creating a calm patch of water space, whereas the hexagonal shapes are more advantageous for wave energy harvesting.

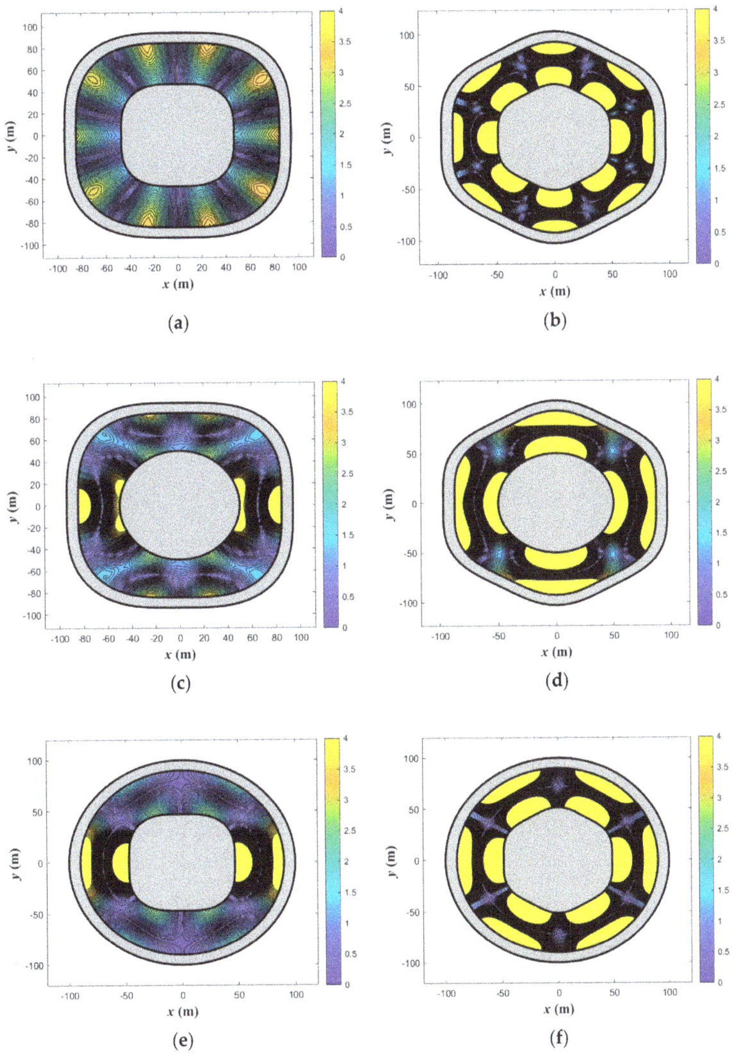

Figure 11. Normalised inner wave fields for various combinations of individually oscillating floating

platforms and ring structures at significant resonance frequencies: (**a**) square platform and square ring ($kh = 3.75$); (**b**) hexagonal platform and hexagonal ring ($kh = 5.25$); (**c**) circular platform and square ring ($kh = 5.25$); (**d**) circular platform and hexagonal ring ($kh = 4.5$); (**e**) square platform and circular ring ($kh = 4.75$); (**f**) hexagonal platform and circular ring ($kh = 4.75$).

10. Concluding Remarks

A semi-analytical method and computer code have been developed for the hydrodynamic analysis of a floating regular polygonal platform that is centrally placed within a regular polygonal ring structure under wave action. The polygonal shapes for equilateral triangle, square, pentagonal, hexagonal platforms and ring structures can be readily generated by using the cosine-type radial perturbation. This shape function is the key to enabling the problem to be solved semi-analytically. Cases involving the two floating bodies oscillating together or individually were considered with the view to understanding the hydrodynamic interactions among the trapped waves, inner platform and outer ring structure. The method has been shown to be able to furnish accurate hydrodynamic quantities such as wave exciting forces, added mass, radiation damping, RAOs, and wave field. The computational speed has been significantly quickened when compared with numerical methods because of the semi-analytical method.

The effects of several parameters such as drafts, radii of platforms and polygonal shapes on major hydrodynamic quantities are investigated by performing parametric studies. When the draft of the ring structure is larger than the floating platform, trapped waves are amplified more in the inner water basin. Additionally, the wave exciting forces increase with increasing radii of platforms. However, the RAOs decrease with the increasing radii of platforms. The inner wave fields for circular shapes tend to spread waves to multiple directions, whereas those for square shapes are relatively mild and their amplified waves are apt to propagate in one direction. For hexagonal shapes, the wave fields are more amplified than other considered shapes and the waves' propagations are in multiple directions. In sum, floating square platform and square ring structure are beneficial for creating a calm patch of water space, whereas the floating hexagonal platform and the hexagonal ring structure are more advantageous for wave energy harvesting.

Author Contributions: Conceptualization, J.C.P. and C.M.W.; methodology, J.C.P.; software, J.C.P.; validation, J.C.P. and C.M.W.; formal analysis, J.C.P.; investigation, J.C.P. and C.M.W.; writing—original draft preparation, J.C.P.; writing—review and editing, C.M.W.; visualization, J.C.P.; supervision, C.M.W.; funding acquisition, C.M.W.; All authors have read and agreed to the published version of the manuscript.

Funding: This research was supported by the Australian Government through the Australian Research Council's Discovery Projects funding scheme (project DP170104546). The views expressed herein are those of the authors and are not necessarily those of the Australian Government or Australian Research Council.

Institutional Review Board Statement: Not applicable.

Informed Consent Statement: Not applicable.

Data Availability Statement: Not applicable.

Conflicts of Interest: The authors declare no conflict of interest.

Appendix A

The divergences between the generalised motion normal $\mathbf{n}_j$ and unit normal vector $\mathbf{n}_{s_l}$ $(l = 1, 2, 3)$ at $r = R_l(\theta)$ are calculated for 6 DOFs in the Cartesian coordinate system as given by

$$\mathbf{n_1}\cdot\mathbf{n}_{s_l} = \mathbf{n}_x\cdot\mathbf{n}_{s_l} = \frac{\cos\theta - \frac{\sin\theta}{R_l}S_{l,\theta}}{\sqrt{1 + \frac{S_{l,\theta}^2}{R_l^2}}} \tag{A1}$$

$$\mathbf{n_2}\cdot\mathbf{n}_{s_l} = \mathbf{n}_y\cdot\mathbf{n}_{s_l} = \frac{\sin\theta + \frac{\cos\theta}{R_l}S_{l,\theta}}{\sqrt{1+\frac{S_{l,\theta}^2}{R_l^2}}} \tag{A2}$$

$$\mathbf{n_3}\cdot\mathbf{n}_{s_l} = \mathbf{n}_z\cdot\mathbf{n}_{s_l} = 0 \tag{A3}$$

$$\mathbf{n_4}\cdot\mathbf{n}_s = -(z-z_G)\mathbf{n}_y\cdot\mathbf{n}_{s_l} + (y-y_G)\mathbf{n}_z\cdot\mathbf{n}_{s_l} = -\frac{(z-z_G)\left(\sin\theta + \frac{\cos\theta}{R_l}S_{l,\theta}\right)}{\sqrt{1+\frac{S_{l,\theta}^2}{R_l^2}}} \tag{A4}$$

$$\mathbf{n_5}\cdot\mathbf{n}_s = (z-z_G)\mathbf{n}_x\cdot\mathbf{n}_{s_l} - (x-x_G)\mathbf{n}_z\cdot\mathbf{n}_{s_l} = \frac{(z-z_G)\left(\cos\theta - \frac{\sin\theta}{R_l}S_{l,\theta}\right)}{\sqrt{1+\frac{S_{l,\theta}^2}{R_l^2}}} \tag{A5}$$

$$\mathbf{n_6}\cdot\mathbf{n}_{s_l} = -(y-y_G)\mathbf{n}_x\cdot\mathbf{n}_{s_l} + (x-x_G)\mathbf{n}_y\cdot\mathbf{n}_{s_l} = \frac{S_{l,\theta}+(y_G-x_G)\left(\cos\theta+\sin\theta-\frac{\sin\theta-\cos\theta}{R_l}S_{l,\theta}\right)}{\sqrt{1+\frac{S_{l,\theta}^2}{R_l^2}}} \tag{A6}$$

If $S_{l,\theta} = 0$, it can be applied to a circular platform or ring breakwater.

Appendix B

The reduced expressions $H_l^{(j)}$, $P_l^{(j)}$, $Q_l^{(j)}$, $\tilde{P}_l^{(j)}$ and $\tilde{Q}_l^{(j)}$ $(j = 0, 1, \cdots, 6; l = 1, 2, 3)$ introduced in Equations (31)–(39) and Equations (50)–(58) are given by

$$\begin{aligned} H_l^{(0)} &= \frac{igA}{\omega\cosh kh}\sum_{m=-\infty}^{\infty} e^{im(\frac{\pi}{2}-\beta)}\mathcal{J}'_{m0}(kR_l)\mathrm{e}^{im\theta}\cosh k(z+h) \\ &= \frac{igA}{\omega\cosh kh}\sum_{m=-\infty}^{\infty} e^{im(\frac{\pi}{2}-\beta)}\sum_{q=-\infty}^{\infty}\sum_{n_r=-\infty}^{\infty}\left[\frac{k}{2}b_{2,0n_r}^{(l)}\left\{f_{m-1,q}^{(l)} - f_{m+1,q}^{(l)}\right\} + mn_r b_{1,0,n_r}^{(l)}f_{m,q}^{(l)}\right]e^{i(m+q+n_r)\theta}\cosh k(z+h) \end{aligned} \tag{A7}$$

$$H_l^{(1)} = R_l^2\cos\theta - R_l S_{l,\theta}\sin\theta = \frac{1}{4}\sum_{n_r=-\infty}^{\infty}\left\{(n_r+1)b_{2,0,n_r-1}^{(l)} - (n_r-1)b_{2,0,n_r+1}^{(l)}\right\}e^{in_r\theta} \tag{A8}$$

$$H_l^{(2)} = R_l^2\sin\theta + R_l S_{l,\theta}\cos\theta = \frac{1}{4i}\sum_{n_r=-\infty}^{\infty}\left\{(n_r+1)b_{2,0,n_r-1}^{(l)} + (n_r-1)b_{2,0,n_r+1}^{(l)}\right\}e^{in_r\theta} \tag{A9}$$

$$H_l^{(3)} = 0 \tag{A10}$$

$$\begin{aligned} H_l^{(4)} &= -\left(z-z_{G_p}\right)\left(R_l^2\sin\theta + R_l S_{l,\theta}\cos\theta\right) \\ &= -\frac{\left(z-z_{G_p}\right)}{4i}\sum_{n_r=-\infty}^{\infty}\left\{(n_r+1)b_{2,0,n_r-1}^{(l)} + (n_r-1)b_{2,0,n_r+1}^{(l)}\right\}e^{in_r\theta} \end{aligned} \tag{A11}$$

$$\begin{aligned} H_l^{(5)} &= \left(z-z_{G_p}\right)\left(R_l^2\cos\theta - R_l S_{l,\theta}\sin\theta\right) \\ &= \frac{\left(z-z_{G_p}\right)}{4}\sum_{n_r=-\infty}^{\infty}\left\{(n_r+1)b_{2,0,n_r-1}^{(l)} - (n_r-1)b_{2,0,n_r+1}^{(l)}\right\}e^{in_r\theta} \end{aligned} \tag{A12}$$

$$H_l^{(6)} = R_l^2 S_{l,\theta} = -\frac{in_r}{3}\sum_{n_r=-\infty}^{\infty} b_{3,0,n_r}^{(l)}e^{in_r\theta} \tag{A13}$$

where $S_{l,\theta}$ denotes $\left.\frac{\partial S_l}{\partial\theta}\right|_{r=R_l(\theta)}$ and z_{G_p} the z-coordinate of the centre of gravity for the oscillating body p.

$$\begin{aligned} \tilde{P}_l^{(0)} &= \frac{igA}{\omega\cosh kh}\sum_{m=-\infty}^{\infty} e^{im(\frac{\pi}{2}-\beta)}\mathcal{J}'_{m0}(kR_l)e^{im\theta}\cosh k(z+h) \\ &= \frac{igA}{\omega\cosh kh}\sum_{m=-\infty}^{\infty} e^{im(\frac{\pi}{2}-\beta)}\sum_{q=-\infty}^{\infty}\sum_{n_r=-\infty}^{\infty}\left[\frac{k}{2}b_{2,0n_r}^{(l)}\left\{f_{m-1,q}^{(l)} - f_{m+1,q}^{(l)}\right\} + mn_r b_{1,0,n_r}^{(l)}f_{m,q}^{(l)}\right]e^{i(m+q+n_r)\theta}\cosh k(z+h) \end{aligned} \tag{A14}$$

$$\tilde{P}_l^{(3)} = -\frac{R_l^3}{2(h-d_1)} = -\frac{1}{2(h-d_1)}\sum_{n_r=-\infty}^{\infty} b_{3,0,n_r}^{(l)}e^{in_r\theta} \tag{A15}$$

$$\begin{aligned}\tilde{P}_l^{(4)} &= \tfrac{1}{8(h-d_1)}\left\{4(z+h)^2\left(R_l^2\sin\theta + R_l S_{l,\theta}\cos\theta\right) - \left(3R_l^4\sin\theta + R_l^3\cos\theta S_{l,\theta}\right)\right\} \\ &= -\tfrac{i}{64(h-d_1)}\sum_{n_r=-\infty}^{\infty}\left[8(z+h)^2\left\{(n_r+1)b^{(l)}_{2,0,n_r-1} + (n_r-1)b^{(l)}_{2,0,n_r+1}\right\}\right. \\ &\quad \left. -\left\{(n_r+11)b^{(l)}_{4,0,n_r-1} + (n_r-11)b^{(l)}_{4,0,n_r+1}\right\}\right]e^{in_r\theta}\end{aligned} \tag{A16}$$

$$\begin{aligned}\tilde{P}_l^{(5)} &= -\tfrac{1}{8(h-d_1)}\left\{4(z+h)^2\left(R_l^2\cos\theta - R_l S_{l,\theta}\sin\theta\right) - \left(3R_l^4\cos\theta - R_l^3\sin\theta S_{l,\theta}\right)\right\} \\ &= -\tfrac{1}{64(h-d_1)}\sum_{n_r=-\infty}^{\infty}\left[8(z+h)^2\left\{(n_r+1)b^{(l)}_{2,0,n_r-1} - (n_r-1)b^{(l)}_{2,0,n_r+1}\right\}\right. \\ &\quad \left. -\left\{(n_r+11)b^{(l)}_{4,0,n_r-1} - (n_r-11)b^{(l)}_{4,0,n_r+1}\right\}\right]e^{in_r\theta}\end{aligned} \tag{A17}$$

$$\tilde{P}_l^{(1)} = \tilde{P}_l^{(2)} = \tilde{P}_l^{(6)} = 0 \tag{A18}$$

$$\begin{aligned}P_l^{(0)} &= \frac{igA}{\omega\cosh kh}\sum_{m=-\infty}^{\infty} e^{im(\frac{\pi}{2}-\beta)} J_m(kR_l)e^{im\theta}\cosh k(z+h) \\ &= \frac{igA}{\omega\cosh kh}\sum_{m=-\infty}^{\infty} e^{im(\frac{\pi}{2}-\beta)}\sum_{q=-\infty}^{\infty} f^{(l)}_{m,q}e^{i(m+q)\theta}\cosh k(z+h)\end{aligned} \tag{A19}$$

$$P_l^{(3)} = \frac{2(z+h)^2 - R_l^2}{4(h-d_1)} = \frac{(z+h)^2}{2(h-d)} - \frac{1}{4(h-d)}\sum_{n_r=-\infty}^{\infty} b^{(l)}_{2,0,n_r}e^{in_r\theta} \tag{A20}$$

$$\begin{aligned}P_l^{(4)} &= \tfrac{4(z+h)^2 - R_l^2}{8(h-d_1)}R_l\sin\theta \\ &= -\tfrac{i}{16(h-d_1)}\sum_{n_r=-\infty}^{\infty}\left[4(z+h)^2\left\{b^{(l)}_{1,0,n_r-1} - b^{(l)}_{1,0,n_r+1}\right\} - \left\{b^{(l)}_{3,0,n_r-1} - b^{(l)}_{3,0,n_r+1}\right\}\right]e^{in_r\theta}\end{aligned} \tag{A21}$$

$$\begin{aligned}P_l^{(5)} &= -\tfrac{4(z+h)^2 - R_l^2}{8(h-d_1)}R_l\cos\theta \\ &= -\tfrac{1}{16(h-d_1)}\sum_{n_r=-\infty}^{\infty}\left[4(z+h)^2\left\{b^{(l)}_{1,0,n_r-1} + b^{(l)}_{1,0,n_r+1}\right\} - \left\{b^{(l)}_{3,0,n_r-1} + b^{(l)}_{3,0,n_r+1}\right\}\right]e^{in_r\theta}\end{aligned} \tag{A22}$$

$$P_l^{(1)} = P_l^{(2)} = P_l^{(6)} = 0 \tag{A23}$$

$$\begin{aligned}\tilde{Q}_l^{(0)} &= \tfrac{igA}{\omega\cosh kh}\sum_{m=-\infty}^{\infty} e^{im(\frac{\pi}{2}-\beta)}\mathcal{J}'_{m0}(kR_l)e^{im\theta}\cosh k(z+h) \\ &= \tfrac{igA}{\omega\cosh kh}\sum_{m=-\infty}^{\infty} e^{im(\frac{\pi}{2}-\beta)}\sum_{q=-\infty}^{\infty}\sum_{n_r=-\infty}^{\infty}\left[\tfrac{k}{2}b^{(l)}_{2,0n_r}\left\{f^{(l)}_{m-1,q} - f^{(l)}_{m+1,q}\right\} + mn_r b^{(l)}_{1,0,n_r}f^{(l)}_{m,q}\right]e^{i(m+q+n_r)\theta}\cosh k(z+h)\end{aligned} \tag{A24}$$

$$\tilde{Q}_l^{(3)} = -\frac{R_l^3}{2(h-d_2)} = -\frac{1}{2(h-d_2)}\sum_{n_r=-\infty}^{\infty} b^{(l)}_{3,0,n_r}e^{in_r\theta} \tag{A25}$$

$$\begin{aligned}\tilde{Q}_l^{(4)} &= \tfrac{1}{8(h-d_2)}\left\{4(z+h)^2\left(R_l^2\sin\theta + R_l S_{l,\theta}\cos\theta\right) - \left(3R_l^4\sin\theta + R_l^3\cos\theta S_{l,\theta}\right)\right\} \\ &= -\tfrac{i}{64(h-d_2)}\sum_{n_r=-\infty}^{\infty}\left[8(z+h)^2\left\{(n_r+1)b^{(l)}_{2,0,n_r-1} + (n_r-1)b^{(l)}_{2,0,n_r+1}\right\} - \left\{(n_r+11)b^{(l)}_{4,0,n_r-1} + (n_r-11)b^{(l)}_{4,0,n_r+1}\right\}\right]e^{in_r\theta}\end{aligned} \tag{A26}$$

$$\begin{aligned}\tilde{Q}_l^{(5)} &= -\tfrac{1}{8(h-d_2)}\left\{4(z+h)^2\left(R_l^2\cos\theta - R_l S_{l,\theta}\sin\theta\right) - \left(3R_l^4\cos\theta - R_l^3\sin\theta S_{l,\theta}\right)\right\} \\ &= -\tfrac{1}{64(h-d_2)}\sum_{n_r=-\infty}^{\infty}\left[8(z+h)^2\left\{(n_r+1)b^{(l)}_{2,0,n_r-1} - (n_r-1)b^{(l)}_{2,0,n_r+1}\right\}\right. \\ &\quad \left. -\left\{(n_r+11)b^{(l)}_{4,0,n_r-1} - (n_r-11)b^{(l)}_{4,0,n_r+1}\right\}\right]e^{in_r\theta}\end{aligned} \tag{A27}$$

$$\tilde{Q}_l^{(1)} = \tilde{Q}_l^{(2)} = \tilde{Q}_l^{(6)} = 0 \tag{A28}$$

$$\begin{aligned}Q_l^{(0)} &= \frac{igA}{\omega\cosh kh}\sum_{m=-\infty}^{\infty} e^{im(\frac{\pi}{2}-\beta)} J_m(kR_l)e^{im\theta}\cosh k(z+h) \\ &= \frac{igA}{\omega\cosh kh}\sum_{m=-\infty}^{\infty} e^{im(\frac{\pi}{2}-\beta)}\sum_{q=-\infty}^{\infty} f^{(l)}_{m,q}e^{i(m+q)\theta}\cosh k(z+h)\end{aligned} \tag{A29}$$

$$Q_l^{(3)} = \frac{2(z+h)^2 - R_l^2}{4(h-d_2)} = \frac{(z+h)^2}{2(h-d_2)} - \frac{1}{4(h-d_2)}\sum_{n_r=-\infty}^{\infty} b^{(l)}_{2,0,n_r}e^{in_r\theta} \tag{A30}$$

$$\begin{aligned} Q_l^{(4)} &= \tfrac{4(z+h)^2-R_l^2}{8(h-d_2)} R_l \sin\theta \\ &= -\tfrac{i}{16(h-d_2)} \sum_{n_r=-\infty}^{\infty} \left[4(z+h)^2 \left\{ b_{1,0,n_r-1}^{(l)} - b_{1,0,n_r+1}^{(l)} \right\} - \left\{ b_{3,0,n_r-1}^{(l)} - b_{3,0,n_r+1}^{(l)} \right\} \right] e^{in_r\theta} \end{aligned} \quad \text{(A31)}$$

$$\begin{aligned} Q_l^{(5)} &= -\tfrac{4(z+h)^2-R_l^2}{8(h-d_2)} R_l \cos\theta \\ &= -\tfrac{1}{16(h-d_2)} \sum_{n_r=-\infty}^{\infty} \left[4(z+h)^2 \left\{ b_{1,0,n_r-1}^{(l)} + b_{1,0,n_r+1}^{(l)} \right\} - \left\{ b_{3,0,n_r-1}^{(l)} + b_{3,0,n_r+1}^{(l)} \right\} \right] e^{in_r\theta} \end{aligned} \quad \text{(A32)}$$

$$Q_l^{(1)} = Q_l^{(2)} = Q_l^{(6)} = 0 \quad \text{(A33)}$$

Appendix C

The reduced expressions $\tilde{a}_{m,n,q}^{(1)}$, $\tilde{b}_{m,n,q}^{(1),(2)}$, $\tilde{c}_{m,n,q}^{(1),(2)}$, $\tilde{d}_{m,n,q}^{(2),(3)}$, $\tilde{e}_{m,n,q}^{(2),(3)}$, $\tilde{f}_{m,n,q}^{(3)}$, $\tilde{a}'^{(1)}_{m,n,q}$, $\tilde{b}'^{(1),(2)}_{m,n,q}$, $\tilde{c}'^{(1),(2)}_{m,n,q}$, $\tilde{d}'^{(2),(3)}_{m,n,q}$, $\tilde{e}'^{(2),(3)}_{m,n,q}$ and $\tilde{f}'^{(3)}_{m,n,q}$ introduced in Equations (50)–(58) are given by

$$\tilde{a}_{m,0,q}^{(1)} = \frac{a_{m,0,q}^{(1)}}{b_1^{|m|}} \ (n=0), \quad \tilde{a}_{m,n,q}^{(1)} = \frac{a_{m,n,q}^{(1)}}{I_m(p_n b_1)} \ (n>0) \quad \text{(A34)}$$

$$\tilde{a}'^{(1)}_{m,0,q} = \frac{1}{b_1^{|m|}} \left(|m| a_{1,0,n_r}^{(1)} a_{m,0,q}^{(1)} + m n_r a_{1,0,n_r}^{(1)} a_{m,0,q}^{(1)} \right) \ (n=0) \quad \text{(A35)}$$

$$\tilde{a}'^{(1)}_{m,n,q} = \frac{1}{I_m(p_n b_1)} \left\{ \frac{p_n}{2} a_{2,0,n_r}^{(1)} \left(a_{m-1,n,q}^{(1)} + a_{m+1,n,q}^{(1)} \right) + m n_r a_{1,0,n_r}^{(1)} a_{m,n,q}^{(1)} \right\} \ (n>0) \quad \text{(A36)}$$

$$\tilde{b}_{m,0,q}^{(1,2)} = b_{m,0,q}^{(1,2)} \ (n=0), \quad \tilde{b}_{m,n,q}^{(1,2)} = \frac{b_{m,n,q}^{(1,2)}}{I_m(k_n b_2)} \ (n>0) \quad \text{(A37)}$$

$$\tilde{b}'^{(1)}_{m,0,q} = \frac{k}{2} a_{2,0,n_r}^{(1)} \left(b_{m-1,0,q}^{(1)} - b_{m+1,0,q}^{(1)} \right) + m n_r a_{1,0,n_r}^{(1)} b_{m,0,q}^{(1)} \ (n=0) \quad \text{(A38)}$$

$$\tilde{b}'^{(2)}_{m,0,q} = \frac{k}{2} d_{2,0,n_r}^{(2)} \left(b_{m-1,0,q}^{(2)} - b_{m+1,0,q}^{(2)} \right) + m n_r d_{1,0,n_r}^{(2)} b_{m,0,q}^{(2)} \ (n=0) \quad \text{(A39)}$$

$$\tilde{b}'^{(1)}_{m,n,q} = \frac{1}{I_m(k_n b_2)} \left\{ \frac{k_n}{2} a_{2,0,n_r}^{(1)} \left(b_{m-1,n,q}^{(1)} + b_{m+1,n,q}^{(1)} \right) + m n_r a_{1,0,n_r}^{(1)} b_{m,n,q}^{(1)} \right\} \ (n>0) \quad \text{(A40)}$$

$$\tilde{b}'^{(2)}_{m,n,q} = \frac{1}{I_m(k_n b_2)} \left\{ \frac{k_n}{2} d_{2,0,n_r}^{(2)} \left(b_{m-1,n,q}^{(2)} + b_{m+1,n,q}^{(2)} \right) + m n_r d_{1,0,n_r}^{(2)} b_{m,n,q}^{(2)} \right\} \ (n>0) \quad \text{(A41)}$$

$$\tilde{c}_{m,0,q}^{(1,2)} = \frac{c_{m,0,q}^{(1,2)}}{H_m(k a_1)} \ (n=0), \quad \tilde{c}_{m,n,q}^{(1,2)} = \frac{c_{m,n,q}^{(1,2)}}{K_m(k_n a_1)} \ (n>0) \quad \text{(A42)}$$

$$c'^{(1)}_{m,0,q} = \frac{1}{H_m(k a_1)} \left\{ \frac{k}{2} a_{2,0,n_r}^{(1)} \left(c_{m-1,0,q}^{(1)} - c_{m+1,0,q}^{(1)} \right) + m n_r a_{1,0,n_r}^{(1)} c_{m,0,q}^{(1)} \right\} \ (n=0) \quad \text{(A43)}$$

$$c'^{(2)}_{m,0,q} = \frac{1}{H_m(k a_1)} \left\{ \frac{k}{2} d_{2,0,n_r}^{(2)} \left(c_{m-1,0,q}^{(2)} - c_{m+1,0,q}^{(2)} \right) + m n_r a_{1,0,n_r}^{(2)} c_{m,0,q}^{(2)} \right\} \ (n=0) \quad \text{(A44)}$$

$$\tilde{c}'^{(1)}_{m,n,q} = \frac{1}{K_m(k_n a_1)} \left\{ -\frac{k_n}{2} a_{2,0,n_r}^{(1)} \left(c_{m-1,n,q}^{(1)} + c_{m+1,n,q}^{(1)} \right) + m n_r a_{1,0,n_r}^{(1)} c_{m,n,q}^{(1)} \right\} \ (n>0) \quad \text{(A45)}$$

$$\tilde{c}'^{(2)}_{m,n,q} = \frac{1}{K_m(k_n a_1)} \left\{ -\frac{k_n}{2} d_{2,0,n_r}^{(2)} \left(c_{m-1,n,q}^{(2)} + c_{m+1,n,q}^{(2)} \right) + m n_r d_{1,0,n_r}^{(2)} c_{m,n,q}^{(2)} \right\} \ (n>0) \quad \text{(A46)}$$

$$\tilde{d}_{0,0,q}^{(2,3)} = d_{0,0,q}^{(2,3)} \ (m=0, n=0), \quad \tilde{d}_{m,0,q}^{(2,3)} = \frac{d_{m,0,q}^{(2,3)}}{b_3^{|m|}} \ (m \neq 0, n=0) \quad \text{(A47)}$$

$$\tilde{d}^{(2,3)}_{m,n,q} = \frac{d^{(2,3)}_{m,n,q}}{I_m(q_n b_3)} \ (n > 0) \tag{A48}$$

$$\tilde{d}'^{(2,3)}_{0,0,q} = d^{(2,3)}_{2,0,n_r} e^{(2,3)}_{1,0,q} \ (m = 0, n = 0) \tag{A49}$$

$$\tilde{d}'^{(2,3)}_{m,0,q} = \frac{1}{b_3^{|m|}} \left\{ |m| d^{(2,3)}_{1,0,n_r} d^{(2,3)}_{m,0,q} + m n_r d^{(2,3)}_{1,0,n_r} d^{(2,3)}_{m,0,q} \right\} \ (m \neq 0, n = 0) \tag{A50}$$

$$\tilde{d}'^{(2,3)}_{m,n,q} = \frac{1}{I_m(q_n b_3)} \left\{ \frac{q_n}{2} d^{(2,3)}_{2,0,n_r} \left(d^{(2,3)}_{m-1,n,q} + d^{(2,3)}_{m+1,n,q} \right) + m n_r d^{(2,3)}_{1,0,n_r} d^{(2,3)}_{m,n,q} \right\} \ (n > 0) \tag{A51}$$

$$\tilde{e}^{(2,3)}_{0,0,q} = e^{(2,3)}_{0,0,q} (m = 0, n = 0), \quad \tilde{e}^{(2,3)}_{m,0,q} = \frac{e^{(2,3)}_{m,0,q}}{a_2^{-|m|}} \ (m \neq 0, n = 0) \tag{A52}$$

$$\tilde{e}^{(2,3)}_{m,n,q} = \frac{e^{(2,3)}_{m,n,q}}{K_m(q_n a_2)} \ (n > 0) \tag{A53}$$

$$\tilde{e}'^{(2,3)}_{0,0,q} = -d^{(2,3)}_{2,0,n_r} e^{(2,3)}_{1,0,q} \ (m = 0, n = 0) \tag{A54}$$

$$\tilde{e}'^{(2,3)}_{m,0,q} = \frac{1}{a_2^{-|m|}} \left(-|m| d^{(2,3)}_{1,0,n_r} e^{(2,3)}_{m,0,q} + m n_r d^{(2,3)}_{1,0,n_r} e^{(2,3)}_{m,0,q} \right) \ (m \neq 0, n = 0) \tag{A55}$$

$$\tilde{e}'^{(2,3)}_{m,n,q} = \frac{1}{K_m(q_n a_2)} \left\{ -\frac{q_n}{2} d^{(2,3)}_{2,0,n_r} \left(e^{(2,3)}_{m-1,n,q} + e^{(2,3)}_{m+1,n,q} \right) + m n_r d^{(2,3)}_{1,0,n_r} e^{(2,3)}_{m,n,q} \right\} \ (n > 0) \tag{A56}$$

$$\tilde{f}^{(3)}_{m,0,q} = \frac{f^{(3)}_{m,0,q}}{H_m(k a_3)} \ (n = 0), \quad \tilde{f}^{(3)}_{m,n,q} = \frac{f^{(3)}_{m,n,q}}{K_m(k_n a_3)} \ (n > 0) \tag{A57}$$

$$\tilde{f}'^{(3)}_{m,0,q} = \frac{1}{H_m(k a_3)} \left\{ \frac{k}{2} d^{(3)}_{2,0,n_r} \left(f^{(3)}_{m-1,0,q} - f^{(3)}_{m+1,0,q} \right) + m n_r d^{(3)}_{1,0,n_r} f^{(3)}_{m,0,q} \right\} \ (n = 0) \tag{A58}$$

$$\tilde{f}'^{(3)}_{m,n,q} = \frac{1}{K_m(k_n a_3)} \left\{ -\frac{k_n}{2} d^{(3)}_{2,n,n_r} \left(f^{(3)}_{m-1,n,q} + f^{(3)}_{m+1,n,q} \right) + m n_r d^{(3)}_{1,n,n_r} f^{(3)}_{m,n,q} \right\} \ (n > 0) \tag{A59}$$

References

1. Wang, C.M.; Wang, B. *Large Floating Structures*; Springer: Singapore, 2015; Volume 3.
2. Wang, C.M.; Lim, S.H.; Tay, Z.Y. *WCFS2019: Proceedings of the the World Conference on Floating Solutions*; Springer: Singapore, 2020.
3. Piątek, Ł.; Lim, S.H.; Wang, C.M.; Graaf-van Dinther, R. *WCFS2020: Proceedings of the the Second World Conference of Floating Solutions, Rotterdam*; Springer: Singapore, 2022.
4. Miloh, T. Wave loads on a floating solar pond. In *The Proceedings, International Workshop on Ship and Platform Motions*; Yeung, R.W., Ed.; University of California: Berkeley, CA, USA, 1983.
5. Wang, C.M.; Chu, Y.I.; Park, J.C. Moving offshore for fish farming. *J. Aquac. Mar. Biol.* **2019**, *8*, 38–39. [CrossRef]
6. McNown, J.S. 18. Waves and Seiche in Idealized Ports. In *Proceedings of NBS Semicentennial Symposium on Gravity Waves Held at the NBS on June 18–20, 1951*; National Bureau of Standards Circular 521: Washington, DC, USA, 1952; pp. 153–164.
7. Miles, J.; Munk, W. Harbor paradox. *J. Waterw. Harb. Div.* **1961**, *87*, 111–132. [CrossRef]
8. Garrett, C. Bottomless harbours. *J. Fluid Mech.* **1970**, *43*, 433–449. [CrossRef]
9. Fernandes, A.C. *Analysis of an Axisymmetric Pneumatic Buoy by Reciprocity Relations and a Ring-Source Method*; Massachusetts Institute of Technology: Cambridge, MA, USA, 1983.
10. Konispoliatis, D.; Mazarakos, T.; Mavrakos, S. Hydrodynamic analysis of three–unit arrays of floating annular oscillating–water–column wave energy converters. *Appl. Ocean Res.* **2016**, *61*, 42–64. [CrossRef]
11. Mavrakos, S. Wave loads on a stationary floating bottomless cylindrical body with finite wall thickness. *Appl. Ocean Res.* **1985**, *7*, 213–224. [CrossRef]
12. Mavrakos, S. Hydrodynamic coefficients for a thick-walled bottomless cylindrical body floating in water of finite depth. *Ocean Eng.* **1988**, *15*, 213–229. [CrossRef]
13. Mavrakos, S.A.; Chatjigeorgiou, I.K. Second-order hydrodynamic effects on an arrangement of two concentric truncated vertical cylinders. *J. Mar. Struct.* **2009**, *22*, 545–575. [CrossRef]

14. Mavrakos, S.A. Hydrodynamic coefficients in heave of two concentric surface-piercing truncated circular cylinders. *Appl. Ocean Res.* **2004**, *26*, 84–97. [CrossRef]
15. Mavrakos, S.A. Hydrodynamic characteristics of two concentric surface–piercing. *Mitochondrial Med.* **2006**, *221–228*.
16. Mavrakos, S.A.; Katsaounis, G.M.; Chatjigeorgiou, I.K. Performance characteristics of a tightly moored piston-like wave energy converter under first-and second-order wave loads. In *International Conference on Offshore Mechanics and Arctic Engineering*; AMSE: Estoril, Portugal, 2008; pp. 783–792.
17. Wetmore, S.B.; Ramsden, H.D. CIDS: A Mobile Concrete Island Drilling System for Arctic Offshore Operations. *Mar. Technol. SNAME News* **1984**, *21*, 1–11. [CrossRef]
18. Fishfarmexpert. FjordMAX Promises More Fish and Smaller Footprint. Available online: https://www.fishfarmingexpert.com/article/fjordmax-promises-more-fish-and-smaller-footprint/ (accessed on 25 July 2022).
19. Wang, P.; Zhao, M.; Du, X.; Liu, J. Analytical solution for the short-crested wave diffraction by an elliptical cylinder. *Eur. J. Mech.-B/Fluids* **2019**, *74*, 399–409. [CrossRef]
20. Park, J.C.; Wang, C.M. Hydrodynamic behaviour of floating polygonal platforms under wave action. *J. Mar. Sci. Eng.* **2021**, *9*, 923. [CrossRef]
21. Park, J.C.; Wang, C. Hydrodynamic behaviour of floating polygonal ring structures under wave action. *Ocean Eng.* **2022**, *249*, 110195. [CrossRef]
22. Mansour, A.M.; Williams, A.N.; Wang, K. The diffraction of linear waves by a uniform vertical cylinder with cosine-type radial perturbations. *Ocean Eng.* **2002**, *29*, 239–259. [CrossRef]
23. Liu, J.; Guo, A.; Li, H. Analytical solution for the linear wave diffraction by a uniform vertical cylinder with an arbitrary smooth cross-section. *Ocean Eng.* **2016**, *126*, 163–175. [CrossRef]
24. Liu, J.; Guo, A.; Fang, Q.; Li, H.; Hu, H.; Liu, P. Investigation of linear wave action around a truncated cylinder with non-circular cross section. *J. Mar. Sci. Technol.* **2018**, *23*, 866–876. [CrossRef]

Journal of
Marine Science and Engineering

MDPI

Article

RANS Prediction of Wave-Induced Ship Motions, and Steady Wave Forces and Moments in Regular Waves

Qingze Gao [1,2], Lifei Song [1,2,*] and Jianxi Yao [1,2,*]

1 Key Laboratory of High-Performance Ship Technology, Wuhan University of Technology, Ministry of Education, Wuhan 430070, China; gao_qz@whut.edu.cn
2 School of Naval Architecture, Ocean and Energy Power Engineering, Wuhan University of Technology, Wuhan 430070, China
* Correspondence: songlifei@whut.edu.cn (L.S.); yao@whut.edu.cn (J.Y.)

Abstract: The wave-induced motions, and steady wave forces and moments for the oil tanker KVLCC2 in regular head and oblique waves are numerically predicted by using the expanded RANS solver based on OpenFOAM. New modules of wave boundary condition are programed into OpenFOAM for this purpose. In the present consideration, the steady wave forces and moments include not only the contribution of hydrodynamic effects but also the contribution of the inertial effects due to wave-induced ship motions. The computed results show that the contribution of the inertial effects due to heave and pitch in head waves is non-negligible when wave-induced motions are of large amplitude, for example, in long waves. The influence of wave amplitude on added resistance in head waves is also analyzed. The dimensionless added resistance becomes smaller with the increasing wave amplitude, indicating that added resistance is not proportional to the square of wave amplitude. However, wave amplitude seems not to affect the heave and pitch RAOs significantly. The steady wave surge force, sway force and yaw moment for the KVLCC2 with zero speed in oblique waves are computed as well. The present RANS results are compared with available experimental data, and very good agreements are found between them.

Keywords: wave-induced ship motions; steady wave force and moment; inertia effects; RANS; OpenFOAM

Citation: Gao, Q.; Song, L.; Yao, J. RANS Prediction of Wave-Induced Ship Motions, and Steady Wave Forces and Moments in Regular Waves. *J. Mar. Sci. Eng.* **2021**, *9*, 1459. https://doi.org/10.3390/jmse9121459

Academic Editors: Carlos Guedes Soares and Serge Sutulo

Received: 26 November 2021
Accepted: 16 December 2021
Published: 20 December 2021

1. Introduction

Nowadays, environmental issues have become more and more prominent. Green, efficient and sustainable development has become the main theme of human development. There is a greater need to reduce pollution emissions and save energy for ships operated in real sea conditions. The International Maritime Organization (IMO) and the Marine Environmental Protection Commission (MEPC) have released the regulations for ship energy efficiency standards, for example, the Energy Efficiency Design Index (EEDI) [1,2], which strictly regulates greenhouse gas emissions for newly-built ships. Ship added resistance is closely related to the calculation of the speed reduction coefficient (f_w) in the estimation formula of EEDI. The IMO has also developed guidelines for minimum propulsion power to ensure ship maneuverability in adverse weather conditions [3] to ensure safe navigation. On one hand, the development of green ship technology requires ships to use equipment with less power to reduce emissions and meet energy efficiency standards. On the other hand, ships will inevitably be affected by various environmental factors in sailing, and ships are required to have sufficient propulsion power to cope with the harsh environment and ensure navigation safety. It needs to seek a balance between the two sides. Therefore, it is necessary to accurately forecast a ship's navigational performance in the actual sea environment at the initial stage of design.

When sailing in waves, the steady forces and moments acting on the ship will increase, for example, the added resistance, compared with that in calm water. The towing tank test is regarded as the most accurate way to determine the added forces and moments due to waves, for example, in the works of Sadat-Hosseini et al. [4] and Lee et al. [5]. In recent years, a large number of towing tank tests have been carried out to investigate the hydrodynamic performance of various ships in waves in the frame of the European Union project SHOPERA [6].

However, the towing tank test is usually of high cost, and needs a long preparation period. In addition, it is not so helpful for a theoretical understanding of the problem. Therefore, numerical methods are more and more popularly applied to studying ship hydrodynamic performance at present. Perhaps the numerical methods used nowadays for ship hydrodynamic performance in waves can be categorized into two methods. One is based on potential flow theory, and the other is based on viscous flow theory. There are two main approaches based on potential flow theory, that is, the near-field method and the far-field method. The near-field method directly integrates pressure on the hull surface. Havelock [7] might be the first one who focused on the problem of ship added resistance. He derived the formula of the added resistance without considering the diffraction action of waves based on the Froude–Krylova hypothesis. Salvesen [8] introduced a simplified asymptotic method based on two-dimensional strip theory to overcome the shortcomings of this method in short waves. Faltinsen et al. [9] improved the near-field method based on the direct pressure integration method. Kim et al. [10] developed a three-dimensional time-domain method to predict ship added resistance based on the near-field method. The far-field method is based on the conservation of energy, by which ship added resistance is calculated by the wave energy and momentum fluxes around the hull. This method was proposed by Mauro [11] at first. Joosen [12] and Newman [13] also used this method to predict the added resistance and wave drift force. Later on, Gerritsma and Beukelman [14] improved the far-field method based on the radiated energy method to predict the added resistance in head waves. Chizhiumov [15] used a boundary element method to model the ship motion in heave sea conditions.

In recent years, the three-dimensional panel method has become popular when using the method based on potential flow theory. Kim et al. [16] computed the added resistance for the container ship S175 by using different numerical methods, that is, strip theory, 3D time-domain panel method, and CFD method. Their results indicated that all the numerical results could predict well ship added resistance, but the results from the 3D time-domain panel method and the CFD method have better accuracy than those with strip theory. Park et al. [17] predicted the added resistance of KVLCC2 under four different draft conditions based on the strip theory and the Rankine panel method. The prediction results of pitch and heave under different draft conditions with the two numerical methods were in good agreement with the experimental data.

The methods based on potential flow theory ignore the viscosity, and may not be appropriate for the strong non-linear problems. Yasukawa et al. [18] carried out an experimental study to investigate the effects of wave amplitude on added resistance, and two amplitudes were considered in the experiment. They found that ship added resistance was not proportional to the square of the wave amplitude, while wave amplitude had very little influence on heave and pitch RAOs (Response Amplitudes Operators). Lee et al. [19] and Yu et al. [20] also found similar conclusions in their studies. With the rapid development of computer performance, the numerical methods based on viscosity flow theory, especially the Reynolds-Average Navier–Stokes (RANS) method, are more and more widely used for the computation of ship hydrodynamic forces and moments in waves.

Guo et al. [21] used the RANS method to numerically predict the wave-induced motions and added resistance for KVLCC2 in head waves, and the numerical results were verified and confirmed by uncertainty analysis. The study showed that when $\lambda/L_{pp} < 0.6$, heave and pitch are of small values. Wu et al. [22] calculated the added resistance for KVLCC2 in head short waves by RANS, and the numerical results are in good agreement with the experimental values. They found that the increase of pressure near the ship's bow contributed the most to the increase of resistance. Kim et al. [23] used both RANS and the potential flow method to predict ship added resistance, and validated the accuracy of the RANS method. They also used the RANS method to study speed drop. Sigmund et al. [24] used the RANS method to predict added resistance, and found that the contribution of viscosity accounted for 30% of the total added resistance in the short-wave conditions. The predicted results were in good agreement with the experimental data. Uharek et al. [25] used RANS code Neptuno to predict steady wave forces and moments for a ship in oblique regular waves. The study proved that the inertia effects due to wave-induced motions could not be ignored, especially in long waves. Lee et al. [26] used the RANS method to calculate the added resistance and propeller wake. Islam et al. [27] predicted ship resistances for four different ship models, and uncertainty analysis was performed. Yao et al. [28] numerically predicted added resistance and ship motion of oil tanker KVLCC2 by using RANS method. The effects of wavelength, wave amplitude and the scales of the ship model on the results were analyzed. Islam et al. [29] determined the propulsion power for an inland container vessel in open and restricted channel for a ship model and a full-scale ship. The simulated results indicated that the channel current was essential while predicting the propulsion power requirement and full-scale simulation was crucial for particular cases. Jiao et al. [30] predicted a nonlinear hydro-elastic response for S-175 by the CFD–FEA two-way coupling method in severe wave conditions. The results indicated that the simulation method was reliable and had broad applicability for ship seakeeping and hydro-elasticity issues. Yao et al. [31] computed the mean forces, moments and wave-induced six-DOF motions for KVLCC2 in the regular head wave and beam waves with the RANS method, and good agreement was shown by comparing the CFD results with the experimental data. The RANS method is gradually being widely used to investigate ship hydrodynamic performance in waves with the development of high-performance computing.

In this work, the wave-induced motions and added resistance are predicted for KVLCC2 performing straight ahead motion in head regular waves by using the expanded RANS solver based on OpenFOAM. Great efforts are devoted to the implement of new modules of wave boundary conditions. The steady wave forces and moments are also computed for the ship with zero speed in oblique waves. The added resistance, steady wave drift forces and moments include the contributions of hydrodynamic effects and inertial effects due to wave-induced motions. The results show reasonable agreements with experimental data. In addition, the effects of wavelength and wave amplitude are investigated. The present works would supply a good reference for ship control [32].

2. Numerical Method

2.1. Governing Equations and Numeric Discretization

A horizontal coordinate system is used to describe the flow around the ship. When the ship is at rest in still water, the origin locates at midship, x-axis towards bow, y-axis towards starboard, and z-axis vertical downwards (see Figure 1). Based on the assumption of incompressible Newtonian fluid, the mass conservation equation and momentum conservation (RANS) equations can be expressed as:

$$\frac{\partial U_i}{\partial x_i} = 0 \tag{1}$$

$$\frac{\partial U_i}{\partial t} + \frac{\partial (U_i U_j)}{\partial x_j} = -\frac{1}{\rho}\frac{\partial p}{\partial x_i} + \frac{\partial}{\partial x_j}\left[v\left(\frac{\partial U_i}{\partial x_j} + \frac{\partial U_j}{\partial x_i} \right) - \overline{U_i' U_j'} \right], \tag{2}$$

where $x_i = (x, y, z)$ are the Cartesian coordinates, $U_i = (U, V, W)$ is flow velocity components, ρ is the fluid density, p is pressure, v is kinematic viscosity coefficient, and $-\rho\overline{U_i' U_j'}$ is the Reynolds stress tensor. The available method of mesh deformation is applied to allow the ship to perform motions in waves. In the present consideration, the ship is free to heave, pitch and roll, whereas surge is restrained by using a spring, and sway and yaw are completely fixed.

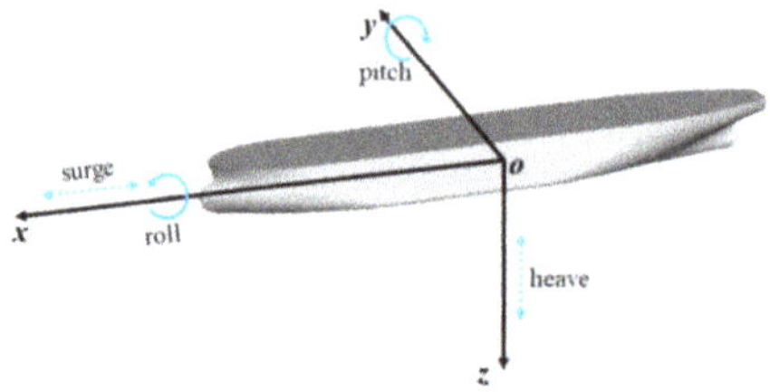

Figure 1. Coordinate system.

According to the Boussinesq hypothesis, the specific Reynolds stress can be expressed as:

$$-\overline{U_i'U_j'} = v_t\left(\frac{\partial U_i}{\partial x_j} + \frac{\partial U_j}{\partial x_i}\right) - \frac{2}{3}\delta_{ij}k, \tag{3}$$

where v_t is the eddy viscosity, k is the turbulent kinetic energy, and δ_{ij} is the Kronecker symbol.

The eddy viscosity in Equation (3) is approximated by the SST $k-\omega$ turbulence model. Since the turbulence transportation equations can be found everywhere, we do not repeat them here.

The Volume of Fluid (VOF) method is used to capture the free surface. The governing equation of volume fraction is:

$$\frac{\partial F}{\partial t} + \frac{\partial (FU_i)}{\partial x_i} = 0, \tag{4}$$

where $F = 0$ means the grid cell is full of air, $F = 1$ means the grid cell is full of water, and $0 < F < 1$ means the grid cell locates at the interface between air and water. The mixed density and viscosity coefficients are expressed as follows

$$\rho = F\rho_w + (1-F)\rho_a \tag{5}$$

$$v = Fv_w + (1-F)v_a \tag{6}$$

where subscripts w and a denote water and air respectively.

The Finite Volume Method (FVM) is used to solve the governing equations of fluid dynamics. The convection terms in the RANS equations and turbulence equations are discretized by Upwind Difference Scheme (UDS), the diffusion term in the RANS equations is discretized using Central Difference Scheme (CDS), and the time term is discretized using a second-order backward scheme. The PIMPLE algorithm is used to correct pressure, where the PIMPLE algorithm is the combination of the Simple (Semi-Implicit Method for Pressure-linked Equations) algorithm and PISO (Pressure Implicit with Split-of-operators) algorithm. For the system of linear equations, the Gauss–Seidel method is used to solve the velocity, turbulent kinetic energy, and dissipation rate iteratively, and the GAMG (Generalized Multi-Grid) method is used to solve the pressure iteratively.

2.2. The Expression of Steady Wave Force and Moment

During the present computations, the heave, pitch and roll are free; however, the surge is restricted by virtual spring, and sway and yaw are fixed, as mentioned. The added resistance, steady sway force and yaw moment are calculated by the following formulas :

$$\overline{F_x} = \left|\overline{X} - X_0\right| + m\overline{wq} + m\overline{\dot{u}} - mg\overline{T}_{31} \tag{7}$$

$$\overline{F_y} = \overline{Y} + m\overline{wp} + mg\overline{T}_{32} \tag{8}$$

$$\overline{M_z} = \overline{N} - (I_y - I_x)\overline{pq}, \tag{9}$$

where $\overline{X}$, $\overline{Y}$ and $\overline{N}$ are the mean force or moment, which are obtained by integrating pressure and viscous stress over hull surface; X_0 is the clam-water resistance; $\left|\overline{X} - X_0\right|$ is added resistance due to the contribution of pure hydrodynamic effects; $m\overline{wq}$ and $m\overline{wp}$ are the mean inertia forces due to the coupled motions of heave, pitch and roll; $m\overline{\dot{u}}$ is mean inertia force due to the surge acceleration; T is the transformation matrix, and T_{31} is the component at the third row of first column; $mg\overline{T}_{31}$ and $mg\overline{T}_{32}$ are the mean inertial forces due to gravity. Their dimensionless forms are expressed as:

$$C_{F_x} = \frac{\overline{F_x}}{\rho g A^2 B^2 / L_{pp}} \tag{10}$$

$$C_{F_x} = \frac{\overline{F_y}}{\rho g A^2 B^2 / L_{pp}} \tag{11}$$

$$C_{M_z} = \frac{\overline{M_z}}{\rho g A^2 B^2}, \tag{12}$$

where A is the wave amplitude; B is the ship beam; L_{pp} is the length between perpendiculars.

2.3. Grid and Boundary Conditions

Due to symmetric flow, only a half ship is considered for head wave cases. The computational domains for head waves and oblique waves are shown in Figures 2a and 3a, respectively. For head wave cases, the velocity inlet boundary condition is set at $x = 1.5\ L_{pp}$, the outlet boundary condition is set at $x = -2.5\ L_{pp}$, the symmetry boundary condition is set at $y = 0$, and others are set as slip boundaries. For oblique wave cases, the boundaries at $x = 1.5\ L_{pp}$ and $y = L_{pp}$ are set as velocity inlet boundaries, the boundaries at $x = -1.5\ L_{pp}$ and $y = -1\ L_{pp}$ are set as an outlet boundary, and others are set as slip boundaries.

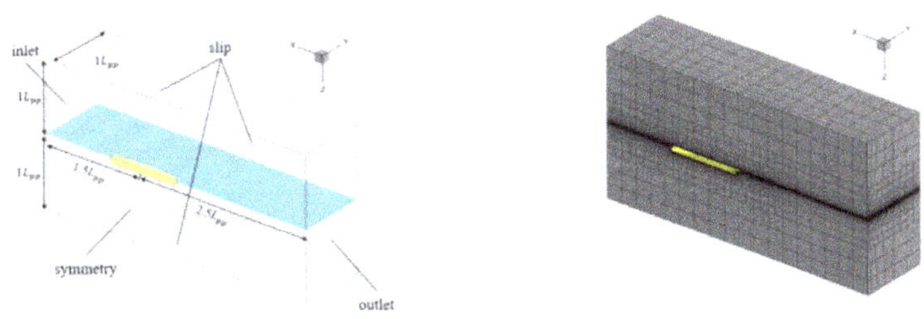

(a) (b)

Figure 2. Computational domain, boundary conditions, and grid arrangement for head wave cases. (**a**) Computational domain and boundary condition; (**b**) Grid arrangement.

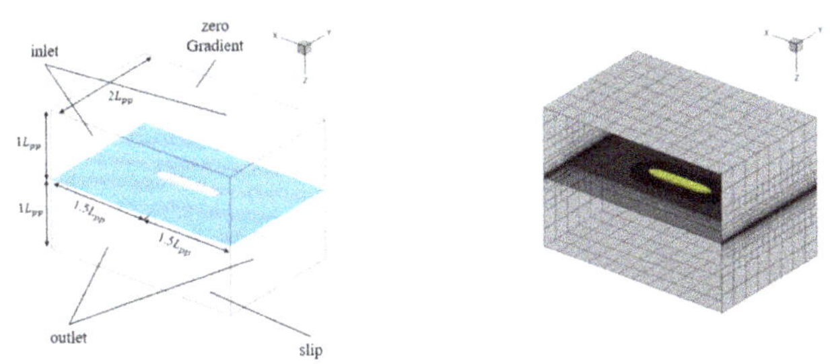

(a) (b)

Figure 3. Computational domain, boundary conditions, and grid arrangement for oblique wave cases. (**a**) Computational domain and boundary condition; (**b**) Grid arrangement.

The software Hexpress is used to generate computational grids. Figures 2b and 3b show the grids for head wave cases and oblique wave cases, respectively. The wall function is used to approximate the flow near the hull surface and the dimensionless distance y^+ is in the range of 30~300.

In this study, wave generation in the computational domain is achieved by setting flow velocity on the inlet boundary. New modules of wave boundary conditions are programed into OpenFOAM. The flow velocity on the inlet boundary is the combination relative flow velocity relative to ship and wave orbital velocity from the solution of the potential theory of linear waves. The components of freestream flow velocity can be expressed as:

$$u_w = -u_0 - A\omega_0 \frac{\cos hk_w(h-z)}{\sin hk_w h} \sin(\omega_e t + k_w x \cos\chi + k_w y \sin\chi)\cos\chi \tag{13}$$

$$v_w = -A\omega_0 \frac{\cos hk_w(h-z)}{\sin hk_w h} \sin(\omega_e t + k_w x \cos\chi + k_w y \sin\chi)\sin\chi \tag{14}$$

$$w_w = -A\omega_0 \frac{\sin hk_w(h-z)}{\sin hk_w h} \cos(\omega_e t + k_w x \cos\chi + k_w y \sin\chi), \tag{15}$$

where u_0 is the ship forward speed; h is water depth; k_w is wave number; A is wave amplitude; ω_0 is natural frequency; χ is the angel of wave direction; ω_e is encounter frequency and it is expressed by:

$$\omega_e = \omega_0 + k_w u_0. \tag{16}$$

The ω_0 is

$$\omega_0 = \sqrt{gk_w \tan hk_w h}. \tag{17}$$

The corresponding freestream wave evaluation is expressed as:

$$\zeta = -A\sin(\omega_e t + k_w x \cos\chi + k_w y \sin\chi). \tag{18}$$

For the cases in head waves, χ equals to zero, so that the Equations (13)–(15) and Equation (18) become:

$$u_w = -A\omega_0 \frac{\cos hk_w(h-z)}{\sin hk_w h} \sin(\omega_e t + k_w x) \tag{19}$$

$$v_w = 0 \tag{20}$$

$$w_w = -A\omega_0 \frac{\sin hk_w(h-z)}{\sin hk_w h} \cos(\omega_e t + k_w x) \tag{21}$$

$$\zeta = -A\sin(\omega_e t + k_w x). \tag{22}$$

2.4. Wave Absorbing

The wave reflection from both ship hull and outlet boundaries usually occur, which will affect numerical accuracy. Thus, wave absorption is often necessary to eliminate reflected waves. During the present simulation, relaxation zones are set near the wave generation boundaries. Within the relaxation zones, the relaxation function $\alpha_R(l_d)$ is expressed as [33]:

$$\alpha_R(l_d) = 1 - \frac{e^{l_d^{3.5}} - 1}{e - 1}, \tag{23}$$

where l_d is the relative distance from 0 to 1. At the interface of relaxed zones and non-relaxed zone, α_R is equal to 1. At the interface of relaxed zone and wave boundary, α_R is equal to 0. Figure 4 presents the sketch of the relaxation zones. Within relaxed zones, the flow velocity and wave evaluation are determined by:

$$\phi = \alpha_R \phi_{computed} + (1 - \alpha_R)\phi_{target}, \tag{24}$$

where ϕ_{target} is the exact freestream value based on the solution of linear wave theory, $\phi_{computed}$ is from RANS.

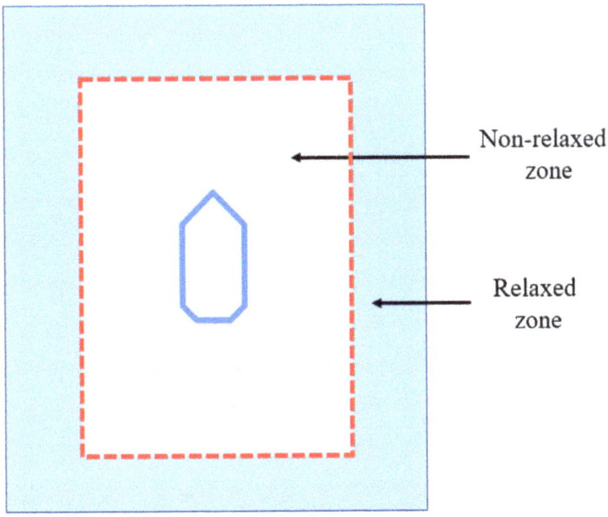

Figure 4. The sketch of wave absorbing zone.

In the case of head waves, because the wave reflection area of ship hull is small, it is necessary to absorb the reflected waves from the downstream boundary. For head wave cases, the relaxation zones may be not necessary. Therefore, the damping term in RANS equations, that is, f_i, is considered to absorb waves downstream. At present, the expression of the damping term is as follows:

$$f_i = (0, 0, d(x)W) \tag{25}$$

$$d(x) = \begin{cases} a(x_s - x) & x \le x_s \\ 0 & x > x_s \end{cases}, \tag{26}$$

where a is the coefficient of wave absorbing, x_s is the starting position of the wave absorbing zone.

2.5. Check of Wave Quality

In order to check the wave quality, RANS simulations are first performed to generate waves in empty computational domains without a ship hull. The domain size is the same as described in Section 2.3. Eight cases of different cell size and time steps are selected. The details are listed in Table 1. The wave parameters are $\lambda = 3.2$ m and $A = 0.016$ m.

Table 1. Cases for the check of wave quality.

Cases	Cell Numbers in a Wavelength	Cell Numbers in a Wave Height	Time Step [s]
C01	36	8	0.001
C02	72	4	0.001
C03	72	8	0.001
C04	72	16	0.001
C05	144	8	0.001
C06	72	8	0.004
C07	72	8	0.002
C08	72	8	0.0005

The computed wave profiles obtained by using different cell numbers in wavelength and height directions are shown in Figure 5. The wave profile based on theory is shown in the figure as well. As observed, with the increase of cell numbers in wavelength and height directions, the wave profile becomes more consistent with the theoretical one. The wave profiles based on cases C03, C04, and C05 are fairly close to the theory one. In order to reduce cell numbers, around 72 cells in wavelength direction and eight cells in wave height direction (i.e., similar with the case C03) are considered when generating waves for the computational domain with ship hull.

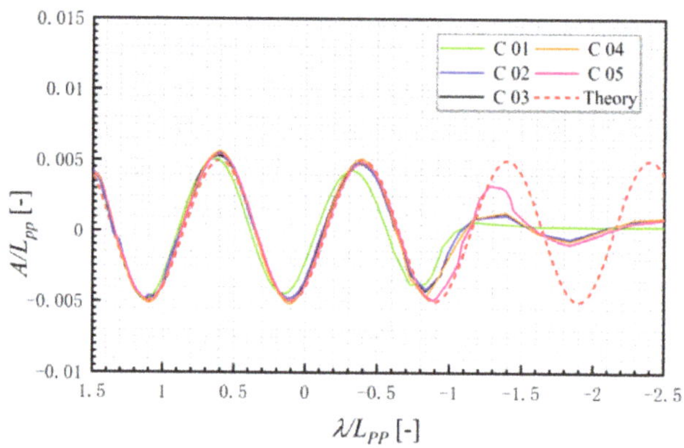

Figure 5. Influence of grid size on wave quality.

Figure 6 presents the computed wave profiles by using different time steps for case C03 as listed in Table 1. It shows that, with the decease of time step, the computed wave profile becomes closer to the theoretical one. When the time step is 0.001 s, the wave quality seems to be acceptable, so that for the following RANS simulations of flow around the ship in waves, the time step 0.001 s is applied.

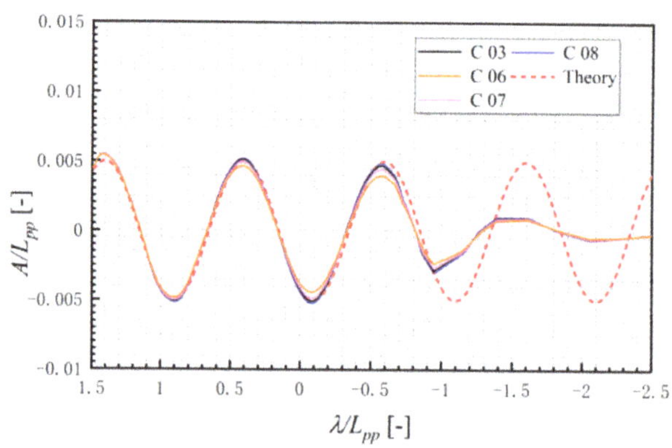

Figure 6. Influence of time step on wave quality.

3. Results and Analysis

3.1. Ship Model and Computational Cases

In this study, the RANS computations are carried out for the naked hull of oil tanker KVLCC2 without propeller and rudder. The principal particulars of the full scale KVLCC2 are listed in Table 2, T is the draft, KG is the height of center of gravity from keel, C_B is the block coefficient. The KVLCC2 geometry is shown in Figure 7.

Table 2. Principal particulars of KVLCC2.

Items	Real Ship
Scale	1
L_{pp} [m]	320
B [m]	58
T [m]	20.8
KG [m]	18.6
C_B [-]	0.81

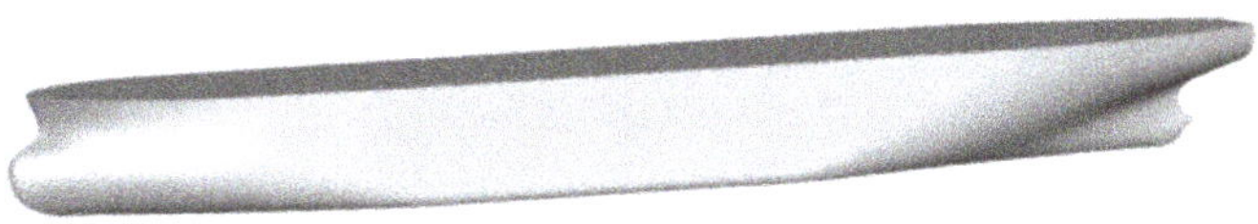

Figure 7. Geometry of the KVLCC2.

In the present consideration, the head wave is defined as a 0° incidence wave angle, and the following waves as 180°, as shown in Figure 8. The computational cases for head waves and oblique waves are listed in Table 3. The experimental data from Seoul National University (SNU), Osaka University (OU), and European project SHOPERA are used to validate the present RANS results. The test conditions are also listed in Table 3.

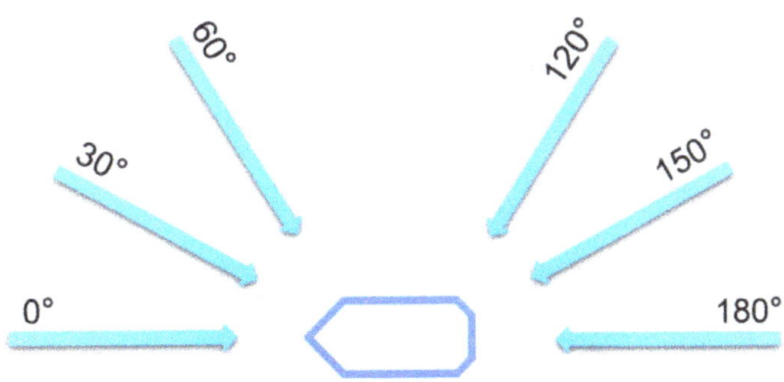

Figure 8. The sketch of wave incident angle.

Table 3. Experimental and CFD cases.

Items	EFD			CFD
	SNU	OU	SHOPERA	
Scale	100	100	80	80, 100
χ [°]	0	0	30~130	0~150
F_r	0.142	0.142	0	0, 0.142
λ/L_{pp}	0.3~2	0.6~2	0.2~1.2	0.5~2
A/L_{pp}	0.005	0.009375	0.007	0.005, 0.007, 0.009375

For head wave cases, to ensure numerical stability, the calm-water resistance is calculated at first. The calm-water computation lasts for 100 s, to obtain the convergence resistance. Afterwards, the wave generation then starts. For each case, more than 20 encounter periods are performed during the computation.

3.2. Grid Dependency Analysis

Grid dependency analysis is carried out at first. For the case of head waves, coarse, medium and fine grids with around 0.29, 0.65 and 1.19 million cells are systematically generated. According to the check of wave quality (see Section 2.5), for the medium grid there are around 72 cells in the wavelength direction and eight cells in wave height direction. The computation is performed for λ/L_{pp} =1.0, A/L_{pp} =0.005, and the ship model is 3.2 m long. The computed results for the head wave case are shown in Figure 9. As seen, the added resistance becomes closer as increasing the grid resolution. The added resistance is computed by subtracting the calm-water resistance from the mean resistance in head waves. The calm-water resistances, which are computed by the coarse, medium, and fine grids are about 4.308 N, 4.067 N, and 3.940 N, respectively. Compared with the SNU experimental data 3.966 N, the computational errors of calm-water resistance are about 8.62%, 2.54%, and 0.66% respectively. The discrepancy between the added resistances based on medium and fine grids is around 5.50%.

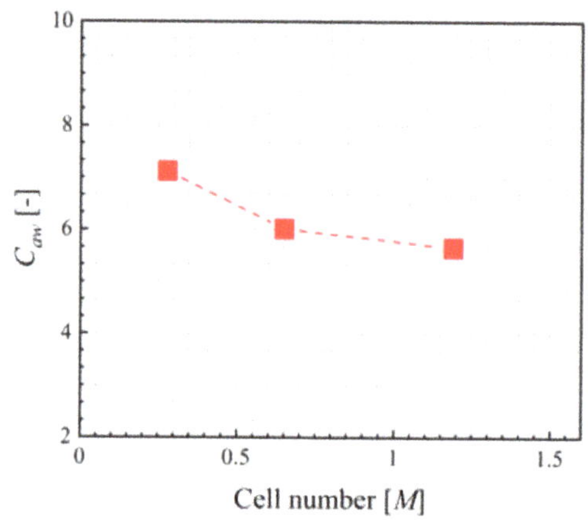

Figure 9. Grid dependency analysis for head waves.

The grid dependency analysis is also carried out for oblique wave cases. Coarse, medium and fine grids are also systematically generated. There are around 0.48, 0.96 and 1.80 million cells, respectively. Similarly, there are around 72 cells in the wavelength direction and eight cells in the wave height direction for the medium grid. Here, wavelength is $\lambda/L_{pp} = 0.635$, wave amplitude is $A/L_{pp} = 0.007$, and wave incidence angle is 30°. For the oblique wave cases, the ship model is 4 m long. In addition, the ship speed is zero for oblique wave cases, as mentioned. The computed steady surge force, sway force and yaw moment are presented in Figure 10. In these figures, the dimensionless steady surge force, sway force, and yaw moment show convergence tendency with the increase of cell numbers, as expected.

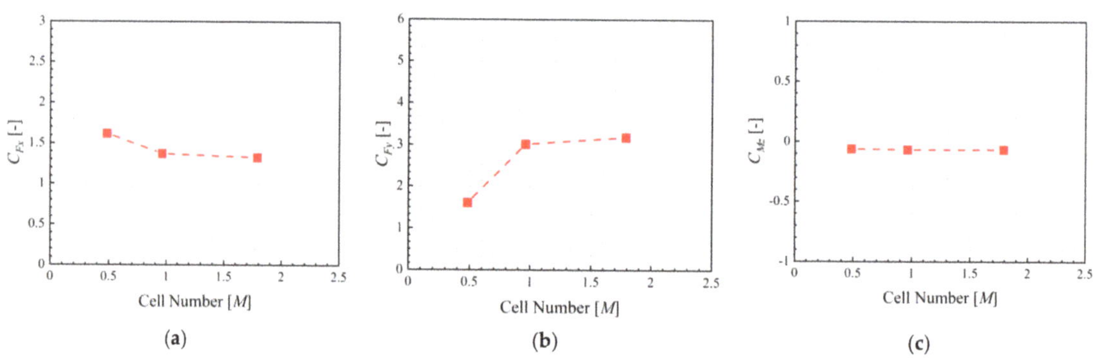

Figure 10. Grid dependency analysis for oblique waves. (**a**) Steady surge force; (**b**) Steady sway force; (**c**) Steady yaw moment.

Based on the above grid dependency analysis, both the medium grids for the head wave case and the oblique wave case seem to be of enough resolution, and they are then used for other computations.

3.3. Results and Analysis

3.3.1. Component Analysis

The head wave cases are computed for nine wave lengths, which are λ/L_{pp} = 0.5, 0.75, 0.9, 1.0, 1.1, 1.2, 1.4, 1.6 and 2.0, and considering two amplitudes A/L_{pp} = 0.005 and 0.009375, where $L_{pp} = 3.2$ m. Figure 11 shows the added resistance in head waves. The mean inertial force generated by longitudinal acceleration is almost zero, and it can be ignored in most cases. However, the mean inertia force generated by the coupling of heave and pitch have a significant contribution, especially in long waves. The comparison conforms the inertia effects of wave-induced ship motions to added resistance.

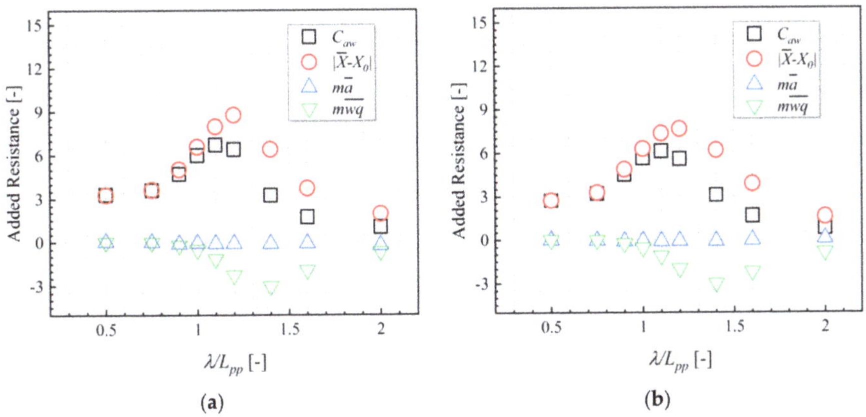

Figure 11. Components of added resistance in head waves. (**a**) $A/L_{pp} = 0.005$; (**b**) $A/L_{pp} = 0.009375$.

The computations in oblique waves are performed for λ/L_{pp} = 0.635 and A/L_{pp} = 0.007, where $L_{pp} = 4$ m. The components of steady wave forces in oblique waves are shown in Figure 12. In oblique wave cases, the ship forward speed equals zero. The inertial effect has small contributions to the steady forces and moment, mainly because pitch, heave and roll are of very small amplitudes.

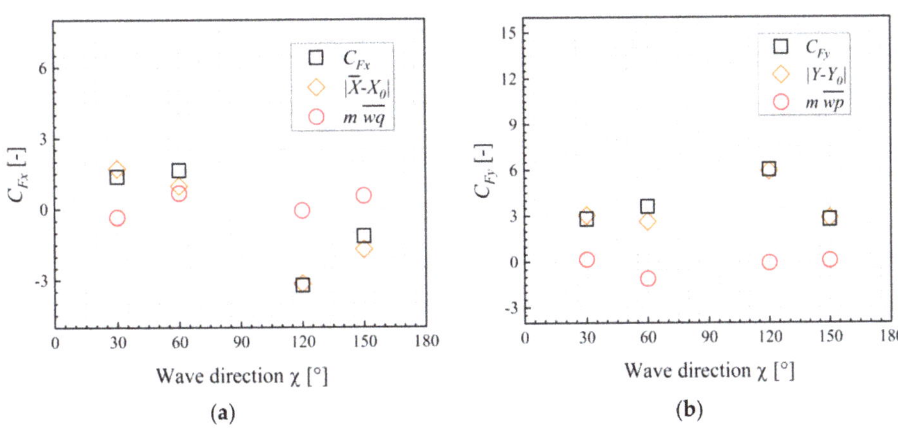

Figure 12. Components of steady surge force and sway force in oblique waves. (**a**) Steady surge force; (**b**) Steady sway force.

3.3.2. Effect of Wave Amplitude on Added Resistance

The results of added resistance, heave, and pitch RAOs at different wave amplitudes are compared in Figure 13. As seen in Figure 14a, for the two wave amplitudes, the computed curves of dimensionless added resistances versus wavelength do not coincide, meaning the added resistance is not proportional to the square of the wave amplitude. With decreasing wave amplitude, the dimensionless added resistances become large, in particular at $\lambda/L_{PP} = 1.1$ and 1.2. In addition, it is also seen from Figure 13b,c that wave amplitude has quite a small influence on the heave and pitch RAOs.

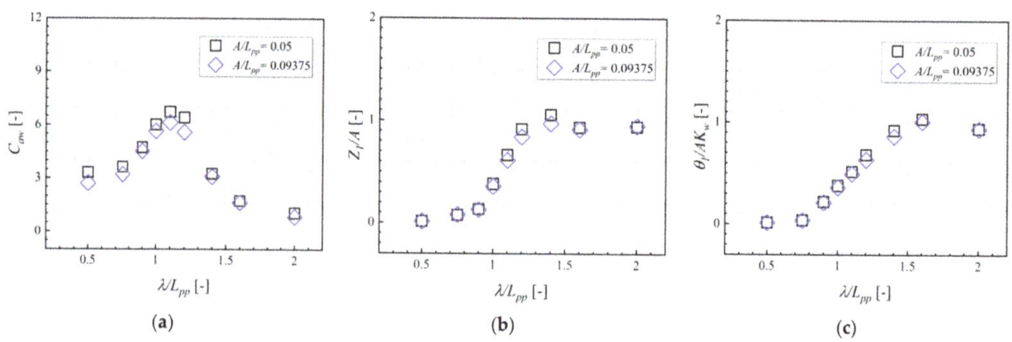

Figure 13. The influence of wave amplitudes. (**a**) Added resistance; (**b**) Heave RAOs; (**c**) Pitch RAOs.

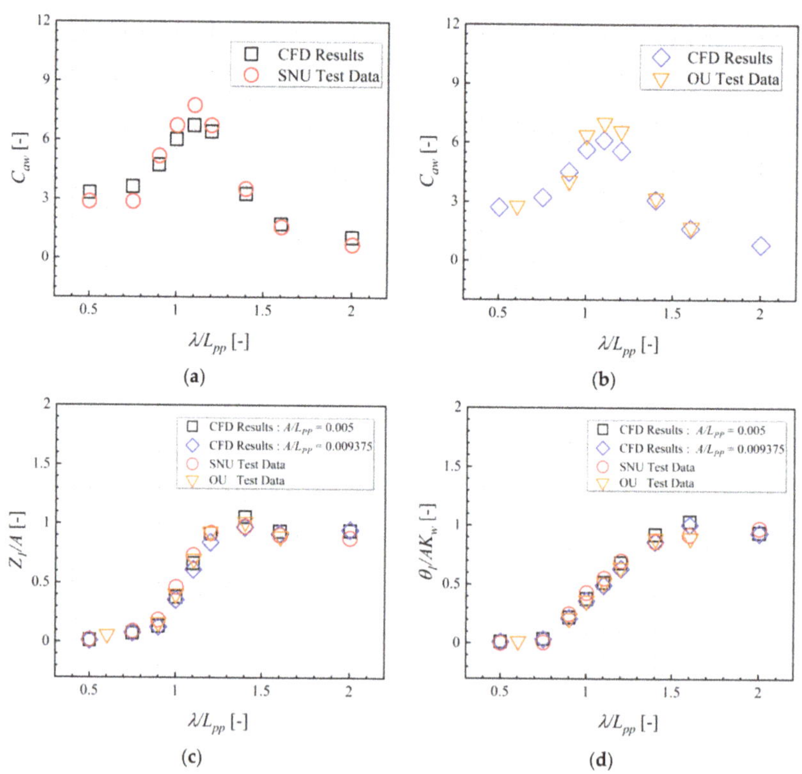

Figure 14. Compare the CFD results with the experiment data for head wave cases. (**a**) $A/L_{PP} = 0.005$; (**b**) $A/L_{PP} = 0.009375$; (**c**) Heave RAOs; (**d**) Pitch RAOs.

3.3.3. Validation of RANS Results

The accuracy of RANS results is validated by comparing them with the experimental data. In Figure 14, the added resistance with inertial effects, heave and pitch RAOs in head waves are compared with the experimental data. It can be observed that the RANS results are in good agreement with the experimental data.

The RANS results in oblique waves are compared with experimental data in Figure 15. For different incidence wave angles, the computed steady wave surge forces, sway forces and yaw moments show acceptable agreements with experimental data. The error for the sway force at an incident angle of 60° is large. The reason is as-yet unknown. The yaw moments are of very small values in the oblique waves as the ship is zero speed.

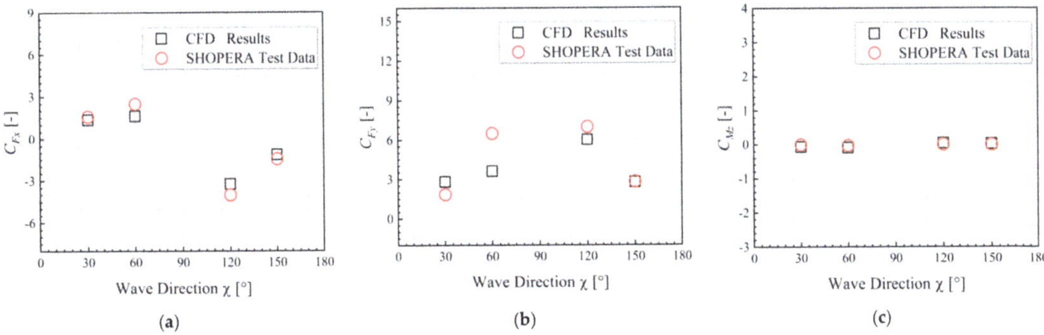

Figure 15. Compare the CFD results with the experiment data for oblique wave cases. (**a**) Steady surge force; (**b**) Steady sway force; (**c**) Steady yaw moment.

The wave snapshots and pressure distribution on the hull surface for a head wave case in an encounter period are shown in Figure 16. It can be observed that the pressure near the bow region changes considerably as the waves propagate from front to back. Green water on the bow deck at $t = 0.75\ T_e$ is observed.

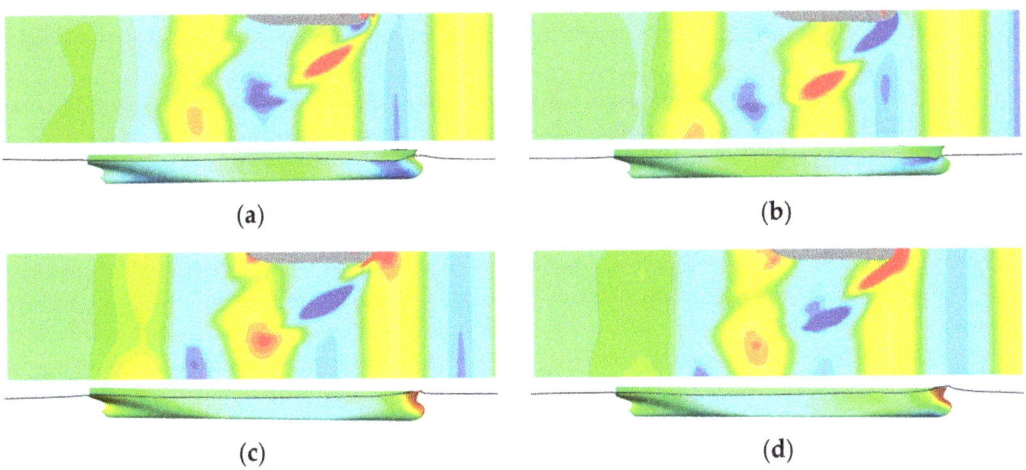

Figure 16. Wave snapshots and hull surface pressure in an encounter period. (**a**) $t = 0$; (**b**) $t = 0.25\ T_e$; (**c**) $t = 0.5\ T_e$; (**d**) $t = 0.75\ T_e$.

The wave snapshots and pressure distribution on the hull surface on the starboard side for four oblique angles are presented in Figure 17.

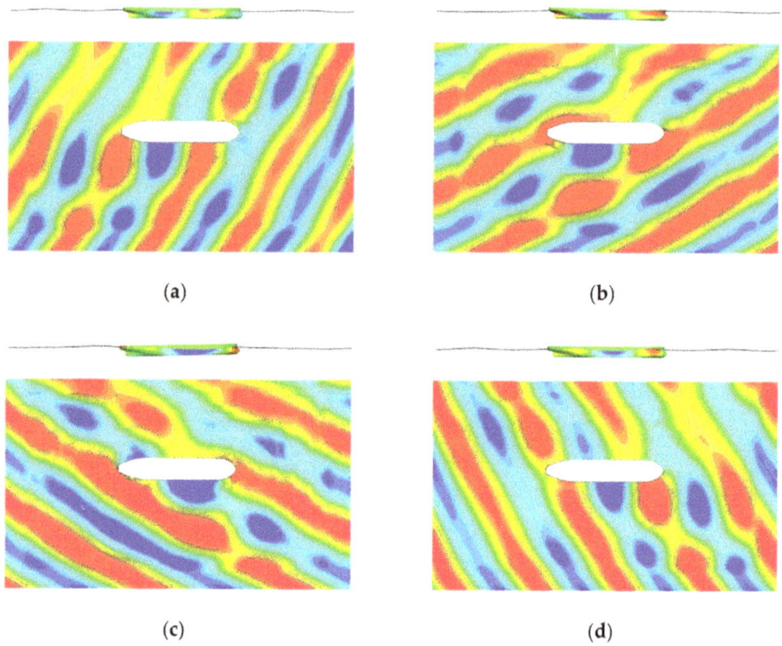

Figure 17. Wave snapshots and hull surface pressure of four incidence angels. (**a**) $\chi = 30^{\circ}$; (**b**) $\chi = 60^{\circ}$; (**c**) $\chi = 120^{\circ}$; (**d**) $\chi = 150^{\circ}$.

4. Conclusions

In this study, the wave-induced ship motions and steady wave force and moment for KVLCC2 are numerically predicted by using the RANS solver on OpenFOAM. New modules of wave boundary conditions are programed into OpenFOAM. The conclusions may be as follows:

(1) The computed added resistance, as well as the steady wave sway force and yaw moment with inertia effects due to the wave-induced motions agrees well with the available experimental data. This confirms the contributions of the inertia effects. The inertia forces cannot be neglected if wave-induced motions are of large amplitudes.
(2) The comparison of the computed resistances using two wave amplitudes indicates that added resistance is not proportional to the square of wave amplitude.
(3) General good agreements between the computed results and the experimental data are observed. This shows that the RANS solver can be used as a tool for ship seakeeping analysis.

Great efforts in future works will be devoted to the cases, considering a full appended ship (propeller and rudder), and will also concern the ship hydrodynamic performances in irregular waves.

Author Contributions: Conceptualization, J.Y.; Data curation, Q.G. and J.Y.; Formal analysis, J.Y.; Funding acquisition, L.S.; Investigation, Q.G. and J.Y.; Methodology, Q.G. and J.Y.; Project administration, J.Y. and L.S.; Resources, Q.G.; Software, Q.G. and J.Y.; Supervision, J.Y.; Validation, Q.G. and J.Y.; Visualization, Q.G. and J.Y.; Writing—original draft, Q.G.; Writing—review & editing, J.Y. All authors have read and agreed to the published version of the manuscript.

Funding: This research was funded by the National Natural Science Foundation of China, grant numbers 51809203, 51720105011, 51609188.

Institutional Review Board Statement: Not applicable.

Informed Consent Statement: Not applicable.

Data Availability Statement: The data used to support the findings of this study are available from the corresponding author upon request.

Conflicts of Interest: The authors declare no conflict of interest.

References

1. International Maritime Organization. *Interim Guidelines for the Calculation of the Coefficient fw for Decrease in Ship Speed in a Representative Sea Condition for Trial Use*; IMO: London, UK, 2012.
2. Specialist Committee on Seakeeping. Final Report and Recommendations to the 27th ITTC. In Proceedings of the the 27th International Towing Tank Conference, Copenhagen, Denmark, 31 August–5 September 2014.
3. International Maritime Organization. *Interim Guidelines for Determining Minimum Propulsion Power to Maintain the Maneuverability in Adverse Conditions*; IMO: London, UK, 2013.
4. Sadat-Hosseini, H.; Wu, P.-C.; Carrica, P.M.; Kim, H.; Toda, Y.; Stern, F. CFD verification and validation of added resistance and motions of KVLCC2 with fixed and free surge in short and long head waves. *Ocean Eng.* **2013**, *59*, 240–273. [CrossRef]
5. Lee, J.; Park, D.-M.; Kim, Y. Experimental investigation on the added resistance of modified KVLCC2 hull forms with different bow shapes. *Proc. Inst. Mech. Eng. Part M J. Eng. Marit. Environ.* **2017**, *231*, 395–410. [CrossRef]
6. Shigunov, V.; Moctar, O.E.; Papanikolaou, A.; Potthoff, R.; Liu, S. International benchmark study on numerical simulation methods for prediction of manoeuvrability of ships in waves. *Ocean Eng.* **2018**, *165*, 365–385. [CrossRef]
7. Havelock, T.H. The resistance of a ship among waves. *Proc. R. Soc. Lond. Ser. A Math. Phys. Sci.* **1937**, *161*, 299–308.
8. Salvesen, N. Second-order steady state forces and moments on surface ships in oblique regular waves. In *Proceedings of the International Symposium on the Dynamics of Marine Vehicles and Structures in Waves*; Naval Ship Research and Development Center: Washington, DC, USA, 1974.
9. Faltinsen, O.M.; Minsaas, K.J.; Liapis, N.; Skjørdal, S.O. Prediction of resistance and propulsion of a ship in a seaway. In Proceedings of the 13th Symposium on Naval Hydrodynamics, Tokyo, Japan, 6–8 October 1980; pp. 505–529.
10. Kim, K.H.; Kim, Y. Numerical study on added resistance of ships by using a time-domain Rankine panel method. *Ocean Eng.* **2011**, *38*, 1357–1367. [CrossRef]
11. Maruo, H. The drift of a body floating on waves. *J. Ship Res.* **1960**, *4*, 1–10.
12. Joosen, W. Added resistance of ships in waves. In Proceedings of the 6th Symposium on Naval Hydrodynamics; National Academy Press: Washington, DC, USA, 1966.
13. Newman, J.N. The drift force and moment on ships in waves. *J. Ship Res.* **1967**, *11*, 51–60. [CrossRef]
14. Gerritsma, J.; Beukelman, W. Analysis of the resistance increase in waves of a fast cargo ship. *Int. Shipbuild. Progress* **1972**, *19*, 285–293. [CrossRef]
15. Chizhiumov, S.D. Numerical Modeling of Ship Motion in Heavy Sea Conditions. In Proceedings of the Sixth ISOPE Pacific/Asia Offshore Mechanics Symposium, Vladivostok, Russia, 12–16 September 2004.
16. Kim, B.Y.; Seo, M.G.; Park, D.M.; Yang, K.K. Numerical and Experimental Analyses of Added Resistance in Waves. In Proceedings of the 29th International Workshop on Water Waves and Floating Bodies, Osaka, Japan, 30 March–2 April 2014.
17. Park, D.-M.; Kim, Y.; Seo, M.-G.; Lee, J. Study on added resistance of a tanker in head waves at different drafts. *Ocean Eng.* **2016**, *111*, 569–581. [CrossRef]
18. Yasukawa, H.; Matsumoto, A.; Ikezoe, S. Wave Height Effect on Added Resistance of Full Hull Ships in Waves. *J. Jpn. Soc. Nav. Archit. Ocean Eng.* **2016**, *23*, 45–54.
19. Lee, J.-H.; Kim, Y. Study on added resistance of a ship under parametric roll motion. *Ocean Eng.* **2017**, *144*, 1–13. [CrossRef]
20. Yu, J.W.; Lee, C.M.; Choi, J.E.; Lee, I. Effect of ship motions on added resistance in regular head waves of KVLCC2. *Ocean Eng.* **2017**, *146*, 375–387. [CrossRef]
21. Guo, B.; Steen, S.; Deng, G. Seakeeping prediction of KVLCC2 in head waves with RANS. *Appl. Ocean Res.* **2012**, *35*, 56–67. [CrossRef]
22. Wu, C.; Yan, D.; Qiu, G.; Ni, Y. CFD Computation of Added Resistance for KVLCC2 Model in Head Short Waves. *J. Ship Mech.* **2015**, *19*, 229–236.
23. Kim, M.; Hizir, O.; Turan, O.; Incecik, A. Estimation of added resistance and ship speed loss in a seaway. *Ocean Eng.* **2017**, *141*, 465–476. [CrossRef]
24. Sigmund, S.; Moctar, O.E. Numerical and experimental investigation of added resistance of different ship types in short and long waves. *Ocean Eng.* **2018**, *147*, 51–67. [CrossRef]
25. Uharek, S.; Cura-Hochbaum, A. The influence of inertial effects on the mean forces and moments on a ship sailing in oblique waves part b: Numerical prediction using a rans code. *Ocean Eng.* **2018**, *165*, 264–276. [CrossRef]
26. Lee, C.M.; Park, S.C.; Yu, J.W.; Choi, J.E.; Lee, I. Effects of diffraction in regular head waves on added resistance and wake using cfd. *Int. J. Nav. Arch. Ocean Eng.* **2019**, *11*, 736–749. [CrossRef]
27. Islam, H.; Guedes Soares, C. Uncertainty analysis in ship resistance prediction using OpenFOAM. *Ocean Eng.* **2019**, *191*, 105805. [CrossRef]

28. Yao, J.; Su, Y.; Song, X.; Liu, Z.; Cheng, X.; Zhan, C. RANS Analysis of the Motions and Added Resistance for KVLCC2 in Head Regular Waves. *Appl. Ocean Res.* **2020**, *105*, 102398. [CrossRef]
29. Islam, H.; Guedes Soares, C.; Liu, J.; Wang, X. Propulsion power prediction for an inland container vessel in open and restricted channel from model and full-scale simulations. *Ocean Eng.* **2021**, *229*, 108621. [CrossRef]
30. Jiao, J.; Huang, S.; Wang, S.; Guedes Soares, C. A CFD–FEA two-way coupling method for predicting ship wave loads and hydroelastic responses. *Appl. Ocean Res.* **2021**, *117*, 102919. [CrossRef]
31. Yao, J.; Cheng, X.; Song, X.; Zhan, C.; Liu, Z. RANS Computation of the Mean Forces and Moments, and Wave-Induced Six Degrees of Freedom Motions for a Ship Moving Obliquely in Regular Head and Beam Waves. *J. Mar. Sci. Eng.* **2021**, *9*, 1176. [CrossRef]
32. Borkowski, P. Numerical Modeling of Wave Disturbances in the Process of Ship Movement Control. *Algorithms* **2018**, *11*, 130. [CrossRef]
33. Jacobsen, N.G.; Fuhrman, D.R.; Fredsone, J. A wave generation toolbox for the open-source CFD library: Openfoam. *Int. J. Numer. Methods Fluids* **2012**, *70*, 1073–1088. [CrossRef]

Journal of
Marine Science and Engineering

MDPI

Article

Assessment of the Roll Derivatives of Different Surface Ships Based on Numerical Pure Roll Simulation

Thi Loan Mai [1], Anh Khoa Vo [1], Hyeon Kyu Yoon [2,*] and Dong Kyou Park [3,*]

[1] Department of Smart Environmental Energy Engineering, Changwon National University, Changwon 51140, Korea
[2] Department of Naval Architecture and Marine Engineering, Changwon National University, Changwon 51140, Korea
[3] Department of Electromechanical Convergence Engineering, Korea University of Technology and Education, Cheonan 31253, Korea
* Correspondence: hkyoon@changwon.ac.kr (H.K.Y.); pdongkyou@koreatech.ac.kr (D.K.P.); Tel.: +82-55-213-3683 (H.K.Y.); +82-41-640-8614 (D.K.P.)

Abstract: Among the 6 degrees of freedom (6-DoF), excessive roll motion is the most dangerous cause of ships capsizing. However, when analyzing the maneuverability of surface ships, the roll components have usually been ignored. It is widely known that the influence of roll moment becomes significant for surface ships with low GM (metacentric height) and high speed. This paper examines the pure roll test for several surface ships to assess the roll-related hydrodynamic derivatives of added mass and damping in maneuvering. The objective ships are the KRISO Container Ship (KCS), David Taylor Model Basin (DTMB), Office of Naval Research Tumblehome (ONRT), and Delft 372 catamaran, where the DTMB and ONRT ships are equipped with complementary bilge keels as damping devices and have a small GM, which the Delft 372 catamaran does not have. The flow during pure roll is analyzed by the Computational Fluid Dynamics (CFD) simulation method that allows the complex flow around ships to be captured, especially when the bilge keel and skeg are considered. The results indicate that the roll moment is greatest in the catamaran. Since the roll moments of the DTMB and ONRT are larger than that of the KCS, bilge keels and surface shape also contribute to increasing roll damping moment. In addition, a comparison of the damping derivatives due to roll rate with results obtained from another method indicates that CFD simulation is capable of accurately predicting the roll-related derivatives, which is difficult to perform by the experiment method.

Keywords: pure roll simulation; small metacentric height; surface ship; bilge keel; skeg; catamaran; roll-related derivatives; CFD

Citation: Mai, T.L.; Vo, A.K.; Yoon, H.K.; Park, D.K. Assessment of the Roll Derivatives of Different Surface Ships Based on Numerical Pure Roll Simulation. *J. Mar. Sci. Eng.* **2022**, *10*, 1702. https://doi.org/10.3390/jmse10111702

Academic Editors: Carlos Guedes Soares and Serge Sutulo

Received: 18 October 2022
Accepted: 6 November 2022
Published: 9 November 2022

1. Introduction

Research related to the maneuvering of the surface ship has been continuously conducted. The roll moment of a ship has been a prime concern due to being the most dangerous among the 6-DoF, which has the possible consequences of instability and ship capsize. In general, the roll moment has usually been ignored when analyzing the maneuverability of the ship with low GM and high-speed ships, such as the KCS, DTMB, and ONRT, according to Hajivand et al. [1], Liu et al. [2], Franceschi et al. [3], and Li et al. [4]. However, it is widely known that the effect of the roll moment becomes significant for surface ships with small GM and high speed. Therefore, some researchers have investigated the effect of roll coupling on ship maneuverability. Kim et al. [5] executed the captive model test that included the roll moment in several tests relative to roll, such as the static heel, combined drift-heel, combined turning-heel, and free roll decay for the KCS model to obtain hydrodynamic derivatives. The maneuvering simulation considering the roll effect was conducted for the 4-DoF motion. The comparison of simulation among 3-DoF, 4-DoF, and the free-running test indicated that the 4-DoF simulation was more accurate than the 3-DoF one. Yoshimura [6] clarified the following principles structure regarding the effect of the roll motion on ship maneuverability. That is, the roll angle became larger by lower GM and a higher Froude number, and the derivatives of Y_φ and N_φ were the key parameter that strongly affected the course-keeping and turning abilities of the ship. Furthermore, the rudder-to-yaw response equation was introduced in this study. Yasukawa and Yoshimura [7] experimented with the oblique towing test and circular motion test with various heel angles for a container ship, pure car carrier, and ferry to obtain hydrodynamic derivatives relative to sway velocity, roll angle, and yaw rate, to analyze the roll-coupling effect. The simplified formulae were proposed for the course stability criterion and steady turning index. By analysis of the course stability criterion, it was concluded that the course stability was worse with increasing forward speed or decreasing GM, whereas the turning performance improved, and the response time until reaching the steady turning increased. Fukui et al. [8] suggested the 4-DoF mathematical model for maneuvering simulation. The effect of roll angle for the KCS and passenger ferry was estimated by the captive model test by setting several roll angles to obtain hydrodynamic derivatives. The proposed mathematical model was validated by comparing the simulation with the free-running test at various GMs and forward speeds. As a result, the turning circle became greater with large GM. Yasukawa et al. [9] proposed a practical maneuvering simulation method considering the roll-coupling effect by adding the motion equation of roll in the 3D-MMG (Maneuvering Modeling Group) model. The roll moment was estimated by multiplying the hull lateral force by the vertical acting point. The empirical formulae of roll-related derivatives and vertical acting points of the lateral force were introduced using the experiment data of four ship models. The validation of the simulation method with free running was implemented for several ships at various GMs, and indicated sufficient accuracy to predict the turning motion considering the roll-coupling effect.

From the literature review just presented, the roll-coupling effect was investigated by roll derivatives with respect to roll angle, which were obtained from the captive model test at various roll angles. The roll derivatives with respect to roll rate and roll acceleration that should be determined in the dynamic test such as the pure roll test have not previously been calculated for a surface ship, due to complex device requirements. Nevertheless, an experimental pure roll test was performed for the underwater vehicle by Kim et al. [10] using an apparatus device. Consequently, roll-added mass and roll damping relative to roll rate were calculated for the underwater vehicle. Additionally, roll damping relative to the roll rate of the submerged body was measured from the pure roll test using Corning motion test equipment by Park et al. [11]. Those devices are difficult to apply to a surface ship.

Another method to estimate the roll derivatives due to roll rate is to apply the system identification (SI) technique. However, this requires data from sea trial tests or free-running model tests. Yoon et al. [12,13] proposed a simpler method to identify the model structure of the hydrodynamic roll moment and its parameter using the data from the turning circle and $20^\circ/20^\circ$ zig-zag tests. Jeon et al. [14] introduced the process of identifying the dynamic models of a combatant in intact and damaged condition, based on the measured data from free-running model tests. The hydrodynamic derivatives constituting 4-DoF were determined through system identification.

In this study, the pure roll simulation is executed for different surface ship types that have a small GM, such as the KCS, DTMB, and ONRT, and a high-speed ship, the Delft 372 catamaran. Although the Delft 372 catamaran has a large GM, it operates at high speed, so the roll moment has a significant change; thus, roll moment is a concern in this ship. Nowadays, numerous investigations of the roll motion on these ships are being conducted, especially the DTMB and ONRT, which are appended with bilge keel and skeg. Khekit et al. [15] and Gokce et al. [16] studied the free roll decay for the DTMB using the CFD simulation method. Meanwhile, the ONRT ship's dynamic instability events and maneuverability were studied with a 4-DoF equation of motion including roll moment by Sadat-Hosseini et al. [17], Araki et al. [18], and Sanada et al. [19]. A commercial CFD code of STAR CCM+ is applied to simulate the pure roll test for different surface ships with given dimensionless angular velocities. The simulation method is verified by conducting the grid and time-step convergence uncertainty analysis. The force and moment of different surface ships indicate the roll moment is greatest in the catamaran. Additionally, the roll moment of the DTMB and ONRT is greater than that of the KCS, due to the bilge keel attached to the DTMB and ONRT. By taking advantage of the CFD method, the flow field as free surface pattern, the pressure distribution, and the vortices field are analyzed during pure roll simulation. Then, roll-related derivatives in pure roll simulation are estimated. A comparison of derivatives due to roll rate demonstrates that the pure roll simulation brings acceptable results. Roll-related derivatives are used to complete the roll equation of motion for 4-DoF maneuverability.

2. Test Overview

2.1. Coordinate System

Figure 1 shows the two kinds of coordinate systems used in this study. The earth-fixed coordinate system (O-xyz), where the O-xy plane coincides with the still-water surface and the z-axis points vertically downward, was used. The body-fixed coordinate system (o-$x_b y_b z_b$) was also employed to define the external forces acting on the ship. For the horizontal plane (o-$x_0 y_0$) located at the still-water surface, the x_0-axis points toward the bow while the y_0-axis points toward starboard. The origin (o) is located at the intersection of midship, centerline, and draft. Considering the vertical plane (o-$y_0 z_0$) for roll motion, which is placed at the midship section, the z_0-axis points downward vertically.

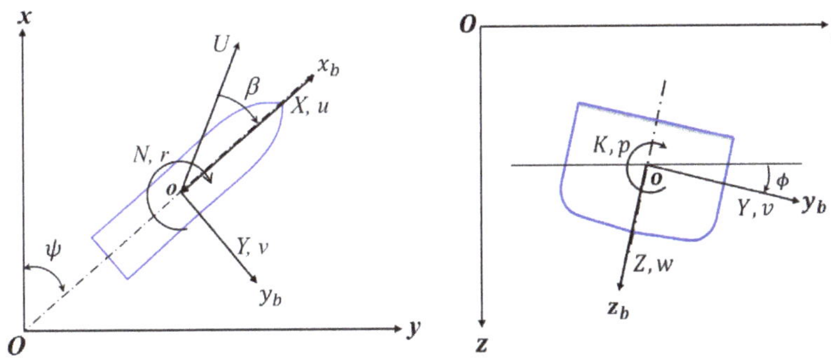

Figure 1. Coordinate system.

2.2. Pure Roll Simulation

Pure roll simulation consists of cyclic oscillation of the model by rotation around the x-axis in the y-z plane and a constant speed along the x-axis. Considering the model is passing a sine path, the roll angle (ϕ), rate (p), and acceleration ($\dot{p}$) of pure roll motion are as follows:

$$\begin{aligned} \phi &= \phi_0 \sin(\omega t) \\ p &= \dot{\phi} = \phi_0 \omega \cos(\omega t) \\ \dot{p} &= \ddot{\phi} = -\phi_0 \omega^2 \sin(\omega t) \end{aligned} \quad (1)$$

where ϕ_0 and ω are the roll angle displacement for the sinusoidal motion and circular frequency of the pure roll motion, respectively. Figure 2 shows an example of sinusoidal motion of pure roll, which describes the relationship between roll angle, rate, and acceleration in Equation (1). The amplitude of roll rate is shifted 90° in phase compared to the roll angle, while the amplitude of the roll acceleration is the same phase as the roll angle, but in the opposite direction.

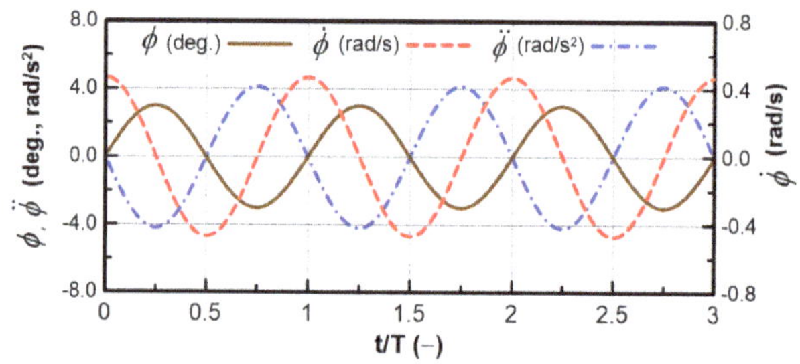

Figure 2. Example of the sinusoidal motion of pure roll: roll angle, rate, and acceleration.

Equation (2) shows an equation of motion for roll moment mentioned by Jeon et al. [14]. Here, z_{HD} is the vertical position where the force acts, and Y_{HD} is the sway hydrodynamic force.

$$I_x\dot{p} - mz_G(\dot{v} + ur) = K_{HD} = K_{\dot{v}}\dot{v} + K_{\dot{p}}\dot{p} + K_{\dot{r}}\dot{r} + K_p p - z_{HD}Y_{HD} \tag{2}$$

Equation (2) is simplified by taking derivative in pure roll test such that $v = \phi = r = 0$. Similarly, the sway force and yaw moment in pure roll simulation are also established as expressed in Equation (3):

$$\begin{aligned} Y_H &= Y_{\dot{p}}\dot{p} + Y_p p \\ K_H &= K_{\dot{p}}\dot{p} + K_p p \\ N_H &= N_{\dot{p}}\dot{p} + N_p p \end{aligned} \tag{3}$$

where the non-dimensional of sway force (Y'), roll moment (K'), and yaw moment (N') and motion variables are calculated based on the water density (ρ), forward speed (U), roll angle amplitude (ϕ_0), roll frequency (ω), and length perpendicular (L_{pp}) as written in Equations (4) and (5):

$$Y' = \frac{F_Y}{0.5\rho U^2 L_{pp}^2};\quad K' = \frac{M_X}{0.5\rho U^2 L_{pp}^3};\quad N' = \frac{M_Z}{0.5\rho U^2 L_{pp}^3} \tag{4}$$

$$p' = \frac{pL}{U} = \frac{\phi_0 \omega L}{U};\quad \dot{p}' = \frac{p'\omega L}{U} \tag{5}$$

Substituting p and $\dot{p}$ from Equation (1) into Equation (3), the following equations are written:

$$\begin{aligned} Y_H &= Y_{\dot{p}}(-\phi_0\omega^2 \sin\omega t) + Y_p(\phi_0\omega\cos\omega t) \\ K_H &= K_{\dot{p}}(-\phi_0\omega^2 \sin\omega t) + K_p(\phi_0\omega\cos\omega t) \\ N_H &= N_{\dot{p}}(-\phi_0\omega^2 \sin\omega t) + N_p(\phi_0\omega\cos\omega t) \end{aligned} \tag{6}$$

The hydrodynamic force and moment measured in pure roll motion can be decomposed by Fourier analysis into the in-phase and out-of-phase components, as in Equation (7). The subscript "in" indicates the in-phase component, which determines the force and moment due to angular acceleration; meanwhile, the subscript "out" refers to the out-of-phase component, which determines the force and moment due to angular velocity. In these, the in-phase component is to obtain the added mass derivative, and the out-of-phase component is to take the damping derivatives.

$$\begin{aligned} Y_H &= Y_{in}\sin\omega t + Y_{out}\cos\omega t \\ K_H &= K_{in}\sin\omega t + K_{out}\cos\omega t \\ N_H &= N_{in}\sin\omega t + N_{out}\cos\omega t \end{aligned} \tag{7}$$

The coefficients of the sine and cosine functions in Equation (7) with Equation (6) are considered to be equal, respectively. Based on the least squares method, the roll-added mass and damping derivatives can be calculated in Equations (8) and (9).

$$Y_{\dot{p}} = \frac{Y_{in}}{-\phi_0\omega^2}; \quad K_{\dot{p}} = \frac{K_{in}}{-\phi_0\omega^2}; \quad N_{\dot{p}} = \frac{N_{in}}{-\phi_0\omega^2} \tag{8}$$

$$Y_p = \frac{Y_{out}}{\phi_0\omega}; \quad K_p = \frac{K_{out}}{\phi_0\omega}; \quad N_p = \frac{N_{out}}{\phi_0\omega} \tag{9}$$

2.3. Ship Models

The roll effect was demonstrated to be significant for surface ships with small GM. Therefore, the four surface ships selected for pure roll simulation were the KCS, DTMB, ONRT, and Delft 372 catamaran, which represent the ship types recently concerned with roll effect. The ships with small GM are KCS, DTMB, and ONRT. KCS was designed by Korea Research Institute of Ships and Ocean Engineering (KRISO) and appended with a propeller and a suspended rudder. The DTMB is known as a preliminary design of the surface combatant introduced in the marine field in 1980 [15,16]. It is an unconventional modern benchmark hull form equipped with two bilge keels, twin propellers, and twin rudders. Similarly, the ONRT is also a preliminary design of a modern surface combatant [17–19] and is appended with skeg, twin-bilge keels, twin propellers, and twin rudders. The Delft 372 catamaran consists of two demi-hulls that were originally designed at the TU-Delft by Veer (1998) [20]; it is a large GM ship, but it operates at high speed, so the roll moment also shows significant change. In addition, due to its use for passengers, the roll moment of the catamaran is of concern. The pure roll simulation is performed in the bare hull condition without propeller and rudder, and with bilge keels for the DTMB, with skeg and bilge keel for the ONRT. Figure 3 shows the geometry of the model ships. Table 1 lists more detail on ship models.

The pure roll simulation was performed for different surface ships with their design speed. Different dimensionless roll angular velocities and roll angle amplitude were given as input parameters of the pure roll simulation. Table 2 depicts the test condition of the pure roll.

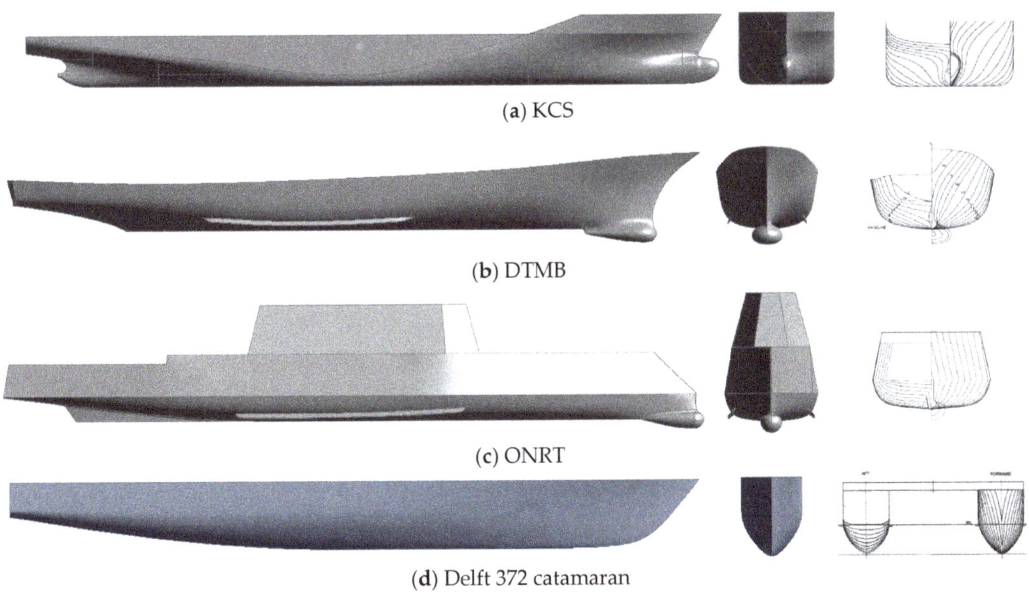

Figure 3. Ship models and body plan: (**a**) KCS; (**b**) DTMB; (**c**) ONRT; and (**d**) Delft 372 catamaran.

Table 1. Main dimension of ship models.

Item (Unit)	KCS	DTMB	ONRT	Delft 372 Cat.
Scale ratio	1/52.667	1/35.480	1/48.935	1/27.571
L_{pp} (m)	4.367	4.002	3.147	3.627
B (m)	0.611	0.537	0.384	1.137
T (m)	0.205	0.173	0.112	0.181
Δ (kgf)	365.0	193.3	72.6	154.0
KG (m)	0.138	0.213	0.162	0.411
GM_T (m)	0.011	0.055	0.042	1.791
U (m/s)	1.701	2.588	1.111	2.685
Fn (-)	0.260	0.413	0.200	0.450

Table 2. Test conditions.

Motion Variable	KCS	DTMB	ONRT	Delft 372 Cat.
Speed (m/s)	1.701	2.588	1.111	2.685
p' (-)	0.2, 0.4, 0.6, 0.8, 1.0, 1.2			
ϕ_0 (°)	0.5, 1.0, 1.5, 2.0, 2.5, 3.0			

3. Numerical Model

3.1. Governing Equation

To model the incompressible flow around the ship, an Unsteady Reynolds-Averaged Navier–Stokes (URANS) approach was applied to solve the governing equation. This includes the mass and momentum conservation equation solved by utilizing the commercial CFD code of STAR CCM+. The average continuity and momentum equations were expressed as follows:

$$\frac{\partial \overline{u}_i}{\partial x_i} = 0 \tag{10}$$

$$\frac{\partial \overline{u}_i}{\partial t} + \overline{u}_j \frac{\partial \overline{u}_i}{\partial x_j} = f_i - \frac{1}{\rho}\frac{\partial \overline{p}}{\partial x_i} + \frac{1}{\rho}\frac{\partial}{\partial x_j}\left(\mu \frac{\partial \overline{u}_j}{\partial x_j} - \rho\overline{u'_i u'_j}\right) \tag{11}$$

where $u_{i,j}$ $(i, j = 1, 2, 3)$ are the averaged velocity vector, $x_{i,j}$ $(i, j = 1, 2, 3)$ are the Cartesian coordinates, $\rho\overline{u'_i u'_j}$ is the Reynolds stress, $\overline{p}$ is the average pressure, and μ is the viscosity coefficient of fluid.

3.2. CFD Simulation

The computation region of the simulation was generated to be sufficiently large to avoid backflow. According to the ITTC recommended procedures [21], the dimension size of the domain was: $-5.0L_{pp} < x < 2.5L_{pp}$, $-2.5L_{pp} < y < 2.5L_{pp}$, $-2.5L_{pp} < z < 1.5L_{pp}$. The physical condition was set respectively for the inlet, sides, top, and bottom of the domain, which were defined as the velocity inlet, while the outlet was set as the pressure outlet. No-slip wall condition was applied on the ship hull. Additionally, a cylindrical cover surrounded the ship as an overset region to execute the pure roll motion. Grid generation applied an automatic meshing generator that used the Cartesian cut-cell method for volume and surface grids. A trimmed cell generated the volume grids, which were mainly composed of hexahedral cells. A surface remesher was applied to generate a high-quality mesh of the surface. The ship hull surface was overlaid with 10 layers of prismatic cells with a growth rate of 1.5 to resolve the surface boundary layer by applying the prism layer mesher. The non-dimensional wall distance y+ value on the ship hull was maintained lower than 1 for all simulations. The grid refinement was generated surrounding the ship hull and free surfaces using volumetric controls to improve the calculation accuracy and capture a proper wave elevation. Figure 4 shows an overview of the computational mesh, boundary condition, and boundary domain.

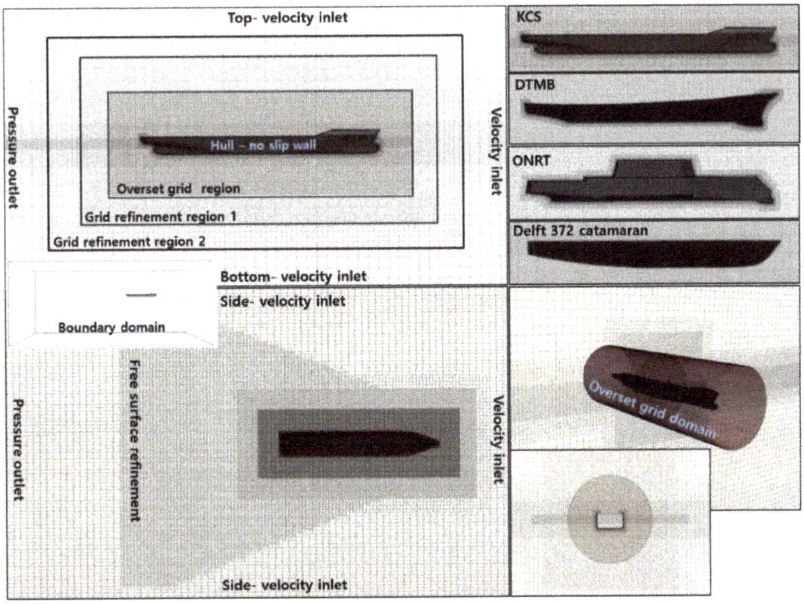

Figure 4. An overview of the computational mesh and boundaries.

To complete the URANS equation, the k-ω SST (Shear Stress Transport) turbulence mode was achieved, which is the most popular model applied in the field of marine maneuvering. The implicit unsteady method was employed throughout all simulations. The Volume of Fluid (VoF) method was used for the Eulerian multiphase to simulate the deformed free surface, so that the flow field characteristics around the ship could be captured. The velocity and pressure were analyzed using the Semi-implicit method for the pressure-linked equation (SIMPLE) algorithm. The diffusion and convection terms of the URANS equations were discretized based on second-order accuracy.

3.3. Numerical Uncertainty Analysis

Before proceeding with the simulation, the convergence study was conducted to verify the present study using the numerical uncertainty method. Figure 5 shows the calculation process of uncertainty estimation proposed in the ITTC recommended procedures [22]. The numerical uncertainty study was applied to the KCS model for the pure roll simulation with p' of 1.2. The computational solutions were in-phase and out-of-phase of the sway force (Y'), roll moment (K'), and yaw moment (N') during pure roll simulation.

We set $S_{i,1}$, $S_{i,2}$, and $S_{i,3}$, respectively as computational solutions of the fine, medium, and coarse. The refinement ratio (r_i) of fine, medium, and coarse was proposed to be $\sqrt{2}$ according to ITTC [22]. The changes in solution between two successive values were calculated as medium-fine ($\varepsilon_{i,21} = S_{i,2} - S_{i,1}$) and coarse-medium ($\varepsilon_{i,32} = S_{i,3} - S_{i,2}$). Then, the convergence ratio (R_i) which indicates the convergence tendency, was defined as follows:

$$R_i = \frac{\varepsilon_{i,21}}{\varepsilon_{i,32}} \tag{12}$$

There are three convergence conditions that are judged based on R_i:

(i) Monotonic convergence (MC): $0 < R_i < 1$;
(ii) Oscillatory convergence (OC): $R_i < 0$;
(iii) Divergence (Div): $R_i > 1$.

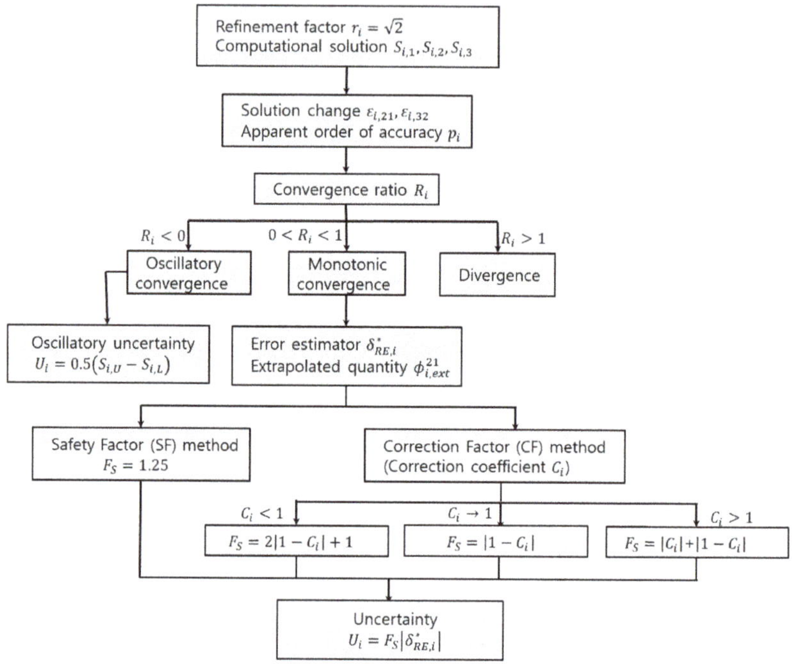

Figure 5. Calculation process of uncertainty estimation.

When oscillatory convergence is achieved, uncertainties are evaluated based on oscillation maxima ($S_{i,U}$) and minima ($S_{i,L}$) as follows:

$$U_i = 0.5(S_{i,U} - S_{i,L}) \tag{13}$$

When monotonic convergence is achieved, the generalized Richardson extrapolation (RE) method is applied to estimate numerical error ($\delta^*_{RE,i}$) and order of accuracy (p_i). Then, the uncertainty error (U_i) is estimated based on $\delta^*_{RE,i}$ and safety factor (F_S) as Equation (16).

$$\delta^*_{RE,i} = \frac{\varepsilon_{i,21}}{r_i^{p_i} - 1} \tag{14}$$

$$p_i = \frac{1}{\ln(r_i)} \ln\left|\frac{\varepsilon_{i,32}}{\varepsilon_{i,21}}\right| \tag{15}$$

$$U_i = F_S|\delta_{RE_i}| \tag{16}$$

There are two methods to determine the safety factor: the safety factor method, and the correction factor method. The correction factor (CF) method is applied for monotonic convergence that is proposed under the asymptotic assumption. The correction factor is calculated in Equation (17), which describes how close the numerical solution is to the asymptotic range. The $p_{i,est}$ is the theoretical accuracy, which is proposed to be 2.0 [22]. The safety factor (F_S) was calculated for each asymptote in Equation (18). For the safety factor (SF) method, the safety factor was proposed to be 1.25 by Roache [23]. This is also called the grid convergence index (GCI). It is applied for both monotonic and oscillatory convergences.

$$C_i = \frac{r_{21}^{p_i} - 1}{r_{21}^{p_{i,est}} - 1} \tag{17}$$

$$F_S = \begin{cases} 2|1 - C_i| + 1, & C_i < 1 \\ |1 - C_i|, & C_i \approx 1 \\ |C_i| + |1 - C_i|, & C_i > 1 \end{cases} \tag{18}$$

In addition, the extrapolated relative error ($\phi^{21}_{i,ext}$) is presented as the expected solution of convergence. The extrapolated relative error is calculated in Equation (19):

$$\phi^{21}_{ext} = \frac{r^{p}_{21} S_{i,1} - S_{i,2}}{r^{p_i}_{21} - 1} \tag{19}$$

here, $e^{21}_{i,a}$ is the approximate relative error, which is calculated in Equation (20):

$$e^{21}_{i,a} = \left| \frac{S_1 - S_2}{S_1} \right| \tag{20}$$

The uncertainty cannot be estimated in divergence.

The convergence study consists of grid convergence and time-step convergence. For the grid convergence study, the grid refinement ratio (r_G) of $\sqrt{2}$ was proposed, according to ITTC [22]. The r_G was to be kept constant by reducing and increasing the cell size on the grid refinement region, free surface, and ship hull. Three mesh sizes of fine, medium, and coarse were generated with grid cells of approximately (8.7, 6.2, and 4.4) M, respectively. For time-step convergence, the size of the grid was medium, and the time-step of 0.005 s was calculated based on the formula $(0.001 \sim 0.0025)L_{pp}/U$ according to ITTC [21]. Then, coarse and fine time-steps were determined to be (0.007 s and 0.0035) s, respectively, based on a refinement ratio of $\sqrt{2}$. Figures 6 and 7 show the time histories of force and moments during pure roll motion with different grid cells and time-steps. Slight discrepancies in the time series of force and moments were observed, but the overall one was largely consistent. The time-step has a more noticeable effect on the convergence study compared with the grid cells, due to a greater difference in the roll moment of the coarse time-step.

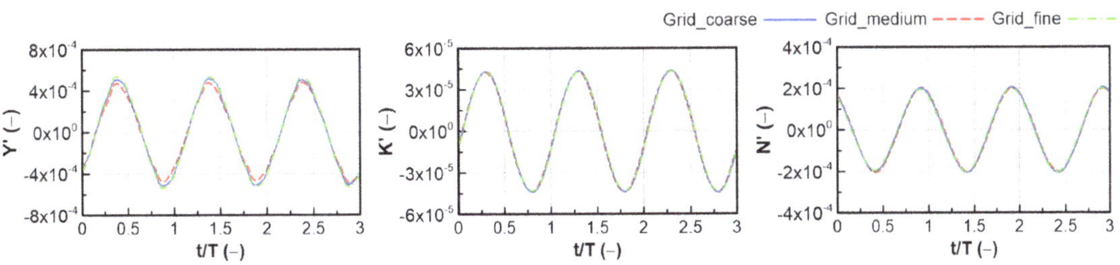

Figure 6. Time histories of force/moment with different grid cells for KCS with p' = 1.2.

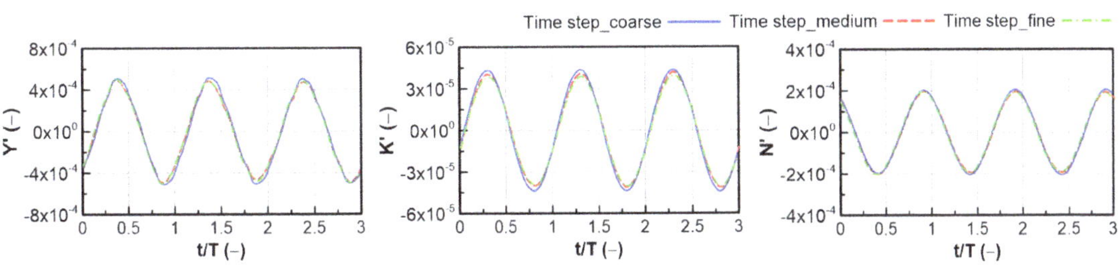

Figure 7. Time histories of force/moment with different time-steps for KCS with p' = 1.2.

Tables 3 and 4 list the grid and time-step uncertainty results. The grid and time-step uncertainty are less than 5%, which indicates that the numerical simulation shows good convergence. To this end, the present simulation method considers the medium grid and time-step of 0.005 s.

Table 3. Grid uncertainty results of pure roll simulation for KCS with $p' = 1.2$.

Mesh Convergence	Y_{in}	Y_{out}	K_{in}	K_{out}	N_{in}	N_{out}
Solution $S_{G,1}$	3.17×10^{-4}	-4.35×10^{-4}	4.13×10^{-5}	-1.24×10^{-5}	-9.93×10^{-5}	1.61×10^{-4}
Solution $S_{G,2}$	3.04×10^{-4}	-4.27×10^{-4}	4.20×10^{-5}	-1.19×10^{-5}	-1.01×10^{-4}	1.72×10^{-4}
Solution $S_{G,3}$	2.78×10^{-4}	-4.10×10^{-4}	4.30×10^{-5}	-1.07×10^{-5}	-9.84×10^{-5}	1.67×10^{-4}
Convergence ratio (R_G)	0.47	0.45	0.72	0.41	−0.64	−2.47
Convergence tendency	MC	MC	MC	MC	OC	OC
p_G	2.16	2.31	0.96	2.21	1.30	2.60
$\phi^{21}_{G,ext}$ ($\%S_{G1}$)	1.29	1.53	0.20	0.96	2.52	2.36
Oscillatory uncertainty U_i ($\%S_{G1}$)	-	-	-	-	1.26	3.6
CF-based uncertainty U_i ($\%S_{G1}$)	1.41	0.78	0.23	1.48	-	-
SF-based uncertainty U_i ($\%S_{G1}$)	1.38	0.91	0.18	1.00	2.84	4.02

Table 4. Time step uncertainty results of pure roll simulation for KCS with $p' = 1.2$.

Mesh Convergence	Y_{in}	Y_{out}	K_{in}	K_{out}	N_{in}	N_{out}
Solution $S_{T,1}$	3.09×10^{-4}	-4.2×10^{-4}	3.70×10^{-5}	-1.13×10^{-5}	-1.06×10^{-4}	1.76×10^{-4}
Solution $S_{T,2}$	3.04×10^{-4}	-4.27×10^{-4}	4.20×10^{-5}	-1.19×10^{-5}	-1.01×10^{-4}	1.72×10^{-4}
Solution $S_{T,3}$	3.15×10^{-4}	-4.32×10^{-4}	3.88×10^{-5}	-1.31×10^{-5}	-1.10×10^{-4}	1.67×10^{-4}
Convergence ratio (R_T)	−0.44	0.81	−1.57	0.51	−0.58	0.80
Convergence tendency	OC	MC	OC	MC	OC	MC
p_T	2.35	0.61	1.30	1.93	1.59	0.63
ϕ^{21}_{ext} ($\%S_{T1}$)	3.05	3.24	3.94	2.93	1.87	1.34
Oscillatory uncertainty U_i ($\%S_{T1}$)	2.35	-	2.50	-	2.60	-
CF-based uncertainty U_i ($\%S_{T1}$)	-	2.33	-	1.94	-	1.56
SF-based uncertainty U_i ($\%S_{T1}$)	1.84	4.45	2.43	1.56	2.64	1.94

Figure 8 shows an example of the comparison of force and moments between CFD simulation and Fourier analysis in Equation (7) for the KCS with the medium grid and time-step of 0.005 s. The Fourier analysis is used to determine in-phase and out-of-phase force and moments that are computational solutions used in numerical uncertainty analysis. It could be observed that the Fourier analysis method is in perfect agreement with the CFD simulation in the roll and yaw moments. Although the response of sway force in the CFD simulation is not smooth as the sinusoidal function perfectly, the Fourier analysis matches well with the CFD simulation.

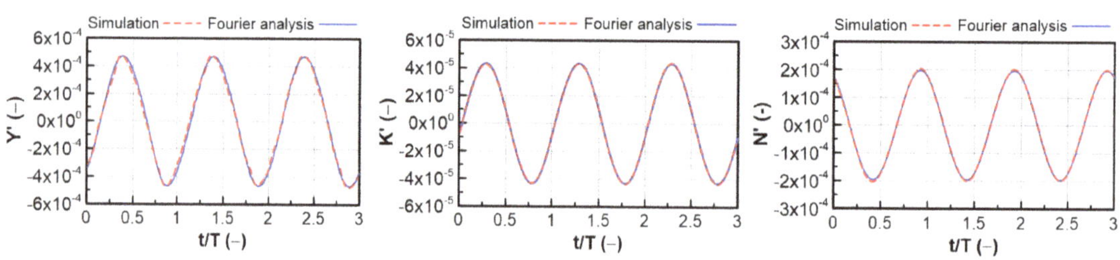

Figure 8. Comparison of force/moment of CFD simulation and Fourier analysis for KCS with medium grid and time-step of 0.005 s.

4. Results and Discussion

Numerical simulation was used to carry out the pure roll simulation for different types of surface ships of the KCS, DTMB, ONRT, and Deft 372 catamaran at their design speed. Hydrodynamic force and moment were taken for three periods after a stable Kelvin wave was generated. Figures 9–12 present the time histories of hydrodynamic force and moment during pure roll simulation of the KCS, DTMB, ONRT, and Delft 372 catamaran at different dimensionless roll angular velocities, respectively. It can be observed in all ships that as the dimensionless roll rate is increased, the amplitude of force and moment increase. Even though it was the pure roll test, the yaw moment was observed to be greater than the roll moment, except for the Delft 372 catamaran. This was demonstrated by Sadat-Hossenini [14] for the ONRT when the yaw moment was larger than the roll moment in the static heel test. The roll moment of the catamaran is greater than the yaw moment, due to the entire demi-hull performing a mission of forcing water when the ship heels to the side, while for a single hull, the starboard or port side performs that mission. Moreover, the length-to-beam (L/B) ratio of the Delft 372 catamaran is small to be 3.2, which is also a reason to explain this problem. On the other hand, Figure 13 compares the amplitude force and moment of the different surface ships. Because the roll moment of the KCS, DTMB, and ONRT is much smaller than that of the Delft 372 catamaran, Figure 13d is plotted to observe clearly the difference in roll moment among single hulls. The comparison indicates that the force and moment are greatest in the Delft 372 catamaran, due to the entire demi-hull being forced to water when the ship heels to the side. For three single hulls, the roll moment of the ONRT and DTMB is observed to be greater than the KCS in Figure 13d, because the bilge keel is a passive anti-roll device appended on the ORNT and DTMB. However, the roll and yaw moment of the DTMB is insignificantly larger than that of KCS, due to the slender body of the DTMB. In addition, a skeg equipped on the ONRT is a cause of the roll moment of the ONRT being larger than that of the DTMB. The yaw moment of different ships tends to be similar to the roll moment, while the sway force of the DTMB is significantly smaller than the KCS, and is seen to be nearly equal between the KCS and ONRT. This demonstrates that not only do the appendages of the bilge keel and skeg contribute to increasing the roll moment, but the surface shape also affects the change of the roll moment.

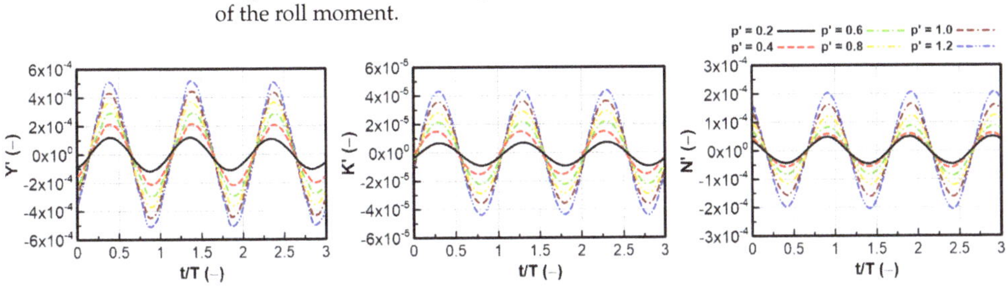

Figure 9. Time histories of force/moment during the pure roll simulation of KCS.

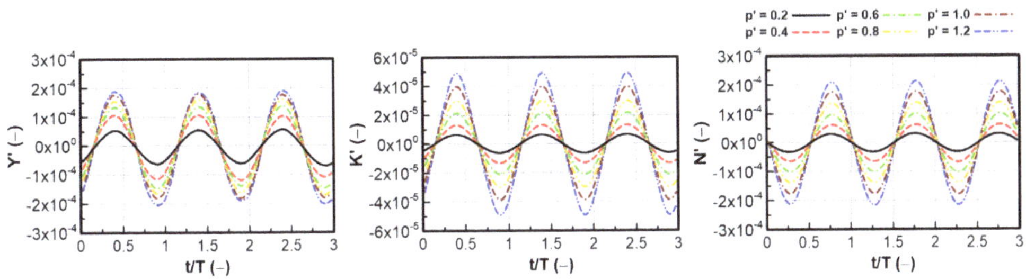

Figure 10. Time histories of force/moment during the pure roll simulation of DTMB.

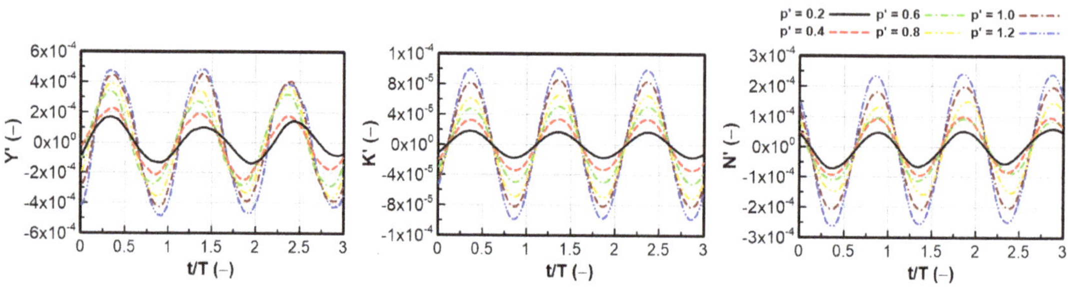

Figure 11. Time histories of force/moment during the pure roll simulation of ONRT.

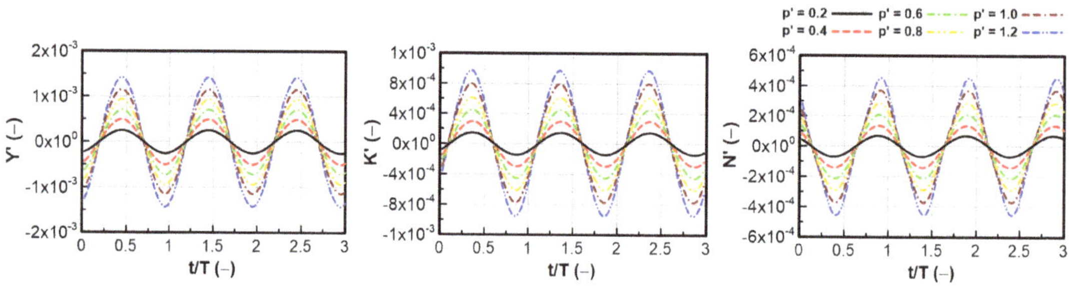

Figure 12. Time histories of force/moment during pure roll simulation of Delft 372 catamaran.

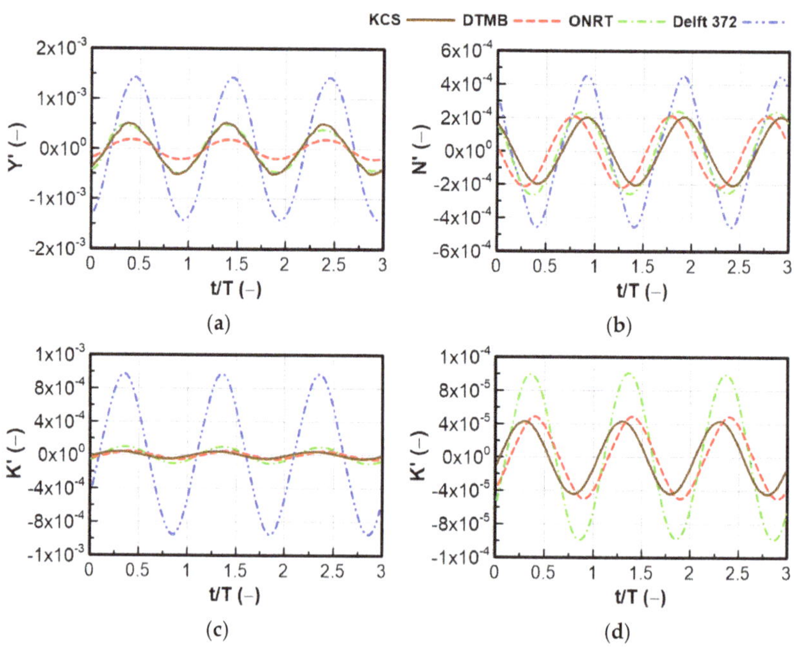

Figure 13. Comparison of amplitude force/moment of different ship types with $p' = 1.2$: (**a**) Sway force; (**b**) Yaw moment; (**c**) Roll moment; (**d**) Roll moment of KCS, DTMB, and ONRT.

Figure 14 displays the Kelvin wave pattern generated around the surface ships during pure roll simulation. The wave pattern was generated to be different due to discrepancies in the surface shape, forward speed, and frequency of pure roll. Figure 15 depicts the dynamic pressure distribution on the bottom of the ships in one period of pure roll simulation. The low-pressure distribution is observed at the side that is inclined, while the other side reaches the greater pressure distribution at that time. For example, when the ship heels to the starboard side ($t/T = 1/4$), the pressure distribution on the starboard is low, while the port side shows higher pressure, and vice versa. The low-pressure distribution at the inclined side is expanded to the middle of the bottom when the period time of t/T reaches 0 and 2/4.

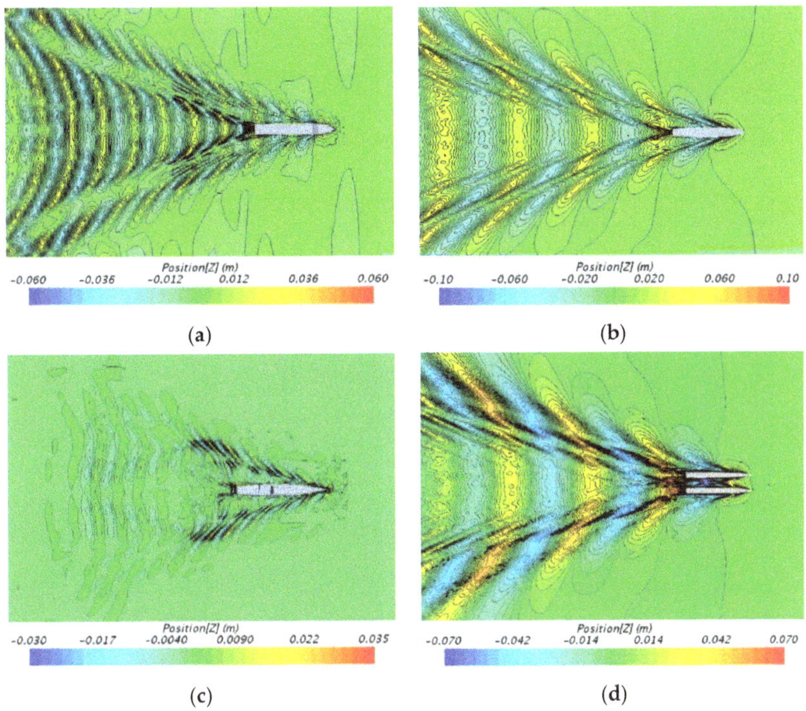

Figure 14. Kelvin wave pattern generated around surface ships during pure roll simulation: (**a**) KCS; (**b**) DTMB; (**c**) ONRT; (**d**) Delft 372 catamaran.

Figures 16 and 17 show the axial vortices around the bilge keel for one period of the DTMB and ONRT at the midship section. The bilge keel is attached to the ship hull as a passive anti-roll device to create extra damping, which leads to diminished roll motion and an increased period of motion due to increasing roll moment. Therefore, during the pure roll simulation, a vortex shedding is generated, which is the main physical phenomenon that affects the flow velocity around the ship. The vortices' filament curls near the ends of the bilge keel, presenting the three-dimensional effect of the flow that can be seen. Moreover, the large and strong vortices are generated at t/T equal to 2/8 and 6/8, where full-grown positive and negative vortices are shed from the bilge keel when the roll angle reaches the peak. Here is also a place to generate the vortex interference that changes the magnitude of the vortices from strong to weaker, and vice versa. This is observed for both DTMB and ONRT. Figure 18 depicts the axial vortices around the skeg for one period at $x/L_{pp} = -0.381$ of ONRT. The vortices are generated near the ends of the skeg. This demonstrates that the skeg also contributes to reducing the roll motion and increasing the roll moment. Therefore, the roll moment of ONRT is greater than that of DTMB as shown in Figure 13.

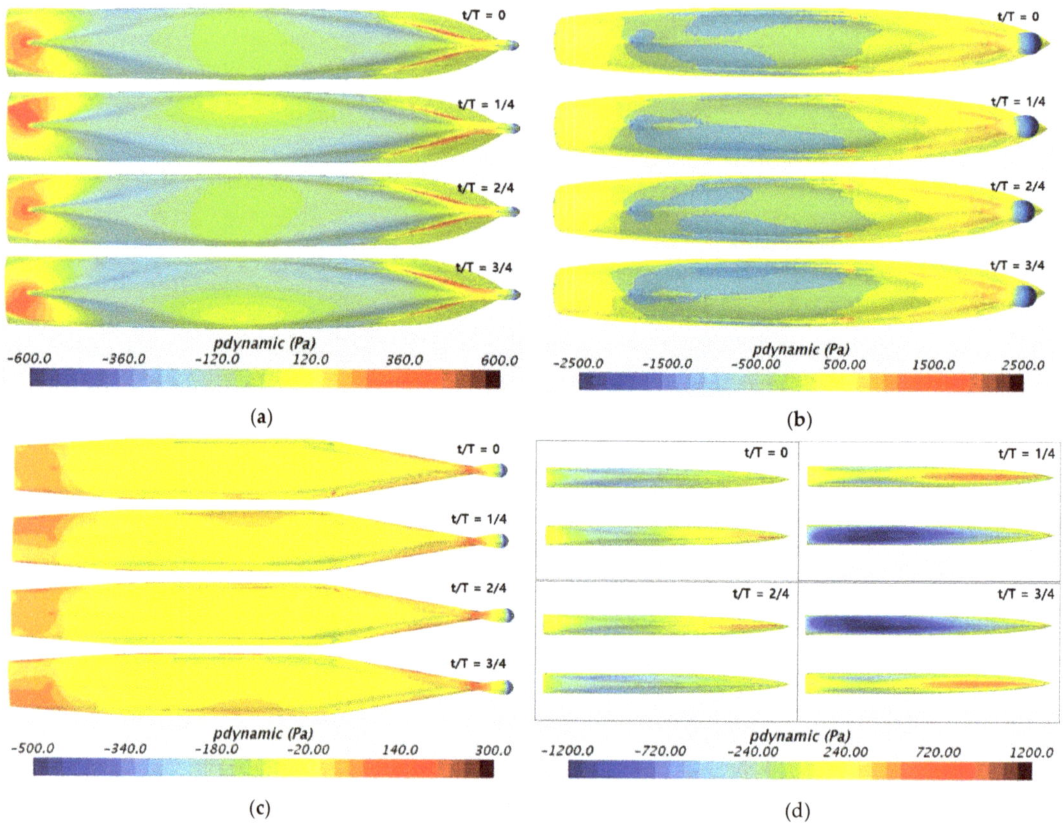

Figure 15. Dynamic pressure distribution of the surface ships for one period during pure roll simulation: (**a**) KCS; (**b**) DTMB; (**c**) ONRT; (**d**) Delft 372 catamaran.

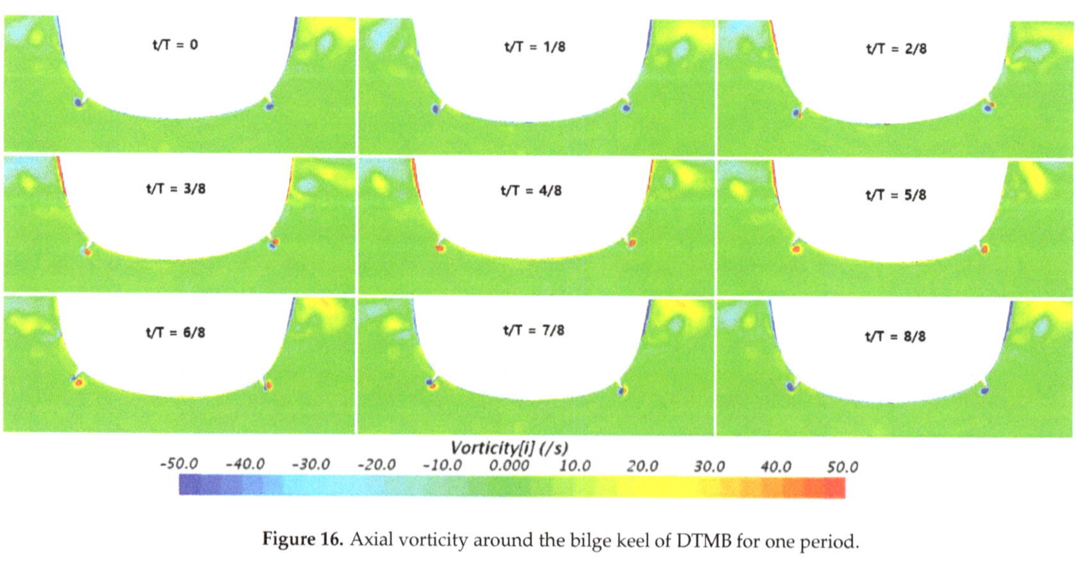

Figure 16. Axial vorticity around the bilge keel of DTMB for one period.

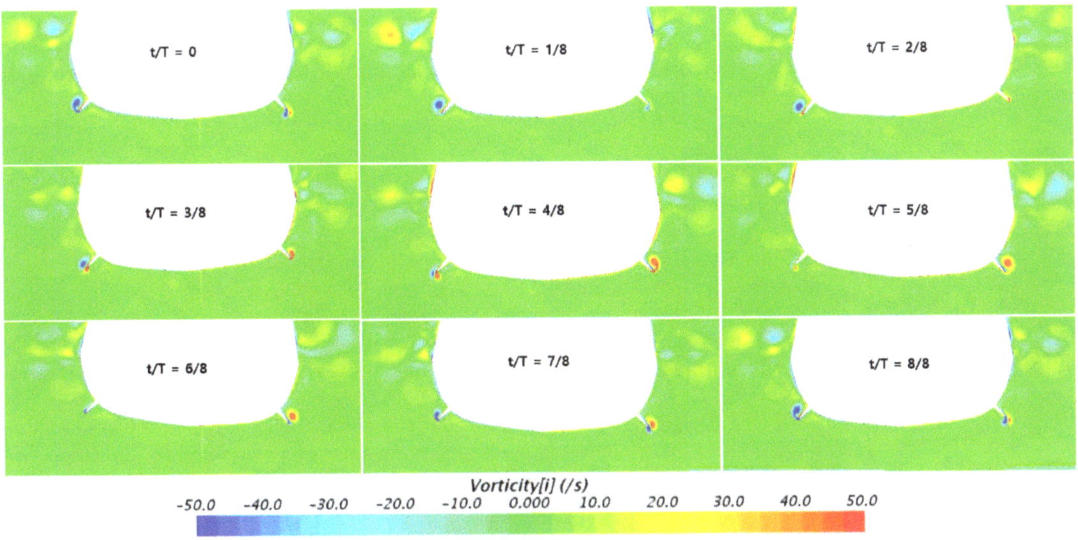

Figure 17. Axial vorticity around the bilge keel of ONRT for one period.

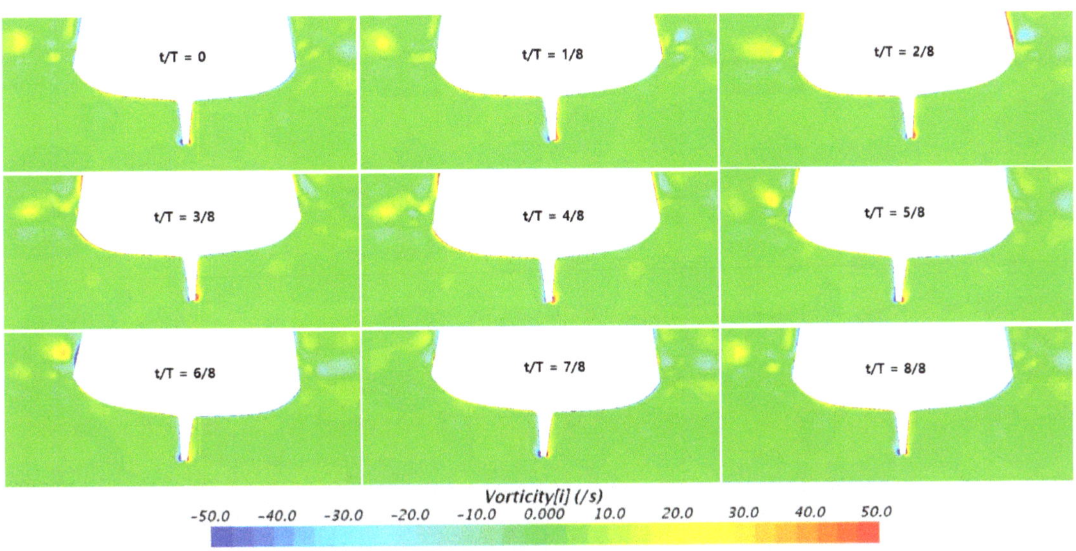

Figure 18. Axial vorticity around the skeg for one period at the $x/L_{pp} = -0.381$ section of ONRT.

The hydrodynamic force and moment of pure roll simulation are distinguished into in-phase and out-of-phase components by Fourier analysis, where the in-phase component determines the added mass due to angular acceleration, while the out-of-phase component determines the damping derivative due to angular velocity. Figures 19 and 20 show the results of the in-phase component and out-of-phase component of different ships, respectively. The in-phase due to roll acceleration and out-of-phase due to roll rate are observed to be linear. Similar to the tendency in hydrodynamic force and moment, the in-phase and out-of-phase of the Delft 372 catamaran are greatest, in which the roll moment shows a significant discrepancy to be 9.6 times in in-phase and 11 times in out-of-phase compared with the ONRT at a dimensionless angular velocity of 1.2. The roll moment of the DTMB is smaller than the KCS at in-phase, but greater than at out-of-phase. For sway force, the in-phase and out-of-phase of the DTMB are the smallest. Although the amplitude of sway force is nearly the same between the KCS and ONRT, the in-phase of the ONRT is shown to be slightly greater than that for the KCS, while out-of-phase shows a smaller amount. For the yaw moment, the in-phase of the DTMB is greater than that of the KCS and ONRT; meanwhile, it is smaller than that of the out-of-phase of the KCS and ONRT in out-of-phase. This demonstrates that both in-phase and out-of-phase have a significant effect on the amplitude of the roll moment. After fitting the in-phase and out-of-phase by linear regression, the roll-related derivatives in pure roll simulation for each ship are calculated based on Equations (5) and (6), respectively. Table 5 lists the roll-related derivatives in pure roll simulation that are then estimated. These derivatives will be used to complement the equation of roll moment in the 4-DOF equation of motion.

Table 6 compares the derivatives obtained from pure roll simulation and system identification performed by Yoon et al. [12] for the container ship and Jeon et al. [14] for the ONRT. The present results show a similar tendency to the derivatives due to roll rate, especially in the container ship, where the present derivatives are in good agreement with the system identification results. Comparing the estimated derivatives due to roll rate indicates that the pure roll simulation yields acceptable results.

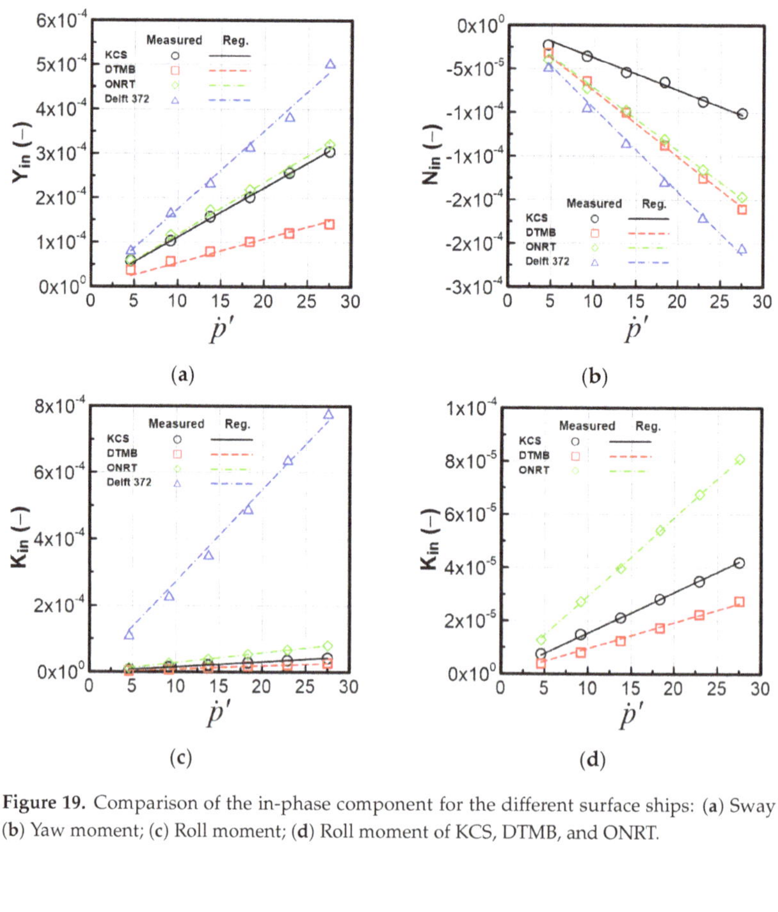

Figure 19. Comparison of the in-phase component for the different surface ships: (**a**) Sway force; (**b**) Yaw moment; (**c**) Roll moment; (**d**) Roll moment of KCS, DTMB, and ONRT.

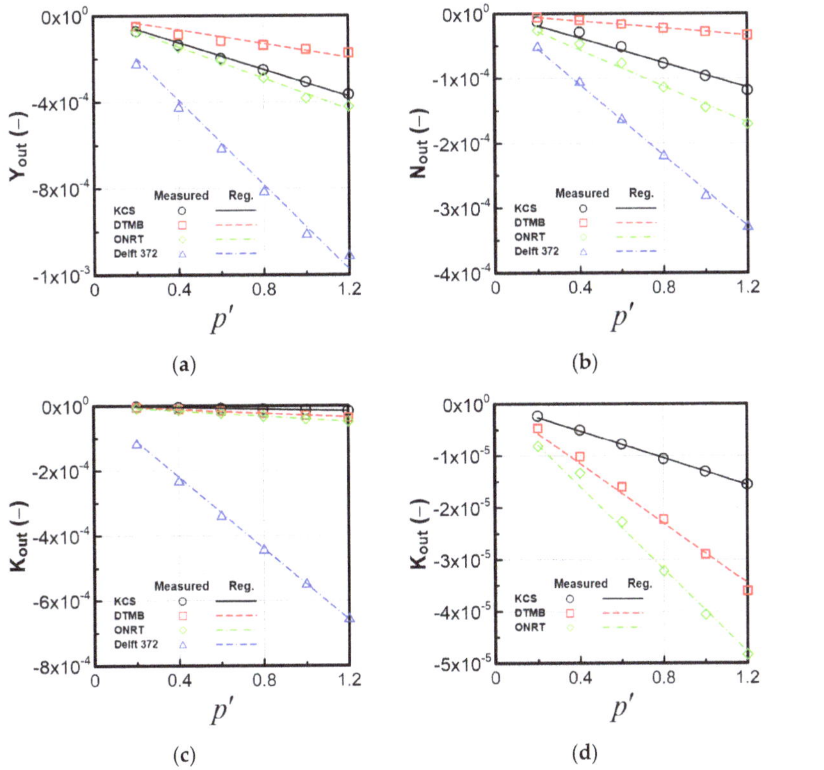

Figure 20. Comparison of the out-of-phase component for the different surface ships: (**a**) Sway force; (**b**) Yaw moment; (**c**) Roll moment; (**d**) Roll moment of KCS, DTMB, and ONRT.

Table 5. Hydrodynamic derivatives estimated from the pure roll simulation.

HD Derivatives	KCS	DTMB	ONRT	Delft 372
$Y_{\dot{p}}$	-7.53×10^{-5}	-1.44×10^{-5}	-9.83×10^{-5}	-3.22×10^{-5}
$K_{\dot{p}}$	-1.03×10^{-5}	-2.22×10^{-6}	-2.34×10^{-5}	-4.82×10^{-5}
$N_{\dot{p}}$	2.62×10^{-5}	1.75×10^{-5}	6.05×10^{-5}	1.80×10^{-5}
Y_p	-9.22×10^{-4}	-2.81×10^{-4}	-7.19×10^{-4}	-1.46×10^{-3}
K_p	-2.64×10^{-5}	-4.30×10^{-5}	-1.41×10^{-4}	-6.75×10^{-4}
N_p	3.21×10^{-4}	4.47×10^{-5}	3.43×10^{-4}	3.43×10^{-4}

Table 6. Comparison of the roll-related derivatives.

HD Derivatives	Container Ship		ONRT	
	Present (KCS)	Yoon et al. (2004) [12]	Present	Jeon et al. (2022) [14]
Y_p	-9.22×10^{-4}	-8.67×10^{-4}	-	-
K_p	-2.64×10^{-5}	-2.85×10^{-5}	-1.41×10^{-4}	-6.60×10^{-4}

5. Conclusions

In this paper, the roll-related derivatives, which were roll-added mass and roll damping due to roll rate, were estimated through pure roll simulation for different surface ships using the CFD simulation method.

The KCS, DTMB, ONRT, and Delft 372 catamaran were selected in this study because they have the characteristics that roll moment becomes dominant, as designed with small GM in the KCS, DTMB, ONRT, and high-speed catamaran. In particular, the passive anti-roll device as bilge keel was appended on the DTMB, and both bilge keel and skeg devices were equipped on the ONRT.

The CFD simulation method was verified by performing the uncertainty analysis of the grid and time-step based on the Richardson Extrapolation method. Generally, the uncertainty error shows less than 5% for both grid and time-step, which indicated that the present method shows good convergence. Therefore, the present numerical simulation was acceptable to execute the pure roll simulation.

Hydrodynamic force and moment throughout pure roll simulation were taken three periods after the wave was generated in stable condition. The force and moment of the Delft 372 catamaran were observed to be greatest, due to the entire demi-hull being forced to water when the ship heels. The roll moment of the ONRT and DTMB was larger than that of the KCS, because they were appended with the bilge keel. Additionally, the skeg was attached on the ONRT, which is the reason the roll moment was larger than for the DTMB. Furthermore, the roll moment of the DTMB was insignificantly larger than that of the KCS, due to the slender shape of the DTMB. Therefore, it demonstrates that not only do the appendages of the bilge keel and skeg contribute to increasing the roll moment, but the surface shape also affects the change of roll moment.

Based on the advantages of the CFD method, the flow field as wave pattern, pressure distribution, and axial vortices around the surface ships were analyzed during pure roll simulation. Low-pressure distribution was observed at the inclined side, while the other side reached greater pressure. In addition, the low-pressure distribution was expanded to the middle of the bottom when the period time t/T reached 0 and 2/4. On the other hand, axial vortices flow was generated near the ends of the bilge keel in the DTMB and ORNT. The full-grown vortices shedding from the bilge keel were observed when the roll angle reached the peak. The generation of vortices demonstrated that bilge keel and skeg contribute to reducing roll motion and increasing roll moment. Flow analysis proved that the obtained force and moment were reasonable.

The hydrodynamic force and moment of pure roll simulation were then distinguished into in-phase and out-of-phase components based on Fourier analysis for different dimensionless angular velocities. Using linear regression to fit the in-phase and out-of-phase, the roll-added mass and damping derivative in the pure roll simulation were estimated. A comparison of the damping derivatives due to roll rate with results obtained from the system identification technique demonstrated that the pure roll simulation using the CFD method brought acceptable results. Pure roll simulation can be applied to determine roll-related derivatives in the early design stage. Those derivatives were used to complete the roll equation of motion for 4-DoF maneuverability.

Author Contributions: Conceptualization, H.K.Y. and D.K.P.; methodology, H.K.Y. and D.K.P.; software, T.L.M.; validation, T.L.M., A.K.V. and H.K.Y.; formal analysis, T.L.M.; investigation, T.L.M. and A.K.V.; resources, T.L.M. and H.K.Y.; data curation, T.L.M.; writing—original draft preparation, T.L.M. and A.K.V.; writing—review and editing, T.L.M., H.K.Y. and D.K.P.; visualization, H.K.Y.; supervision, H.K.Y. and D.K.P.; project administration, H.K.Y.; funding acquisition, H.K.Y. All authors have read and agreed to the published version of the manuscript.

Funding: This research was supported by the 'Development of Autonomous Ship Technology (PJT201313, Development of Autonomous Navigation System with Intelligent Route Planning Function)' funded by the Ministry of Oceans and Fisheries (MOF, Korea).

Institutional Review Board Statement: Not applicable.

Informed Consent Statement: Not applicable.

Data Availability Statement: The data presented in this study are available in this article (Tables and Figures).

Conflicts of Interest: The authors declare no conflict of interest.

Nomenclature

Symbol	Unit	Description
B	m	Breadth of ship
Fn	-	Froude number
GM_T	m	Transverse metacentric height
K	Nm	Roll moment
KG	m	Height center of gravity above keel
L_{pp}	m	Ship length between perpendiculars
N	Nm	Yaw moment
p	rad/s	Roll rate
$\dot{p}$	rad/s^2	Roll acceleration
p'	-	Nondimensionless roll rate
r	rad/s	Yaw rate
T	m	Draft
U	m/s	Ship speed
u	m/s	Surge velocity
v	m/s	Sway velocity
w	m/s	Heave velocity
X	N	Surge force
Y	N	Sway force
Z	N	Heave force
β	deg.	Drift angle
ρ	kg/m^3	Water density
ϕ	deg.	Roll angle
ψ	deg.	Heading angle
ω	rad/s	Frequency
Δ	kgf	Displacement weight

References

1. Hajivand, A.; Mousavizadegan, H. Virtual simulation of maneuvering captive tests for a surface vessel. *Int. J. Nav. Archit. Ocean Eng.* **2015**, *7*, 848–872. [CrossRef]
2. Liu, Y.; Zou, L.; Zou, Z.; Guo, H. Prediction of ship maneuverability based on virtual captive model tests. *Eng. Appl. Comput. Fluid Mech.* **2018**, *12*, 334–353. [CrossRef]
3. Framceschi, A.; Piaggio, B.; Tonelli, R.; Villa, D.; Viviani, M. Assessment of the maneuverability characteristics of a twin shaft naval vessel using an open-source CFD code. *J. Mar. Sci. Eng.* **2021**, *9*, 665. [CrossRef]
4. Li, S.; Liu, C.; Chu, X.; Zheng, M.; Wang, Z.; Kan, J. Ship maneuverability modeling and numerical prediction using CFD with body force propeller. *Ocean Eng.* **2022**, *264*, 112454. [CrossRef]
5. Kim, Y.G.; Yeo, D.J.; Son, N.S.; Kim, S.Y.; Yun, K.H.; Oh, B.I. Prediction of Maneuverability of KCS with 4 Degrees of Freedom. *J. Soc. Nav. Archit. Korea* **2011**, *48*, 267–274. [CrossRef]
6. Yoshimura, Y. Effect of Roll Motion on Ship Manoeuvrability by a Rudder to Yaw Response Equation. *J. Jpn. Soc. Nav. Archit. Ocean Eng.* **2011**, *13*, 11–18. [CrossRef]
7. Yasukawa, H.; Yoshimura, Y. Roll-coupling effect on ship maneuverability. *Ship Technol. Res.* **2014**, *21*, 16–32. [CrossRef]
8. Fukui, Y.; Yokota, H.; Yano, H.; Kondo, M.; Nakano, T.; Yoshimura, Y. 4-DOF Mathematical Model for Manoeuvering Simulation including Roll Motion. *J. Jpn. Soc. Nav. Archit. Ocean Eng.* **2016**, *24*, 167–179. [CrossRef]
9. Yasukawa, H.; Sakuno, R.; Yoshimura, Y. Practical maneuvering simulation method of ships considering the roll-coupling effect. *J. Mar. Sci. Technol.* **2019**, *24*, 1280–1296. [CrossRef]
10. Kim, D.H.; Baek, H.M.; Lee, S.K.; Kim, E.S. Development of Apparatus for Pure Roll-Motion Test of Underwater Vehicles. *J. Navig. Port Res.* **2021**, *45*, 16–25.
11. Park, J.Y.; Kim, N.; Rhee, K.P.; Yoon, H.K.; Kim, C.; Jung, C.; Ahn, K.; Lee, S. Study on Coning Motion Test for Submerged Body. *J. Ocean Eng. Technol.* **2015**, *29*, 436–444. [CrossRef]
12. Yoon, H.K.; Son, N.S. Estimation of Roll Related Coeficients of a Ship by Using the System Identification Method. *J. Soc. Nav. Archit. Korea* **2004**, *41*, 53–58.
13. Yoon, H.K.; Son, N.S.; Lee, G.J. Estimation of the Roll Hydrodynamic Moment Model of a Ship by Using the System Identificaiton Method and the Free Runing Model Test. *IEEE J. Ocean. Eng.* **2007**, *32*, 798–806. [CrossRef]

14. Jeon, M.; Yoon, H.K.; Park, J.; Rhee, S.H.; Seo, J. Identification of 4-DoF Maneuvering Mathematical Models for a Combatant in Intact and Damaged Condition. *Int. J. Nav. Archit. Ocean Eng.* **2022**, *14*, 100480. [CrossRef]
15. Bekhit, A.; Popescu, F. URANSE-Based Numerical Prediction for the Free Roll Decay of the DTMB Ship Model. *J. Mar. Sci. Eng.* **2021**, *9*, 452. [CrossRef]
16. Gokce, M.K.; Kinaci, O.K. Numerical simulations of free roll decay of DTMB 5415. *Ocean Eng.* **2018**, *159*, 539–551. [CrossRef]
17. Sadat-Hosseini, H.; Carrica, P.; Stern, F.; Umeda, N.; Hashimoto, H.; Yamamura, S.; Mastuda, A. CFD, system-based and EFD study of ship dynamic instability events: Surf-riding periodic motion, and broaching. *Ocean Eng.* **2011**, *38*, 88–110. [CrossRef]
18. Araki, M.; Sadat-Hosseini, H.; Sanada, Y.; Tanimoto, K.; Umeda, N.; Stern, F. Estimating maneuvering coefficients using system identification methods with experimental, system-based, and CFD free-runing trial data. *Ocean Eng.* **2012**, *51*, 63–84. [CrossRef]
19. Sanada, Y.; Tanimoto, K.; Takagi, K.; Gui, L.; Toda, Y.; Stern, F. Trajectories for ONR Tumblehome maneuvering in calm water and waves. *Ocean Eng.* **2013**, *72*, 45–65. [CrossRef]
20. Veer, V. *Experimental Results of Motions, Hydrodynamic Coefficients and Wave Loads on the 372 Catamaran Model*; Report 1129; Delft University of Technology: Delft, The Netherlands, 1998.
21. ITTC. Practical Guidelines for Ship CFD Application. In Proceedings of the 26th International Towing Tanks Conference, Rio de Janeiro, Brazil, 28 August–3 September 2011; Number 7.5-03-02-03; ITTC: 2011. Available online: https://ittc.info/media/1357/75-03-02-03.pdf (accessed on 17 October 2022).
22. ITTC. Uncertainty Analysis in CFD Verification and Validation Methodology and Procedures. In Proceedings of the 29th International Towing Tanks Conference, Virtual, 13–18 June 2021; 7.5-03-01-01; ITTC: 2021. Available online: https://www.ittc.info/media/9765/75-03-01-01.pdf (accessed on 17 October 2022).
23. Roache, P.J. *Verification and Validation in Computational Science and Engineering*; Hermosa: Albuquerque, NM, USA, 1998.

Journal of
Marine Science and Engineering

MDPI

Article

Aerodynamic Load Prediction on a Patrol Vessel Using Computational Fluid Dynamics

Hafizul Islam, Serge Sutulo and C. Guedes Soares *

Centre for Marine Technology and Ocean Engineering (CENTEC), Instituto Superior Técnico, Universidade de Lisboa, 1049-001 Lisbon, Portugal; hafizul.islam@centec.tecnico.ulisboa.pt (H.I.); serge.sutulo@centec.tecnico.ulisboa.pt (S.S.)
* Correspondence: c.guedes.soares@centec.tecnico.ulisboa.pt

Abstract: Aerodynamic loads and moments on a naval patrol vessel are investigated using computational fluid dynamic simulations based on the OpenFOAM solver. After the initial turbulence, time, and grid dependency study, model scale simulations were performed for a wide range of inflow angles to predict aerodynamic forces and moments acting on the vessel at different heading conditions. For validation, model scale results were compared with wind tunnel data for similar hull forms. Finally, full-scale simulations were performed for a few cases to investigate possible scale effects on simulation results. The revealed scale effect turned out significant only for the yaw moment response. In this study, we aimed to produce reliable aerodynamic load data for the high-speed vessel, which is essential to developing reliable manoeuvring models. We conclude that Computational Fluid Dynamics is capable of providing reliable aerodynamic load predictions for high-speed vessels with sophisticated superstructures, in an economical manner.

Keywords: aerodynamic loads; CFD; OpenFOAM; scale-effect; patrol vessel; turbulence model dependency

Citation: Islam, H.; Sutulo, S.; Guedes Soares, C. Aerodynamic Load Prediction on a Patrol Vessel Using Computational Fluid Dynamics. *J. Mar. Sci. Eng.* **2022**, *10*, 935. https://doi.org/10.3390/jmse10070935

Academic Editor: Decheng Wan

Received: 15 June 2022
Accepted: 5 July 2022
Published: 7 July 2022

1. Introduction

Although aerodynamic loads on surface displacement ships are typically viewed as secondary ones, their prediction is important for analysing the intact stability and simulating manoeuvring motion in wind [1,2]. Reliable information on aerodynamic loads can also be potentially used in the inverse problem of identification of the mathematical model parameters using full-scale ship data typically obtained in the presence of significant wind [3–8]. Another important application of the aerodynamic forces is to assess the ship speed loss due to wind and its effect on voyage duration and ship emissions [9–11].

As direct full-scale measurements of aerodynamic loads are practically impossible, these are typically estimated either through wind-tunnel tests with scaled models of the above-water part of the hull or using CFD codes. Several experimental or numerical results can be found in the available literature and part of these results are presented in the form of systematised databases suitable for a quick and rough estimation of aerodynamic forces and moments. Detailed wind tunnel test data for different vessel types were obtained by Owens and Palo [12], and somewhat later by Blendermann [13,14]. Andersen [15] provided some benchmark data for a post-Panamax container ship and made recommendations to make the container stack profiling smoother. As an alternative to using these databases of wind load data, approximate methods based on neural networks have also been adopted by Haddara and Guedes Soares [16].

With the development of computational fluid dynamics (CFD) and computational resources, several researchers have also started focusing on CFD for reliable prediction of wind loads. Hassan et al. [17] used CFD to investigate wind loads on a full-scale 2800 TEU container ship and discussed the optimization of stack arrangements. Koop et al. [18] investigated five different ship models and compared wind tunnel test results with those obtained with CFD. Luquet et al. [19] performed CFD simulations for an F70 frigate in heeled conditions and estimated the aero- and hydrodynamic coefficients. Wnęk and Guedes Soares [20] conducted a CFD study of the wind loads on an LNG shuttle tanker in the vicinity of an offshore LNG production platform and compared the results with wind tunnel test data obtained by Wnęk et al. [21]. Watanabe et al. [22] and Nguyen et al. [23] investigated wind resistance and moment acting on a 20,000 TEU container ship using CFD and discussed possibilities of reduction of the air drag. Majidian and Azarsina [24] investigated a 9000 TEU container ship following the Andersen [15] model. The same authors also investigated a 2748 TEU container vessel using CFD and proposed a wind resistance model based on statistical data [25]. Wang et al. [26] studied a cruise ship's aerodynamics, experimentally and using CFD, and discussed design optimization possibilities through analysis of flow separation regions.

Apart from wind load studies, another group of researchers was involved in studies of ship air wakes, which are important for modelling helicopter landings and propagation of exhaust gases [27–32]. In particular, Linton and Thornber [33] presented a comprehensive review on CFD practices for ship air wake simulations, together with a discussion on a parameter dependency study and a method describing uncertainty quantification in associated turbulent flows. However, air wake studies are more focused on accurate descriptions and reproduction of the flow contours, including replication of the observed turbulence intensity and flow propagation. Whereas, in the case of load prediction, some compromise may be accepted as long as the average turbulence stress is well captured. Aerodynamic load studies are more focused on associated averaged Reynolds stress; thus, the total involved forces and moments, rather than the detailed flow field. As such, the study target and approach are quite different compared to air wake studies.

Although the studies performed for the determination of wind loads are numerous, they do not cover all existing types of ships and the available databases are not always sufficient for reliable predictions, which explains the necessity for individual studies contributing to the knowledge in ship aerodynamics. Practically, most of the existing aerodynamic load studies have been carried out in a model scale and the possible scale effect on forces and moments has rarely been discussed. The present study focuses on studying wind loads on a relatively small patrol ship of the Portuguese Navy with the help of the open-source Reynolds-averaged Navier–Stokes (RANS)-based CFD toolkit, OpenFOAM. While the main purpose of the study was to obtain a set of data on aerodynamic coefficients for systematically varying inflow angles for the model scale, the scale effect was also studied by means of a limited number of full-scale computations for a subset of the inflow angles. Similarly, the influence of the type of the involved turbulence model was investigated.

2. CFD Computational Layout

2.1. Ship Model

The ship under study belonged to the Portuguese Navy and is typically referred to as the Tejo class vessel. Historically, this design belonged to the family of Flyvefisken-class [34,35] patrol vessels primarily designed for the Royal Danish Navy.

The main particulars of the ship are length overall $L_{OA} = 54$ m, beam $B = 9$ m, and the full displacement draught $T = 2.5$ m. Displacement of the vessel can vary from 320 m^3 to 450 m^3 and the maximum design speed reaches 30 knots. The image of the actual vessel (a) and the simplified model (b) used for simulations are shown in Figure 1. Since the purpose of the study was to estimate the involved aerodynamic loads and not the turbulence intensity, a simplified model was assumed sufficient for the purpose. Furthermore, being a multipurpose vessel, the deck arrangement of the vessel changed based on the intended function. Thus, the simplified model helped provide a generalised prediction for different deck arrangements or functionalities.

Figure 1. The original (**a**) and simplified model (**b**) of the Tejo class vessel.

2.2. Mesh

For the study, single-phase simulations were performed only for the above-water part (freeboard) of the vessel with a somewhat simplified geometry of the superstructure. Thus, the mesh was generated only for the portion above the water, with the waterplane being a flat surface.

The simulation domain was generated following the general ITTC-2011 guidelines [36], with the inlet placed one ship length before the bow, the outlet placed two ship lengths after the stern, and the side boundaries at one ship length in each lateral direction from the sides of the hull (to the starboard and port). Although the ITTC guidelines are for hydrodynamic studies, a literature review by Linton and Thornber [33] revealed that the domain size has limited influence on aerodynamic simulations, unless the domain is too small. Furthermore, flow contours from the simulations did not reveal any obvious boundary interactions. Thus, the adopted domain size was considered sufficient.

The bottom of the domain was modelled as a rigid wall fixed at the waterplane, and the upper boundary (atmosphere) was located at about half the ship length above the bottom boundary. This upper boundary and the side boundaries were treated as local symmetry planes. The symmetry boundary condition was applied at a mirror surface, where the fluxes and normal components of all variables across the symmetry were set to zero. The symmetry in this case works as a non-reflecting boundary condition that ignores the interaction of the flow field with the sides and the top. The pressure boundary condition at the inlet is a zero-gradient (the OpenFOAM function zeroGradient: Neumann boundary condition), and the outlet boundary supplies a constant pressure condition (fixedValue: Dirichlet boundary condition). Regarding the velocity field, the inlet was set as a constant velocity condition (fixedValue), and the outlet was set to be a generic outflow condition with zero return inflow (inletOutlet). Initial values for the turbulence parameters were calculated following the Reynolds number [37]. The OpenFOAM built in a rough wall function was used to reduce the mesh dependency near the wall to capture the boundary layer. The airflow at the inlet was assumed to be uniform. The domain size and the domain boundaries are shown in Figure 2.

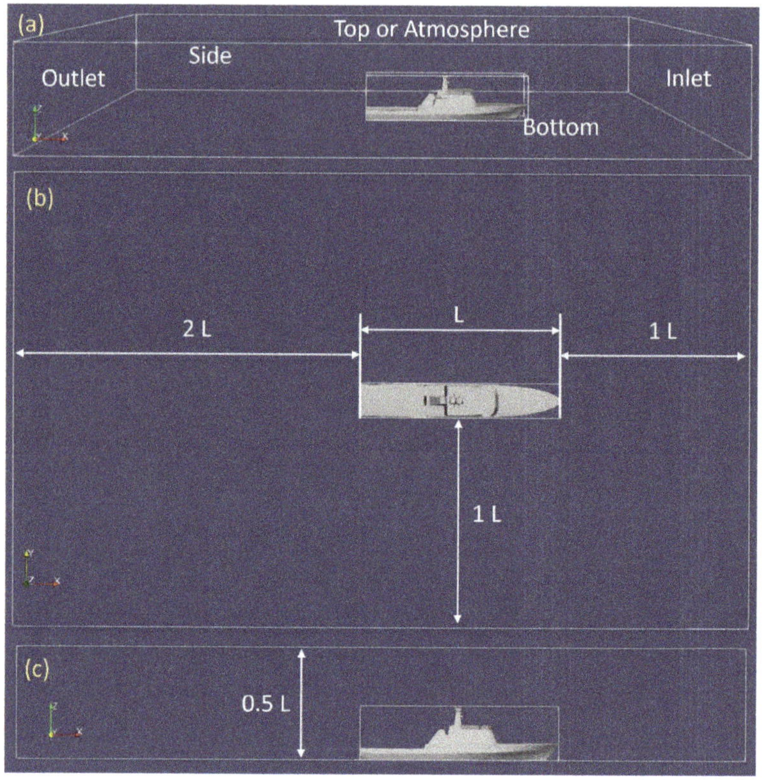

Figure 2. The domain size and the general boundaries used in the simulations, showing (**a**) the name of each boundary, (**b**) the dimensions in the x and y directions, (**c**) the dimension in the z-direction of the domain.

The simulation domain was created using the blockMeshDict utility from OpenFOAM, which generates a block of a defined dimension with structured grids. The block was refined consecutively four times to obtain a higher mesh resolution around the vessel. Separate refinement blocks were used for the hull up to the deck and for the superstructure above. This was done to avoid unnecessary refinement in empty areas and to reduce the total mesh size. Finally, the hull was integrated into the domain using the snappyHexMesh utility, which uses the Boolean approach for merging the solid object into the grid space. Five layers around the hull were generated using the function snappyHexMesh to ensure better capturing of the viscous resistance. The refinement area and the hull position were adjusted by changing the inflow angle while keeping the simulation domain unchanged. In the domain, the ship's longitudinal direction was in a positive x-axis (stern to bow), the lateral section was in the y-axis and the vertical direction was in the positive z-axis.

For the model scale simulations (1:10 scale), an average mesh resolution of 3.95 million cells was used, with a non-dimensional wall distance, $y^+ \approx 87$, where y^+ was used to describe the height of the first grid element next to the wall. Further details about the model scale mesh resolution are provided in the verification study section. The full-scale simulations were performed with a similar mesh topology with a substantially higher mesh resolution of around 80 million cells. A general image of the simulation mesh and refinement around the vessel is shown in Figure 3.

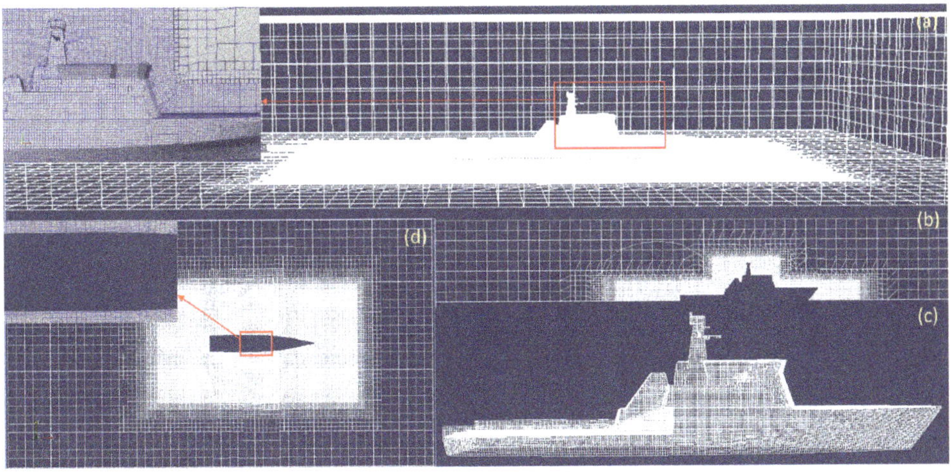

Figure 3. The mesh distribution around the above-water part of the hull for the headwind condition, where (**a**) is the top view of the domain, (**b**) is the side view at the centre of the mesh, (**c**) is the mesh distribution on the ship hull, and (**d**) is the mesh distribution at the bottom plane.

2.3. The Solver and Computational Resource

As previously mentioned, the study is based on the OpenFOAM library, whose solvers have been elaborately described by Weller et al. [38] and Jasak [39]. For the presented study, the OpenFOAM version 2006, managed by the ESI group, (https://www.openfoam.com/openfoam-is-open-source) (accessed on 15 August 2021) was used.

The governing equations for the solver applied here are the Reynolds-averaged Navier–Stokes (RANS) equations coupled with the continuity equation. In the vector–tensor form for an incompressible, turbulent, single-phase unsteady flow, these equations are written as:

$$\frac{\partial \mathbf{v}}{\partial t} + \mathbf{v}\cdot\nabla\mathbf{v} - \nabla\cdot\ (\nu\nabla\mathbf{v}) = \frac{1}{\rho}\nabla p, \nabla\cdot\mathbf{v} = 0 \tag{1}$$

Here, t is the time, ∇ is the Hamilton operator describing the gradient or divergence, $\mathbf{v}$ is the fluid velocity, ρ is the fluid density, p is the pressure, ν is the effective viscosity. The finite volume method (FVM) is used to discretise the governing equations. The pressure–velocity coupling is handled through the PIMPLE algorithm (a combination of PISO and SIMPLE algorithms). OpenFOAM incorporates several turbulence models, and for the present paper, the SST $k-\omega$ model is used for most computations. All computations are in a single phase and the free surface was modelled as a flat surface with no friction.

For the simulations, a desktop computer with Intel i9 processors with 18 cores, 128 GB RAM, and a 2TB SSD disk was used. The model scale simulations were performed using 12 processors in parallel, while the full-scale simulations were performed using 18 processors. Each model scale simulation was run for at least 25 simulation seconds (30 flow-through times) to reach stable outputs; that took roughly 40 physical hours. As for the full-scale simulations, each case was simulated for roughly 40 s of simulation time (15 flow-through times); that took roughly 650 h. For this case, the simulation time was reduced due to large resource dependency and because the solution converged (steady response) earlier due to smaller time steps.

3. Results

The OpenFOAM PIMPLE (single-phase) solver was used for both the model- and full-scale simulations. Only the slightly simplified above-water vessel geometry was involved and it was assumed symmetric with respect to the centre plane. Simulations were performed on the vessel model remaining fixed at each value of the inflow angle, assuming zero heel and fixed trim. The air at the inlet was assumed to have constant velocity without any pre-defined gust or flow oscillation.

A 1:10 model (5.4 m in length) was used for the model scale simulations. Simulations were performed at the inflow velocity of 20 m/s for the full-scale and 6.3 m/s for the model scale. The corresponding values of the Reynolds number based on the overall ship and model length were 7.12×10^7 and 2.28×10^6, respectively, assuming an air temperature of 20 °C.

In experimental aerodynamic studies, the focus is mostly given to maintaining Reynolds and Strouhal similarity, since it is easier and less time-consuming to reach high Reynolds numbers in laboratories [38]. However, a compromise had to be made in the CFD study considering resource limitations. Even though boundary layer and, thus, frictional resistance, have limited impacts on the total aerodynamic load, maintaining a reasonable y^+ value near the wall also ensures that the meshing follows reliable guidelines and mesh resolution around the hull is sufficient to properly capture the vortices and pressure forces. To maintain Reynold's similarity, the mesh requirement for the model scale becomes even higher compared to a full-scale simulation. Thus, to maintain the practicality of the study, Froude similarity has been applied, ignoring the wind tunnel practice of maintaining Reynold's similarity.

The CFD computations were performed in the following steps:

1. Turbulence model dependency study for a mesh with 3.95 million cells using three different approaches, by modelling with SST K-Omega and K-Epsilon two-equation turbulence models, and by resolving with the Delayed Detached Eddy Simulation.
2. Time convergence study for three inflow angles (0°, 45°, 90°, 135°, and 180°) using a mesh with 3.95 Mcells and the selected (SST K-Omega) turbulence model.
3. Grid and time step dependency study for a 45° inflow angle using three mesh resolutions and the selected turbulence model.
4. Study of aerodynamic forces and moments encountered by the vessel while facing wind from different inflow angles. In total, the resulting database includes wind loads for 24 inflow angles ranging from head to stern flow.
5. Study of scale effects on aerodynamic load prediction by simulating five selected cases in full-scale and comparing with model scale results.

The results are presented in the form of several visualisations (flow contours and pressure distributions) to show some peculiarities of the flow, by plotted time histories reflecting convergence studies and by plotted and tabulated results for the aerodynamic forces, moments, and their normalised values. All time histories are shown from 2 to 20 s, to show the initial oscillation before reaching a steady response. A longer time history was avoided to make the oscillations clearly visible, which becomes difficult to represent with longer time histories (especially since some figures had to be presented in smaller dimensions). Primary results for the forces were obtained in the velocity axes in form of the drag F_x and lateral force F_y, and were used directly in the convergence studies. The final study results are represented in body axes by means of evident transformations.

Direct validation of the results for the ship under study has not been possible but comparisons with somewhat similar shapes from Blendermann's database [13] were performed.

3.1. Preliminary Studies

3.1.1. Turbulence Models

The interaction between the vessel's upper structures and the high-velocity flow is a highly turbulent phenomenon. For accurate modelling of the forces involved in the interaction, reliable modelling of the turbulence is essential. In air wake studies, mostly the Large Eddy Simulation (LES) or Detached Eddy Simulation (DES) are used [33]. However, LES and DES are computationally too expensive and normally not practical for series computations. In the present study, two simpler and commonly known turbulence models and a numerical approach are tested. These are the two-equation turbulence models

(SST K-Omega [40] and K-Epsilon [41,42]) and the Delayed Detached Eddy Simulation (DDES) [43]. Among the three approaches, the turbulence models model the turbulence, whereas, the DDES attempts to resolve the turbulence.

Simulations were run for the 45°, 90°, and 135° air drift angles with the three approaches and with the same mesh resolution of 3.95 million. All simulations were performed for at least 30 flow-through times.

The body axis-based non-dimensional surge and sway force coefficients C_{XA} and C_{YA} (as defined in Section 3.3) and roll and yaw moment coefficients C_{KA} and C_{NA} (as defined in Section 3.3) are presented in Table 1. The results show somewhat oscillating behaviours with different models predicting higher values in different cases. Nevertheless, the overall differences among the results remain minimum for most cases. Because of the absence of experimental data, it is difficult to conclude which model predicts better. However, the relatively low mesh resolution for the DDES cases might have contributed to the observed over-predictions.

Table 1. The surge and sway force and heel and yaw moment results for the aerodynamic load-predicted using model scale simulations in OpenFOAM.

Drift Angle (Deg)	Turbulence Model	C_{XA}	C_{YA}	C_{KA}	C_{NA}
45	SST K-Omega	−0.22	0.85	−0.522	0.16
	K-Epsilon	−0.19	0.85	−0.515	0.16
	DDES	−0.13	0.92	−0.564	0.17
90	SST K-Omega	−0.04	1.10	−0.599	0.06
	K-Epsilon	0.03	1.14	−0.587	0.06
	DDES	−0.12	1.17	−0.627	0.06
135	SST K-Omega	0.36	0.93	−0.458	−0.06
	K-Epsilon	0.29	1.00	−0.484	−0.08
	DDES	0.29	0.98	−0.471	−0.06

Considering that the same mesh resolution was applied, the time required for performing the simulations was roughly similar. However, in general, simulations with the K-Epsilon model were faster compared to the other two, and the model with a relatively slower solution was SST K-Omega.

The force and moment time history results from the simulations are presented in Figures 4 and 5, respectively. The time histories shown are based on the velocity axis. The figure shows drag F_x and transverse force F_y histories, and the roll, pitch, and yaw moments ($M_x = K$, $M_y = M$, and $M_z = N$) time histories.

The images show that for all three turbulence models, maximum oscillation in the force and moment was observed in the 90° drift simulation. For all drift angles, the K-Epsilon model showed the most stable results with almost a linear flat line for forces and moments. This might suggest that the K-Epsilon turbulence model oversimplifies the turbulence in this case and provides average stress results instead of considering the oscillating turbulent phenomenon. The SST K-Omega model shows better capturing of turbulence, which is represented by the repeated oscillations in the time history within certain bounds. As for the DDES model, it shows the highest oscillation for the force and moment recordings compared to the other two models. In theory, DDES is supposed to better represent the turbulence phenomenon, which explains the higher oscillation. However, DDES requires a substantially high mesh resolution, both near and far from the body (vessel), and low CFL conditions to properly capture the turbulent behaviour of the flow. Considering the applied mesh resolution, DDES results would most probably reveal high uncertainty in results, thus reducing the reliability of the study. Moreover, the application of a higher mesh resolution is not feasible considering the large number of simulations planned for the study.

Turbulence modelling is a complex issue and assessing different turbulence models without direct comparison with experimental data is challenging. Different models might be better suited for different types of fluid–structure interactions and different regions. As such, making conclusive statements about the better suitability of different turbulence models for a particular study is very difficult with an extensive investigation. As mentioned earlier, the purpose of the present study was not to observe or measure the turbulence intensity involved in the interaction, but rather to study the forces and moments involved. As such, the SST K-Omega turbulence model was selected for the study, which seemed most reliable for this study after the investigation.

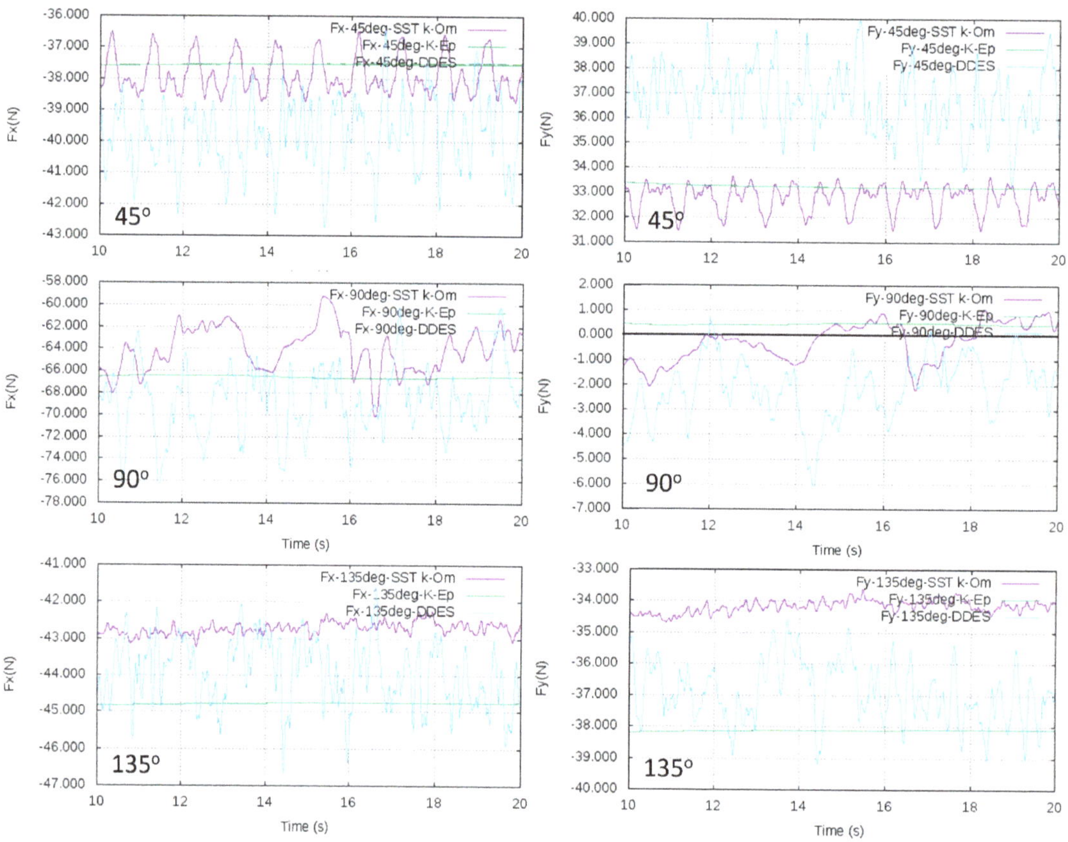

Figure 4. Relative comparison among force time histories for the Tejo vessel model simulated using three different turbulence models. In the figure, the purple colour represents the SST K-Omega model, green represents the K-Epsilon model, and blue represents the DDES model.

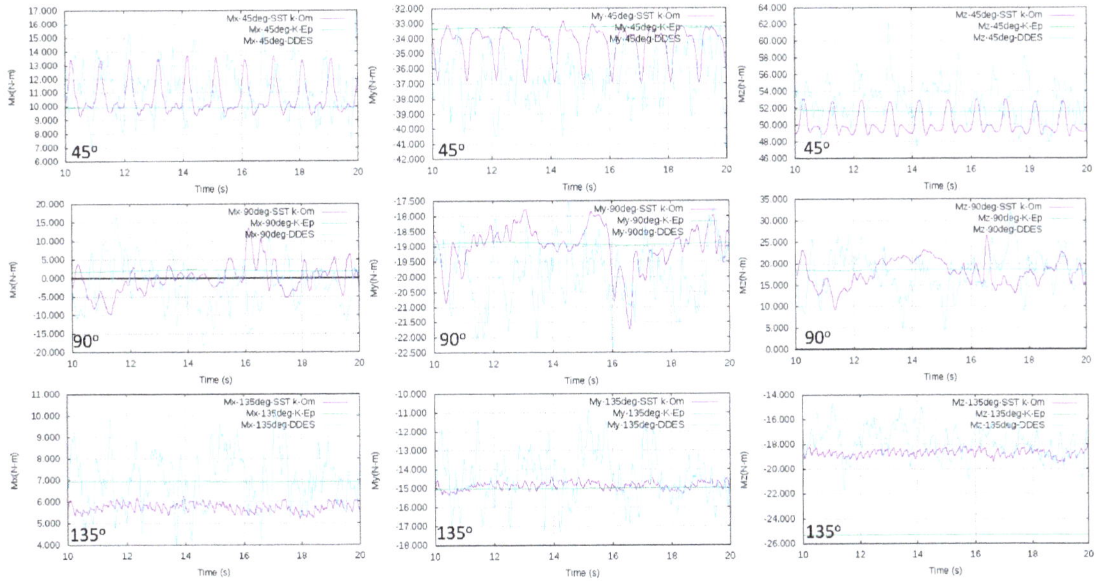

Figure 5. Relative comparison among moment time histories for the Tejo vessel model simulated using three different turbulence models. In the figure, the purple colour represents the SST K-Omega model, the green represents the K-Epsilon model, and blue represents the DDES model.

3.1.2. Time Domain Settling

CFD simulations in the time domain are performed until some steady values of the forces are obtained or a regime of somewhat stationary oscillations is reached. The latter may occur when non-streamlined shapes are placed into a steady flow. In those cases, averaging over the last 10 s (12 flow-through times) of simulation time, after reaching a repeated steady response, is performed to obtain the mean values of the loads. For full-scale simulations, a larger flow-through time is considered for averaging. All simulations after the initial turbulence study are performed using the SST K-Omega turbulence model.

The drag and the transverse force time histories for five simulation cases (0°, 45°, 90°, 135°, and 180°) are shown in Figure 6. The roll, pitch, and yaw moment histories are shown in Figure 7, for the same cases. However, as before, the time histories shown are based on the velocity axis.

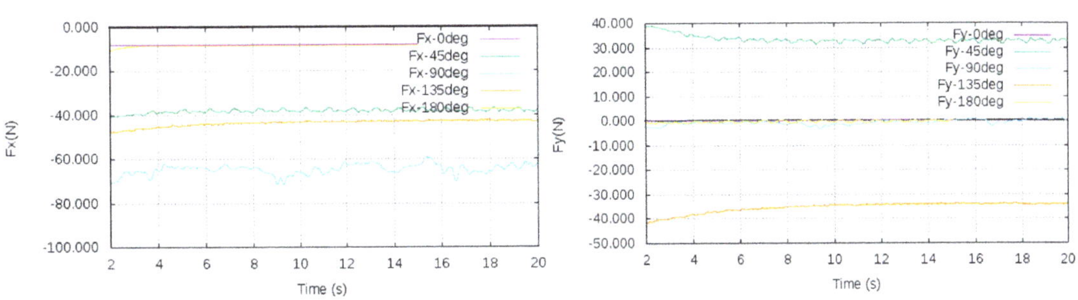

Figure 6. Drag and transverse force time history for bow, bow quartering, beam, stern quartering, and stern wind inflow conditions. In the figure, purple represents 0°, green 45°, blue 90°, orange 135°, and yellow 180° of the wind direction.

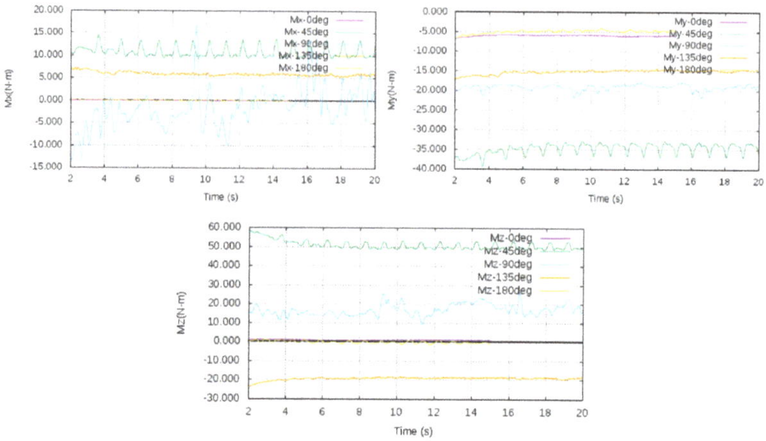

Figure 7. Roll, pitch, and yaw moment time histories for bow, bow quartering, beam, stern quartering, and stern inflow conditions. In the figure, purple represents 0°, green 45°, blue 90°, orange 135°, and yellow 180° of the wind direction.

Although the time history for the first 20 s of the simulation (24 flow-through times) is shown in the figures, simulations were run for a longer period, especially for the cases with high oscillation (e.g., the 90 deg case). The figures show that relatively high oscillation is observed for the beam wind case. The oscillation can be partly explained by the high turbulence associated with this case, as can be seen in Figure 8.

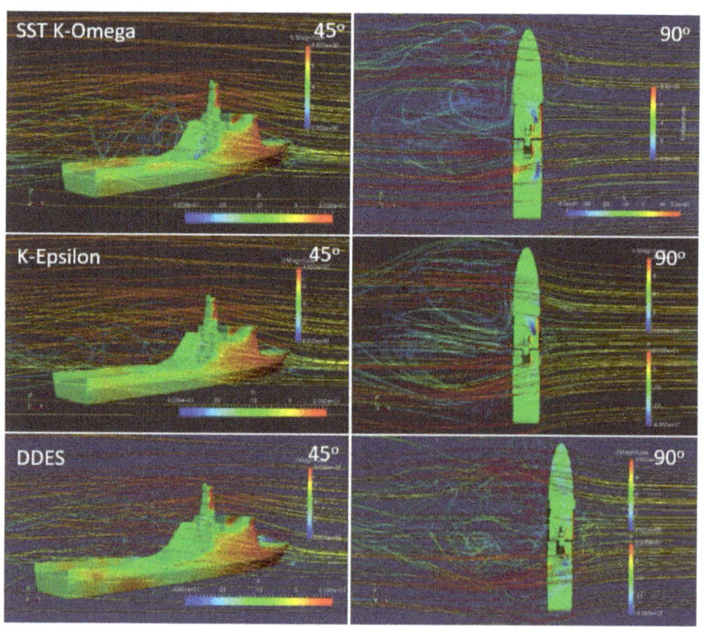

Figure 8. Relative comparison among the flow contours, for the vessel at 45° (**left**) and 90° (**right**) drift angle, generated using SST K-Omega, K-Epsilon, and DDES turbulence modelling, respectively (velocity scale: 0 to 8; Pr. scale: −40 to 20; model scale simulations).

3.1.3. Grid Convergence Study

Next, a systematic verification study was performed for the simulation settings and mesh configuration to assess the associated numerical uncertainty. Three different mesh configurations were used to perform a systematic verification study following the processes suggested by ITTC-2008 [44] guidelines, namely the factor of safety or the F_s-based approach [45]. However, contrary to the ITTC guidelines, instead of conducting an independent time and grid-based study, a constant Courant–Friedrichs–Lewy (CFL)-based discretization approach is used with equal refinement for both the grid and the time step. The justification for this approach has been discussed by Islam and Guedes Soares [46,47] and Wang et al. [48]. The details of the mesh resolutions used for the model scale uncertainty study are shown in Table 1. The table shows the total cell number in the domain and the general cell dimensions for each mesh in x, y, and z directions. The minimum layer thickness represents the minimum cell dimension in the direction normal from the hull surface). The layer is used to ensure the minimum required cell size near the hull surface to meet the y^+-criterion for boundaries with wall function (30 < y^+ < 300, following the law of wall). The refinement ratio is defined as: $r_{n+1,n} = h_{n+1}/h_n$, where h is the dimension of a cell in any particular direction (x, y, and z), and the subscript $n = 1, 2$ is the mesh number, according to Table 2.

Table 2. Mesh resolutions used for the verification study for the model scale simulations.

Mesh	Total Number of Cells	Dimensions x (m)	Of y (m)	Cells z	Minimum Layer Thickness	Non-Dimensional Wall Distance, y^+	Coarsening Ratio
1	8.7×10^6	0.01953	0.01875	0.01875	0.00375	65	1.00
2	3.95×10^6	0.02500	0.02500	0.02344	0.00500	87	1.30
3	2.25×10^6	0.03125	0.03125	0.03125	0.00625	108	1.25

For the uncertainty study, all computations were performed for the inflow angle of 45 degrees. The uncertainty was quantified for the surge (X_A) and sway (Y_A) force, and roll (K_A) and yaw (N_A) moments. The results are shown in Table 3, where $S_{1,2,3}$ stands for the computation results, per se. The absolute value of the convergence factor indicates whether the solution is converging (<1) or diverging (>1). For predicting the apparent order of accuracy, p, the following equation was used by Celik et al. [45]:

$$
\begin{aligned}
p &= \frac{1}{\ln(r_{21})}\left|\ln\left|\frac{\varepsilon_{32}}{\varepsilon_{21}}\right| + q(p)\right|, \\
q(p) &= \ln\left(\frac{r_{21}^p - s}{r_{32}^p - s}\right), \\
s &= \mathrm{sgn}\left(\frac{\varepsilon_{32}}{\varepsilon_{21}}\right).
\end{aligned}
\tag{2}
$$

Celik's paper also proposed the following extrapolated values for the solutions:

$$
s_{ext}^{21} = \left|\frac{r_{21}^p S_1 - S_2}{r_{21}^p - 1}\right| \tag{3}
$$

and the estimates of the errors and the extrapolated errors are:

$$
\begin{aligned}
e_a^{21} &= \left|S_1 - \frac{S_2}{S_1}\right|, \\
e_{ext}^{21} &= \left|S_{ext}^{21} - \frac{S_1}{S_{ext}^{21}}\right|
\end{aligned}
\tag{4}
$$

Table 3. Uncertainty analysis for the aerodynamic forces and moments.

Property		X_A (N)	Y_A (N)	K_A (Nm)	N_A (Nm)
Simulation results	S_1 (fine)	−3.73	50.18	−16.45	53.25
	S_2 (mid)	−3.65	49.95	−16.68	50.20
	S_3 (coarse)	−4.63	49.84	−16.47	48.48
Refinement ratio	$r_{21} = h_2/h_1$	1.3	1.3	1.3	1.3
	$r_{32} = h_3/h_2$	1.25	1.25	1.25	1.25
Difference of estimation	$\varepsilon_{21} = S_2 - S_1$	0.0778	−0.233	−0.233	−3.050
	$\varepsilon_{32} = S_3 - S_2$	−0.985	−0.114	0.208	−1.720
Convergence	$\varepsilon_{21}/\varepsilon_{32}$	−0.079	2.050	−1.122	1.773
Order of accuracy	p	11.25	1.96	0.41	1.45
Extrapolated values	S_{ext}^{21}	−3.73	50.53	−14.39	59.84
	S_{ext}^{32}	−3.56	50.16	−18.85	54.70
Approximate relative error	e_a^{21}	−0.0209	−0.0046	0.0142	−0.0573
	e_a^{32}	0.270	−0.0023	−0.0125	−0.0343
Extrapolated relative error	e_{ext}^{21}	−0.00115	−0.0069	0.1428	−0.1101
	e_{ext}^{32}	0.0245	−0.0041	−0.1151	−0.0823
Grid convergence index (GCI)	GCI_{21}	−0.0014	−0.0086	0.15616	−0.15467
	GCI_{32}	0.0298	−0.0052	−0.1626	−0.1121
Corrected uncertainty	U_{c1}	0.0288%	0.1729%	3.1231%	3.0933%
	U_{c2}	0.5968%	0.1039%	3.2522%	2.2421%

The Grid Convergence Index (GCI), which is the uncertainty estimate U_i, is defined as a product of the Richardson normalised discretization error ($\delta^*_{RE,1}$) by F_s (factor of safety). The value F_s is suggested to be 1.25 for the systematic parameter refinement study with at least three inputs, or 3 for simple convergence studies with two values of the input parameter.

$$GCI_{21} = \frac{1.25\,\delta^a_{21}}{r^p_{21} - 1} = F_s|\delta^*_{RE,1}| \quad (5)$$

$$\delta^a_{21} = \left|\frac{S_2 - S_1}{S_1}\right| \quad (6)$$

Following the ITTC-2008 guidelines, the corrected uncertainty (U_c), is defined as:

$$U_c = (F_s - 1)|\delta^*_{RE,1}| \quad (7)$$

The uncertainty study shows (Table 3) oscillatory convergence for the surge force and the roll moment and divergence for the sway force and yaw moment results. In general, for high Reynolds number simulations with complex structures, the involved turbulence makes it very difficult to have monotonously converging results. Furthermore, the study angle of 45 degrees represents a very turbulent condition. Nevertheless, in all the cases, the estimated corrected uncertainty is less than 3%, except for the rolling moment. Direct validation of the results could not be attempted due to the absence of experimental data for the same ship model. Nevertheless, a comparison is shown in Section 3.4 for the simulated results and experimental results for similar hull forms to assess the reliability of the predictions.

3.2. Flow Field Visualization Results

A detailed study of the aerodynamic loads on the Tagus class vessel was performed using a large number of model scale simulations. Single-phase simulations were performed for the vessel facing wind at 6.3 m/s velocities from different directions. Assuming symmetry of the above-water part of the vessel, the wind inflow direction was changed from the ship bow to the stern with the 10 deg interval, but some additional values of the inflow angle were also investigated.

Initially, flow field visualizations from the turbulence model study were presented in Figure 8, for 45- and 90-degree drift angles. The flow field contours further confirmed the findings discussed in Section 3.1.1, based on the force and moment results. The figure shows higher chaotic behaviour of the flow in the case of DDES and relatively more streamlined flow for the K-Epsilon model.

Next, flow field simulation results are presented for some specific airflow angles, performed using the defined mesh resolution (3.95 mil) and turbulence model (SST K-Omega), to understand how wind interacts with the vessel above the water structure and how forces and moments are influenced.

A flow field visualization for the wind interacting with the model scale vessel for the head and the following wind is shown in Figure 9. The figure shows a vortex formation after (for headwind) and before (for the following wind) the cabin structure. The streamlines show changes when the inflow angle is changing, and the colour on the hull (above the water part) shows pressure distribution due to wind interaction. The high-pressure region at the front of the superstructure in the head waves indicates relatively higher wind resistance.

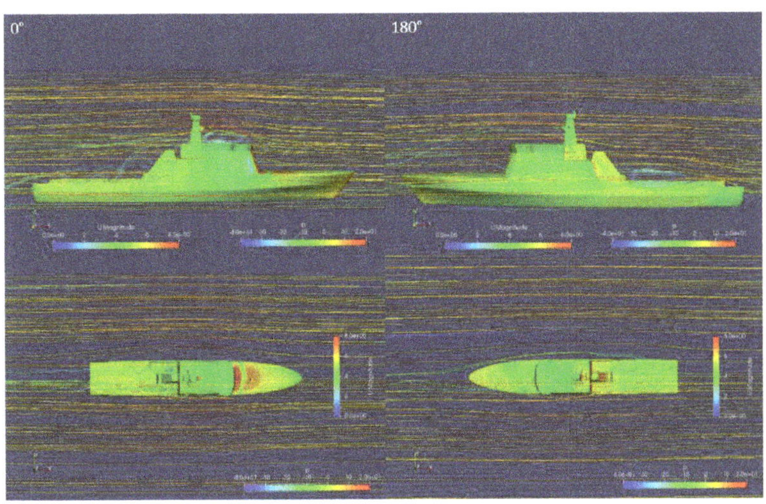

Figure 9. Free stream velocity and pressure distribution on the hull (above the waterline), left for headwind and right for the following wind (velocity scale: 0 to 8; Pr. scale: −40 to 20), after reaching the steady-state.

To further illustrate the results, Figure 10 shows the flow stream interacting with the vessel at different angles. The contours show the change in wind direction and vorticity formation after interacting with the vessel. The streamlines represent the change in free stream velocity and pressure distribution on the hull superstructure (as shown). The turbulent airflow in the case of the beam sea indicates a higher roll motion. The pressure distribution, in this case, is also different because of local recirculation regions. Notable turbulence is also observed in 45° and 135° cases.

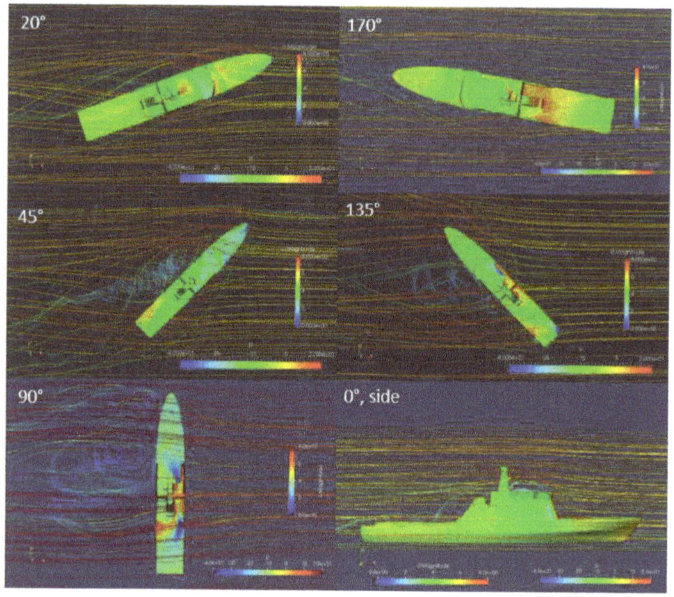

Figure 10. Velocity stream and pressure distribution on the model scale hull form (above the water part) while interacting with the wind at various angles (velocity scale: 0 to 8; Pr. scale: −40 to 20), after reaching a steady-state.

After model scale simulations, full-scale simulations were performed from the head to the following wind with a 45° interval. The number of simulation cases was reduced considering the substantial amount of computational resources required for each full-scale simulation.

Pressure distribution on the hull form (above water) and free stream flow around the hull are shown in Figure 11 for the full-scale simulations. The figures show notably higher turbulence compared to the model scale simulations. The variation mostly comes from the difference in the Reynolds scale, since in full-scale simulations, turbulence is fully developed. This enhanced turbulence would lead to a higher roll moment prediction in a full-scale simulation compared to the model scale.

3.3. Force and Moment Results

As mentioned earlier, the forces and moments were brought to the right-handed Cartesian body frame with the origin in the intersection of the midship plane, the centre plane, and the waterplane; with the x-axis pointing forward, y-axis—to the starboard, and the z-axis—downwards. The aerodynamic forces of surge X_A and sway Y_A, and moments of roll K_A and yaw N_A defined in body axes can be represented as:

$$\begin{aligned} X_A &= C_{XA}(\beta_A)\frac{\rho V_A^2}{2}A_T,\ Y_A = C_{YA}(\beta_A)\frac{\rho V_A^2}{2}A_L, \\ K_A &= C_{KA}(\beta_A)\frac{\rho V_A^2}{2}A_L L_{OA},\ N_A = C_{NA}(\beta_A)\frac{\rho V_A^2}{2}A_L L_{OA}, \end{aligned} \quad (8)$$

where C_{XA}, C_{YA}, C_{KA}, C_{NA} are the corresponding aerodynamic force/moment coefficients; β_A is the air drift or inflow angle,A_T and A_L are, respectively, the frontal (transverse) and the lateral projected area of the above-water hull; L_{OA} is the length overall.

The model scale simulation results of aerodynamic loads are shown in Table 4. The results show both the dimensional values for the model scale simulations and the non-dimensional coefficient.

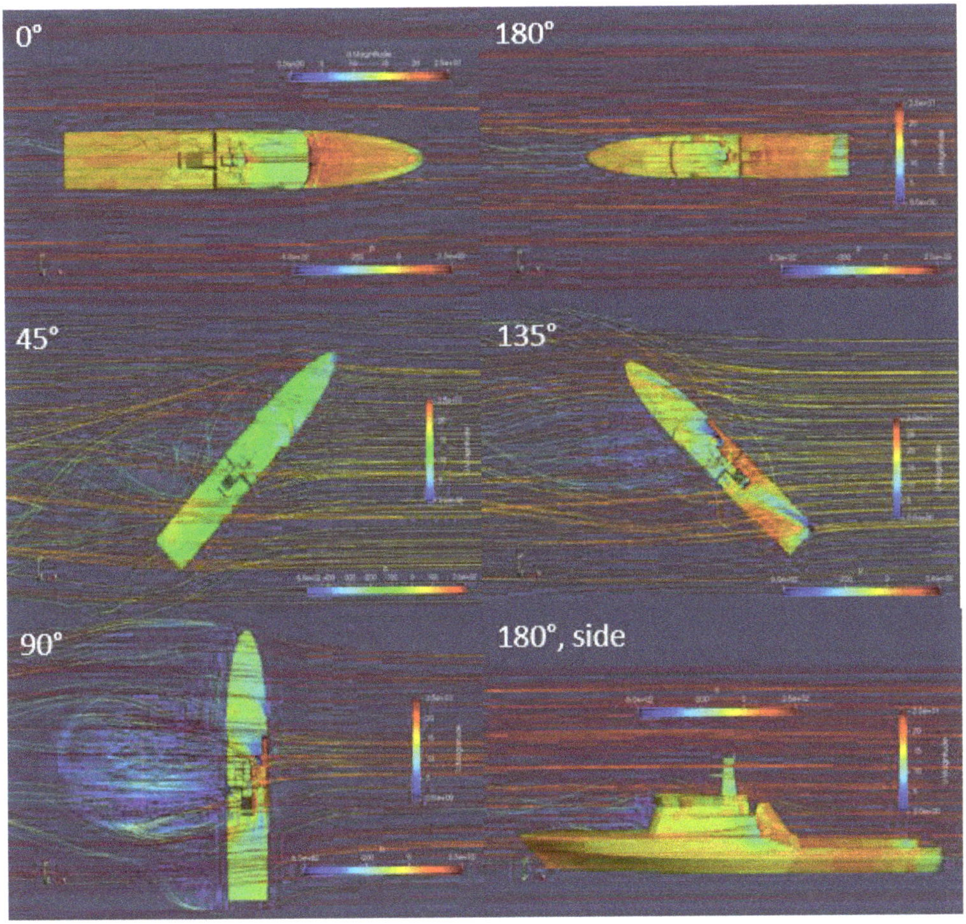

Figure 11. Velocity stream and pressure distribution on the full-scale hull form (above water part) while interacting with the wind at various angles (velocity scale: 0 to 25; Pr. scale: −500 to 250), after reaching the steady-state.

The negative values represent the drag. According to the results, the vessel faces a maximum air drag at bow wind, which keeps decreasing as it approaches the beam wind conditions. After the beam, the drag continues for a while, and then it starts facing a lift force from the following wind. The sway force coefficient shows a gradual rise from the headwind to beam condition, reaches the maximum at the beam, and then starts decreasing again as it approaches following wind conditions. The roll moment results show that the maximum moment is observed close to the beam wind condition. As for the moment, the result shows notable asymmetry around the beam condition. The yaw moment observed is significantly higher for headwind conditions compared to the following wind cases.

The full-scale simulation results are shown in Table 5 with the predicted forces, moment data, and coefficients. The dimensional values from the full-scale simulations can serve as a reference to understand the encountered aerodynamic loads by the vessel.

Table 4. The computed aerodynamic forces/moments in the model scale and their coefficients.

Drift Angle	X_A (N)	Y_A (N)	K_A (Nm)	N_A (Nm)	C_{XA}	C_{YA}	C_{KA}	C_{NA}
0	−7.90	0.17	−0.07	1.00	−0.48	0.00	−0.002	0.00
10	−9.19	9.33	−2.39	15.94	−0.56	0.16	−0.075	0.05
20	−8.68	24.02	−7.49	29.93	−0.53	0.41	−0.234	0.09
30	−6.75	39.57	−12.41	43.33	−0.41	0.68	−0.388	0.14
40	−4.71	45.88	−15.05	51.10	−0.29	0.78	−0.471	0.16
45	−3.65	49.95	−16.70	50.20	−0.22	0.85	−0.522	0.16
50	−2.82	52.97	−17.44	49.76	−0.17	0.90	−0.545	0.16
60	−3.71	56.42	−16.75	43.31	−0.23	0.96	−0.524	0.14
70	−3.38	62.12	−17.53	37.22	−0.21	1.06	−0.548	0.12
80	−3.15	66.06	−19.85	27.79	−0.19	1.13	−0.621	0.09
85	0.22	63.21	−19.62	20.00	0.01	1.08	−0.614	0.06
90	−0.62	64.51	−19.15	18.41	−0.04	1.10	−0.599	0.06
95	−2.40	66.04	−19.21	7.71	−0.15	1.13	−0.601	0.02
100	−2.15	64.86	−20.09	4.56	−0.13	1.11	−0.628	0.01
110	0.63	63.21	−17.03	−9.72	0.04	1.08	−0.533	−0.03
120	2.19	62.15	−16.69	−10.20	0.13	1.06	−0.522	−0.03
130	5.00	55.55	−15.20	−15.32	0.30	0.95	−0.475	−0.05
135	5.89	54.58	−14.66	−18.81	0.36	0.93	−0.458	−0.06
140	6.37	51.20	−13.93	−21.91	0.39	0.87	−0.436	−0.07
145	7.70	46.99	−12.60	−22.67	0.47	0.80	−0.394	−0.07
150	7.73	41.70	−11.20	−21.95	0.47	0.71	−0.350	−0.07
160	8.83	28.57	−6.99	−14.12	0.54	0.49	−0.219	−0.04
170	9.97	13.03	−2.80	−7.38	0.61	0.22	−0.087	−0.02
180	8.25	0.20	−0.02	−0.27	0.50	0.00	−0.001	0.00

Table 5. The computed aerodynamic forces/moments in full-scale and their coefficients.

Drift Angle	X_A (N)	Y_A (N)	K_A (Nm)	N_A (N-m)	C_{XA}	C_{YA}	C_{KA}	C_{NA}
0	−7396.12	−174.56	700.00	−1227	−0.44	0.00	0.00	0.00
45	−4201.27	48,412.00	−210,030.51	535,125	−0.25	0.82	−0.65	0.17
90	−1338.89	64,698.98	−251,004.00	135,944	−0.08	1.10	−0.78	0.04
135	5146.32	57,890.83	−221,197.14	−264,855	0.31	0.98	−0.69	−0.08
180	9579.00	−272.46	1490.47	−2939.8	0.58	0.00	0.00	0.00

The non-dimensional results for forces and moments from both the model scale and full-scale simulations are also shown in Figures 12 and 13.

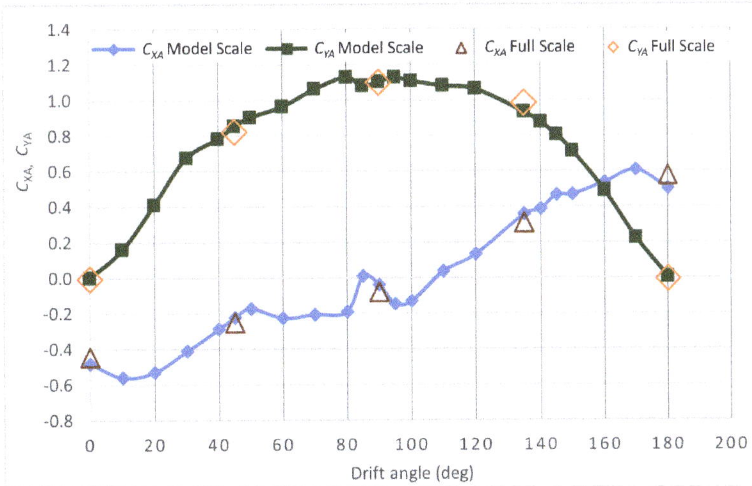

Figure 12. Surge and sway force coefficients (aerodynamic) for the Tejo class vessel.

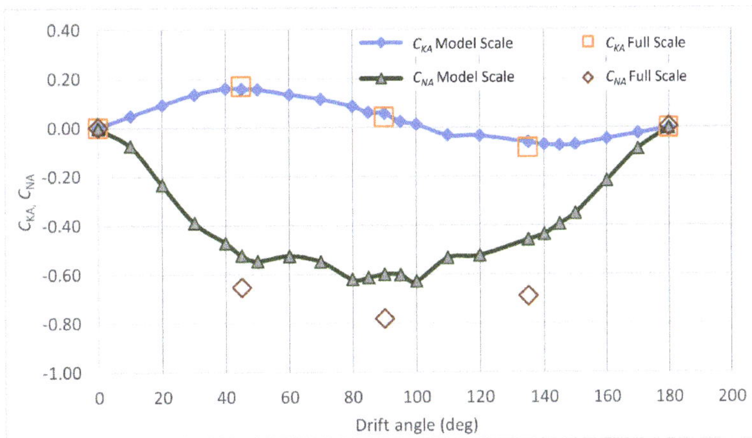

Figure 13. Roll and yaw moment coefficients (aerodynamic) for the Tejo class vessel.

Relative comparison between the full-scale and model scale results (Figures 12 and 13) suggests that the non-dimensional force results agree quite well with each other. Although slight over-predictions are observed in the case of model scale results for surge force, sway force results agree very well. Relatively notable variations are observed in the case of moment prediction, especially at the 135° drift angle.

The results suggest that the model scale simulations slightly underpredict the yaw moments in most cases. However, in the case of rolling moments, a notable difference is observed. This might be attributed to the fully developed turbulence in the actual Reynolds number, which was missing in the model scale. The variation, in this case, is mostly related to the difference in the Reynolds number (scale effect).

However, some contributions in the deviation also come from the simulation meshes, since the y^+ value for the model scale and full-scale simulations could not be maintained at the same level. Nevertheless, the overall difference between the model and full-scale results remains limited. This suggests that aerodynamic load coefficients show a minor impact from the simulation scale factor. This is mostly because aerodynamic loads are pressure dominant with a minor contribution from viscous forces. As such, scaling of the results mostly depends on Reynolds scaling and does not require empirical formulation for scaling of viscous forces, which often provide over prediction [49].

3.4. Comparison of Results with Blendermann Models

Direct validation of the results was not possible due to the absence of experimental or sea-trial data. Nevertheless, a general comparison with the results for a Destroyer (DES0101BN), a research vessel (RES0101BN), and a speedboat (SPE0102BN) model from Blendermann [13] wind load dataset suggests that, in general, OpenFOAM overpredicts the sway force and yaw moment.

The vessel outlines are shown in Figure 14. The navy destroyer (DES0101BN) is a 133.8 m vessel, with a breadth of 13.40 m, and with an above-water lateral area of 1342.07 m^2. The experiment was performed on a 1:100 scale. The research vessel had an overall length of 82.05 m, a breadth of 13.50 m, and the reference area was 670.27 m^2. The speedboat (SPE0102BN) had an overall length of 53.60 m, a breadth of 9.20 m, and a reference area of 317.63 m^2. As can be observed, the parameters and free-board geometry of the vessels considered here are notably different from the Tejo class vessel. Nevertheless, it provides a general understanding of the results.

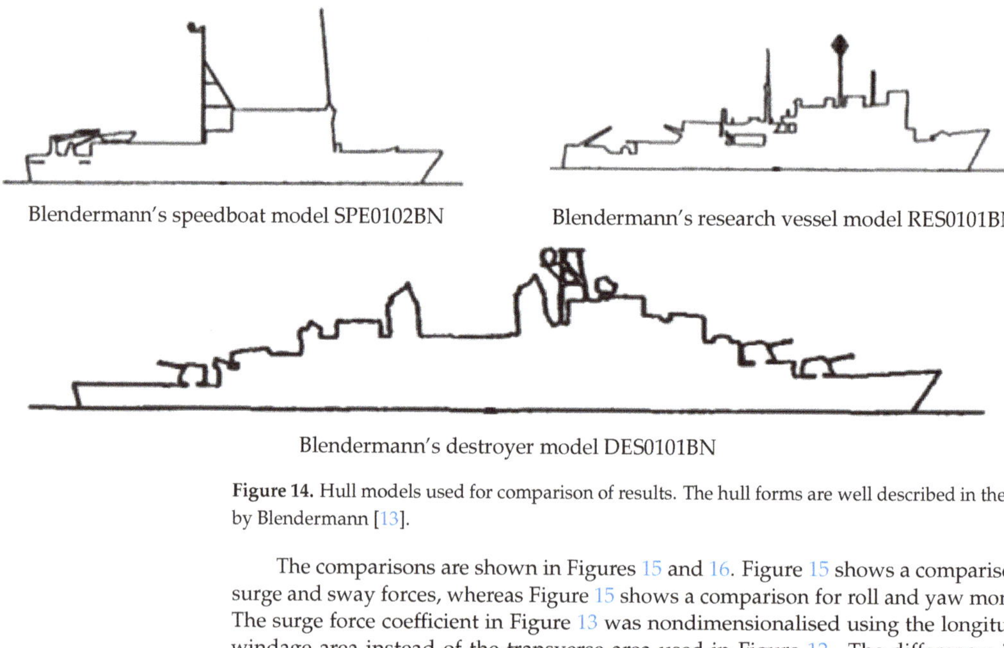

Figure 14. Hull models used for comparison of results. The hull forms are well described in the paper by Blendermann [13].

The comparisons are shown in Figures 15 and 16. Figure 15 shows a comparison for surge and sway forces, whereas Figure 15 shows a comparison for roll and yaw moments. The surge force coefficient in Figure 13 was nondimensionalised using the longitudinal windage area instead of the transverse area used in Figure 12. The differences in the responses seem to be quite adequate regarding notable differences in the shape of the above-water part of the hull.

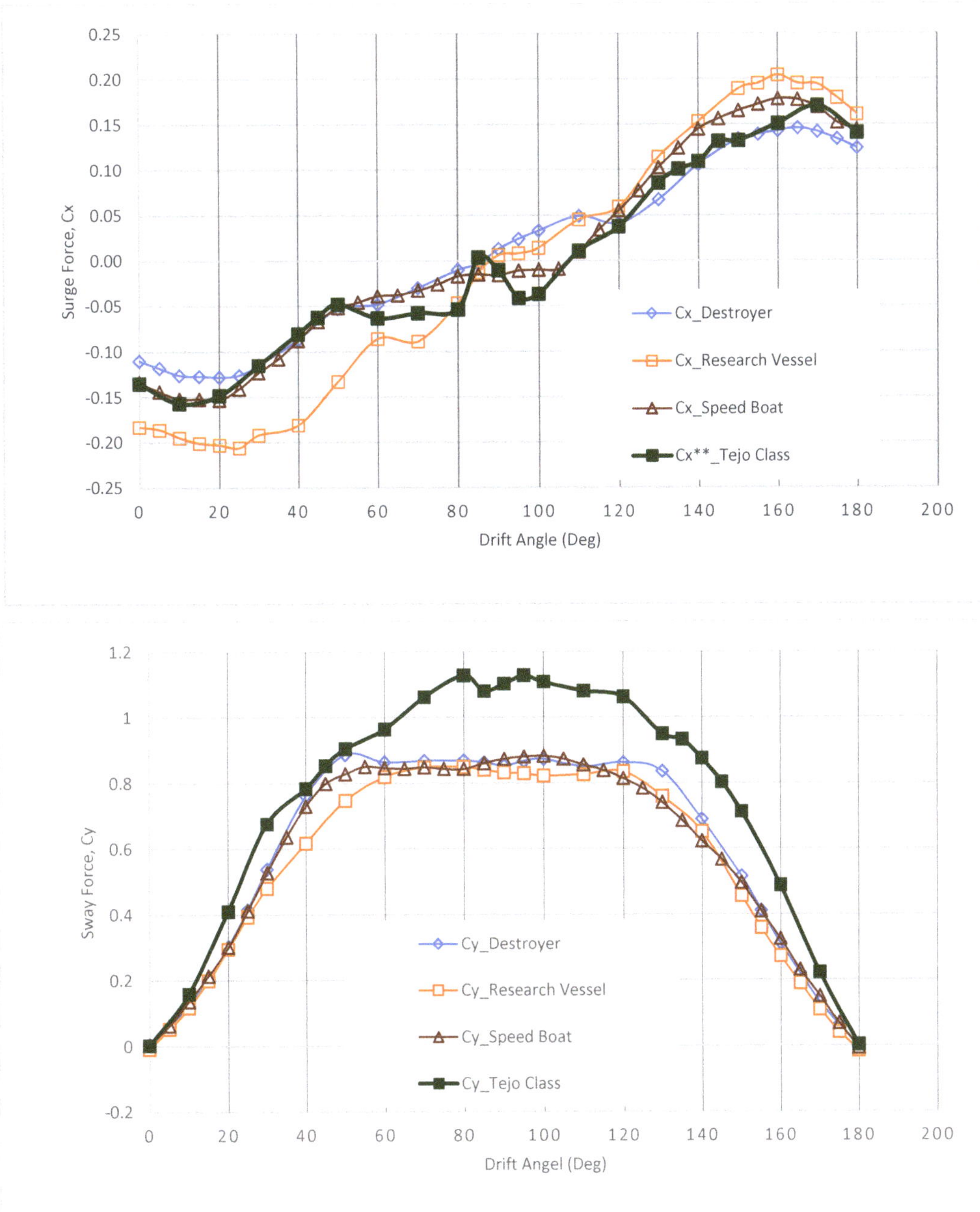

Figure 15. Relative comparison among force results for the Tejo vessel and three other ship models from Blendermann [13]. For the non-dimensionalizaiton of the surge force (Cx), a different formulation has been used here, as used by Blenderman. Thus, it has been marked as Cx **.

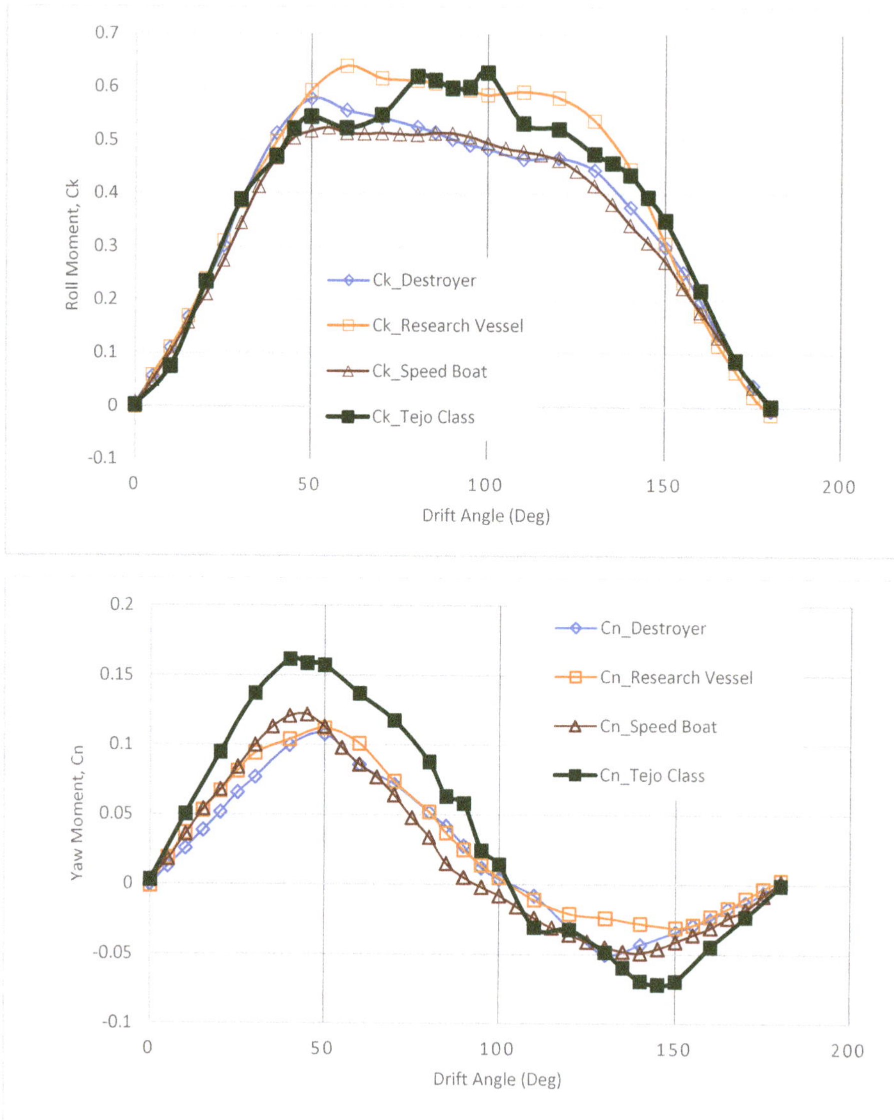

Figure 16. Relative comparison among moment results for the Tejo vessel and three other ship models from Blendermann [13].

4. Conclusions

In the study, RANS simulations for a navy vessel were performed to predict the aerodynamic drag of the vessel at different heading angles. Both model scale and full-scale simulations were performed using OpenFOAM, and drag coefficients were predicted, ignoring the free surface effect. Initially, a turbulence model dependency and a verification study were performed for model scale simulations to understand related dependencies. Next, 24 model scale simulations were performed for various heading angles. Finally, full-scale simulations were performed for five heading positions. Direct validation of the results was not possible due to the absence of experimental or sea trial data. However, a relative comparison of results with similar hull forms confirmed the reliability of the results.

The study provides surge and sway force and roll and yaw moment coefficients for the aerodynamic loads on Tejo class vessels. The result curves show reasonable agreement with similar numerical studies available in the literature. The model scale and full-scale simulations show acceptable agreement in most cases. The comparison of the model and full-scale results show that the scale effect is insignificant for the surge and sway forces and more significant for the yaw moment. Reasons for such differences remain subject to further investigation.

In general, following the dependency studies and comparison with available data, it may be concluded that CFD is well capable of providing reliable data for the aerodynamic loads acting on the studied vessel, which can be used in further studies to improve the manoeuvring trajectories of the vessel under the action of wind. The paper contributes by providing reliable quantitative predictions of aerodynamic loads and moments for a high-speed vessel. It also demonstrates where scale effects are more prominent. However, the paper adopts a relatively low mesh resolution and skips detailed capturing of turbulence intensity and propagation, which are important for several operational aspects of a navy ship. Thus, to assess the difference, a more detailed approach following the air wake study group for simulators might be a future study target.

Author Contributions: Conceptualization, S.S. and C.G.S.; methodology, H.I., S.S. and C.G.S.; software, H.I.; validation, H.I., S.S. and C.G.S.; formal analysis, H.I. and S.S.; writing—original draft preparation, H.I.; writing—review and editing, S.S. and C.G.S.; visualization, H.I.; supervision, S.S. and C.G.S.; project administration, S.S. and C.G.S.; funding acquisition, S.S. and C.G.S. All authors have read and agreed to the published version of the manuscript.

Funding: The work was performed within the NAVAD project "Simulation of manoeuvrability of ships in adverse weather conditions", which is co-funded by the European Regional Development Fund (Fundo Europeu de Desenvolvimento Regional—FEDER) and by the Portuguese Foundation for Science and Technology (Fundação para a Ciência e a Tecnologia—FCT) under contract 02/SAICT/032037/2017. This work contributes to the Strategic Research Plan of the Centre for Marine Technology and Ocean Engineering (CENTEC), which is financed by the Portuguese Foundation for Science and Technology (Fundação para a Ciência e Tecnologia—FCT) under contract UIDB/UIDP/00134/2020.

Institutional Review Board Statement: Not applicable.

Informed Consent Statement: Not applicable.

Data Availability Statement: Not applicable.

Conflicts of Interest: The authors declare no conflict of interest.

References

1. Sutulo, S.; Moreira, L.; Guedes Soares, C. Mathematical Models for Ship Path Prediction in Manoeuvring Simulation Systems. *Ocean. Eng.* **2002**, *29*, 1–19. [CrossRef]
2. Sutulo, S.; Guedes Soares, C. Mathematical models for simulation of manoeuvring performance of ships. In *Marine Technology and Engineering*; Guedes Soares, C., Garbatov, Y., Fonseca, N., Teixeira, A.P., Eds.; Taylor & Francis Group: London, UK, 2011; pp. 661–698.
3. Guedes Soares, C.; Sutulo, S.; Francisco, R.A.; Santos, F.M.; Moreira, L. Full-Scale Measurements of the Manoeuvring Capabilities of a Catamaran. In Proceedings of the International Conference on Hydrodynamics of High Speed Craft, London, UK, 24–25 November 1999; Royal Institution of Naval Architects: London, UK, 1999; pp. 1–12.
4. Guedes Soares, C.; Francisco, R.A.; Moreira, L.; Laranjinha, M. Full-Scale Measurements of the Manoeuvering Capabilities of Fast Patrol Vessels, Argos Class. *Mar. Technol.* **2004**, *41*, 7–16.
5. Sutulo, S.; Guedes Soares, C. An Algorithm for Offline Identification of Ship Manoeuvring Mathematical Models after Free-Running Tests. *Ocean. Eng.* **2014**, *79*, 10–25. [CrossRef]
6. Wang, Z.; Guedes Soares, C.; Zou, Z.J. Optimal design of excitation signal for identification of nonlinear ship manoeuvring model. *Ocean. Eng.* **2020**, *196*, 106778. [CrossRef]

7. Dubbioso, D.; Viviani, M. Aspects of twin screw ships semi-empirical maneuvering models. *Ocean. Eng.* **2012**, *48*, 69–80. [CrossRef]
8. Sutulo, S.; Guedes Soares, C. On the application of empiric methods for prediction of ship manoeuvring properties and associated uncertainties. *Ocean. Eng.* **2019**, *186*, 106111. [CrossRef]
9. Vettor, R.; Prpić-Oršić, J.; Guedes Soares, C. The effect of wind loads on the attainable ship speed on seaways. In *Towards Green Marine Technology and Transport*; Guedes Soares, C., Dejhalla, R., Pavletic, D., Eds.; Taylor and Francis: London, UK, 2015.
10. Vettor, R.; Prpić-Oršić, J.; Guedes Soares, C. Impact of wind loads on long-term fuel consumption and emissions in trans-oceanic shipping. *Brodogradnja* **2018**, *69*, 15–28. [CrossRef]
11. Moreira, L.; Vettor, R.; Guedes Soares, C. Neural network approach for predicting ship speed and fuel consumption. *J. Mar. Sci. Eng.* **2021**, *9*, 119. [CrossRef]
12. Owens, R.; Palo, P. *Wind-Induced Steady Loads on Ships*; Technical Note; Naval Civil Engineering Laboratory: Port Hueneme, CA, USA, 1982; p. 93043.
13. Blendermann, W. *Wind Loadings of Ships—Collected Data from Wind Tunnel Tests in Uniform Flow*; Report 574; Institute of Naval Architecture: London, UK, 1996.
14. Blendermann, W. *Practical Ship and Offshore Structure Aerodynamics*; Technische Universität Hamburg-Harburg: Hamburg, Germany, 2013.
15. Andersen, I.M.V. Wind Loads on post-Panamax container ship. *Ocean. Eng.* **2013**, *58*, 115–134. [CrossRef]
16. Haddara, M.; Guedes Soares, C. Wind Loads on Marine Structures. *Mar. Struct.* **1999**, *12*, 199–210. [CrossRef]
17. Hassan, K.; White, M.F.; Ciortan, C. Effect of Container Stack Arrangement on the Power Optimization of a Container Ship. *J. Ship Prod. Des.* **2012**, *28*, 10–19. [CrossRef]
18. Koop, A.; Rossin, B.; Guilherme, V. Predicting wind loads on typical offshore vessels using CFD. In Proceedings of the ASME 31th International Conference on Ocean, Offshore and Arctic Engineering (OMAE2012), Rio de Janeiro, Brazil, 1–7 July 2012.
19. Luquet, R.; Vonier, P.; Prior, A.; Leguen, J.F. Aerodynamic Loads on a Heeled Ship. In Proceedings of the 12th International Conference on the Stability of Ship and Ocean Vehicles, Glasgow, UK, 14–19 June 2015.
20. Wnęk, A.D.; Guedes Soares, C. CFD assessment of the wind load on an LNG carrier and floating platform models. *Ocean. Eng.* **2015**, *97*, 30–36. [CrossRef]
21. Wnęk, A.D.; Paço, A.; Zhou, X.Q.; Sutulo, S.; Guedes Soares, C. Experimental Study of Aerodynamic Loads on an LNG Carrier and Floating Platform. *Appl. Ocean. Res.* **2015**, *51*, 309–319. [CrossRef]
22. Watanabe, I.; Nguyen, T.V.; Miyake, S.; Shimizu, N.; Ikeda, Y. A Study on Reduction of Air Resistance Acting on A Large Container Ship. In Proceedings of the 8th Asia-Pacific Workshop on Marine Hydrodynamics in Naval Architecture, Ocean Technology and Constructions, Hanoi, Vietnam, 20–23 September 2016.
23. Nguyen, T.V.; Kinugawa, A.; Shimizu, N.; Ikeda, Y. Studies on Air Resistance Reduction Methods for a Large Container Ship (Part 1). In Proceedings of the Japan Society of Naval Architects and Ocean Engineers, Autumn Meeting, Okayama, Japan, 21–22 November 2016.
24. Majidian, H.; Azarsina, F. Aerodynamic Simulation of a Containership to Evaluate Cargo Configuration Effect on Frontal Wind Loads. *China Ocean. Eng.* **2018**, *32*, 196–205. [CrossRef]
25. Majidian, H.; Azarsina, F. Numerical simulation of container ship in oblique winds to develop a wind resistance model based on statistical data. *J. Int. Marit. Saf. Environ. Aff. Shipp.* **2018**, *2*, 67–88. [CrossRef]
26. Wang, P.; Wang, F.; Chen, Z. Investigation on aerodynamic performance of luxury cruise ship. *Ocean. Eng.* **2020**, *213*, 107790. [CrossRef]
27. Forrest, J.; Owen, I. An investigation of ship airwakes using Detached-Eddy simulation. *Comput. Fluids* **2010**, *39*, 656–673. [CrossRef]
28. Thornber, B.; Starr, M.; Drikakis, D. Implicit large eddy simulation of ship airwakes. *Aeronaut. J.* **2010**, *114*, 715–736. [CrossRef]
29. Forrest, J.S.; Kääriä, C.; Owen, I. Evaluating ship superstructure aerodynamics for maritime helicopter operations through CFD and flight simulation. *Aeronaut. J.* **2016**, *120*, 1578–1603. [CrossRef]
30. Yuan, W.; Wall, A.; Lee, R. Combined numerical and experimental simulations of unsteady ship airwakes. *Comput. Fluids* **2018**, *172*, 29–53. [CrossRef]
31. Thedin, R.; Kinzel, M.; Horn, J.; Schmitz, S. Coupled simulations of atmospheric turbulence-modified ship airwakes and helicopter flight dynamics. *J. Aircr.* **2019**, *56*, 812–824. [CrossRef]
32. Watson, N.A.; Kelly, M.F.; Owen, I.; Hodge, S.J.; White, M.D. Computational and experimental modelling study of the unsteady airflow over the aircraft carrier HMS Queen Elizabeth. *Ocean Eng.* **2019**, *172*, 562–574. [CrossRef]
33. Linton, D.; Thornber, B. Quantifying uncertainty in turbulence resolving ship airwake simulations. *Ocean. Eng.* **2021**, *229*, 108983. [CrossRef]
34. Flyvefisken-Class Patrol Vessel. 2021. Available online: https://en.wikipedia.org/wiki/Flyvefisken-class_patrol_vessel (accessed on 15 August 2021).
35. Flyvefisken Class (SF 300). Available online: https://www.naval-technology.com/projects/fly/ (accessed on 15 August 2021).
36. *ITTC-7.5-03-02-03*; ITTC-2011—Recommended Procedures and Guidelines. Practical Guidelines for Ship CFD and Application. Curran Associates, Inc.: Red Hook, NY, USA, 2012.

37. Labanti, J.; Islam, H.; Guedes Soares, C. CFD assessment of Ropax hull resistance with various initial drafts and trim angles. In *Maritime Technology and Engineering 3*; Guedes Soares, C., Santos, T.A., Eds.; Taylor & Francis Group: London, UK, 2016; pp. 325–332.
38. Weller, H.G.; Tabor, G.; Jasak, H.; Fureby, C. A tensorial approach to computational continuum mechanics using object-oriented techniques. *Comput. Phys.* **1998**, *12*, 620–631. [CrossRef]
39. Frisch, U. *Turbulence: The Legacy of A.N. Kolmogorov*, 1st ed.; Cambridge University Press: Cambridge, UK, 1995.
40. Menter, F.R. Zonal Two Equation k-ω Turbulence Models for Aerodynamic Flows. In Proceedings of the 23rd Fluid Dynamics, Plasma Dynamics, and Lasers Conference, Orlando, FL, USA, 6–9 July 1993; AIAA Paper: Reston, VA, USA, 1993; p. 2906.
41. Launder, B.E.; Spalding, D.B. The numerical computation of turbulent flows. *Comput. Methods Appl. Mech. Eng.* **1974**, *3*, 269–289. [CrossRef]
42. Wilcox, C.D. *Turbulence Modeling for CFD*, 2nd ed.; DCW Industries: La Canada, CA, USA, 1998.
43. Gritskevich, M.S.; Garbaruk, A.V.; Schütze, J.; Menter, F.R. Development of DDES and IDDES Formulations for the k-w Shear Stress Transport Model. *Flow Turbul. Combust.* **2012**, *88*, 431–449. [CrossRef]
44. *ITTC-7.5-03-01-01*; ITTC-2008—Recommended Procedures and Guidelines. Uncertainty Analysis in CFD Verification and Validation Methodology and Procedures. Curran Associates, Inc.: Red Hook, NY, USA, 2009.
45. Celik, I.B.; Ghia, U.; Roache, P.J.; Freitas, C.J.; Coleman, H.; Raad, P.E. Procedure for Estimation and Reporting of Uncertainty Due to Discretization in CFD Applications. *J. Fluids Eng. Trans. ASME* **2008**, *130*, 078001.
46. Islam, H.; Guedes Soares, C. Uncertainty analysis in ship resistance prediction using OpenFOAM. *Ocean. Eng.* **2019**, *191*, 105805. [CrossRef]
47. Islam, H.; Guedes Soares, C. Assessment of uncertainty in the CFD simulation of the wave-induced loads on a vertical cylinder. *Mar. Struct.* **2021**, *80*, 103088. [CrossRef]
48. Wang, S.; Islam, H.; Guedes Soares, C. Uncertainty due to discretization on the ALE algorithm for predicting water slamming loads. *Mar. Struct.* **2021**, *80*, 103086. [CrossRef]
49. Islam, H.; Guedes Soares, C.; Liu, J.; Wang, X. Propulsion power prediction for an inland container vessel in open and restricted channel from model and full-scale simulations. *Ocean. Eng.* **2021**, *229*, 108621. [CrossRef]

Journal of
Marine Science and Engineering

MDPI

Review

Review on Ship Manoeuvrability Criteria and Standards

Serge Sutulo * and **C. Guedes Soares**

Centre for Marine Technology and Ocean Engineering (CENTEC), Instituto Superior Técnico, Universidade de Lisboa, Av. Rovisco Pais 1, 1049-001 Lisbon, Portugal; c.guedes.soares@centec.tecnico.ulisboa.pt
* Correspondence: sutulo@centec.tecnico.ulisboa.pt

Abstract: Possible reduction of the installed power on newly designed merchant ships triggered by requirements of the Energy Efficiency Design Indices (EEDI) raised concern in possible safety degradation and revived interest in manoeuvrability standards to make them capable to compensate for negative effects of underpowering. A substantial part of the present article presents a detailed analytical review of general principles laid in the foundation of consistent safety standards in the naval architecture and analysis of the existing IMO manoeuvrability criteria and standards. Possible ways of extension of the existing standards to embrace situations associated with adverse sea and wind conditions are discussed and modification of the present standards related to the directional stability is considered as one of the possible solutions. At the same time, it was found that introduction of additional standards for the ship controllability in wind is justified, and the second part of the contribution is dedicated to developing a theoretical basis useful for devising such standards. This includes obtaining a set of analytical solutions related to the steady motion in wind and analysis of wind-tunnel data which resulted in simple equations for conservative generalized envelopes for the aerodynamic forces which are especially convenient for standardizing purposes. Possible design decisions aimed at augmentation of the ship's capacity to resist adverse environmental factors are outlined.

Keywords: manoeuvrability standards; standardizing methodology; EEDI; influence of wind and sea waves; controllability in wind; ship aerodynamic characteristics

Citation: Sutulo, S.; Guedes Soares, C. Review on Ship Manoeuvrability Criteria and Standards. *J. Mar. Sci. Eng.* **2021**, *9*, 904. https://doi.org/10.3390/jmse9080904

Academic Editor: Michele Viviani

Received: 24 July 2021
Accepted: 17 August 2021
Published: 21 August 2021

1. Introduction

There are no doubts that since ancient times the seafarers were able to discriminate between "good" and "not so good" vessels albeit this was only possible on a purely qualitative and intuitive level. Such discrimination was necessary and sufficient for a slow but steady evolution of various types of ships. As result, it became possible for ships to reach a certain undisputable level of perfection without any scientific support: medieval sailing ships were capable to perform very long voyages in rough sea waves including round-the-world trips. Of course, development based exclusively on traditions and some empiric rules was extremely slow.

Acceleration of development was not possible without certain scientific contributions unthinkable without quantification of the ship performance and seaworthiness. It can be noticed that the tangible raise of the "quantitative" phase of development of the naval architecture approximately coincided with the transition from sailing ships to mechanically powered ones. Likely, this was not only a coincidence associated with concurrent development of the naval architecture and ship theory but was also facilitated by the fact that any mechanically driven ship is, in many respects, a much simpler technical system than a sailing vessel. In particular, while the speed of a powered ship can be measured rather reliably at some specified standard testing conditions (deep water, calm sea, specified engine rpm) this is practically impossible for a tall ship whose speed depends substantially on the available wind. Further, as the aerodynamic configuration of a sailing ship is extremely variable, it is difficult to quantify its seaworthiness. Finally, it is practically impossible to establish any reasonable measures of manoeuvring qualities of such ships as, for instance, a sail-driven ship is not able to perform steady turning circles and it is very difficult to reasonably define parameters of any turn executed through rather complex coordination between the rudder deflection and handling of the sails.

However, nowadays the sailing ships occupy a relatively narrow niche of recreational and training vessels and can be treated in some special way. As to the powered ships, all quantitative measures of their performance can be divided into two groups:

1. Contractual parameters whose values are established, first, as the customer's requirements and, further, as a compromise and agreement between the customer and the builder specified in the contract.
2. The parameters whose acceptable values are typically specified by various rules and regulations. Most of such regulated parameters are related to maritime safety.

All these parameters may be related to the ship hydrodynamics or the strength and reliability of a ship hull and other elements or other operational qualities. Only hydrodynamics-related parameters will be considered further. The most common parameter of the first group is the contractual or design speed of the ship. The inability of a newly built ship to reach the specified speed value in the trials is typically treated as incomplete fulfilment of the contract and, unless the problem is fixed, is penalized. Most parameters of the second group are related to seaworthiness and unsinkability although in the several last decades manoeuvring performance of ships also became subject to substantial efforts aimed at the development and implementation of manoeuvrability criteria and standards.

The present article contains an analysis of existing requirements to the manoeuvring performance of civil displacement ships with emphasis given to the well-known manoeuvrability standards of the International Maritime Organization (IMO) [1]. All studies related to the introduction of any standards are characterized by many poorly formalizable issues and they strongly depend on subjective opinions of numerous experts. Final solutions are typically reached after long discussions in form of a more or less acceptable compromise, a retrospective analysis of existing approved standards and criteria and various alternative proposals is rather important.

The authors deliberately abstain from formulating any completed specific proposals complementing or modifying the existing recognized system of manoeuvring criteria but rather focus on discussion of possible approaches and methodology. This discussion constitutes the main content of the present article.

Section 2 of the article is dedicated to the formulation of general requirements to any system of criteria and standards related to maritime safety with a brief review of the existing intact-stability criteria. In addition, relatively recent minimum powering criteria are briefly commented. Section 3 contains a brief review of some existing calm-water manoeuvring criteria and of some earlier published alternative proposals. The influence of accounting for adverse factors and external disturbances on the composition and acceptable values of manoeuvrability criteria is discussed in Section 4, where also some possible novel approaches are outlined.

While the first four sections are mainly of descriptive character, Section 5 is of a more specific nature presenting some analytical developments that can be useful for devising simplified manoeuvring standards accounting for the ship aerodynamic characteristics and wind action. Finally, Section 6 contains a concluding discussion.

2. General Principles of the Criteria and Standards in Naval Architecture

Most theoretical developments and methods in naval architecture are associated with problems of analysis or design. However, it is reasonable also to talk about the standardizing problem characterised by some specifics which justify placing this problem into a separate class.

Most standards in ship design are associated with various aspects of the ship safety depending on the global and local strength, seaworthiness and manoeuvrability where the latter area was embraced by standardizing activities relatively recently and possesses a shorter historical record. At the same time, studies on the intact stability standards have been actively carried out since the 1930s and they provide rather rich material for comparative analysis of various approaches [2–4]. As a result of such analysis, it was possible to formulate the following reasonable requirements to any consistent set of safety standards independently of the application area.

A. **Simplicity:** Criteria must be as simple as possible and preferably described by relatively simple analytical formulae.
B. **Testability:** Criteria must be verifiable both in operation and in the design stage.
C. **Clearness:** Criteria must be crisp, prescriptive and preferably of "pass" vs "non-pass" binary type.
D. **Physical consistency:** Criteria must be based on transparent and consistent physical (mechanical) principles. This makes undesirable recourse to criteria based exclusively on failure statistics although verification on statistical data is indispensable.
E. **Realism:** No matter how solidly based is any new proposed standard, it should not be implemented if the fraction of the existing fleet for which it fails, exceeds a certain value: typically 5 or 10 per cent. Similarly, no one standard is acceptable if it leads to evidently excessive requirements.

Apparently, the most straightforward and intuitively transparent criterion of a ship's seaworthiness could be the expected duration of the ship's survival in certain severe weather conditions characterized by parameters of the sea and wind spectra but in practice, some better defined simplified scenarios are needed.

The so-called Second Generation Intact Stability Criteria [2,5–7] envisage 5 different basic failure scenarios including parametric roll and broaching to. However, the existing IMO Code [2] still in effect is based, besides direct requirements to the righting-arm curve, almost exclusively on the so-called weather criterion, i.e., on the estimation of dynamic stability of a dead ship under the action of resonant roll and an unfavourable wind gust. Despite its simplicity, this criterion is quite reasonable and reliable as in fact it resulted from several decades of evolution based on intensive research studies carried out in many countries. The criterion still survives despite considerable efforts aimed at accounting for, say, parametric rolling, broaching or just involving more sophisticated and accurate mathematical models [5,6]. It is also remarkable that while the direct statistical approach introduced by Rahola is considered outdated and no longer acceptable, statistical analysis keeps playing an important role [2,8].

Of course, all safety-aimed criteria and standards have a clear trend to involve more and more sophisticated mathematical models, scenarios and schemes which sometimes seems to be more driven by the interests of developers than by clear practical demands. In particular, it is envisaged that in the mentioned "Second-generation intact-stability criteria", see, e.g., [9], 3 levels or tiers of criteria are to be introduced with the level 1 corresponding to the most simple and conservative criteria while higher-level criteria presume involvement of rather sophisticated schemes and methods. It is supposed that a higher-level criterion is only invoked if the corresponding lower-level criterion is not satisfied. Whether such multi-level system of standards will be definitely beneficial for the overall safety of shipping is not so evident and some specialists reserved a somewhat sceptical attitude [10]. The authors of the present contribution tend to believe that any practical system of criteria and standards should be kept as simple and transparent as possible and, regarding the classification mentioned above, all ideas presented in this paper should be viewed as corresponding to the lowest level.

In addition to the aforementioned standardizing areas, relatively recently a completely new problem emerged—that of minimum installed power requirements [11]. It was triggered by the so-called Energy Efficiency Design Index (EEDI) requirements, see [12] for more details. These requirements encouraged reduction of the installed main engine power as the most effective mean of decreasing the fuel consumption and, concurrently, emissions of all kinds. As such a trend obviously leads to reduction of the capability of ships to withstand adverse conditions thus impairing navigational safety, it gave birth to rather natural concerns. As result, considerable efforts have already been dedicated to working out certain minimum-power standards limiting the EEDI-driven reduction of the installed power [13].

The minimum-power requirements are connected with the seakeeping and manoeuvring qualities of ships. However, in this article, the authors avoid detailed discussion of the minimum-power problem limiting themselves to the short comments that follow.

Since the appearance of powered ships the required power of the main engine(s) has always been estimated as a compromise between a natural desire to reach the highest possible operation speed and safety and a desire to build a more economical vessel. It is obvious that the latter encourages the designer to reduce the ship's speed and the engine's rating. In some special cases (tugs, fishing vessels, icebreakers) the engine power is governed not by the design speed but by the required bollard pull. While the bollard pull-based power can be determined relatively straightforwardly, a more common task of setting the design speed is more complex and fuzzy. The value of this speed depends on the type of the ship, its size, actual and envisaged fuel prices, traditions and expectations of ship owners and seafarers.

A kind of natural selection during many decades resulted in several typical values of the design speed for various types and classes of ships although those typical values were subject to relatively small variations and corrections especially when new types of ships, engines and propulsors were introduced. However, so far they have never been conditioned by some legal requirements. Thus, in some sense, the introduction of EEDIs is breaking the existing dynamic equilibrium exerting substantial legal pressure that encourages the ship designers to decrease the power of main engines. It was soon understood that in some cases the newly built ships could become underpowered and excessively prone to exogenous factors. As result, the problem of standardizing also some minimum safe power of ships was formulated with the purpose of legal counterbalancing the unfavourable impact of EEDI requirements.

In addition to vulnerability to environmental factors, the underpowering may have one more negative effect related to the stopping distance in crash stop. While the stopping distance or the track reach from the speed at maximum continuous rating in calm water, as specified by the current IMO standards, may remain unchanged as long as poorer reversing capabilities come in certain harmony with the reduced speed, this is not so certain for partial speed and/or in presence of unfavourable wind, sea and current. This issue, closely related to the powering requirements is also dropped here.

As the danger of underpowering was realised, lately a concept of the reserved power to only be used in extreme conditions was discussed [13]. It may happen that this situation will finally lead to the appearance of more adaptable and flexible power plants.

Returning to the general standardization problem, it can be observed that there are three main different standardizing schemes:

1. **Explicit criteria scheme.** It is presumed that the limiting values of certain criteria are explicitly specified and the compliance to them is to be verified by some certified or generally recognized procedures which may include full-scale trials, model-scale tests and Computational Fluid Dynamics (CFD) modelling. For instance, regarding some manoeuvring criterion such as, e.g., the minimum turning radius, its value can be specified explicitly and their fulfilment is then verified by estimating the corresponding measures for the actual ship or design. This approach was realized in IMO standards [1,14] and STANAG standards [15]. The advantage of this scheme is that the standards in form of natural and meaningful manoeuvring qualities are verified directly and the estimation methods can be gradually improved in course of natural scientific progress without affecting the standards themselves. The evident disadvantage is the necessity to select and use some methods not incorporated into the text of the standards: this complicates application of the standards and may introduce additional uncertainty.

2. **Implicit scheme.** Alternatively, the standards can be formulated in form of a set of direct requirements to the rudder effectiveness. The latter must be defined within the standards in some unambiguous way allowing immediate verification at least for the most typical steering devices. This approach was followed in the implicit manoeuvring standards implemented within the Rules of the Russian Maritime Register of Shipping [8], see also [12,16]. In contrast with the IMO standards which were developed and adopted 14 years later, the Russian standards do account for aerodynamic loads albeit in a somewhat hidden way. However, although implementation of these standards was by all means somewhat useful, their straightforward reproduction in the part concerning controllability in wind cannot be recommended: the exploited mathematical models and prediction methods were too simplistic by all reasonable standards and even suffered from some inconsistencies. Even though these inconsistencies were traced by some experts, they practically could not be fixed as all the models and methods are embedded in the standards and their correction, modification and improvement is unthinkable without revision of the corresponding part of the Rules which is impossible without complicated bureaucratic procedures. This is one of the main reasons why the mentioned "hidden" manoeuvring standards remained unchanged since their embedment into the text of the Rules in 1982. Even though these Rules will keep these standards for an indefinite time, it is obvious that any new developments should follow a more "explicit" approach and it makes sense to realize them as some modifications and extensions of the existing IMO standards.

The task of specifying limiting values for criteria involved in any system of standards is also important and non-trivial. The following approaches can be followed:

1. **Statistical approach.** This approach is based on the assumption that for the existing successful ships values of the considered criteria are acceptable while those for ships that went through some kind of failure are not. In its direct form, this method had been applied by Rahola [2,3] for devising first-generation intact stability standards but rather soon this method was abandoned in favour of "mechanical" schemes. At the same time, statistical checks and corrective actions are considered mandatory for any set of standards. For instance, regarding any manoeuvrability standards, Norrbin [17] proposed the acceptability level as being worse by half of the standard deviation than the observed average value of this or that criterion over a representative set of vessels. Koyama and Kose [18] also have explicitly indicated that manoeuvring qualities of any ship must not deviate too much from average or typical which is especially important for marine pilots normally possessing very limited time to become acquainted with specifics of every piloted ship.
2. **Scenario approach.** This method is based on the faith that it is possible to formulate one or more operation scenarios that can be considered critical beyond any reasonable doubt and on the assumption that for every such scenario it is possible to develop sufficiently reliable prediction methods. Typically, simpler and more conservative methods are preferred to keep with the simplicity principle but also a hierarchy of methods with various degrees of accuracy and complexity may be envisaged by the already mentioned multi-level approach. A good example of the scenario approach for the turning ability standard was given by Mastushkin [19] who considered a scenario of avoiding the collision with a suddenly detected obstacle in form of an infinite wall crossing the ship's path. Starting from some assumed radar detection distance it was assumed that the evasion manoeuvre would be a sharp turn at full helm (this was found to be more efficient than a crash stop). The upper boundary for the advance in turning manoeuvre was defined setting in this way a standard for the turning ability. This and other considered scenarios were inevitably rather artificial and schematized. In addition, broad application of simplified mathematical models and empiric methods further increased uncertainties and final adjustment was performed "statistically" using data for the then existing fleet. The same procedure was applied to all scenarios considered. As result, the following general conclusion

can be drawn: although any of the meaningful scenarios provides reasonable and useful indicators, no one of those scenarios is guaranteed to be "played" successfully in real world even if the ship meets the finally adjusted standards.

3. **Ergatic approach.** This approach is very common in aeronautics [20] where standardization focuses on ergatic issues, i.e., on the man–machine interaction presuming a definition of acceptable values of parameters of mathematical models of aircraft. Nowadays, this has become a rather routine practice and is performed with the help of interactive flight simulators operated by representative groups of pilots. Probably, for the first time, the idea of applying a similar approach to ship manoeuvring was expressed by Segel [21] and much later, though independently and in a more elaborated form, by Sutulo [22]. In fact, elements of this approach combined with the scenario of steering a ship in a complex bent canal were also used by Nobukawa et al. [23] to establish an acceptable level of directional stability and finally these results contributed to the IMO standards. However, in general, the significance of methods based on interactive simulations in ship manoeuvrability is much inferior to that in the flight dynamics because of substantial differences in dynamic properties, in particular in the values of time lags, between the surface displacement ships and aircraft. In addition, this approach is hardly suitable for standardizing any properties of the craft related to exogenous perturbations.

3. On Calm-Water Manoeuvrability Criteria and Corresponding Standards

3.1. General Specifics of Manoeuvring Standards

It should be noted that the parameters defining the manoeuvring performance of ships for a long time belonged to the first "contractual" group that is at best they were specified in the contract but it was clear that this situation could not last forever. As long as the manoeuvring qualities were related to the navigational safety, in the middle of the 20th century first attempts to establish manoeuvring criteria and to set corresponding standards were undertaken.

Even though ship seakeeping and ship manoeuvrability in some sense can be viewed as two faces of ship dynamics, the differences between possible standardizing strategies in both fields are, however, substantial:

1. It is possible to speak about manoeuvring qualities of a ship in still water, i.e., without any excitation at all while any seakeeping criterion is associated with a certain level of external excitation caused by wind and waves.
2. The tighter is any intact stability standard, the higher will be the safety level, which is always welcome unless the implementation costs are too high. The situation is more delicate with the manoeuvring performance as, for instance, an extremely good turning or stopping ability can increase the probability of collision as the behaviour of some surrounding vessels may become too unusual and unpredictable. This does not present any problems in deep-sea conditions and in absence of other craft but is highly undesirable in dense traffic conditions. In particular, that means that a rating-based system of standards encouraging the highest possible performance such as those suggested in [24,25] could produce unpredictable effects regarding the overall navigational safety.
3. The human factor is much more important when dealing with manoeuvring performance. It is even possible to say that, unless the manoeuvring qualities are extremely poor, competent operators after suitable training are capable to compensate for imperfections in manoeuvring properties and successfully and safely handle practically any ship although this adaptability is achieved at a price: difficulties and strain in controlling the ship may be very different and not always acceptable. In general, the human factor is very difficult to be accounted for but the ergatic approach mentioned in the previous section probably can be useful in handling this issue. Of course, in the seakeeping the human factor is also present in the sense that a high seasickness level may affect selection of the speed and of the heading but the difference is obvious.

4. While ship mathematical models serving as a basis for standardizing the intact stability may only consider one degree of freedom, i.e., accounting for the roll motion only, this cannot be allowed for the manoeuvring motion where a mathematical model must have at least two degrees of freedom, i.e., the coupled sway and yaw and often also the surge motion must be included. To prevent possible misunderstandings, it should be emphasized that the often-used Nomoto equations, contrary to their appearance, implicitly embrace two degrees of freedom.

Here the verifiability requirement may be very difficult to meet as long as it presumes the possibility to check any manoeuvring criterion in full-scale trials. This is not so easy even with calm-water criteria: high-quality measurements require practical absence of wind which is a rather rare situation and typically full-scale data suffer from considerable uncertainties. The difficulties are becoming even more serious for criteria related to the waves and wind as it is necessary to fix a certain level of external perturbations. This problem was discussed by Dand [26] who expressed the opinion that the requirement of the possibility of a full-scale verification of any particular standard could be lifted. He suggested that "a combination of a simulation model and suitable indices derived from past best practice, be used as a way of assessing manoeuvrability of vessels".

While the number of publications on the intact stability and unsinkability criteria is definitely superior to that on manoeuvring standards, the latter is also quite significant. Rather complete reviews on proposed manoeuvring criteria and standards can be found in [12,22] and the following text is mainly focusing on the sources related to the development and validation of the existing IMO standards [1].

There exists a more or less clear understanding of which calm-water manoeuvring qualities should be standardized although, e.g., Quadvlieg [27] tried to extend the list of standard criteria, aiming at the assessment of manoeuvring prediction methods. In the following, only dlassic, commonly recognized criteria will be discussed.

It is recognized that regarding manoeuvring qualities of surface displacement ships in calm water the following properties are of primary interest:

1. Turning ability
2. Directional stability

A ship is assessed as perfectly controllable if it can be characterized by a high level of these two properties at the same time although often one of them may be partly sacrificed. For all purposes, it is important to establish reasonable measures for these qualities.

3.2. Turning Ability

This quality can be subdivided into the turning ability per se which is also called sometimes the ultimate turning ability and the so-called initial turning ability (ITA).

The (ultimate) turning ability of a ship is its capability to describe a highly curved trajectory in absence of external factors. When the moderate manoeuvring (excluding crabbing and rotation with the help of side therusters) is considered, a very natural measure of the turning ability can be introduced which is the steady turning diameter at full helm non-dimensionalized by the ship length. In reality, the easier to register "tactical diameter" is usually preferred. It is, however, well correlated with the steady turning diameter.

According to the IMO standards, the tactical diameter is required to not exceed 5 ships length independently of the value of that length and of the speed of the ship.

Despite a certain terminological similarity, the initial turning ability is rather different from the ultimate turning ability discussed above: not the path curvature is here considered but the ability of a ship to rapidly change the heading enough. In most cases, the 10 deg heading change is considered, and no helm check is performed. The latter indicates that it does not go about a true course-changing manoeuvre. The meaning of standardizing the initial turning ability consists in preventing temptations to assure good turning ability at large helms at the expense of the inherent stability of the ship and without increasing the effectiveness of the rudder: despite technically good turning abilities such ships will be slow in responses to the rudder deflections or, in other words, they will be not well controllable. The ITA is correlated with both the ultimate turning ability and with the ship directional stability although not in a quite straightforward way. It is clear, however, that a ship with a sufficiently large (say 3–4% of the centreplane area) rudder working outside of the propeller slipstream (a rather typical configuration for naval ships) is very likely to possess, at the same time, perfect directional stability combined with the good turning ability both ultimate and initial. With a smaller rudder in the slipstream the combination of manoeuvring qualities may turn out less favourable and then invoking the ITA criterion can serve as a good remedy against unsatisfactory controllability of the ship.

3.3. Directional Stability

3.3.1. Definitions and Criteria

The directional stability is, in practical terms, the capability of a ship to follow a straight path with moderate yawing, at a low average intensity of rudder orders and without an excessive strain of the human operator (helmsman) [22]. The ship is treated here as a closed-loop system and the course-keeping process is interpreted as continuing mitigation of deviations from the ordered heading caused by low-level excitations from sea waves. On directionally unstable ships this process can be accompanied by self-sustained oscillations whose admissible frequency and amplitude are to be standardized. Even though criteria of this type had been suggested, see, e.g., [28], they appeared to be impractical as the mentioned parameters of oscillations depend also on the parameters of the controller represented in this case by the helmsman. The undesirable influence of the controller is eliminated if the "inherent" stability of the ship with the rudder fixed in its neutral position is considered. Both the practical stability and the inherent stability are correlated: the higher is the degree of the inherent directional stability, the easier and better is the course keeping. However, quantitative criteria of the inherent directional stability are not obvious. The most natural criteria, at first sight, must be associated with the eigenvalues of the mathematical model of the sway-yaw motion linearized around the straight run. It is known that both eigenvalues are always real, and one of them has a relatively small absolute value and can be either positive or negative for different ships. The positive eigenvalue correspond to inherently unstable ships, the negative one—to inherently stable craft and it is obvious that the absolute value can serve as a measure of the stability/instability degree. Even though proposals to use the critical eigenvalue or the corresponding time lag as a stability criterion were made [29], its inconvenience became clear very soon: there is no way to check the value of this criterion directly from the results of full-scale or model-scale trials. As result, most of the proposals were exploiting more practical indirect measures.

3.3.2. Connection with Parameters of the Spiral Curve

The first of such indirect criteria is based on the configuration of the static characteristic (spiral curve) of a ship, which represents the dependence $r'(\delta_R)$ of the non-dimensional rate of yaw or the path curvature r' on the rudder angle δ_R in a steady turn. When the ship is directionally unstable, this curve possesses a hysteresis loop around the origin and the dimensions of this loop, i.e., its height and/or width can serve as measures of the degree of instability [12,22]. Some specialists believed that only some admissible degree of instability must be specified while all inherently stable ships are equally and unconditionally acceptable [30,31]. However, the stability margin can be very different and, aiming at standardizing directional stability of naval combatants and trying to keep the spiral curve as the main source of information on manoeuvring qualities, it was proposed by Sutulo [22] to standardize the value of the spiral curve derivative at the origin, i.e., $\left.\frac{\mathrm{d}r'}{\mathrm{d}\delta_R}\right|_{\delta_R=0}$ or of its normalized modification $\frac{\delta_m}{r'(\delta_m)} \cdot \left.\frac{\mathrm{d}r'}{\mathrm{d}\delta_R}\right|_{\delta_R=0}$, where δ_m is the maximum rudder angle.

There are reasons to believe that the optimal value of the latter normalized parameter should be 1.0 which corresponds to an absolutely straight spiral curve. This can be viewed as an educated guess but this guess is, however, indirectly confirmed by the fact that all high-speed moving objects, e.g., steady-wing aircraft, are highly linear. The aircraft are not only always inherently stable but typically a certain margin of the so-called static directional stability is required [20]. It is interesting to note that Oltmann [32] also considered a similar though inversed criterion $\left.\frac{d\delta_R}{dr'}\right|_{\delta_R=0}$ and he demonstrated invoking multiple full-scale data for unstable ships that this criterion is only weakly correlated with the hysteresis loop width.

However, such "initial tangent" criteria have not gained recognition. One of the causes is that it is difficult to estimate these quantities from full-scale trials that must include the Bech reverse spiral test being rather difficult to execute let alone that the obtained estimates of the derivatives can be highly uncertain. Regarding the fact that most civil ships indeed remain well steerable even at some moderate degree of inherent instability, the dimensions of the hysteresis loop can serve as suitable indicators of acceptable directional stability at least for merchant ships.

However, the analysis based on the spiral curve configuration also has inconveniences related to full-scale trials. The Dieudonné spiral manoeuvre is not the easiest manoeuvre to execute as it requires substantial water area and is long in duration. The presence of wind is especially unfavourable for this manoeuvre as it makes practically impossible reaching the steady turn regime at each rudder deflection. To obtain more or less reliable averaged results in wind, it is necessary to perform at least two full turns at each helm which would increase the duration of the trials to unacceptable values. That is why, while requirements to the parameters of the spiral manoeuvre are formulated in the Explanatory Notes to the IMO manoeuvring standards [33] they were not even mentioned in the main text of the Standards [1] where an alternative in form of overshoot angles in zigzag tests was preferred [34]. However, Norrbin [35] characterized the fact that the spiral characteristic finally did not enter the main text of the IMO standards as "unfortunate".

3.3.3. Connection with Parameters of the Zigzag Manoeuvre

As long as there is no way to establish an analytical connection between the overshoot angles and, say, traditional stability indices, this solution raises, however, the following questions:

1. Whether both the width and the height of the hysteresis loop influence the actual steerability and the overshoot angles of a ship or one of them can be neglected?
2. Are the overshoot angles highly correlated with the parameters of the loop or they should be treated as additional and independent criteria?

As will be seen from the review below, the existing data are rather contradictory and this fact alone was sufficient for stimulating some criticism with respect to the existing IMO standards [1,23,33]. For instance, the data presented in [23] indicated that the relation between the loop width and the overshoot angles is rather rigid: the larger is the former, the larger are the latter. Norrbin [17] discussed the relative importance of the loop width and height and pointed out that the overshoot angles are better correlated with the loop width than with its height. Nikolayev et al. [36] discovered that the correspondence between the overshoot angles and the loop width alone can be very poor but can be improved if the loop height is used as an additional factor.

It is generally recognized that the following 3 overshoot angles adopted as standard measures are significant: the first overshoot angle in the 10°–10° zigzag (OS1), the second overshoot angle in the same zigzag (OS2) and the first overshoot angle in the 20°–20° zigzag (OS3). It is not possible to link analytically the parameters of Kempf's zigzag manoeuvres to other measures of the inherent directional stability, multiple simulations and trials confirm that typically the higher is the directional stability, the smaller are values of the overshoot angles. There is also a direct evidence that reduction of OS1 is beneficial for steering. In particular, such a conclusion was obtained in [37] based on multiple interactive simulations in the conditions of a straight and double-bent channel under influence of current. Many specialists suspected that these three criteria are not of equal significance.

For instance, Ræstad [38] came to the conclusion that the necessity of inclusion of OS3 into the Standards was not confirmed by analysis of full-scale data and by polling the seafarers although OS3 is typically considered important as a measure of the turn checking ability. At least, Yoshimura et al. [39], on the other hand, concluded that the OS1 requirement is less restrictive than OS3 while the least restrictive is OS2. At the same time, according to [40] OS1 is the strictest criterion.

Yoshimura et al. [39] showed that the correlation between the loop width and the overshoot angles OS1 and OS2 was confirmed by full-scale trials carried out with 50 ships. It can be noticed, however, that the shown correlation was better than in most other sources. Rhee et al. [41] performed a rather detailed study of the width–height correlations. Simulations showed a rather strong correlation with the width and a somewhat weaker one with the height of the loop. At the same time, the full-scale data demonstrated substantially inferior correlation in general and with OS3 it was even practically absent. As to the 10°–10° zigzag overshoots, OS1 correlated better with the loop height while OS2—with its width. Haraguchi and Nimura [42] have found, performing simulations based on tank data, a definitely positive width–angles correlation for pod-driven ships. A good correlation between the loop width and the overshoot angles was obtained by Sohn et al. [43] for a number of simulated ships but Yamada [44] pointed out that his data do not confirm such correlation.

The brief review above might leave an impression that the correlation between the parameters of the hysteresis loop is uncertain as the available data are controversial especially regarding full-scale sea trials data. However, it must be taken into account that such data are always given without uncertainty estimates while the uncertainties may be considerable for measurements taken in real sea conditions. In addition, there is sufficient evidence that the hysteresis loop height though ignored by the existing standards also influences the ship's behaviour in steering and the values of the overshoot angles thus impairing their correlation with the loop width. It must be admitted that detailed study of the influence of the loop width–height relation is still the matter of future research and so far even qualitative discussions such as that by Norrbin [17] are very rare.

However, in practice, there is no doubt that both the loop width and the overshoot angles can serve as reasonable measures of the directional stability of ships and the overshoot angles have the advantage to suit not only unstable but also directionally stable vessels. Remarkably, Ræstad [38] demonstrated using full-scale measurements and questionnaires distributed among seafarers that ships with subjectively abnormal behaviour indeed exhibit especially large overshoot angles.

3.4. Analysis of IMO Criteria of Directional Stability

The measures of the ship's directional stability used in the IMO standards [1,33] are reviewed and analysed in this subsubsection. It is important to keep in mind that these standards had been devised for calm-water steering.

The measures for the turning ability in IMO standards in terms of the ship length are fixed for all vessels but the required degree of the directional stability depends on the ship reference time $T_{\mathrm{ref}} = L/V$, where V is the speed specified in the standards. The stability requirements can be formulated either in terms of the hysteresis loop width (Figure 1) or in terms of the overshoot angle OS1 (Figure 2). The IMO requirements are shown on the plots as a solid line (graphs plotted in dotted and dash-dot lines and the meaning of $L_{1,2,3}$ will be discussed below, in the next section). If the reference time is smaller than 9 or 10 s, the requirement becomes flat. In particular, every inherently stable ship unconditionally satisfies the spiral loop requirement although further tightening of the inherent stability standard for small values of the reference time would seem more natural. However, any extrapolation on the plot for the loop width can only result in unrealizable negative values, see the dotted line in Figure 1 and application of alternative parameters of the spiral curve, such as its initial slope, discussed in the previous section is not compatible with the IMO standard. At the same time, a reasonable extrapolation can be performed on the overshoot angle plot, see the dashed line in Figure 2.

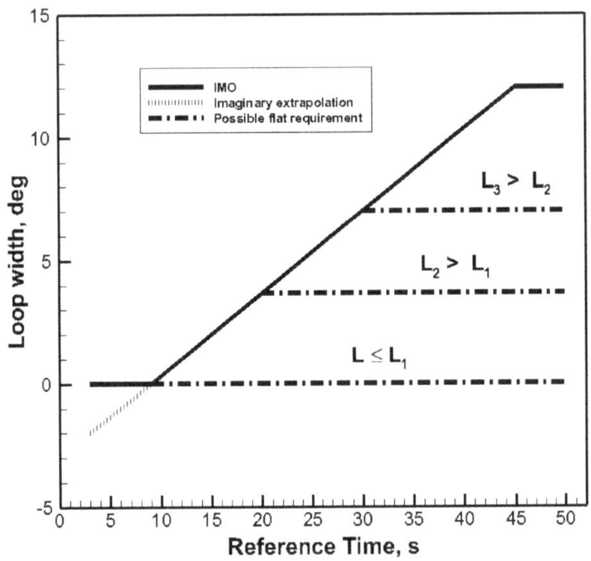

Figure 1. Standards for hysteresis loop width according to IMO standards and alternative requirements.

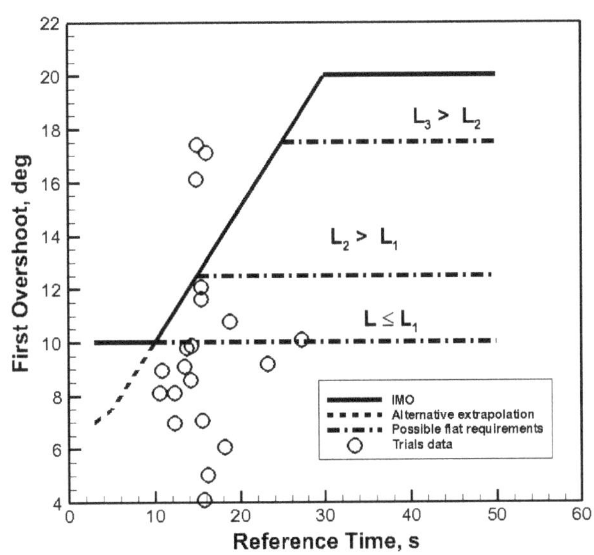

Figure 2. Existing and proposed standards for the first overshoot angle in 10°–10° zigzag.

As to the reference time exceeding 9 s, such ships are allowed to be directionally unstable and only the degree of instability is limited. As the reference time is increasing further, the inherent stability requirements according to the existing IMO standards are becoming looser and looser which looks consistent for manoeuvring in calm water. Constant limiting values applied again for $T_{ref} > 30$ s are explained by the necessity to have some helm margin at large loop width.

Dependence of the stability criteria on the ship reference time looks quite logical from the viewpoint of the ergatic approach: the smaller and faster is the craft, the higher is the required degree of directional stability. This is explained by the obvious fact that smaller time lags of the controlled object require also smaller time lags of the controller. Even though the human operator is adaptable in a rather wide range of dynamic properties of the controlled object, steering faster and smaller craft requires from the operator a higher degree of concentration and tension which cannot be completely eliminated even after intense and long training.

The IMO-adopted standardizing numerical values of the manoeuvring criteria are a result of intense and long research work and of multiple discussions carried out over many years. So, it can be saidthat they represent some acceptable compromise [34]. However, they should not be viewed as a kind of dogma and alternative values of the manoeuvrability measures were proposed not only before but also after the interim and even definitive standards were approved. For instance, in one of the earlier proposals [45] it was suggested to allow the hysteresis loop width not exceeding 4 degrees as a measure of an acceptable degree of instability for all ships independently of the reference time. A number of values of the parameters of the hysteresis loop proposed from 1959 to 1987 were listed by Norrbin [17]. Li and Wu [41] suggested to consider manoeuvring criteria as fuzzy variables but this suggestion did not have any impact apparently because it was unclear how to use fuzzy requirements in a document of legal nature.

Yoshimura et al. [39] analysed full-scale data for 50 ships and came to the conclusion that OS2 should be increased by 5 deg and OS3 should be defined as doubled OS2 while in fact it was fixed at 25 deg which looked too strict for ships with $T_{ref} > 12$ s. On the other hand, Rhee et al. [41] suggested easing the overshoot angles requirements for small T_{ref} having pointed out that the smaller ships are easier to operate in narrow waterways without considering, however, external disturbances. Sohn et al. [43] performed interactive simulations in a narrow waterway under the action of wind and current and discovered that in discordance with the IMO standards ships with larger T_{ref} and with the same loop width were more difficult to steer (Nishimura and Kobayashi [46] came to similar conclusions). Alternative flat standards, 20° for OS1 and 40° for OS2 independently of the T_{ref} value, were proposed in [43] on the basis of subjective assessments and data on mean rudder deflections. Another flat standard for only OS1 was proposed by Oltmann [32]: the suggested value of 17° was close to the average of the IMO standard over the whole T_{ref} interval. In theory, a lower speed must not impair either the turning ability or the directional stability in calm conditions although such effect was traced by Yasukawa et al. [47] in the case when the propeller loading was substantially reduced. However, as the reference time of the ship $T_{ref} = L/V$ at a lower operational speed V becomes larger, the standard related to the stability will be easier to meet according to the existing IMO requirements (Figures 1 and 2).

It is worth noting that the ship reference time can serve, to some extent, as an indirect measure of the degree of powering: the smaller is T_{ref}, the better powered is the ship. Hence, according to the existing IMO standards, the less powered is the ship, the less directionally stable it may be. Of course, such simple and general parameter as T_{ref} is not capable to account for all specifics reflecting the capability of this or that ship to withstand adverse conditions but in general it captures that capability fairly enough to be used in first-level standards: the smaller is T_{ref} the faster is the ship at the same length and, in general, the higher is the disposable power. All specifics related to the characteristics of the propulsion complex (different configurations can reveal very different reactions to augmentation of the resistance) should be accounted for in higher-level criteria or, even preferably, within special design and simulation procedures.

Looser directional stability requirements for larger and slower ships have always been considered beneficial for the designers as in this case good turning ability can be achieved with less effective steering devices. However, when external loads of any nature, roughly independent of the ship speed, are taken into consideration, it can be expected quite the opposite: as the proper hydrodynamic forces on the ship hull and the rudder are roughly proportional to the square of the speed, a more effective steering device must be required. In other words, if one needs to keep the resisting capacity of a ship to given external factors at a lower speed, one must stimulate installation of a more effective steering device. Stimulation can be achieved by augmenting the directional stability requirements without easing (or maybe even with tightening) those for the turning ability. This task is always feasible although not always very easy.

Possible solutions for a single-screw vessel could be:

1. Installation of sufficiently large twin rudders working outside of the propeller slipstream.
2. An azipod arrangement with enlarged skeg and/or installation of additional stabilizing fins.
3. A steering nozzle with additional stabilizers.
4. A high-lift rudder (flapped, Schilling) with additional stabilizing fins.

It is expected that all these solutions may presume augmentation of maximum deflection angles as compared to the orthodox 35 degrees. Of course, these solutions make the steering complex heavier, bulkier and more expensive but this is inevitable as a definite improvement of the steering ability has its price.

4. Accounting for Adverse Conditions and External Perturbations in Manoeuvrability Criteria

4.1. Earlier Proposals

As has already been noticed, the existing IMO manoeuvrability standards [1,33] had been devised without account for external disturbances.

The authors are not aware of any reasons for such a decision and of any related discussions. In fact, it is possible to avoid including requirements into official standards presuming that the environmental factors can be taken into account for each design individually using available prediction methods whenever it is considered necessary by the customer. In particular, this viewpoint was discussed and recognized with some hesitation as preferable by the first author 25 years ago [48].

However, the current trend is exactly to complement the existing IMO set of standards with additional environment-dependent requirements [9,30,49,50].

Even though the existing IMO standards are formulated for calm water, the importance of accounting for environmental factors had been discussed in the literature more than once. These factors, first of all, are disturbances caused by wind, sea waves and current although in a more general sense variations of the ship draught and trim, shallow water effects and hydrodynamic interaction with other borders and bodies can also be interpreted as perturbations.

Norrbin [17] underlining the importance of accounting for the wind action pointed out that "some passenger ferries have the relative windage area comparable to that of the old windjammers". Spyrou and Vassalos [51] remarked that while a ship may have good manoeuvring performance in calm sea and even keel, its behaviour may become unacceptable under external disturbances and/or in a trim-by-the-bow condition.

Lowry [52] formulated a general definition that "a vessel is controllable if the forces and moments generated by the vessel can overcome the [external] aero/hydrodynamic forces acting on the vessel". As the "forces generated by the vessel" are also hydrodynamic, this statement may look not very exact but it is clear that the author meant here disturbances from sea waves.

Li and Wu [37] also paid attention to the necessity of the introduction of criteria related to environmental factors. Koyama and Kose [18] stated that in developing manoeuvring standards the current should not be considered at all as long as strong tidal currents presumably can be avoided while the wind effects can not and must be taken into account. It was proposed to consider the wind speed 15 m/s for the course keeping at 5 kn speed of advance. A similar steering-in-the wind formulation was also discussed in [40]. Zaky and Yasukawa [30], studying the effects of the load condition for the virtual tanker KVLCC2, discovered that the wind balancing rudder angle becomes smaller in ballast condition with a stern trim than in full load in spite of a substantially larger relative windage area. Some researchers, including the first author of the present paper, see [48], tended to believe that the capacity of a ship to withstand external factors in the sense of manoeuvrability should not be standardized but, in any necessary case, must be subject to design specifications and special analysis.

Latest studies on manoeuvrability in adverse weather conditions mainly focused on second-order wave excitation effects including added resistance in waves [13]. The latter is really important in the formulation of minimum-power requirements which include the ability of the ship to keep controllability under adverse conditions characterised by the significant wave height varying from 4.5 to 6.0 m and for the wind speed from 19 to 22.6 m/s depending on the ship's length. The controllability must be demonstrated using some trusted prediction method in the sense that at some small speed, e.g., 2 knots, the ship is capable to maintain heading advancing into head seas [13]. Regarding that "trusted method," it is clear that at present the only reliable method for predicting a ship's behaviour in waves is an application of CFD codes which are, however, prohibitively slow for performing systematic calculations, especially in irregular seas. However, as will be discussed below, there is a possibility to bypass this difficulty.

The latest case of interest in manoeuvring criteria accounting for external factors is associated with expected underpowering caused by EEDI requirements while "low-powered" tankers designed to minimize the costs of construction, operation and maintenance had been already mentioned by Norrbin [17]. It still does not mean, however, that EEDI-compliant ships require some special sets of criteria and standards but they are more likely to become vulnerable to external factors and the existing standards are supposed to be revised and extended in such a way that this vulnerability is limited to acceptable levels.

Of course, any kind of hydrodynamic and aerodynamic perturbations must be somehow checked and counteracted and the effectiveness of the ship control device must be sufficient for that purpose. The hydrodynamic interaction issues (such as the ship-to-ship interaction) may be quite important per se but they are hardly associated with the underpowering as in this case all categories of forces will be simultaneously proportional to the squared ship speed.

If the minimum-power requirements lead in this or that case to the reduction of the installed power and of the ship design speed as compared to those selected without the influence of such requirements, it may happen that either the IMO standards must be somehow modified or complemented with standards based on additional criteria.

In general, the following external factors can be considered:

1. Hydrodynamic interaction with other vessels, banks, steady and floating structures and the seabed.
2. Excitation from the sea waves.
3. Aerodynamic loads caused by wind.
4. Additional forces that may be significant for certain special ships: reaction from the towline for tugs, reaction from towed acoustic arrays, trawls, sweeps and other towed devices.

It is interesting to note that all (!) these factors had been considered during the development of earlier Russian national manoeuvring standards as presented by Mastushkin [19] who led and supervised that development. As those national standards, see [8], in no way influenced the development of the IMO criteria, the issue of the external factors practically had never been discussed on the international level before studies on the STANAG standards [15]. These standards, however, were driven by a desire to account for all possible navigational situations and missions sensitive to the manoeuvring performance of naval combatants and auxiliary ships. While this approach is understandable, it is hardly rational as long as the simplicity principle is definitely neglected.

4.2. Standards Accounting for Disturbances from Sea Waves

The task of developing manoeuvrability criteria accounting for the wave action is more complex and contains more uncertainties than it could seem at first sight. It is only evident and known from the seamanship practice that less powered (slower) ships are more sensitive to the action of the sea. The scenario approach followed by some researchers [49,50] looks quite natural and suitable. However, the selection of critical scenarios is not easy. Probably, an adequate understanding of the complexity of the task of handling a ship in rough sea waves can be obtained from the following abridged quotation [53], (pp. 118–121):

> "The great wave flung the bows up, pushed the *Ulysses* far over to starboard, then passed under. The *Ulysses* staggered over the top, corkscrewed wickedly down the other side, her masts, great gleaming tree trunks thick and heavy with ice, swinging in a great arc as she rolled over, burying her port rails in the rising shoulder of the next sea.
>
> 'Full ahead port!' ... 'Starboard 30!'
>
> The next sea, passing beneath, merely straightened the *Ulysses* up. Next, at last, Nicholls understood. Incredibly, because it had been impossible to see so far ahead, Carrington had known that two opposing wave systems were due to interlock in an area of comparative calm: how he had sensed it, no one knew, would ever know, not even Carrington himself: but he was a great seaman and he had known. For 15, 20 s, the sea was a seething white mass of violently disturbed, conflicting waves—of the type usually found, on a small scale, in tidal races and overfalls—and the *Ulysses* curved gratefully through. Then another great sea, towering almost to bridge height, caught her on the far turn of the quarter circle. It struck the entire length of the *Ulysses* for the first time that night—with tremendous weight. It threw her far over on her side, the lee rails vanishing. [...]
>
> And still, the great wave had not passed. It towered high above the trough into which the *Ulysses*, now heeled far over to 40°, had been so contemptuously flung, bore down remorselessly from above and sought, in a lethal silence and with an almost animistic savagery, to press her under. The inclinometer swung relentlessly over—45°, 50°, 53° and hung there an eternity, while men stood on the side of the ship, braced with their hands on the deck, numbed minds barely grasping the inevitable. This was the end. The *Ulysses* could never come back. [...] ... the *Ulysses* shuddered, then imperceptibly, then slowly, then with vicious speed lurched back and whipped through an arc of 90°, then back again. [...] 'Slow ahead both! Midships!' 'Steady as she goes!' The *Ulysses* was round."

The quoted fragment is an artist's description of a turn in heavy seas of a cruiser participating in a WW2 arctic convoy. However, this description is based on personal experience of its author and in fact it well agrees with recommendations given in seamanship manuals, see, e.g., [54]. This indicates that the selection of ship handling scenarios and formulation of explicit criteria for manoeuvrability in sea waves may become a prohibitively difficult task with plenty of uncertainties. However, even if some reasonable scenarios are established, it will be very difficult to verify compliance of a given ship with the corresponding criteria: there is no chance to perform full-scale trials, scaled model experiments of this nature are very complex and the existing theoretical and numerical methods are not matured enough to provide reliable estimates. Of course, there are some ways to bypass those difficulties.

The first way was de facto followed in several later proposals for manoeuvrability criteria in adverse conditions and consists, first, in lifting the requirement of direct verifiability of each criterion in the spirit of the philosophy proposed in [26] and, second, in simplifying the model for the wave action, which is very complex. Considering the hull forces, it is possible to separate the first-order and second-order effects of sea waves. The first-order excitation forces are of primary importance in ship seakeeping and also are the main cause of the ship's yawing in a seaway. In regular waves, they result in ship responses with the wave encounter frequency and are roughly proportional to the wave amplitude at least when the nonlinear effects are not very pronounced. From the point of view of manoeuvrability, the most unfavourable is the case of long following waves. Possible loss of controllability in extreme seas and situations similar to that described in the quotation above are also caused by first-order effects. The second-order reactions result in regular waves in time-averaged steady forces and moments which, of course, will affect the motion of the ship in the horizontal plane. In particular, they will result in the added resistance in waves but also in the additional constant sway force and yaw moment. These forces and moment are roughly proportional to the square of the wave amplitude and reach their maxima in relatively short waves. In irregular seas the second-order excitation is more complex and, besides the steady component, results in low-frequency excitation which can cause resonant motions of moored or anchored ships.

Most latest proposals on manoeuvring criteria and standards are based on the methods predicting only second-order wave forces possibly combined with the steady current and wind action while the first-order effects are accounted for indirectly by setting some sea margin to the maximum rudder deflection angle, see, e.g., [9,11,49,50,55]. Apparently, for the power standardization problem, this approach is completely justified but it is less evident for the ship controllability in waves. In particular, the extreme situation described above cannot be captured in this way.

However, there are reasons to believe that it is possible to bypass the difficulties related to accounting for wave disturbances. Such alternative is based on two rather evident observations concerning possible outcomes of the analysis of each possible sea waves scenario:

1. The rudder effectiveness required by the scenario is superior to that conditioned by the existing standards for the turning ability and directional stability in calm water.
2. The effectiveness of the rudder required by the sea scenario is inferior to or is absorbed by the effectiveness required by the calm-water criteria.

It is clear that in the latter case any additional criterion turns out excessive and can be ignored but in the former case, a more effective rudder should be required.

Another evident observation is that such a more effective rudder will also improve the indices of the turning ability and directional stability. This means that this additional requirement can be accounted for implicitly, through appropriate tightening of the main manoeuvring standards, i.e., decreasing the allowed values of the advance, tactical diameter and zigzag overshoot angles. Does that mean that no additional separate or explicit criteria accounting for the seas and hydrodynamic interaction would then be necessary? The answer would be definitely "no" if and only if it were possible to reduce the impact of the mentioned external factors without enhancing the effectiveness of the steering device. This looks, however, highly unlikely.

Of course, the shape of the hull and especially its fullness may influence the interaction and wave excitation forces but this shape is almost entirely determined by resistance, propulsion and other design considerations. Similarly, it is practically impossible to alter the gyration radius in yaw motion through redistribution of mass along the hull. The only realistic modification consists of increasing the size of the stabilizing skeg or installation of additional stabilizers which improves the directional stability but inevitably at the expense of the turning ability. An objection may arise that, for instance, ships with different values of the block coefficient are subject to different levels of excitation loads and would require different requirements to the rudder effectiveness but this difference can be accounted for both by explicit or implicit criteria.

Hence, it can be assumed that the resistive capacity of a ship with respect to the action of hydrodynamic exogenous factors (sea waves) can only be augmented by increasing the effectiveness of the rudder which, in its turn, is guaranteed if the turning ability and the directional stability are increased simultaneously. It means that the existing standards of the turning ability and stability are to be revised and maybe tightened but no additional special explicit standards are necessary. It must be emphasized that ships with a good turning ability alone can easily show poor resistance against external factors including sea waves if these ships are not directionally stable or even stable but with insufficient stability margin. This may happen with full-bodied vessels and, as long as the external factors are ignored, can be even viewed as beneficial because a less effective small rudder is then sufficient for good turning qualities but such rudder will not be able to appropriately counteract external forces especially at low speed.

4.3. Standards for Controllability in Wind

The same principle of increasing the rudder effectiveness through tightening the calm-water standards could be also applied to the aerodynamic (wind) factors but here the situation looks somewhat different because, contrary to hydrodynamic loads, it is possible to define several very evident and significant parameters affecting the level of sensitivity of a ship to wind. Since long ago, it has been established with certainty, see, e.g., [12,56–58], that the manoeuvring sensitivity to the wind depends on the following four fundamental dimensionless parameters: the relative density of the air $\overline{\rho}_A = \rho_A/\rho$, where ρ_A is the air density and ρ is the water density; the relative lateral windage area $A'_L = A_L/(LT)$, where A_L is the lateral windage area, L is the ship waterline length, T is the ship draught; the relative longitudinal position of the centroid of A_L defined as $x'_A = x_A/L_{OA}$, where x_A is the abscissa of the centroid and L_{OA} is the length overall; and the relative wind speed $\overline{V}_w = V_w/V$, where V_w is the absolute wind speed and V is the actual speed of the ship with respect to water. The air relative density is practically constant and its variability can be ignored while the remaining parameters can vary in a wide range and are extremely important as they influence the mentioned sensitivity substantially. This fact is quite transparent intuitively and well familiar to all seafarers. In particular, it has always been known that slow vessels with a large windage area and small draught are especially difficult for handling in wind and even the controllability can be lost completely.

It seems that contrary to the criteria driven by sea waves excitation it is easier to introduce a separate standard based on the criterion of controllability in the wind than to establish some additional rules for tightening calm-water criteria, which must capture dependence on the windage area parameters mentioned above. In fact, two additional wind-controllability requirements were developed by Mastushkin [19] and these are up to present embedded into the Rules of the Russian Maritime Register of Shipping [8], see also [12,16]. Multiple test calculations performed with more than 100 ships showed that after appropriate adjustment this criterion became non-trivial: some existing ships that did not satisfy it did really suffer from controllability problems in wind. Of course, for the majority of ships that criterion looked excessive, as the requirements of turning ability/stability in calm water were for them more strict. A similar situation will be with any consistent wind criterion. In particular, it will be certainly redundant for ships with high draught and low windage area: for such ships, second-order wave loads may often exceed aerodynamic ones. On the other hand, many ships in ballast condition, passenger ships and ferries will be more prone to aerodynamic loads than to any other type of excitation and then that additional requirement turns out significant. For instance, it was found by Zaky and Yasukawa [30] that while the second-order wave forces for a typical tanker in full load exceeded by 50% the wind forces, for the same ship in ballast the wave force was 10% lower than the wind force.

4.4. Modification of Calm-Water Standards

4.4.1. Simplified Approach

Recognizing that simultaneous improvement of the turning ability and directional stability definitely improves the ship's capability to counteract external factors, it can be sufficient to modify the required (standardizing) values of the turning ability and directional stability criteria. Regarding the turning ability, as a zero approximation the existing values (5 ship lengths for the tactical diameter and 4.5 lengths for the advance, both independent of the reference time) can be conserved or somewhat tightened regarding that they are satisfied with the majority of existing ships with a tangible margin [29]. However, the values of the measures of directional stability apparently must be revised. Given the previous discussion in Section 3, it goes about the dependence of the width of the hysteresis loop or of the zigzag 1st overshoot on the ship reference time T_{ref}. Again, as a zero approximation, it could be recommended to eliminate this dependence by introducing flat standards, see Figures 1 and 2. This modification seems quite appropriate as most existing ships already satisfy the thus modified standard: a sampling of existing vessels presented in [40] demonstrated compliance with the "flat" standard for most ships and did not show definite dependence on T_{ref} while the scatter is substantial.

It is remarkable that such deep revision and tightening of IMO standards apparently would not result in any serious problems regarding compliance difficulties as multiple sources (see [40,59–61] and Figure 2 where full-scale data from [40] are plotted as circle symbols) indicate that while the scatter of full-scale data is substantial, the dependence of the overshoot angles and/or loop width on T_{ref} is weak and often even absent. In the majority of cases, the standards are satisfied with a considerable margin. Exceptions are mainly associated with very specific vessels, such as hopper dredgers [62] but these hardly can be treated in a common way.

While Gong et al. [59] detected a larger scatter of trials data at low Froude numbers, the data in the cited source indicate that OS1 rarely exceeds 10–11 degrees. The same authors have also found by means of interactive simulation that reduction of OS1 from 13 to, say, 8 degrees with simultaneous reduction of the tactical diameter from 4.3 to 3.7 lengths definitely resulted in safer and more accurate steering.

Even though such modifications of the IMO standards would be, most likely, beneficial even in this form, it would be hardly the best possible solution.

4.4.2. Improved Modifications

Before an extended numerical study, see the next Subsubsection, is undertaken some heuristic improvements can be made. In particular, the same value of the reference time can correspond to larger and faster vessels or to smaller and slower ones. It is, however, evident that in the first case sensitivity to waves at the same sea state will be weaker, which means that the directional stability requirement can be eased (but not necessarily!) for larger ships and probably the ship length can then be introduced as additional parameter although for each L the requirement may remain flat as far as it is stricter than the existing IMO standard. This is illustrated by additional dash-dot lines in Figures 1 and 2 where it is supposed that the base requirement described by the lowest dash-dot line must be applied for all ships with the length not exceeding some value L_1, which must be specified, while the two remaining dash-dot lines show possible looser requirements for some larger ship lengths L_2 and L_3. These eased requirements can be viewed as a compromise between the existing and completely "flat" standards. The flatness of the standard, at least above some value of the reference time, makes less critical selection of the speed V: the IMO standards define it as 0.9 of the speed corresponding to 0.85 of the maximum continuous rating of the engine(s) and the same definition can be kept although the actual reachable speed in waves may be lower than that.

4.4.3. Supporting Studies

Regarding a better account for sea waves, development of more sophisticated and better substantiated requirements to the measures of directional stability, multiple simulations in waves must be performed for a set of generic ship mathematical models with varying indices of the turning ability and inherent directional stability must be performed and followed by appropriate analysis. Moreover, any representative set of mathematical models must embrace not only various values of the shaft power but also different types of propulsors (fixed- or controllable-pitch propellers) and various typical engine characteristics as, e.g., the loss of actual power will be very different for a DC electric motor and a common turbocharged 2-stroke Diesel engine. It does not mean that the resulting standards must be differentiated for all possible configurations but differences in behaviour must be captured and analysed. Such numerical study must be, however, preceded by the development of a sufficiently accurate, fast and commonly recognized program for simulating the arbitrary motion of a ship in waves. Existing codes of this kind can hardly be applied as they suffer from various limitations and uncertainties. Application of Reynolds-Averaged Navier–Stokes Equations (RANSE) CFD codes or model tests looks unrealistic because of unacceptable time requirements: at least real-time or even accelerated-time simulations must be feasible. Unfortunately, no one of faster existing available codes can guarantee capturing all substantial effects accompanying manoeuvring in waves. For instance, one of the latest and relatively consistent developments [63] is fully linear (except for the account for 2nd-order excitation forces) and neglects many coupling effects, let alone variations of the hydrodynamic characteristics of propellers and rudders and interaction coefficients. The 2nd-order wave forces combined with the linear seakeeping model were used by Chillece and el Moctar [55] and by Kobayashi [64]. Only second-order forces were modelled for a tanker by Zaky and Yasukawa [30] who combined the sea with the wind action.

4.4.4. Design Implications for Steering Arrangement

Even though most ships apparently will satisfy even modified and tighter manoeuvring criteria, situations in which design decisions aimed at strengthening the ship's controllability are required may happen and then various solutions are possible. It is important to emphasize one more time that the directional stability must be augmented together (concurrently) with the turning ability of the ship design in consideration as only in this case the ship capacity to resist adverse factors will be improved. Augmentation of the rudder area is the most evident and straightforward solution. Simulations under wave and wind disturbances performed by Ohtagaki and Tanaka [44] demonstrated that the mean course deviation could be reduced from approximately 10 to 5 degrees with simultaneous reduction of the square-root mean rudder deflections from 13 to 8 degrees when the rudder relative area was increased from 1.2% to 2.5%. Norrbin [17] suggested using skegs or fixed fins to improve the directional stability. Haraguchi and Nimura [42] demonstrated direct reduction of the OS1 and OS2 overshoot angles by the factor of 3 and reduction of the OS3 angle by the factor of 2 when the relative rudder area was increased from 1% to 1.5%.

It must be said that in a rather typical situation of a normal rudder working in the propeller slipstream augmentation of the rudder area will definitely increase the turning ability while improvement of the directional stability may become much less pronounced. In this case, simultaneous augmentation of the rudder and the skeg areas may be recommended [38]. Similar solutions can be very promising in the case of pod-driven ships [42,65]. However, it makes sense to keep in mind the natural improvement of controllability in some defining loading conditions. For instance, Zaky and Yasukawa [30] obtained that a typical ballast condition often results in simultaneous growth of the turning ability (due to larger relative rudder area) and of the directional stability (due to trim by the stern). Finally, the so-called high-lift rudders and other steering devices can be applied. For instance, Yamada [44] indicated that the application of the high-lift Schilling rudder permits considerable reduction of the OS2 angle. Eda and Numano [66] demonstrated the effectiveness of twin flap rudders.

5. Manoeuvrability Criteria in Wind: Theoretical Support

5.1. Ship Wind Resisting Capacity Criterion and Standardization Scheme

5.1.1. Preliminary Comments

As was discussed above, although in theory the approach based on adjustment of the existing IMO criteria can also be applied to account for wind perturbations, the special nature of aerodynamic exogenous factors and, in particular, the existence of few well-defined parameters characterizing aerodynamic peculiarities of ships makes reasonable the introduction of an additional criterion. Even though a straightforward approach has inclined many researchers to require direct demonstration of the capability of a ship to perform arbitrary manoeuvres in a given wind [67], since long ago it has been understood that the ship remains controllable in wind of some specified velocity if it is capable to maintain any required course in straight run [56]. It was also demonstrated with the help of direct computations [58] that the ability to maintain an arbitrary course is equivalent to the ability of maintaining an arbitrary heading or arbitrary air drift angle. This scenario had been used by Mastushkin [19] for developing wind controllability criteria implicitly contained in the Rules [8], see also their concise description in [12]. Those criteria were applied using some simplified manoeuvring models under the assumption of the presence of constant wind with a certain specified speed. In general, the scheme turned out reasonable and practical although certain additional simplifications, such as the assumption that the rudder effectiveness is unconditionally higher for rudders with a higher aspect ratio, somewhat impair the quality and consistency of the standard. The approach followed here is similar in some respects but the authors tried to avoid evident drawbacks of earlier suggestions and to leave more flexibility in the standardizing scheme. Any wind controllability standard is supposed to be applied indiscriminately to all ships but in the case of underpowered ships, the corresponding criterion is more likely to become critical, i.e., leading to augmentation of the rudder effectiveness. The following material assembles practically all theoretical elements necessary for devising a rational and relatively simple standard for ship resistance to wind action, but the corresponding standard itself presumes specification of the wind speed value through a statistical adjustment to some representative subset of the existing fleet of vessels is left beyond the scope of the present article.

5.1.2. Formulation and Main Relations

The gustiness of the wind can be neglected in manoeuvring problems as the ship in the horizontal plane represents an aperiodic system with relatively large time lags and possessing no resonant frequencies. Similarly, self-sustained oscillations of the loads observed on non-steamlined bodies even in a sready flow will not have any significant effect. Assuming also that the wind-induced roll does not affect significantly the behaviour of the ship in the horizontal plane, the most general modular mathematical model for the steady ship motion in wind is described by the following set of nonlinear algebraic equations:

$$\begin{aligned} &X_H(u,v) + X_P(u,v,n) + X_R(u,v,n,\delta_R) + X_A(u_A,v_A) = 0, \\ &Y_H(u,v) + Y_R(u,v,n,\delta_R) + Y_A(u_A,v_A) = 0, \\ &N_H(u,v) + N_R(u,v,n,\delta_R) + N_A(u_A,v_A) = 0, \\ &Q_P(u,v,n) + Q_E(n^*,n) = 0, \end{aligned} \tag{1}$$

where X, Y, N, Q stand for the surge force, sway force, yaw moment and the shaft torque, respectively; the subscripts H, P, R, A, E correspond to the hull, propeller, rudder, air (or aerodynamic) and engine, respectively; u and v are the velocities of surge and sway; n is the actual propeller rotation frequency (rps), and n^* is the ordered rotation frequency.

The kinematic parameters present in the set (1) are most convenient for the description of involved forces and are related to alternative parameters, which may be more suitable for human perception:

$$\begin{aligned} &u = V\cos\beta,\ v = -V\sin\beta, \\ &u_A = V_A\cos\beta_A,\ v_A = -V_A\sin\beta_A, \end{aligned} \tag{2}$$

where $V = \sqrt{u^2+v^2}$ is the ship speed relative to the water, β is the drift angle, $V_A = \sqrt{u_A^2+v_A^2}$ is the airspeed of the ship or the relative wind speed and β_A is the air drift angle or the apparent wind angle.

In the case of absence of current, the following relations are valid:

$$\begin{aligned} u_A &= u + V_w \cos(\chi_w - \psi), \\ v_A &= v - V_w \sin(\chi_w - \psi), \end{aligned} \tag{3}$$

where V_w is the absolute wind speed, χ_w is the absolute wind direction angle, ψ is the ship heading angle.

The aerodynamic surge force X_A, sway force Y_A and yaw moment N_A are represented in the standard way:

$$X_A = C_{XA}(\beta_A)\frac{\rho V_A^2}{2}A_T,\ Y_A = C_{YA}(\beta_A)\frac{\rho V_A^2}{2}A_L,\ N_A = C_{NA}(\beta_A)\frac{\rho V_A^2}{2}A_L L_{OA}, \tag{4}$$

where C_{XA}, C_{YA}, C_{NA} are the force/moment coefficients, A_T is the transverse projection of the above-water part of the hull, A_L is its longitudinal projection and L_{OA} is the length overall.

The dependencies of the forces and moments in (1) are closing the ship mathematical model and can be specified in various ways. In general, these dependencies can be very complicated and highly nonlinear. So far, the only simplifying assumption was that the propeller sway force and yaw moment could be neglected. A comprehensive nonlinear mathematical model is suitable and even desirable for simulating the ship manoeuvring motion in wind but seems to be excessively complicated for standardizing purposes. The following simplification seems to be adequate for that task: the speed of the ship with respect to water V is fixed and specified. In this case, the equations for the surge forces and shaft torques can be separated from the remaining two and are often ignored.

The assumption about the fixed speed V is sometimes doubted as in reality it also depends on the wind and in more detailed mathematical models used in simulation systems, it is continuously estimated and depends not only on the relative wind but also on the sea state, type and limiting characteristics of the main engine, type of the propulsor and on the throttle settings while the latter may be conditioned by voluntary reduction of speed to avoid resonances, reduce slamming and green water on deck. However, in simplified models oriented at the definition of manoeuvring criteria not only the assumption of a fixed ship speed is acceptable but the specific value of this speed is less important than it could seem because what matters is its magnitude relative to the wind speed V_w and the standardizing value for the latter is supposed to be adjusted to the existing fleet.

Most empiric methods use linearized or relatively easily linearizable rudder models [56,68,69] which is sufficiently appropriate for a standardizing problem. Then the sway–yaw equilibrium equations can be represented as

$$\begin{aligned} &Y_H'(v')\cdot\tfrac{1}{2}\rho V^2 LT - C_{RN}^{\alpha}\cdot\tfrac{1}{2}\rho V_R^2 A_R \alpha_R \cos\delta_R + C_{AY}(\beta_A)\cdot\tfrac{1}{2}\rho_A V_A^2 A_L = 0, \\ &N_H'(v')\cdot\tfrac{1}{2}\rho V^2 L^2 T - C_{RN}^{\alpha}\cdot\tfrac{1}{2}\rho V_R^2 A_R \alpha_R x_R \cos\delta_R + C_{AN}(\beta_A)\cdot\tfrac{1}{2}\rho_A V_A^2 A_L L_{OA} = 0, \end{aligned} \tag{5}$$

where Y_H' and N_H' are the hull sway force and yaw moment coefficients, $v' = v/V$ is the dimensionless sway velocity, V_R is the magnitude of the rudder velocity with respect to water, A_R is the rudder area, $\alpha_R = \delta_R - \kappa\beta$ is the rudder attack angle, κ is the flow straightening factor, x_R is effective rudder abscissa in the body frame, ρ and ρ_A are the water and air density, respectively. It is assumed here that the linear rudder model is characterized by the normal force coefficient gradient C_{RN}^{α}. As the minimum value of the function $\cos\delta_R$ is 0.85 in most cases, this factor can be dropped. In addition, it can be assumed that $L_{OA} \approx L$.

The set (5) can then be represented in the following non-dimensional form:

$$\begin{aligned} &Y_H'(v') - C_{RN}^{\alpha}\overline{V}_R^2 A_R'(\delta_R + \kappa v') + C_{AY}(\beta_A)\overline{\rho}_A \overline{V}_A^2 A_L' = 0, \\ &N_H'(v') - C_{RN}^{\alpha}\overline{V}_R^2 A_R'(\delta_R + \kappa v')x_R' + C_{AN}(\beta_A)\overline{\rho}_A \overline{V}_A^2 A_L' = 0, \end{aligned} \tag{6}$$

where $\overline{V}_R = V_R/V$, $A'_R = A_R/(LT)$ is the relative rudder area, $\overline{\rho}_A = \rho_A/\rho$, $\overline{V}_A = V_A/V$ and $A'_L = A_L/(LT)$.

The difference between V_R and V is caused by the influence of the wake behind the hull and by the influence of the propeller slipstream. The total effect can be positive (more typical for merchant ships) or negative depending on which part of the rudder is working in the propeller race and on the configuration of the stern. The straightening coefficient κ depends on the same factors but never exceeds 1.0.

The ideal propulsor model typically used by empiric methods, see e.g., [69,70], for modelling the influence of the propeller slipstream on the rudder will yield:

$$\begin{aligned} \overline{V}_R^2 &= \overline{A}_{R0} + \overline{A}_{RP}(1 + C_{TA}), \\ \kappa &= \kappa_H \frac{\overline{A}_{R0} + \overline{A}_{RP}\sqrt{1 + C_{TA}}}{\overline{A}_{R0} + \overline{A}_{RP}(1 + C_{TA})}, \end{aligned} \tag{7}$$

where $\overline{A}_{R0} = A_{R0}/A_R$, A_{R0} and A_{RP} are the parts of the rudder area A_R outside and inside the slipstream, respectively, κ_H is the hull straightening factor varying from 0.3 to 1.0, C_{TA} is the propeller loading coefficient defined as:

$$C_{TA} = \frac{8T_P}{\rho \pi D_P^2 V^2}, \tag{8}$$

where T_P is the propeller thrust and D_P is the propeller diameter.

The hull straightening factor is assumed following the general recommendations [57,69] or it can be adjusted. The propeller thrust and the ship speed in (8) should correspond to the actual ship motion in wind and be obtained from the surge equation but, in first approximation and for standardizing purpose, it is possible to assume that the thrust corresponds to the free run with the speed V, which is then a free parameter.

As the drift angle of the ship advancing along a straight path in wind practically never exceeds 30 degrees, polynomial approximations for the functions $Y'_H(v')$ and $N'_H(v')$ will be adequate. It makes sense to apply the simplest approximations used in the Pershitz empiric method [12,56]:

$$\begin{aligned} Y'_H(v') &= Y'_v v' + Y'_{vv} v' |v'|, \\ N'_H(v') &= N'_v v', \end{aligned} \tag{9}$$

where the coefficients ("hydrodynamic derivatives") can be estimated using the procedure described in [57,69]. This model exploits the fact that the drift angle nonlinearity is much weaker for the yaw moment than for the sway force and is the nonlinear model of minimum complexity. A fully linearized model is also possible and was used in [58] but, as the substantially more accurate model (9) still permits analytical solutions, its application may be preferable. If the primary mathematical model for the sway force and yaw moment has a different structure, it always can be reasonably approximated with (9). For instance, suppose that the coefficients of a very common cubic model

$$\begin{aligned} Y'_{H3}(v') &= Y'_{v3} v' + Y'_{vvv} v'^3, \\ N'_{H3}(v') &= N'_{v3} v' + N'_{vvv} v'^3, \end{aligned} \tag{10}$$

are known. Then, requiring direct matching at the values of the dimensionless sway velocity v'_1 and v'_2 (for instance, they may correspond to 10 deg and 20 deg values of the drift angle) for the sway force and only at v'_2. it is possible to obtain the approximating formulae:

$$\begin{aligned} Y'_v &= Y'_{v3} - Y'_{vvv} v'_1 v'_2, \\ Y'_{vv} &= Y'_{vvv} (v'_1 + v'_2), \\ N'_v &= N'_{v3} + N'_{vvv} v'^2_2. \end{aligned} \tag{11}$$

Of course, the approximation can be performed in different ways including the application of the least-square method.

Substituting (9) into (6) and performing obvious transformations it is possible to bring the equilibrium equations to the following form:

$$\begin{aligned} C_Y^v v' + Y'_{vv} v'|v'| + C_Y^\delta \delta_R &= f_Y(\overline{V}_A, \beta_A), \\ C_N^v v' \qquad\qquad\quad + C_N^\delta \delta_R &= f_N(\overline{V}_A, \beta_A), \end{aligned} \tag{12}$$

where

$$\begin{aligned} C_Y^v &= Y'_v - \kappa E_R, & C_Y^\delta &= -E_R, & f_Y(\,) &= -K_{AY}(\beta_A)\overline{V}_{AL}^2, \\ C_N^v &= N'_v - \kappa E_R x'_R, & C_N^\delta &= -E_R x'_R, & f_N(\,) &= -K_{AN}(\beta_A)\overline{V}_{AL}^2, \end{aligned} \tag{13}$$

where

$$E_R = C_{Rn}^\alpha \overline{V}_R^2 A'_R;\; K_{AY}(\beta_A) = C_{AY}(\beta_A)\overline{\rho}_A A'_L;\; K_{AN}(\beta_A) = C_{AN}(\beta_A)\overline{\rho}_A A'_L, \tag{14}$$

and where E_R has the meaning of the rudder effectiveness index.

5.1.3. Useful Analytical Solutions

Standard analysis based on the Equation (12) presumes computation of equilibrium values of v' (or β) and δ_R as functions of β_A at some fixed value of $\overline{V}_A$. The solution always exists for this model but when the values of the parameters A'_L and $\overline{V}_A$ are large enough, the required equilibrium rudder deflection angle can exceed the maximum possible value δ_m within a certain interval of the air drift angle β_A and this is then interpreted as a loss of controllability in wind. This analysis can be complemented by an investigation of the inherent local stability of attainable equilibrium regimes, see, e.g., [71–74] and such analysis is of considerable theoretical interest. However, it is not relevant for estimation of controllability in the sense formulated above as the steered ship represents a closed-loop system and can be stabilized on any reachable course with the helmsman or autopilot. Necessary allowance for the rudder oscillatory motions (sea margin) is accounted for in setting the value of δ_m which is assumed to be smaller by 5–7 degrees than the maximum deflection angle of the rudder.

The set (12) can be solved analytically. First, physical considerations and computations performed with the linearized model show that at positive β_A the drift angle β must be negative, which corresponds to $v' > 0$ which makes possible substituting $v'|v'|$ with v'^2 as, due to symmetry, it is sufficient to consider only non-negative values of the air drift angle. Then, expressing the rudder angle from the second equation and substituting it into the first one the following quadric equation is obtained:

$$Y'_{vv} v'^2 + (C_Y^v - C_N^v / x'_R) v' + f_N / x'_R - f_Y = 0. \tag{15}$$

The assumption that always $x'_R = -\frac{1}{2}$ will not lead to substantial errors and is justified in approximate analysis. Then, the non-negative solution to this equation is given by the formulae:

$$\begin{aligned} v' &= \frac{-2C_N^v - C_Y^v - \sqrt{(C_Y^v + 2C_N^v)^2 + Y'_{vv}[f_Y(\beta_A) + 2f_N(\beta_A)]}}{2Y'_{vv}}, \\ \delta_R &= [f_N(\beta_A) - C_N^v v'] / C_N^\delta. \end{aligned} \tag{16}$$

These formulae should be applied for a set of values of $\beta_A \in [0, \pi]$ to check whether the controllability can be lost or not. However, the equilibrium Equation (12) can also be interpreted in other ways.

First, assuming that the ship sails at the controllability limit, i.e., $\delta_R = \delta_m$ and that the value of $\overline{V}_A$ is fixed, it will be possible to solve the set (12) with respect to v' and β_A. However, due to nonlinear dependencies $C_{AY}(\beta_A)$ and $C_{AN}(\beta_A)$ this can only be performed numerically and it is clear that the solution does not exist if at a given $\overline{V}_A$ the ship is controllable with some margin.

In addition, under the same assumption $\delta_R = \delta_m$ it is possible to eliminate v' and to obtain a single equivalent equation which can be represented in one of the following two forms:

$$C_{VVVV}\overline{V}_A^4 + C_{VV}\overline{V}_A^2 + C_{V0} = 0 \tag{17}$$

or,

$$C_{EE}E_R^2 + C_E E_R + C_{R0} = 0, \tag{18}$$

where,

$$\begin{aligned}
&C_{VVVV} = Y'_{vv}K_{AN}^2,\ C_{VV} = A_{VVEE}E_R^2 + A_{VVE}E_R + A_{VV},\ C_{V0} = A_{EE}E_R^2 + A_E E_R,\\
&C_{EE} = A_{VVEE}\overline{V}_A^2 + A_{EE},\ C_E = A_{VVE}\overline{V}_A^2 + A_E,\ C_{R0} = C_{VVVV}\overline{V}_A^4 + A_{VV}\overline{V}_A^2,\\
&A_{VVEE} = \kappa^2 x'_R(K_{AY}x'_R - K_{AN}),\\
&A_{VVE} = \kappa K_{AN}(N'_v + Y'_v x'_R) - 2x'_R(Y'_{vv}K_{AN}\delta_m + \kappa N'_v K_{AY}),\\
&A_{VV} = N'_v(K_{AY}N'_v - K_{AN}Y'_v),\ A_{EE} = x'_R\delta_m(Y'_{vv}x'_R\delta_m - \kappa Y'_v x'_R + \kappa N'_v),\\
&A_E = N'_v\delta_m(Y'_v x'_R - N'_v).
\end{aligned} \tag{19}$$

The bi-quadric Equation (17) can be easily solved analytically with respect to the relative critical airspeed $\overline{V}_A(\beta_A)$, which corresponds to the maximum ship air speed at which no loss of controllability occurs for the current ship configuration. It is often assumed [19] that the maximum sensitivity to the wind is observed at $\beta_A \approx \frac{3}{4}\pi$. However, while on average this assumption is justified, the most "dangerous" air drift angle can be substantially different and it makes sense to search for $\beta_A = \text{argmin}\overline{V}_A$ which is rather simple in the one-dimensional problem and can be performed, e.g., by means of a dichotomy algorithm. Once the critical values of V_A and β_A are determined, it is possible to determine the minimum value of the absolute wind speed V_w at which controllability can be lost.

A solution to the quadric Equation (18) at fixed values of $\overline{V}_A$ and β_A will provide the minimum required rudder effectiveness index E_R which is always positive. Once the required value of E_R is fixed, the designer can adjust the actual rudder effectiveness mainly through variating the effective rudder area A'_{RE} or the rudder normal force gradient C_{RN}^{α}. The latter can be increased by increasing the rudder aspect ratio but this must be performed with care as the higher is the aspect ratio, the lower is the stall angle and this highly undesirable stall may then occur at moderately large rudder attack angles.

To close the schemes described above it is necessary to define the dimensionless aerodynamic characteristics of the ship hull $C_{XA}(\beta_A), C_{YA}(\beta_A), C_{NA}(\beta_A)$.

5.2. Aerodynamic Characteristics of Ships for Standardizing Purposes

While ship mathematical models used in ship handling simulators require the most accurate prediction of aerodynamic loads to obtain a more realistic reaction of the model to the simulated wind, the logic of developing manoeuvring standards often permits some conservative approximate estimates which somewhat facilitates the task. In this section, methods for estimating the aerodynamic characteristic of ships are mainly discussed from this viewpoint.

In contrast with the submerged part of the hull of a surface displacement ship, the above-water part is rarely streamlined and very often possesses a rather complicated, edged and peculiar shape especially bearing in mind that not only it includes the hull per se but also superstructures, deckhouses and, possibly, other structures and equipment.

As result, during several decades the aerodynamic characteristics of ships were mainly determined using wind-tunnel tests although lately application of CFD methods is gaining popularity and it is recognized that the credibility of physical and numerical modelling has become comparable [75].

Fortunately, the dependence of the aerodynamic characteristics on the angular velocity of yaw can be neglected and, besides the Reynolds number, the aerodynamic force coefficients will only depend on the air drift (or attack, or sideslip) angle. As most above-water shapes are edged, the separation lines will be fixed and dependence on the Reynolds number is then insignificant. On the other hand, the following circumstances lead to various complications and uncertainties:

1. The natural wind over the sea surface is always more or less gusty which is difficult to reproduce in physical or numerical modelling. However, the frequency of fluctuations either is too high to trigger a more or less significant response of the vessel or they can be handled on the basis of the quasi-steadiness principle. Moreover, while in the classic intact stability weather criterion the gustiness factor is accounted for as the duration of the gust can be comparable with the natural period of roll, the filtering property of the ship in the horizontal plane is much superior as the ship represents then an aperiodic system without any resonant frequencies, the responses will be weaker and the gusts are typically neglected. Of course, in any realistic simulation, at least relatively low-frequency gusts must be modelled and even the quasi-steady part of the load must be complemented with the inertial one [75]. This can be performed using various wind spectra such as those suggested by Davenport, and Harris and Kaimal for gusts along the main wind direction and the spectrum proposed by Lumly and Panofsky for the transverse components, see [75]. However, as mentioned above, the account for realistic air velocity fluctuations is hardly important for manoeuvring standardizing purposes when just some constant wind velocity can be specified.
2. When a typical above-water shape is subject to an absolutely steady and uniform flow, the resulting perturbed flow will not remain steady and the observed loads will not be constant because of unsteady vortex shedding. However, the resulting fluctuations are always of relatively high frequency and will not affect the ship motion. In wind-tunnel experiments and in the CFD modelling, the observed time series are filtered and averaged resulting in average steady values of the forces.
3. The full-scale Reynolds number cannot be reached in atmospheric wind tunnels used in ship hydrodynamics applications. On edged forms typical for ship superstructures the scale effect is negligible what was directly confirmed by comparative CFD computations performed at model and full-scale values of the Reynolds number. However, if a ship contains substantial non-edged elements similar to cylinders or spheres, application of appropriately placed turbulence stimulators is recommended. In general, the results of well performed wind-tunnel tests can be taken with sufficient credibility and applied for estimating loads on full-scale ships.
4. When the wind blows over the sea surface, a boundary layer is formed resulting in a sheared flow with the velocity distribution typically following the 1/7 power law [75]. If the wind tunnel tests are carried out using a ground plate, the situation is similar but the relative thickness of the boundary layer is different. Typically, the wind-tunnel data are brought to the uniform flow and the apparent wind in simulations is also supposed to be uniform with the equivalent speed corresponding to the 10 m height above the sea level. In most cases, this will yield somewhat conservative estimates that seem to be beneficial for standardizing purposes.
5. The apparent wind is composed of the true wind and and of that originating from the motion of the ship with respect to the still air. In the second case, the flow is not sheared and the resulting relative flow can have different directions on various heights forming the so-called "twisted wind". This effect practically cannot be reproduced in the wind tunnel but can be modelled with CFD. However, as long as so far no data on this phenomenon were reported, the common practice is to neglect the effect.

It can be assumed, that as long as results of appropriate wind-tunnel tests or CFD computations for a given configuration are available, they can be effectively applied for simulating the manoeuvring motion of the corresponding ship in wind or for checking compliance with some wind related criteria. The experimental option is nowadays relatively costly not because of wind-tunnel tests per se, which are in this case simple and not excessively time-consuming, but because of the necessity of manufacturing required scaled models although maybe rapid development of the 3D printing will reduce this cost considerably.

If the above-water configuration of a ship is close to one of the models tested before and the corresponding aerodynamic coefficients are available, these can be used with a rather high degree of certainty. However, the prototype approach is not always possible and several attempts to devise "universal" methods were undertaken. For instance, the method developed by Melkozerova and Pershitz, see [57], provides approximations for the sway force and moment coefficients for single-island and twin-island superstructures whose relative lengths and positions of the centroids serve as defining factors.

Later two alternative empiric methods were developed on the basis of systematic wind-tunnel tests: the method of parametric loading functions developed by Blendermann and the method of trigonometric series. Both methods are described in detail by Blendermann [75] and are only outlined below.

The method of loading functions is a specific approximation of tabulated aerodynamic characteristics performed over 15 ship types and allowing corrections for the longitudinal position of the centroid of the lateral windage area. The method of trigonometric series was developed by Fujiwara and it presumes approximation of the aerodynamic coefficients with 3 or 4-terms trigonometric polynomials over the whole range of the sideslip angle. The coefficients of these trigonometric polynomials are algebraic polynomial regressions on a set of geometric parameters. Specifics of the latter is the separate treatment of the geometry of the above-water hull per se and of the superstructures. Recently, an interesting and promising method based on the combination of elliptic Fourier descriptors with artificial neural networks was proposed [76] but still not brought to the practical tool level.

The methods outlined above are more suitable for simulation applications requiring accurate predictions of aerodynamic loads. As was mentioned earlier, standardization tasks can be often effectively handled using some approximate or upper-end conservative estimates and a simple method aiming at possessing such properties is proposed below.

A simple generic model for the surge and sway force coefficients is based on the evident fact that all loads are 2π-periodic. Then, accounting for the evident symmetry properties and retaining only the first non-zero terms in the Fourier expansion of the coefficients it is obtained:

$$C_{AX} = -C_{AX0} \cos \beta_A,\ C_{AY} = C_{AY0} \sin \beta_{A,} \quad (20)$$

where C_{AX0} and C_{AY0} are some positive constant parameters representing the magnitude of the aerodynamic force in head/stern or beam apparent wind, respectively.

The average recommended values of these parameters are [57]:

$$C_{AX0} = 1.0,\ C_{AY0} = 1.05. \quad (21)$$

Of course, these values are approximate, suggested many decades ago and reflect the averaged properties of the then existing hull shapes. In addition, an alternative and even more obsolete value $C_{AY0} = 1.2$ is mentioned in the book [56].

To estimate possible uncertainties associated with approximations (20), it makes sense to check them against recent experimental data. The richest available database of this kind is that collected by Blendermann on the basis of his own wind tunnel tests. Blenedermann's data can be found in 3 sources: in the reference book [77], in the report [78] and the manual [75]. The cited report describes the most complete database containing data for 48 configurations (in some cases two different configurations correspond to the same ship but at different loading conditions). A thorough comparison carried out by the authors showed that 21 of these configurations were also presented in [77], and 21—in [75]. All data from the report [78] were checked against alternative sources and digitized by the authors using the available tables of values. The book [77] contained one additional configuration not present in the report and it was decided to ignore it.

The experimental data for C_{AX} are plotted as symbols in Figure 3 where also the response provided by (20) is shown as a dashed line. It is evident that this simple model provides a fair averaged estimation for the set of empiric data but in many cases, the magnitude of the surge force is substantially underestimated. To provide more conservative estimates, the following two-term equation can be proposed:

$$C_{AX} = -C^{M}_{AX0} \cos \beta_A + C^{5}_{AX0} \cos 5\beta_A \tag{22}$$

with $C^{M}_{AX0} = 1.35$ and $C^{5}_{AX0} = 0.2$, and the corresponding response is shown in Figure 3 with a solid line. It is obvious that the formula (22) does really provide an upper limit for the absolute value of the surge force for most configurations, The experimental points more or less significantly dropping outside the envelope at air drift angles around 30 and 150 degrees correspond not exactly to a ship but to an empty floating dock with rather high walls but with a relatively small transverse section area. So, Equation (22) with the indicated values of the constant parameters can be recommended if really conservative estimates are desirable. A similar analysis was also performed for the sway force coefficient in Figure 4 where the dashed line represents the response provided by the Equation (20) and the solid line—by the following model:

$$C_{AY} = C^{M}_{AY0} \sin \beta_A + C^{3}_{AY} \sin 3\beta_A \tag{23}$$

with $C^{M}_{AY0} = 1.17$ and $C^{3}_{AY} = 0.15$.

The latter formula produces a rather clear envelope for almost all experimental data except those corresponding to the same floating dock and, at $\beta_A = 60^\circ$, to a car carrier with an exceptionally high freeboard.

Analysis of data for the yaw moment coefficient turns out somewhat more complicated as an additional parameter comes into play. The generic model for the yaw moment coefficient is [57]:

$$C_{AN}(\beta_A) = C_{AY}(\beta_A) \cdot x'_A(\beta_A), \tag{24}$$

where x'_A is the relative lever of the sway force which, in its turn, is represented as

$$x'_A(\beta_A) = x'_w + \Delta x'_A(|\beta_A|), \tag{25}$$

where $x'_w = x_w / L_{OA}$ is the relative abscissa of the centroid of the lateral windage area and the function $\Delta x'_A()$ accounts for the displacement of the sway force application point and can be approximated as

$$\Delta x'_A = k_x \left(\frac{1}{4} - \frac{|\beta_A|}{2\pi} \right), \tag{26}$$

which is generalization of the equation recommended in [56,57] where it is assumed that $k_x \equiv 1.0$.

Blendermann's data on the sway force and yaw moment coefficients were used to generate experimental values of $\Delta x'_A$ using the evident relation:

$$\Delta x'_A = C_{AN}(\beta_A) / C_{AY}(\beta_A) - x'_w \tag{27}$$

and the resulting values are plotted as symbols in Figure 5 except for the values corresponding to the drift angles close to 0 or 180 degrees as the absolute values of C_{AY} and C_{AN} are very small which may lead to unnaturally high relative errors in $\Delta x'_A$.

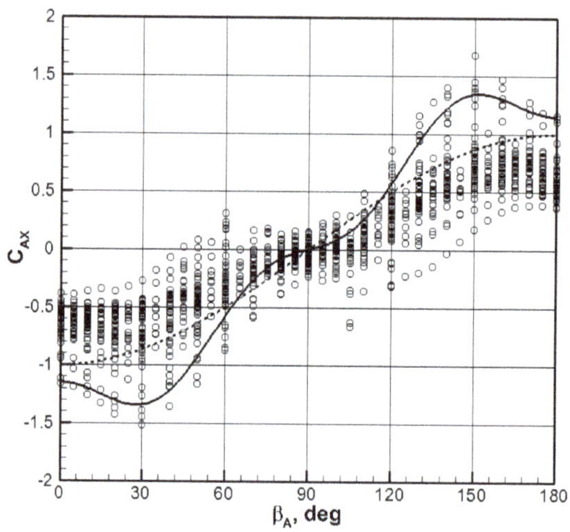

Figure 3. Experimental (circles) and analytical (lines) data for surge force coefficient.

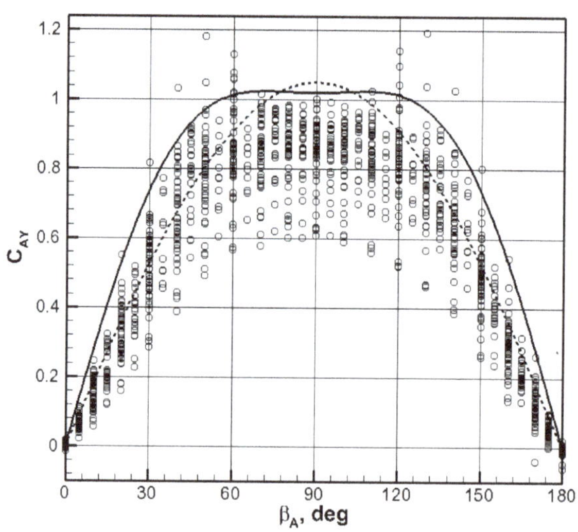

Figure 4. Experimental (circles) and analytical (lines) data for sway force coefficient.

The responses provided by (24) at $k_x = 1.0$ (dashed line) and $k_x = 0.8$ (solid line) are also plotted there. It is obvious that the scatter of empiric data is considerable, and in any particular case, the error caused by using (24) can be very large. However, in general, the linear approximation looks reasonably adequate and the value $k_x = 1$ leads to somewhat more conservative estimates. At the same time, the value $k_x = 0.8$ worked somewhat better delimiting the stripe of the x'_A values for various x'_w. The lower line in Figure 6 was obtained from (24) at $k_x = 0.8$ and $x'_w = -0.11$ which is the lowest value of the centroid's abscissa for all configurations presented in [78] while the upper line corresponds to $x'_w = +0.13$ which is the highest value.

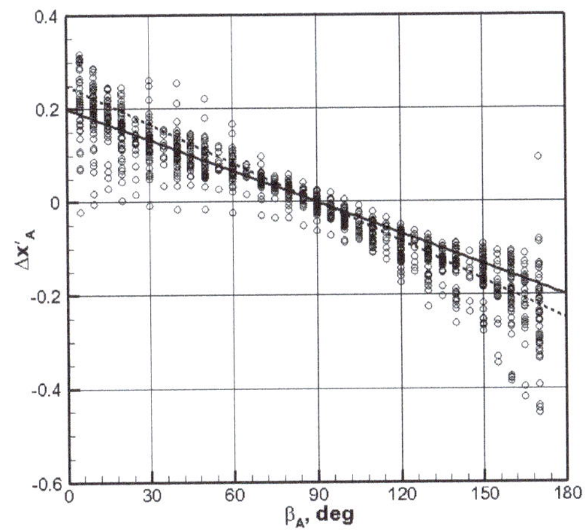

Figure 5. Increment of sway force lever as a function of the air drift angle.

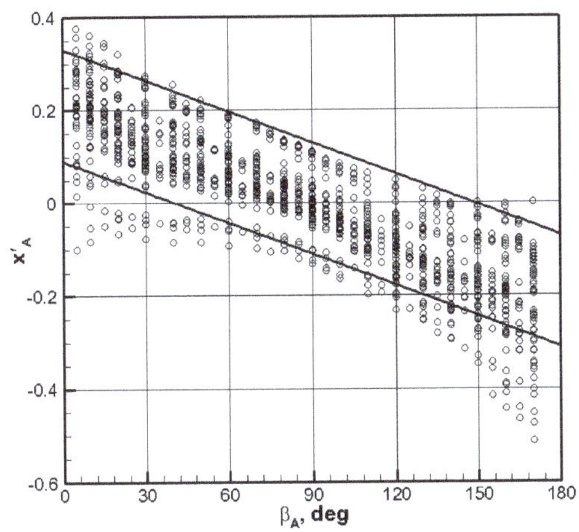

Figure 6. Sway force lever as a function of the air drift angle.

Finally, the measured values of the yaw moment coefficients C_{AN} are plotted in Figure 7 together with the predictions obtained with the formula (24) at $x'_w = -0.11$ (lower line), $x'_w = 0$ (middle line) and $x'_w = 0.13$ (upper line). The outlying empiric values correspond to a ro-ro/lo-lo ship, to a tanker with the aft superstructures and to the floating dock mentioned above.

Regarding some possible future developments, it makes sense to note that Dand [26] proposed two additional parameters describing the windage area geometry: the lateral aspect ratio $k_{AL} = 2A_L/L_{OA}^2$ and the transverse aspect ratio $k_{AT} = 2A_T/L_{OA}^2$. Unfortunately, the authors are not aware of any attempts to use these parameters for the classification and clusterization of ship aerodynamic data.

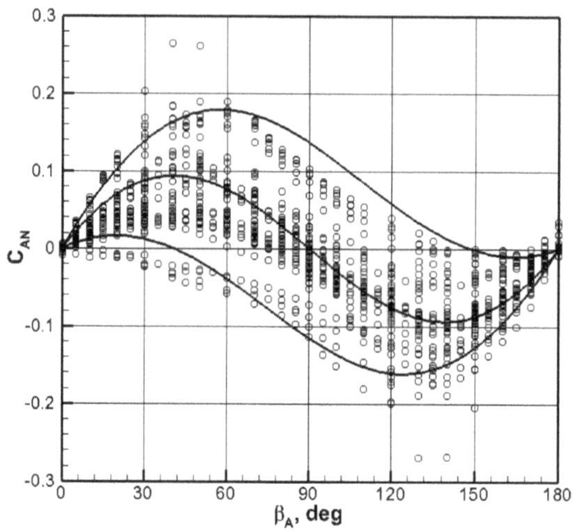

Figure 7. Experimental (circles) and analytical (lines) data for yaw moment coefficient.

6. Discussion and Conclusions

An attempt to work out rational approaches to the development of manoeuvrability criteria and standards was undertaken and is described in the present article. A definite emphasis was put on additional "environmental" criteria, which aim at complementing the existing IMO set of standards. Even though primarily it was found more rational to avoid criteria related to the resisting capacity of a ship to exogenous factors leaving this issue to performance specifications supposed to be formulated by the customer, the recently emerged danger of the appearance of underpowered and more prone to external factors ships stimulated revision of that attitude. As long as the introduction of the Energy Efficiency Design Indices stimulating the under-powering was performed on the level of officially introduced IMO regulations, it is quite natural to balance those indices with some counter-weighting requirements in order to prevent unacceptable reduction of safety.

A natural solution could be to accompany the reduction of the engine power and the design speed with appropriate augmentation of the effectiveness of the main control device in such a way that the resistance capacity of the ship to external factors remains unchanged. In some respects, such formulation of the manoeuvrability standardization problem is not typical and requires a special analysis, which is presented in this article. The main result of this analysis is an outline of a procedure for modification and extension of the IMO set of manoeuvring standards which seems to be relatively simple and promising. The procedure presumes execution of the following steps:

1. The existing turning ability standards, i.e., the limiting values for the relative advance and tactical diameter must be somewhat tightened on the basis of statistical data corresponding to the vessels currently in operation. Collection of these data may present certain organizational difficulties but, as long as all ships are now equipped with GPS units, it is, likely, possible, first, to create appropriate guides for shipmasters and to find means for stimulating them to perform necessary tests which are rather simple. In addition to the objective numerical data, it is also important to collect information about the subjective linguistic evaluation of manoeuvring qualities of ships by their operators.
2. The same as stated in the previous item applies to the directional stability measured by the values of the overshoot angles in zigzag manoeuvres which are relatively simple to be carried out during the operation of ships. It is expected that the existing IMO requirements could be re-considered and tightened.

3. The ship wind resisting capacity should be standardized separately on the basis of an additional criterion and some suitable standardization schemes must be adopted. It is suggested to use the classic criterion, i.e., the ability of a ship to sail with an arbitrary course (or heading, or apparent wind angle) under the action of wind with a certain specified speed. The wind controllability problem was formulated in the general form and it was further subject to reduction aiming at simplification of the standardizing procedure while still capturing correctly dependence on the most important parameters, i.e., on the ship speed and on its relative windage area. Likely, the most practical approach should be based on Equation (18) whose solution will provide a requirement to the relative rudder effectiveness defined by (14). The actual rudder effectiveness index is supposed to be estimated using any approved or commonly acknowledged method. Of course, the term "rudder" must be interpreted in the generalized sense embracing also steering nozzles and azipods. The standardizing value of the wind speed should be set from statistical considerations and, judging from the earlier standardization experience, its expected value may lie in the interval from 15 to 30 m/s.
4. Estimation of aerodynamic loads is part of any wind controllability standardization scheme. While the wind tunnel or CFD data are preferable when the ship configuration is fixed with certainty, simpler empiric methods can be recommended for early design stages. A very simple conservative method especially suitable for standardizing tasks and based on envelopes constructed on the wind tunnel data is suggested in the present article.
5. Reduction of the available engine power of newly built ships stimulated by the EEDI requirements can and must be compensated by increasing the effectiveness of the main steering device in order to keep the same level of resistibility of ships to the external factors. A direct appropriate augmentation of the rudder area in theory always solves the problem but may become unacceptable from the viewpoint of arrangement of the steering complex. In such cases, implementation of rudders with increased effectiveness such as the Schilling or flap rudder possibly combined with the installation of additional skegs or fixed fins may become a more suitable alternative. Other novel developments such as the so-called gate rudders [79] may be considered. Less evident possible solutions could be based on arrangements used on inland (river-going) ships that can include multiple rudders, combinations of rudders with steering nozzles, flanking rudders etc., see, e.g., [80]. However, these "inland" arrangements must be adapted with caution as a rather high degree of directional instability is often tolerable for river-going vessels rarely subject to perturbations from waves and the inherent stability of the resulting configuration must be verified with care.
6. It is evident that compensation of negative effects of the reduced powering through augmentation of the effectiveness of the steering device always comes at some cost. It is not clear whether these extra costs were considered when the EEDI requirements were worked out. However, much earlier Lowry [35] discussed possible costs of improving controllability and remarked that they will be hardly acceptable if exceed the overall construction and operation costs by more than 1–2%. Of course, this estimate is very crude and may be reconsidered.

It can be concluded from the present study that future progress in developing consistent and practical manoeuvring standards depends on development of a fast, reliable and validated method for simulating manoeuvring motion in waves. Systematic simulations with various hull forms are expected to provide sufficient information for confirming or disproving the hypothesis of concurrency of manoeuvring qualities in still water and of behaviour of a ship in waves. If the hypothesis is confirmed, modification of IMO standards can be carried out. Of course, such modification, as well as development of additional standards for controllability in wind, will require adjustment to statistical data on the existing fleet.

Author Contributions: Conceptualization, S.S. and C.G.S.; methodology, S.S.; validation, S.S.; formal analysis, S.S.; investigation, S.S.; resources, C.G.S.; data curation, S.S.; writing—original draft preparation, S.S.; writing—review and editing, S.S. and C.G.S.; visualization, S.S.; supervision, C.G.S.; project administration, C.G.S.; funding acquisition, C.G.S. Both authors have read and agreed to the published version of the manuscript.

Funding: This work was supported by the project SHOPERA—Energy Efficient Safe SHip OPERAtion, which was partially funded by the EU under contract 605221 and by the project IC&DT–AAC n.º02/SAICT/2017 "Simulation of manoeuvrability of ships in adverse weather conditions" funded by the Portuguese Foundation for Science and Technology (FCT). This work contributes to the Strategic Research Plan of the Centre for Marine Technology and Ocean Engineering (CENTEC), which is financed by the Portuguese Foundation for Science and Technology (Fundação para a Ciência e Tecnologia—FCT) under contract UIDB/UIDP/00134/2020.

Institutional Review Board Statement: Not applicable.

Informed Consent Statement: Not applicable.

Data Availability Statement: Not applicable.

Acknowledgments: The authors appreciate support obtained from Blendermann and Soeding in relation to the wind tunnel data resulted from studies carried out at the Institute of Shipbuilding of the University of Hamburg.

Conflicts of Interest: The authors declare no conflict of interest. The funders had no role in the design of the study; in the collection, analyses or interpretation of data; in the writing of the manuscript or in the decision to publish the results.

References

1. *Standards for Ship Manoeuvrability*; International Maritime Organization (IMO); Resolution MSC.137(76)4. Available online: https://wwwcdn.imo.org/localresources/en/KnowledgeCentre/IndexofIMOResolutions/MSCResolutions/MSC.137(76).pdf (accessed on 16 August 2021).
2. Biran, A.; Pulido, R.L. *Ship Hydrostatics and Stability*; Butterworth–Heinemann: Amsterdam, The Netherlands, 2003.
3. Lugovsky, V.V. *Theoretical Foundation of Standardization of Intact Stability of Sea-Going Vessels*; Sudostroyeniye Publ.: Leningrad, Russia, 1971. (In Russian)
4. The Stability in Waves Committee: Final Report and Recommendations to the 29th ITTC. 2021. 57p. Available online: https://ittc.info/media/9110/stability-in-waves-committee.pdf (accessed on 16 August 2021).
5. Jensen, J.J.; Choi, J.-H.; Kristinsen, H.O.; Nielsen, H.O.; Nielsen, U.D.; Erichsen, H.; Tvedt, E.I. An attempt to define critical wave and wind scenarios leading to capsize in beam sea. In Proceedings of the 13th International Symposium on Practical Design of Ships and Other Floating Structures (PRADS'2016), Copenhagen, Denmark, 4–8 September 2016; 8p.
6. Mironiuk, W. A research model of a ship's angle of heel. *Zeshyty Naukowe Akademii Morskiej w Szczecinie* **2016**, *46*, 43–48.
7. IMO Resolution MSC.267(85). Adoption of the International Code on Intact Stability, 2008 (2008 IS Code). Adopted on 4 December 2008. Available online: https://wwwcdn.imo.org/localresources/en/KnowledgeCentre/IndexofIMOResolutions/MSCResolutions/MSC.267(85).pdf (accessed on 16 August 2021).
8. Rules for Classification and Construction of Sea-Going Ships. In *Part III: Equipment, Arrangements and Outfit*; ND No. 2-020101-138-E; Russian Maritime Register of Shipping: St. Petersburg, Russia, 2021.
9. Shigunov, V. Direct stability assessment pragmatic solutions. *Ship Technol. Res.* **2017**, *64*, 144–162. [CrossRef]
10. Kobyliński, L. Stability criteria—Present status and perspectives of improvement. *TransNav Int. J. Mar. Navig. Saf. Sea Transp.* **2014**, *8*, 281–286. [CrossRef]
11. Shigunov, V.; Papanikolaou, A. Criteria for minimum powering and maneuverability in adverse weather conditions. *Ship Technol. Res.* **2015**, *62*, 140–147. [CrossRef]
12. Sutulo, S.; Guedes Soares, C. Analysis of manoeuvrability criteria and standards in view of environmental factors and EEDI impact. In *Maritime Technology and Engineering 3*; Guedes Soares, C., Santos, T.A., Eds.; Taylor & Francis Group: London, UK, 2016; pp. 371–383.
13. The Specialist Committee on Manoeuvring in Waves: Final Report and recommendations to the 29th ITTC. 2021. 33p. Available online: https://ittc.info/media/9106/sc-manoeuvring-in-waves.pdf (accessed on 16 August 2021).
14. Sutulo, S.; Guedes Soares, C. Mathematical models for simulation of manoeuvring performance of ships. In *Marine Technology and Engineering*; Guedes Soares, C., Garbatov, Y., Fonseca, N., Teixeira, A.P., Eds.; Taylor & Francis Group: London, UK, 2011; pp. 661–698.
15. Quadvlieg, F.H.H.A.; Armaoğlu, E.; Eggers, R.; van Coevorden, P. Prediction and verification of the maneuverability of naval surface ships. *Trans. Soc. Nav. Archit. Mar. Eng.* **2010**, *118*, 180–197.
16. Baquero, A. Consideration of ship manoeuvrability in the preliminary design phase (in Spanish). *Ing. Nav.* **1982**, *50*, 284–295.

17. Norrbin, N.H. Ship handling standards—Capabilities and requirements. In Proceedings of the MARSIM and ICSM 90 (Joint International Conference on Marine Simulation and Ship Manoeuvrability), Tokyo, Japan, 4–7 June 1990; pp. 37–50.
18. Koyama, T.; Kose, K. Recent studies and proposals of the manoeuvring standards. In Proceedings of the MARSIM and ICSM 90 (Joint International Conference on Marine Simulation and Ship Manoeuvrability), Tokyo, Japan, 4–7 June 1990; pp. 545–553.
19. Mastushkin, Y.M. *Manoeuvrability of Fishing Vessels*; Lyogkaya i Pishchevaya Promyshlennost Publ.: Moscow, Russia, 1981. (In Russian)
20. Stevens, B.; Lewis, F.L.; Johnson, E.N. *Aircraft Control and Simulation*; John Wiley and Sons: Hoboken, NJ, USA, 2016.
21. Segel, L. Ship maneuverability as influenced by the transient response to the helm. In Proceedings of the First Symposium on Ship Maneuvering, David Taylor Model Basin, Washington, DC, USA, 24–25 May 1960; pp. 151–177.
22. Sutulo, S.V. On the ergonomic approach to evaluation of ship manoeuvring criteria, Trans. In Proceedings of the International Symposium Manoeuvrability'95", Iława, Poland, 16–19 October 1995; pp. 93–118.
23. Nobukawa, T.; Kato, T.; Motomura, K.; Yoshimura, Y. Studies on manoeuvrability standards from the viewpoint of marine pilots. In Proceedings of the MARSIM and ICSM 90 (Joint International Conference on Marine Simulation and Ship Manoeuvrability), Tokyo, Japan, 4–7 June 1990; pp. 59–66.
24. Biancardi, C.G.; Cavazzi, D.; Grazziano, G.; Mosullo, M.T. Full scale measurements of a set of yaw/sway manoeuvring Q-indices. In Proceedings of the International Conference on Marine Simulation and Ship Manoeuvrability 1996 (MARSIM'96), Copenhagen, Denmark, 9–13 September 1996; pp. 635–642.
25. Belenky, V.; Falzarano, J. Rating-based maneuverability standards. *ABS Tech. Pap.* **2006**, 227–246. Available online: https://citeseerx.ist.psu.edu/viewdoc/download?doi=10.1.1.432.5803&rep=rep1&type=pdf (accessed on 16 August 2021).
26. Dand, I. Low speed manoevring criteria: Some considerations. In Proceedings of the International Conference on Marine Simulation and Ship Maneuverability, MARSIM'03, Kanazawa, Japan, 25–28 August 2003; pp. KN-4-1–KN-4-8.
27. Quadvlieg, F. Current status of manoeuvring prediction methods, their mutual pros and cons and ways to quantify the capabilities. In Proceedings of the Yücel Odabaşı Colloquium Series—2nd International Meeting "Recent Advances in Prediction Techniques for Safe Manoeuvring of Ships and Submarines", Istanbul, Turkey, 17–18 November 2016; pp. 1–9.
28. McCallum, I.R. "Which model?"—A critical survey of ship simulator mathematical models. In Proceedings of the Third International Conference on Marine Simulation (MARSIM'84), Rotterdam, The Netherlands, 19–22 June 1984; pp. 375–387.
29. Bassin, A.M. *Ship Propulsion and Manoeuvrability: Manoeuvrability*; Transport Publ.: Moscow, Russia, 1968. (In Russian)
30. Zaky, M.; Yasukawa, H. Effect of load condition on zig-zag manoeuvres of a VLCC in adverse weather conditions. In Proceedings of the International Conference on Ship Manoeuvrability and Maritime Simulation (MARSIM'2018), Halifax, NS, Canada, 12–16 August 2018; Paper 2, 10p.
31. Yudin, E.B. Measure of operational directional stability of ships. *Trans. Krylov Ship Res. Inst.* **1967**, *239*, 72–81. (In Russian)
32. Oltmann, P. Identification of hydrodynamic damping derivatives—A pragmatic approach. In Proceedings of the International Conference on Marine Simulation and Ship Manoeuvrability MARSIM'03, Kanazawa, Japan, 25–28 August 2003; pp. 575–583.
33. International Maritime Organization: Explanatory notes to the standards for ship manoeuvrability. MSC/Circ. 1053. 16 December 2002. 41p. Available online: https://www.register-iri.com/wp-content/uploads/MSC.1-Circ.1053.pdf (accessed on 16 August 2021).
34. Srivastava, S.D. IMO's activities on ship manoeuvring. In Proceedings of the International Conference on Marine Simulation and Ship Manoeuvrability (MARSIM'93), St. John's, NL, Canada, 26 September–2 October 1993; pp. 527–541.
35. Norrbin, N.H. Ship geometry, a dynamic stability parameter and standard test analysis. In Proceedings of the International Conference on Marine Simulation and Ship Manoeuvrability (MARSIM'93), St. John's, NL, Canada, 26 September–2 October 1993; pp. 447–456.
36. Nikolayev, E.; Inutina, T.; Lebedeva, M. On ship manoeuvrability estimation based on IMO resolution No. A.751(18). In Proceedings of the International Conference on Marine Simulation and Ship Manoeuvrability 1996 (MARSIM'96), Copenhagen, Denmark, 9–13 September 1996; pp. 303–307.
37. Li, M.; Wu, X. Simulation calcu;ation and comprehensive assessment of ship manoeuvrabilities in wind, wave, current and shallow water. In Proceedings of the MARSIM and ICSM 90 (Joint International Conference on Marine Simulation and Ship Manoeuvrability), Tokyo, Japan, 4–7 June 1990; pp. 403–411.
38. Ræstad, A.E. A classification society's experience with IMO resolution No. A.751(18). In Proceedings of the International Conference on Marine Simulation and Ship Manoeuvrability 1996 (MARSIM'96), Copenhagen, Denmark, 9–13 September 1996; pp. 309–318.
39. Yoshimura, Y.; Kose, K.; Haraguchi, T. Criteria for yaw-checking and course-keeping abilities in IMO's Interima standards for ship manoeuvrability. In Proceedings of the International Conference on Marine Simulation and Ship Manoeuvrability 2000 (MARSIM 2000), Orlando, FL, USA, 8–12 May 2000; pp. 367–378.
40. Quadvlieg, F.H.H.A.; van Coevorden, P. Manoeuvring criteria: More than IMO A751 requirements alone! In Proceedings of the International Conference on Marine Simulation and Ship Manoeuvrability MARSIM'03, Kanazawa, Japan, 25–28 August 2003; Volume 2, pp. 291–298.
41. Rhee, K.P.; Kim, S.Y.; Sung, Y.J.; Son, N.S. On the correlation between the overshoot angles in zigzag tests and the navigational difficulty. In Proceedings of the International Conference on Marine Simulation and Ship Manoeuvrability MARSIM'03, Kanazawa, Japan, 25–28 August 2003; Volume 2, pp. 299–307.

42. Haraguchi, T.; Nimura, T. A study on manoeuvrability standards for a ship with a POD propulsion system. In Proceedings of the International Conference on Marine Simulation and Ship Manoeuvrability MARSIM'03, Kanazawa, Japan, 25–28 August 2003; pp. 308–315.
43. Sohn, K.; Yang, S.; Lee, D.; Bae, J. A simulator study on yaw-checking and course-keeping ability of unstable ships. In Proceedings of the International Conference on Marine Simulation and Ship Manoeuvrability MARSIM'03, Kanazawa, Japan, 25–28 August 2003; Volume 2, pp. 327–333.
44. Yamada, K. Application of ship-handling/manoeuvrability criteria to stern configuration and rudder area design. In Proceedings of the International Conference on Marine Simulation and Ship Manoeuvrability MARSIM'03, Kanazawa, Japan, 25–28 August 2003; Volume 2, pp. 316–326.
45. Ohtakagi, Y.; Tanaka, M. Application of real-time ship manoeuvring simulator in the past ten years. In Proceedings of the Third International Conference on Marine Simulation (MARSIM'84), Rotterdam, The Netherlands, 19–22 June 1984; pp. 25–37.
46. Nishimura, T.; Kobayashi, H. The estimation of ship manoeuvrability for controlling position of unstable ships. In Proceedings of the International Conference on Marine Simulation and Ship Manoeuvrability MARSIM'03, Kanazawa, Japan, 25–28 August 2003; pp. 334–343.
47. Yasukawa, H.; Zaky, M.; Yonemasu, I.; Miyake, R. Effect of engine output on maneuverability of a VLCC in still water and adverse weather conditions. *J. Mar. Sci. Technol.* **2017**, *22*, 574–586. [CrossRef]
48. Sutulo, S.V. On the manoeuvring standards for sea-going vessels. *Trans. Russ. Marit. Regist. Shipp.* **1996**, *19*, 109–121. (In Russian)
49. Shigunov, V. Norming manoeuvrability in adverse conditions. *J. Offshore Mech. Arct. Eng.* **2017**, *139*, 1–10. [CrossRef]
50. Shigunov, V. Manoeuvrability in adverse conditions: Rational criteria and standards. *J. Mar. Sci. Technol.* **2018**. [CrossRef]
51. Spyrou, K.; Vassalos, D. Recent advances in the development of ship manoeuvring standards. In Proceedings of the MARSIM and ICSM 90 (Joint International Conference on Marine Simulation and Ship Manoeuvrability), Tokyo, Japan, 4–7 June 1990; pp. 51–58.
52. Lowry, I.J. Ship of Japan controllability: An industrial perception. In Proceedings of the MARSIM and ICSM 90 (Joint International Conference on Marine Simulation and Ship Manoeuvrability), Tokyo, Japan, 4–7 June 1990; pp. 333–340.
53. MacLean, A. *HMS "Ulysses"*; Harper Collins Publ.: London, UK, 1994.
54. Alekseyev, L.L. *Practical Manual on Handling Seagoing Vessels*; CNIIMF Publ.: St. Petersburg, Russia, 2003. (In Russian)
55. Chillece, G.; el Moctar, O. A numerical method for manoeuvring simulation in regular waves. *Ocean. Eng.* **2018**, *170*, 434–444. [CrossRef]
56. Pershitz, R.Y. *Ship Manoeuvrability and Steering*; Sudostroyeniye Publ.: Leningrad, Russia, 1983. (In Russian)
57. Voytkunsky, Y.I. (Ed.) *Handbook on Ship Hydrodynamics*; Sudostroyeniye Publ.: Leningrad, Russia, 1985. (In Russian)
58. Sutulo, S.; Guedes Soares, C. Preliminary analysis of ship manoeuvrability criteria in wind. In *Maritime Technology and Engineering*; Guedes Soares, C., Santos, T.A., Eds.; Taylor & Francis Group: London, UK, 2015; pp. 933–946.
59. Gong, I.-Y.; Yang, Y.-H.; Yun, K.-H.; Park, S.-K.; Oh, J.-Y.; Kim, S.-Y.; Jeong, M.-H.; Seo, T.-H.; Baek, M.-S. The influence of ship maneuverability modelling on the navigational safety assessment result. In Proceedings of the International Conference on Marine Simulation and Ship Manoeuvrability 2012 (MARSIM'2012), Singapore, 23–27 April 2012; pp. 740–748.
60. Kang, C.-G.; Kim, J.-H. Assessment of ship manoeuvrability based on IMO resolution No. A.751. In Proceedings of the International Conference on Marine Simulation and Ship Manoeuvrability 1996 (MARSIM'96), Copenhagen, Denmark, 9–13 September 1996; pp. 293–302.
61. Capurro, G.; Sodomaco, P. Applications of the manoeuvring prediction program SIMSUP to meet the new IMO standards. In Proceedings of the International Conference on Marine Simulation and Ship Manoeuvrability 1996 (MARSIM'96), Copenhagen, Denmark, 9–13 September 1996; pp. 563–576.
62. Jurgens, A.J.; de Jager, A. Controllability at too high speeds in too shallow water. In Proceedings of the International Conference on Marine Simulation and Ship Manoeuvrability MARSIM 2006, Terschelling, The Netherlands, 25–30 June 2006; pp. 393–404.
63. Zhang, W.; Zou, Z.-J.; Deng, D.-H. A study on prediction of ship maneuvering in regular waves. *Ocean. Eng.* **2017**, *137*, 367–381. [CrossRef]
64. Kobayashi, E. Development of a simulation system to evaluate ship manoeuvrability in waves. In Proceedings of the International Conference on Marine Simulation and Ship Manoeuvrability (MARSIM'93), St. John's, NL, Canada, 26 September–2 October 1993; pp. 295–303.
65. Woodward, M.D.; Clarke, D.; Atlar, W. On the manoeuvring prediction of pod driven ships. In Proceedings of the International Conference on Marine Simulation and Ship Manoeuvrability MARSIM'03, Kanazawa, Japan, 25–28 August 2003; pp. 344–352.
66. Eda, H.; Numano, M. Analysis of major marine disasters. In Proceedings of the International Conference on Marine Simulation and Ship Manoeuvrability MARSIM'03, Kanazawa, Japan, 25–28 August 2003; pp. 547–554.
67. Martin, L.L. Ship Maneuvering and control in wind. *SNAME Trans.* **1980**, *88*, 257–281.
68. Yasukawa, H.; Yoshimura, Y. Introduction of MMG standard method for ship maneuvering predictions. *J. Mar. Sci. Technol.* **2015**, *20*, 37–52. [CrossRef]
69. Sutulo, S.; Guedes Soares, C. On the application of empiric methods for prediction of ship manoeuvring properties and associated uncertainties. *Ocean. Eng.* **2019**, *186*, 106111. [CrossRef]

70. Sutulo, S.; Guedes Soares, C. Comparative simulation of definitive manoeuvres of the KVLCC2 benchmark ship using different empiric mathematical models. In *Maritime Transportation and Harvesting of Sea Resources*; Guedes Soares, C., Teixeira, A.P., Eds.; Taylor & Francis Group: London, UK, 2018; Volume 1, pp. 571–579.
71. Eda, H. Low-speed controllability of ships in wind. *J. Ship Res.* **1968**, *80*, 181–200. [CrossRef]
72. Benedict, K. Wind influence on directional stability and turning of ships. In Proceedings of the 14th Scientific and Methodological Seminar on Ship Hydrodynamics (SMSSH), Varna, Bulgaria, 24–27 September 1985; Volume 1, pp. 11.1–11.11.
73. Sutulo, S.V. A study on the course keeping of a ship in wind. In Proceedings of the 14th Scientific and Methodological Seminar on Ship Hydrodynamics (SMSSH), Varna, Bulgaria, 24–27 September 1985; Volume 1, pp. 12.1–12.8. (In Russian)
74. Spyrou, K. Yaw stability of ships in steady wind. *Schiffstechnik/Ship Technol. Res.* **1995**, *42*, 21–30.
75. Blendermann, W. *Practical Ship and Offshore Structure Aerodynamics*; Technische Universität Hamburg-Harburg: Hamburg, Germany, 2013.
76. Prpić-Oršić, J.; Valčić, M. Sensitivity analysis of wind load estimation method based on elliptic Fourier descriptors. In *Maritime Technology and Engineering 3*; Guedes Soares, C., Santos, T.A., Eds.; Taylor & Francis Group: London, UK, 2016; pp. 151–160.
77. Brix, J. (Ed.) *Manoeuvring Technical Manual*; Seehafen Verlag: Hamburg, Germany, 1993.
78. Blendermann, W. *Wind Loading of Ships—Collected Data from Wind Tunnel Tests in Uniform Flow*; Technical Report Nr. 574; Institut für Schiffbau der Universität Hamburg: Hamburg, Germany, 1996.
79. Carchen, A.; Turkmena, S.; Piaggiob, B.; Shia, W.; Sasakic, N.; Atlarc, M. Investigation of the manoeuvrability characteristics of a Gate Rudder system using numerical, experimental, and full-scale techniques. *Appl. Ocean. Res.* **2020**, *106*, 102419. [CrossRef]
80. Quadvlieg, F.H.H.A. Mathematical models for the prediction of manoeuvres of inland ships: Does the ship fit in the river. In Proceedings of the "Smart Rivers 2013", Liège, Belgium, 23–27 September 2013; Paper 187, pp. 1–9.

Journal of
Marine Science and Engineering

MDPI

Article

Mesh Properties for RANS Simulations of Airfoil-Shaped Profiles: A Case Study of Rudder Hydrodynamics

Suli Lu [1], Jialun Liu [2,3,*] and Robert Hekkenberg [4]

[1] School of Naval Architecture, Ocean and Energy Power Engineering, Wuhan University of Technology, Wuhan 430063, China; sulilu@whut.edu.cn
[2] Intelligent Transportation Systems Research Center, Wuhan University of Technology, Wuhan 430063, China
[3] National Engineering Research Center for Water Transport Safety, Wuhan 430063, China
[4] Maritime and Transport Technology, Delft University of Technology, Mekelweg 2, 2628 CD Delft, The Netherlands; R.G.Hekkenberg@tudelft.nl
* Correspondence: jialunliu@whut.edu.cn

Abstract: A good mesh is a prerequisite for achieving reliable results from Computational Fluid Dynamics (CFD) calculations. Mesh properties include mesh types, computational domain sizes, and node distributions. However, in literature, we found no clear consensus about what these properties should be. In this article, we performed a case study on ship rudders to determine what the suitable mesh properties are for airfoil-shaped profiles. A classic NACA 0012 profile is chosen as an example, and commercial packages ANSYS ICEM are applied for meshing with an ANSYS Fluent solver. With a strategy in consideration of relationships among different mesh properties, a comprehensive parametric investigation is conducted to study the impacts of these properties on the accuracy of rudder hydrodynamic coefficients obtained by CFD methods. The step-by-step study outputs recommended Reynolds numbers, domain sizes, and near- and far-field node distributions for mesh types with distinct topology structures, i.e., C-mesh, O-mesh, H-mesh, and Hybrid-mesh. Specifically, the study shows that a critical Reynolds number is needed for the perspective of efficiency, while a domain extending 60 times of the chord length enables the boundary effects to be negligible. As for node distributions, the near-field nodes should be treated carefully, compared with those in the far-field. After that, corresponding mesh properties for different calculation objectives are illustrated in detail based on the characteristics of mesh types mentioned above. With the proposed strategy for mesh refinements, impacts of different mesh properties on rudder hydrodynamics are clarified and recommended settings are applicable for other airfoil-shaped profiles such as wind turbines and marine propellers.

Keywords: rudder hydrodynamics; mesh properties; mesh independence; computational fluid dynamics

Citation: Lu, S.; Liu, J.; Hekkenberg, R. Mesh Properties for RANS Simulations of Airfoil-Shaped Profiles: A Case Study of Rudder Hydrodynamics. *J. Mar. Sci. Eng.* **2021**, *9*, 1062. https://doi.org/10.3390/jmse9101062

Academic Editors: Carlos Guedes Soares and Serge Sutulo

Received: 31 July 2021
Accepted: 5 September 2021
Published: 28 September 2021

1. Introduction

Stimulated by developments in computer power and mechanics theories, Computational Fluid Dynamic (CFD) methods were widely applied in ocean engineering to study the hydrodynamic performance of marine structures. Generally, computation domains containing target structures need to be discretized into multiple mesh elements to solve differential equations numerically. In ship maneuvering simulations, the presence of CFD-based tools enables researchers to obtain more precise hydrodynamic coefficients compared with that of traditional empirical methods, which is beneficial to establish individualized and accurate mathematical maneuverability models for different objectives. As the most widely used steering devices for ships, rudders play an important role in the maneuvering of ships [1], and the accuracy of lift coefficients has a great impact on motion predictions. Many CFD simulations were conducted to study rudder hydrodynamics (Liu et al. [2], Badoe et al. [3], Van Nguyen and Ikeda [4]), but few concentrate on mesh generations for rudders. In this paper, the study on mesh properties for RANS simulations of rudder hydrodynamics is performed to present detailed insight into meshes around rudders.

A good mesh is fundamental to the convergence and the accuracy of solutions, while a poor one leads to nonconverging results, inaccurate solutions, and an increase in computation time. Due to the complexities of geometries in structural surfaces and the lack of automatic generation methods, it costs plenty of time to generate meshes in calculation processes. On the one hand, mesh quality is influenced by various factors, but on the other hand, it is unrealistic to adjust all of them to the best states in practice, which requires more computation power and computation time.

A good quality mesh should have a sufficiently large number of cells, implying that a further increase in the number of cells will not lead to significant changes in the solutions. The mesh needs to be fine enough in the regions where fluid characteristics change rapidly. Therefore, mesh properties need to be coordinated to make a trade-off between the accuracy and the time scale. In literature, different mesh properties, i.e., mesh types, domain sizes, and node distributions, are applied in RANS simulations of airfoil-shaped profiles. Lutton [5] indicated that the O-mesh is superior to the C-mesh in determining the pressure coefficients in the vicinity of leading and trailing edges, while the C-mesh shows a better resolution in the wake. Basha [6] constructed a structured O-mesh, a structured C- mesh, and a hybrid mesh on a NACA 0012 airfoil, while the hybrid mesh shows better results in drag coefficient predictions due to finer mesh resolutions in the boundary layer domain. Stuck et al. [7] selected a hybrid mesh with quadrilateral unstructured cells rather than triangular ones to achieve smoother transitions from the boundary mesh to the unstructured outer mesh for RANS simulations on a rudder profile. Parametric grid independence studies are conducted by varying chord-wise and layer-wise grids, respectively, and optimum values are chosen based on variations of integral force coefficients. Turnock et al. [8] carried out decoupled studies on the boundary location and mesh node distributions in different directions for a NACA 0012 section, while related parameters are determined by convergence study.

As no clear consensus about mesh properties was found in current studies, systematic parametric investigations on mesh properties are performed in this paper. As naval architects, our primary interest in CFD lies in applications related to the flow around ships and resultant hydrodynamic forces. In this particular case, we focus on the flow around the rudder and the resulting rudder performance. Kim et al. [9] stated that the NACA series are the most widely used economic rudder profiles. Thus, a classic NACA 0012 airfoil-shape rudder profile is utilized to study mesh properties for its simplicity in the shape, which makes mesh properties easier to change and enhances the applicability of obtained results for other complicated structures.

This paper analyzes the impacts of mesh properties on RANS simulations and proposes suitable meshing strategies for rudder hydrodynamics. Following this introduction, Section 2 further discusses the reason why a 2D rudder profile is selected as the research object in this paper. Section 3 explains definitions of mesh properties in this study. Section 4 illustrates applied research methods in detail, and Section 5 discusses the impacts of different mesh properties based on simulated cases. Finally, a recommended meshing strategy is presented to provide guidance for hydrodynamic analysis in Section 7.

2. The Selection of the Research Object

Bare hulls, propellers, rudders, or full-appendages ships were all studied in CFD methods in literature, though not all of them are suitable for mesh properties studies. There are great curvature changes on the surface of hulls and propellers, and that is why mixture mesh topologies and millions of cells are desired to capture their flow features. The complex nature makes it difficult to perform parametric investigations on a single mesh property economically. What's more, the geometries of hulls or propellers in the marine industry vary from each other significantly, and rules derived from limited cases may not be applicable for those with different configurations.

Although special rudders drew a growing interest, conventional spade rudders, or those modified but still geometrically similar versions like flap rudders, are still mainstream maneuvering devices, however. Besides, the meshing strategy aiming at a NACA profile can be easily applied to its geometrically similar counterparts. Force coefficients of rudders during ship maneuverings can be obtained based on 2D profiles in open water conditions, which also makes the investigation of 2D profiles beneficial to maneuverability studies. For all these reasons, in this paper, a 2D profile, in other words, a rudder with an infinite aspect ratio, is determined as the research object. According to Fujii [10], Fujii and Tsuda [11,12], the normal force coefficient of actual rudders whose aspect ratios are limited can be expressed as:

$$C_N = 6.13 \sin \alpha_R \frac{\Lambda_G}{\Lambda_G + 2.25} \tag{1}$$

where C_N is the rudder normal force coefficient, and 6.13 is an approximate constant to estimate the partial derivative of C_N to $\sin \alpha_R$ of 2D sections. Liu et al. [2] calculated $\frac{\partial C_N}{\partial \sin \alpha_R}$ for different rudder profiles by performing 2D CFD simulations, which can modify the original transformation relation in Equation (1) and obtain force coefficients for 3D rudders in open water conditions. Since rudders generally operate in the propeller slipstream, rudder hydrodynamics affected by the propeller can be estimated based on series model tests by Nienhuis [13], and the experimental data are shown in Figure 1. Compared with that of open water conditions, the stall angle for the rudder increased to about 35° with propeller impacts, while $\frac{\partial C_N}{\partial \sin \alpha_R}$ was not significantly changed. Hence, based on 2D simulation results, corrected formulations for a rudder with an aspect ratio Λ_G considering propeller impacts can be expressed as:

$$\left.\begin{aligned} C_L^{\Lambda_G} &= k_p \left(\frac{\partial C_L^{2D}}{\partial \sin \alpha_R} \sin \alpha_R + C_{L_0}^{2D} \right) \frac{\Lambda_G}{\Lambda_G + 2.25} \\ C_D^{\Lambda_G} &= k_p \left(\frac{\partial C_D^{2D}}{\partial \sin \alpha_R} \sin \alpha_R + C_{D_0}^{2D} \right) \frac{\Lambda_G}{\Lambda_G + 2.25} \end{aligned}\right\}, (\alpha_R \leqslant 35^\circ) \tag{2}$$

where $\frac{\partial C_L^{2D}}{\partial \sin \alpha_R}$, $C_{L_0}^{2D}$, $\frac{\partial C_D^{2D}}{\partial \sin \alpha_R}$, $C_{D_0}^{2D}$ can be obtained from 2D CFD simulations, and k_p can be estimated experimentally. The availability of Equation (2) in ship maneuvering studies was validated by Liu et al. [2,14] against two typical inland vessels. Apart from the transformation relation, another factor that makes the 2D study effective is that span-wise mesh generations can be omitted and cells in the profile section can be focused on. Therefore, 2D rudder profiles are set as the research object in following studies.

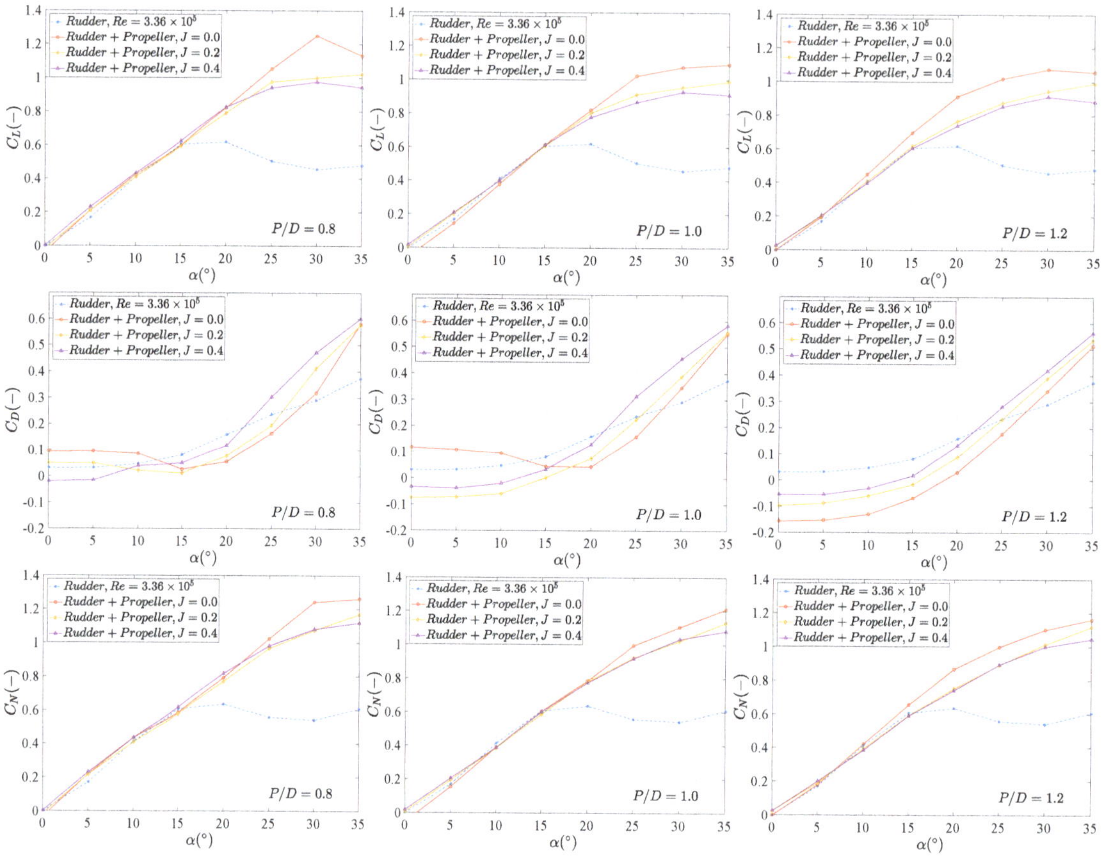

Figure 1. Comparison of the hydrodynamic coefficients of a rudder in open water conditions to those of the rudder in propeller slipstream. Data are adopted from Nienhuis [13].

3. Mesh Properties

In literature, mesh types and computation domain sizes are generally determined before studying grid independence on certain mesh parameters, and some settings are shown in Tables 1 and 2. Mesh properties in grid independence study are denoted as near-field and far-field parameters according to locations of distributions. All of these mesh properties are studied in this paper. The detailed descriptions of mesh properties above are in Sections 3.1–3.3.

Table 1. Mesh properties studies in literature: objects, Reynolds numbers, mesh types, and domain sizes.

No.	Settings / Literature	Object	*Re*	Mesh Type: Tested	Mesh Type: Selected	Domain Size
1	Lutton [5]	NACA 0012	2.00×10^6	C, O	\	\
2	Turnock et al. [8]	NACA 0012	6.00×10^6	\	\	Outlet 15 c
3	Stuck et al. [7]	NACA 0012	6.00×10^6	Hybrid C, Hybrid triangle and quadrilateral	Hybrid quad	Inlet 15 c Outlet 15 c
4	Basha [6]	NACA 0012	2.00×10^6	O, C, Hybrid	Hybrid	\
5	Wasberg and Reif [15]	NACA 0009, NACA 65209	7.04×10^6	H	H	Inlet 29 c Outlet 30 c
6	Tveiterås [16]	NACA 0012	2.00×10^6, 3.00×10^6	Hybrid, C	Hybrid	Outlet 19 c
7	Narsipur et al. [17]	Multi-element airfoils	2.51×10^6	Hybrid	Hybrid	\
8	Eleni et al. [18]	NACA 0012	3.00×10^6	C	C	Side 20 c Outlet 30 c
9	Van Nguyen and Ikeda [4]	NACA 0024	6.00×10^6	Hybrid	Hybrid	
10	Langley Research Center [19]	NACA 0012	6.00×10^6	C	C	Outlet 500 c
11	Nordanger et al. [20]	NACA 0012	3.00×10^6	Multiple patch	Multiple patch	Inlet 15 c Outlet 20 c
12	Siddiqui et al. [21]	NACA 0015	1.96×10^6	C	C	Inlet 8 c Outlet 20 c

Table 2. Mesh properties studies in literature: node distributions and turbulence models.

No.	Settings / Literature	Nodes Distribution: Chord	Nodes Distribution: Boundary Layer	Nodes Distribution: Far Field	Turbulence Model
1	Lutton [5]	\	\	\	\
2	Turnock et al. [8]	244 nodes	$y^+ = 30$	49 radial cells 99 wake cells	k-ϵ standard and RNG
3	Stuck et al. [7]	200 nodes	$t \geq 3\delta$	\	k-ϵ RNG
4	Basha [6]	60 nodes	$y^+ < 5$	\	SA
5	Wasberg and Reif [15]	\	$y^+ \approx 1$ $\Delta s = 1.50{\sim}2.00 \times 10^{-5}$	\	SA
6	Tveiterås [16]	340 nodes	$y^+ = 1.7{\sim}3$ $\Delta s = 3.00 \times 10^{-5}, r = 1.14$	\	Transition SST
7	Narsipur et al. [17]	\	$y^+ \approx 1$	\	k-ω SST
8	Eleni et al. [18]	\	$y^+ \approx 0.2$ $\Delta s = 1.00 \times 10^{-5}$	\	k-ω SST
9	Van Nguyen and Ikeda [4]	\	$y^+ \approx 1$	\	k-ω SST
10	Langley Research Center [19]	\	\	\	\
11	Nordanger et al. [20]	127 nodes	$y^+ = 30$	\	SA
12	Siddiqui et al. [21]	\	\	\	k-ω SST

3.1. Mesh Types

Based on the connectivity of elements, meshes are identified as structured, unstructured, and hybrid. As regular connectivity can be expressed as a two/three-dimensional array, the structured model is highly efficient in the usage of computational resources. However, regular connectivity restricts the element type to quadrilateral in 2D and hexahedral in 3D, which inherently limits applications of the structured mesh for complex geometries. For simple geometries, a structured mesh may have better convergence and higher resolution than an unstructured mesh.

Instead of a single-block structured mesh, a block-structured or multiple-structured mesh is more widely applied. Three commonly used topologies, C-mesh, H-mesh, and O-mesh, are shown in Figure 2. It is superior in computational efficiency to the unstructured mesh and more flexible in handling the complex geometries than the single-block structured mesh. The choice of the mesh topology has a considerable influence on mesh quality [22], which depends on the domain geometry, the structure of the solution, and the topology in the adjacent block. According to Liseikin [22], in H-mesh, the computational domain is square and two singularities exist in the interior boundary. O-mesh and C-mesh are both in solid squares, while the interior boundary for the former one is smooth and the latter has a singularity the same type as the H-mesh. In this particular case, some modifications are made based on basic topologies above to accurately capture the characteristics of the flow near the rudder, as shown in Figure 2. For brevity, modified types are still marked as C-mesh, H-mesh, and O-mesh, respectively.

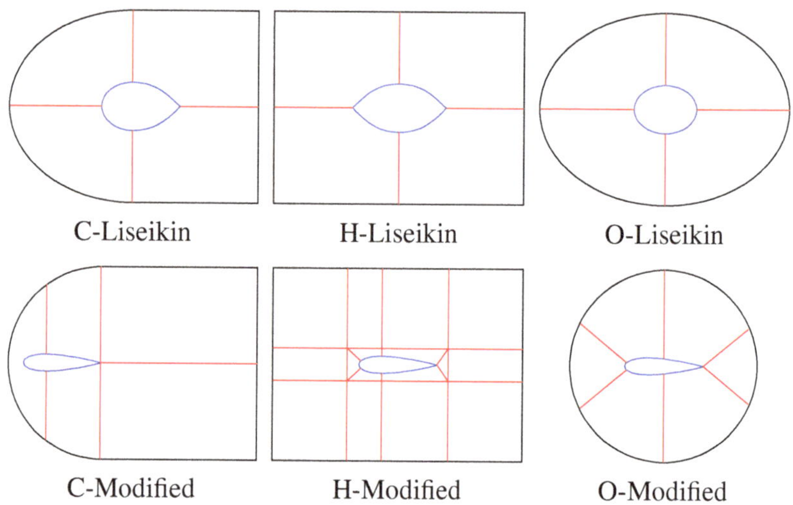

Figure 2. Original and modified mesh types for the block-structured mesh.

The irregular connection of unstructured meshes allows for more freedom in element choices, typically triangular in 2D and tetrahedral in 3D. Compared to that of the structured mesh, the unstructured mesh is more suitable for complex configurations, but it can be highly inefficient. Aftosmis et al. [23] reported that unstructured triangular meshes are 50 times more expensive in both memory and time than that of structured quadrilateral meshes with nearly the same accuracy. Unstructured meshes, however, have advantages over structured meshes in mesh refinement and generation time. In practice, hybrid meshes are applied instead of pure unstructured meshes. Hybrid meshes commonly consist of a structured portion near wall surfaces to capture the boundary layer and an unstructured portion that fills the domain. Thus, hybrid meshes inherit the advantages of both structured and unstructured meshes, such as good orthogonality to wall surfaces, suitability for mesh refinement, flexibility for complex geometries, and fast generation. Figure 3 demonstrates a hybrid triangle mesh, which has quadrilateral elements for the boundary layer and triangular elements for the remainder of the domain.

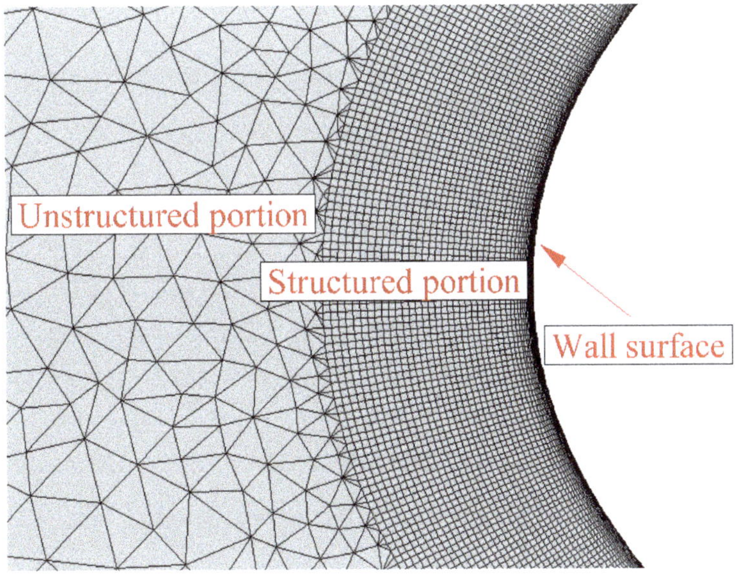

Figure 3. The mesh type of Hybrid-mesh.

3.2. Domain Sizes

The computational domain is the geometrical region that bounds the numerical simulation [17]. To obtain highly accurate solutions, the position of boundaries has to be discussed to demonstrate that the interior flow field is not disturbed. Thus, there is the need to study the domain size, which not only minimizes the influences of the boundaries, but also prevents an excessively large domain. Typical mesh types correspond to related topology structures, which lead to different shapes of calculation domains. The geometric parameters for different mesh types are shown in Figure 4, while subscripts, i.e., in, out, and $side$, stand for distances from the rudder to the inlet, outlet, and side boundary of corresponding mesh types, and O_r is the radius of the domain of O-mesh.

3.3. Node Distributions

For structured meshes, since cell distributions vary with node distributions in different directions of the computation domain, convergence studies need to be done by varying related mesh parameters. In this study, node distributions are divided into two categories, including near- and far-field distributions.

Near-field distributions are located in the boundary layer regions near the wall, which are generally refined to capture the flow around the structure. The chord-wise spacing is the distance between two nodes along with the profile, and the layer-wise growth rate is the distances between adjacent mesh points along a mesh line, which determine the boundary layer mesh along with the first mesh height.

Different mesh types share similar near-wall distributions, while it is difficult to define far-field distributions owing to differences in mesh topologies. In this paper, far-field distributions are defined as node distributions that do not contain near-field distributions.

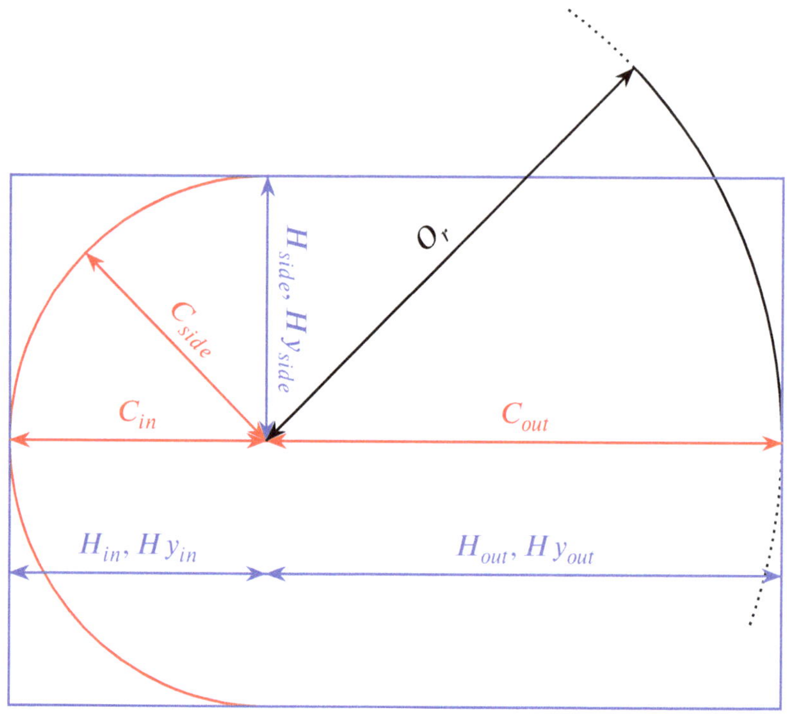

Figure 4. The geometric parameters for different mesh types.

4. Research Approach

A series of parametric studies are conducted to observe the impacts of mesh properties on the prediction of the accuracy of solutions. The classic validation profile NACA 0012 is tested through commercial CFD code ANSYS Fluent. Strategies for investigating mesh properties mentioned in Section 4.1. Numerical methods utilized to study the hydrodynamic performance of the rudder profile are described in Section 4.2, and the error calculation method is mentioned in Section 4.3.

4.1. Strategies

Different mesh properties contribute to mesh quality in varying degrees, and it is the combination of these properties that determines the accuracy of calculation results. Instead of rough global refinement of cells [24], several main impact factors on mesh fineness and the required number of cells are illustrated with simulations. This parametric mesh independence study is chosen because optimization on individual parameters specifically addresses the crucial impact factor of mesh quality, avoiding the waste of cells [7]. The global refinement is more frequently applied for unstructured meshes than structured meshes. However, due to the lack of specific classification on mesh properties, mesh properties above are not independent of each other strictly. Cases with different mesh types also differ from each other in domain sizes, node distributions, and numbers of mesh elements, etc.

As shown in Figure 5, the typical mesh type corresponds to a related topology structure, which leads to different shapes of calculation domains and connection patterns between the profile and domain boundaries. With domain size varying, node distribution in different directions changes. Further, the number of mesh elements is concerned with all mesh properties above.

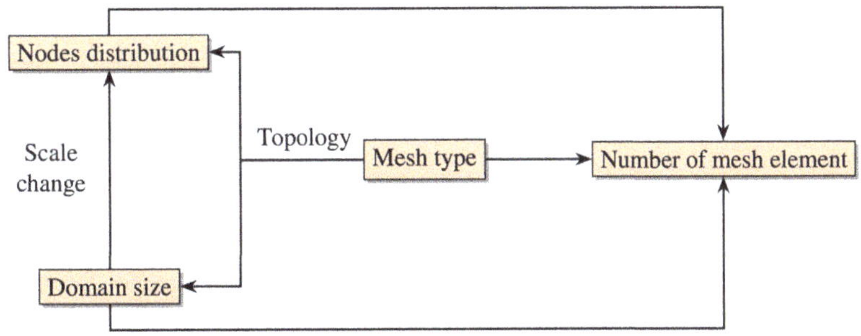

Figure 5. Relationship between different mesh properties.

The strategy in Figure 6 is applied to investigate how different mesh properties affect the calculation results of rudder hydrodynamics. Firstly, cases of C-mesh type with Reynolds numbers in a wide range are tested to obtain the critical Reynolds number ($Re_{critical}$), above which hydrodynamic coefficients tend to be stable. Secondly, the domain sizes of the C-mesh are varied to obtain the proper boundary sizes that are large enough to eliminate boundary effects for the following simulations. Thirdly, cases are divided into 4 groups classified by 4 different mesh types. Since it is inappropriate to compare mesh qualities of different mesh types when other mesh properties are not determined yet, the impacts of node distribution on each mesh type should be studied independently. In this part, the node distribution is the only factor that affects mesh quality for the typical mesh type, and the effects of node distributions are studied based on $Re_{critical}$ and the domain size obtained in the previous stage. Therefore, grid independence studies are conducted on parameters in the near-field, i.e., chord-wise spacings and layer-wise growth rates, and the far-field, i.e., radial nodes, wake nodes, and element sizes. Finally, four mesh types with optimal mesh properties are compared to give recommended mesh settings for rudder hydrodynamics.

4.2. Numerical Methods

The RANS methods presented in this paper use a time-average Reynolds decomposition, which assumes that all the components of the flow velocity and pressure consist of a mean value and a bounded fluctuation to represent turbulence. The governing equations for incompressible viscous flow are as follows:

$$\frac{\partial \bar{u}_i}{\partial x_i} = 0 \tag{3}$$

$$\rho\frac{\partial \bar{u}_i}{\partial t} + \rho\bar{u}_j\frac{\partial \bar{u}_i}{\partial x_j} = f_i - \frac{\partial \bar{p}}{\partial x_i} + \frac{\partial}{\partial x_j}\left(\mu\frac{\partial \bar{u}_i}{\partial x_j} - \rho\overline{u_i'u_j'}\right), \quad (i = 1,2,3) \tag{4}$$

where $\bar{u}_i$ is the mean velocity component in the x direction, ρ and μ are the density and viscosity coefficient of the fluid, $\bar{P}$ is the mean pressure, f_i is the body force, while $\rho\overline{u_i'u_j'}$ is the Reynolds stress. In RANS simulations, turbulence models are introduced to close equations above. The k-ω SST model was designed to give highly accurate predictions of the onset and the amount of flow separation, which provides a better description of flow around the rudder. Therefore, the k-ω SST model is applied for simulations in the following parts. Numerical settings for ANSYS Fluent is shown in Table 3.

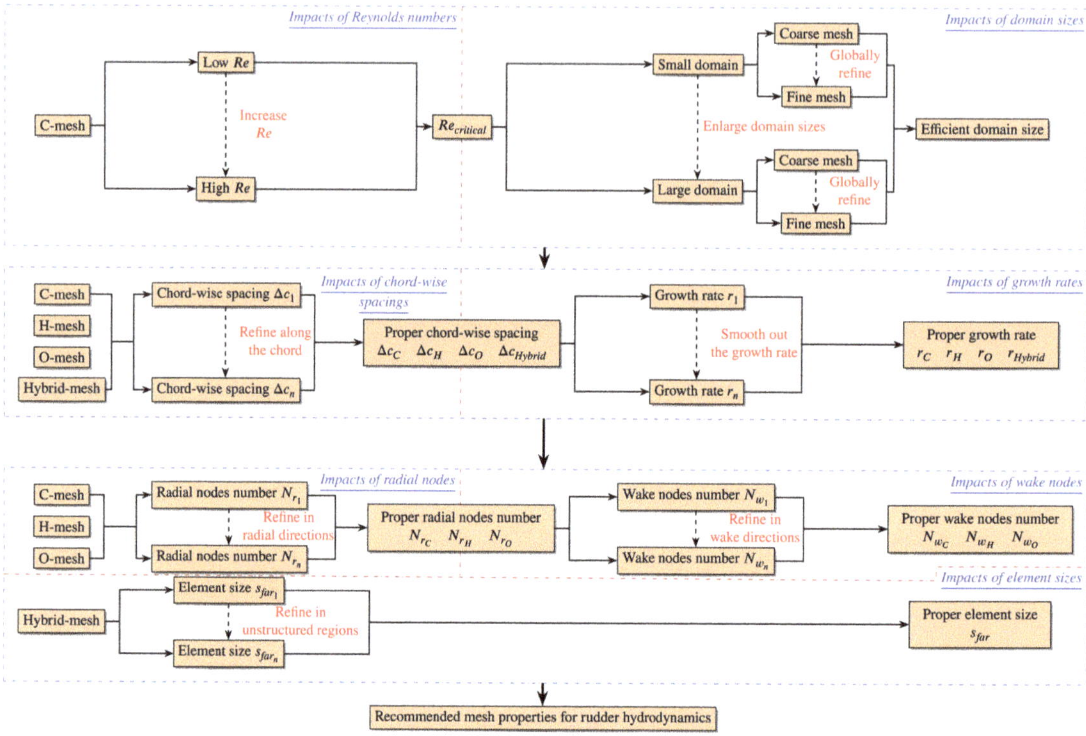

Figure 6. Strategy applied to study different mesh properties.

Table 3. Numerical settings for RANS simulations with ANSYS.

Property	Setting
Turbulence model	k-ω SST
Pressure-velocity coupling	Coupled
Gradient	Least squares cell based
Pressure	Second order
Momentum	Second order upwind
Turbulent kinetic energy	Second order upwind
Specific dissipation rate	Second order upwind

4.3. Error Calculation

Routinely, nondimensional coefficients are used to compare the rudder hydrodynamic performance with that of various design choices, while main hydrodynamic characteristics are the lift coefficient C_L, the drag coefficient C_D, and the moment coefficient C_M ([1]), which are expressed as:

$$C_L = L_R / \left(0.5 \rho V_R^2 A_R\right) \quad (5)$$

$$C_D = D_R / \left(0.5 \rho V_R^2 A_R\right) \quad (6)$$

$$C_M = M_R / \left(0.5 \rho V_R^2 A_R c\right) \quad (7)$$

where L_R, D_R and M_R are the lift force, drag force, and moment, respectively, ρ is water density, α is the angle of attack, V_R is the rudder inflow speed, A_R is the rudder area, and

c is the chord length of the rudder profile. The center of pressure is set as 0.25 c from the rudder's leading edge, based on engineering practice and experimental data [25].

To evaluate the mesh quality, wind tunnel results of the NACA 0012 profile, which are tested by Ladson [25], are taken as the benchmark case. Moreover, three independent compressible CFD codes [19], i.e., CFL3D (NASA LaRC, USA), FUN3D (NASA LaRC, USA), and NTS (NTS, Russia), are cited to make a crosswise comparison with the Fluent solutions. Results of the benchmark and the CFD codes are achieved in essentially incompressible air with a Mach number of 0.15. Thus, they are comparable with the results obtained in incompressible water in this paper. Presented results are compared to that of the benchmark cases in relative differences under various angles of attack at a critical Reynolds number as follows:

$$\Delta C_{\alpha} = \frac{C_{\alpha Fluent}}{C_{\alpha Benchmark}} - 1 \quad (8)$$

$$\Delta C(\mathrm{AVE}) = \frac{\sum_{i=1}^{n} |\Delta C_{\alpha_i}|}{n}, \quad i = 1, 2, \ldots, n \quad (9)$$

where ΔC_{α} is the relative differences of the calculated lift, drag, or moment coefficients achieved by Fluent and provided by the benchmark data, respectively, and $\Delta C(\mathrm{AVE})$ is the average relative differences for a series of α.

4.4. Uncertainty Estimation

The uncertainty estimation is important to evaluate the reliability of data in CFD calculations. According to International Towing Tank Conference [26], simulation numerical uncertainty U_{SN} and simulation modeling uncertainty U_{SM} need to be assessed respectively. In this paper, only the numerical errors are considered since modeling errors are difficult to quantify in practical applications [27]. Generally, numerical errors relate to grid size, time step, and other parameters. This paper performs steady simulations and focuses on mesh properties. Accordingly, only the grid uncertainty U_G is calculated for uncertainty estimations. With validation uncertainty U_V, which is simplified as U_G in this study, obtained, U_V and the comparison error $|E|$ can be compared. If $|E| < U_V$, the validation is achieved at U_V level.

The Grid Convergence Index (GCI) method [28,29] and the correction factor (CF) based method [30] remain popular in uncertainty estimations, however, it is deficient that both methods fail to give reliable estimations when the order of grid convergence p is too high (i.e., $p > 2$) or too small (i.e., $p < 0.5$). In other words, GCI and CF methods require data in the so-called asymptotic range, which means that data on grids are fine enough to give a single dominant term in a power series expansion of the error [31]. Considering that practical flow phenomenons are generally complex, and it is inevitable that p may be not in the range of $[0.5, 2]$, Eça and Hoekstra [31] proposed another GCI method based on least-square roots (LSR-GCI). Such a method allows for error estimations even if the monotonic convergence is not achieved, and it is proven to be more reliable compared with GCI and CF based methods by De Luca et al. [32]. Therefore, the LSR-GCI method [31] is applied for uncertainty estimations in this paper.

For four solutions $\phi_i, (i = 1, 2, 3, 4)$ obtained from systematically refined cases and characterized by typical cell sizes $h_i, (i = 1, 2, 3, 4)$, the discretization error can be estimated with one of the four following equations:

$$\varepsilon_{\phi} \simeq \phi_i - \phi_0 = \alpha h_i^p \quad (10)$$

$$\varepsilon_{\phi} \simeq \phi_i - \phi_0 = \alpha h_i \quad (11)$$

$$\varepsilon_{\phi} \simeq \phi_i - \phi_0 = \alpha h_i^2 \quad (12)$$

$$\varepsilon_{\phi} \simeq \phi_i - \phi_0 = \alpha_1 h_i + \alpha_2 h_i^2 \quad (13)$$

where ϕ_0 is the estimate of the exact solution, α is an undetermined constant, and p is the observed order of grid convergence. Besides, the following functions are proposed to select the error estimator:

$$S_{RE}(\phi_0, \alpha, p) = \sqrt{\sum_{i=1}^{n_g} \left(\phi_i - \left(\phi_0 + \alpha h_i^p\right)\right)^2} \tag{14}$$

$$S_{RE}^{w}(\phi_0, \alpha, p) = \sqrt{\sum_{i=1}^{n_g} w_i \left(\phi_i - \left(\phi_0 + \alpha h_i^p\right)\right)^2}, (w_i = \frac{\frac{1}{h_i}}{\sum_{i=1}^{n_g} \frac{1}{h_i}}) \tag{15}$$

$$S_1(\phi_0, \alpha) = \sqrt{\sum_{i=1}^{n_g} (\phi_i - (\phi_0 + \alpha h_i))^2} \tag{16}$$

$$S_1^{w}(\phi_0, \alpha) = \sqrt{\sum_{i=1}^{n_g} w_i (\phi_i - (\phi_0 + \alpha h_i))^2}, (w_i = \frac{\frac{1}{h_i}}{\sum_{i=1}^{n_g} \frac{1}{h_i}}) \tag{17}$$

$$S_2(\phi_0, \alpha) = \sqrt{\sum_{i=1}^{n_g} \left(\phi_i - \left(\phi_0 + \alpha h_i^2\right)\right)^2} \tag{18}$$

$$S_2^{w}(\phi_0, \alpha) = \sqrt{\sum_{i=1}^{n_g} w_i \left(\phi_i - \left(\phi_0 + \alpha h_i^2\right)\right)^2}, (w_i = \frac{\frac{1}{h_i}}{\sum_{i=1}^{n_g} \frac{1}{h_i}}) \tag{19}$$

$$S_{12}(\phi_0, \alpha_1, \alpha_2) = \sqrt{\sum_{i=1}^{n_g} \left(\phi_i - \left(\phi_0 + \alpha_1 h_i + \alpha_2 h_i^2\right)\right)^2} \tag{20}$$

$$S_{12}^{w}(\phi_0, \alpha_1, \alpha_2) = \sqrt{\sum_{i=1}^{n_g} w_i \left(\phi_i - \left(\phi_0 + \alpha_1 h_i + \alpha_2 h_i^2\right)\right)^2}, (w_i = \frac{\frac{1}{h_i}}{\sum_{i=1}^{n_g} \frac{1}{h_i}}) \tag{21}$$

Details in selecting the error estimator are shown in Figure 7 according to Eça and Hoekstra [31]. $\phi_{0,RE}^*, \alpha_{RE}^*, p_{RE}^*$ are variables that make $S_{RE}(\phi_0, \alpha, p)$ minimum, which can be denoted as $\left(\phi_{0,RE}^*, \alpha_{RE}^*, p_{RE}^*\right) = S_{RE}^{-1} \min\{S_{RE}(\phi_0, \alpha, p)\}$ with the inverse function symbol S_{RE}^{-1}. Similarly, we have $\phi_{0,RE}^{w^*}, \alpha_{RE}^{w^*}, p_{RE}^{w^*}, \phi_{0,1}^*, \alpha_1^*, p_1^*, \phi_{0,1}^{w^*}, \alpha_1^{w^*}, p_1^{w^*}, \phi_{0,2}^*, \alpha_2^*, p_2^*, \phi_{0,2}^{w^*}, \alpha_2^{w^*}, p_2^{w^*}$, $\phi_{0,12}^*, \alpha_{12}^*, p_{12}^*, \phi_{0,12}^{w^*}, \alpha_{12}^{w^*}, p_{12}^{w^*}$. Corresponding standard deviations can be expressed as:

$$\sigma_{RE} = \sqrt{\frac{\sum_{i=1}^{n_g} nw_i \left(\phi_i - \left(\phi_0 + \alpha h_i^p\right)\right)^2}{(n_g - 3)}} \tag{22}$$

$$\sigma_1 = \sqrt{\frac{\sum_{i=1}^{n_g} nw_i (\phi_i - (\phi_0 + \alpha h_i))^2}{(n_g - 2)}} \tag{23}$$

$$\sigma_2 = \sqrt{\frac{\sum_{i=1}^{n_g} nw_i \left(\phi_i - \left(\phi_0 + \alpha h_i^2\right)\right)^2}{(n_g - 2)}} \tag{24}$$

$$\sigma_{12} = \sqrt{\frac{\sum_{i=1}^{n_g} nw_i \left(\phi_i - \left(\phi_0 + \alpha_1 h_i + \alpha_2 h_i^2\right)\right)^2}{(n_g - 3)}} \tag{25}$$

For no-weighted cases, $nw_i = 1$. Following criteria are applied in selecting the error estimators:

- **Criterion 1** (S_{RE}, S^*_{RE} in Figure 7)
  Select p^* corresponding to $\min \sigma^*_{RE}, \sigma^{w^*}_{RE}$
- **Criterion 2** (S_2, S^w_2 in Figure 7)
  When $p^* > 2$, small uncertainty quantities can be obtained due to overestimates of the order of accuracy. As recommended by Eça and Hoekstra [31], when $p^* > 2$, the error estimator is chosen by comparing the standard deviations resulting from Equations (16)–(19). Then, ε_ϕ is calculated with the selected estimator. This procedure substitutes the estimated order of accuracy with 1 or 2 because the exponential power of error estimators to be selected (Equations (11) and (12)) is either 1 or 2. Slightly different from Eça and Hoekstra [31], De Luca et al. [32] substitute p as the theoretical order of accuracy p_{th}, i.e., $p = p_{th} = 2$, when $p^* > 2$. In fact, the method by De Luca et al. [32] is more conservative, and it is applied in following calculations. After that, the error estimator can be selected according to $\min \sigma^*_2, \sigma^{w^*}_2$
- **Criterion 3** ($S_1, S^w_1, S_2, S^w_2, S^*_{12}, S^{w^*}_{12}$ in Figure 7)
  Select the error estimator according to $\min \sigma^*_1, \sigma^{w^*}_1, \sigma^*_2, \sigma^{w^*}_2, \sigma^*_{12}, \sigma^{w^*}_{12}$

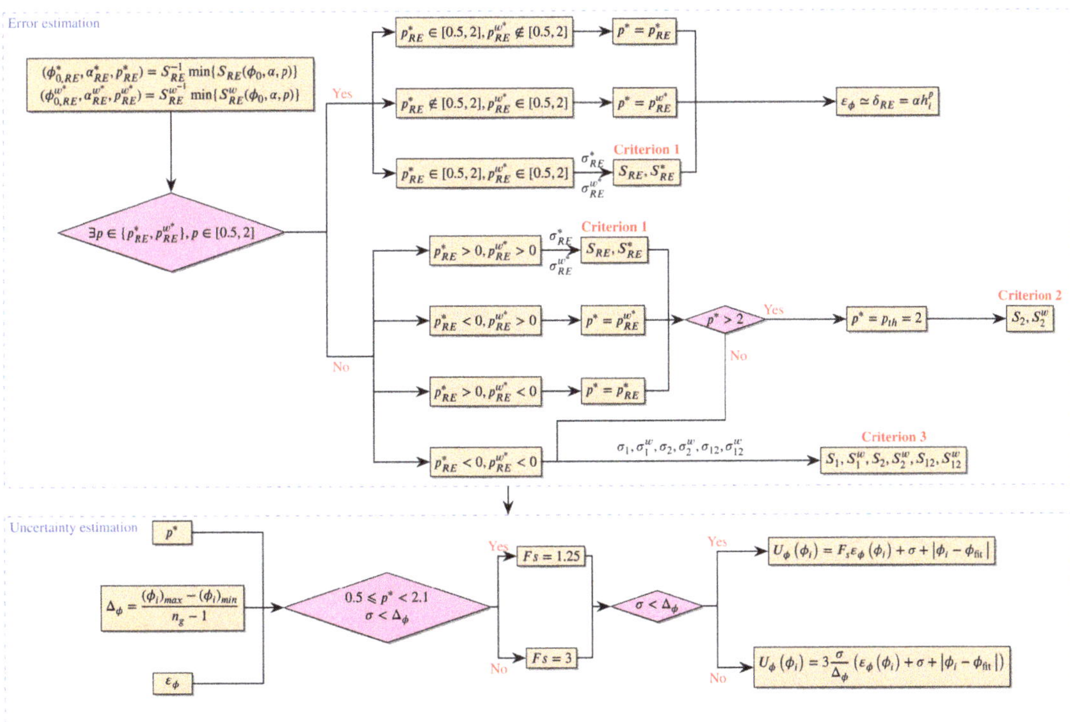

Figure 7. Procedure for error and uncertainty estimations by GCI-LSR method recommended by Eça and Hoekstra [31], De Luca et al. [32].

5. Results and Discussions

With the strategies and numerical methods illustrated in Section 4, systematic simulations are conducted to study the impacts of mesh properties on hydrodynamic performance of the NACA 0012 rudder profile in this section. For each case with different mesh settings, errors with respect to benchmark data [19,25] are displayed in tabular form. Besides, uncertainty estimations are conducted for cases with varying cell sizes. For colored error tables in this section, the color legends are marked based on the values of relative differences. The smallest difference is the darkest green and the largest one is the darkest red for each mesh group, while other colors between the darkest green to red show the transition. The impacts of Reynolds numbers are presented in Section 5.1, and Section 5.2 discusses the selection of domain sizes taking the C-mesh as an example, and impacts of node distributions and mesh types are discussed in Sections 5.3 and 5.4.

5.1. Impacts of Reynolds Numbers

Before conducting parametric investigations, the Reynolds number should be checked first. The Reynolds number (*Re*) is defined as the ratio of inertial forces to viscous forces, which represents the viscous similarity among flow patterns and can be expressed as:

$$Re = \frac{\rho U_\infty L}{\mu} \tag{26}$$

where U_∞ is the inflow velocity, L is the characteristic length, and μ is the viscosity coefficient. For rudder hydrodynamics investigations, due to the limited model size and capacity of the test facility, *Re* of tests in wind tunnels and towing tanks has to be scaled. Analogously, model-scale rudders are preferred by CFD methods since a fine mesh and a large domain are needed for full-scale simulations. In most cases, the similarity law of viscosity and gravity can not be satisfied simultaneously. Generally, experiments or numerical simulations are conducted under a *Re* which is large enough to guarantee that the turbulence develops sufficiently. A high *Re* ensures a fully turbulent flow and reduces the effect of the laminar-turbulent transition, achieving an increase of prediction accuracy compared to that of an analysis at low *Re* with the same mesh.

Cases with larger *Re* show less calculation efficiency seeing that more cells around boundaries are required. The height of the first row of cells Δs varies with *Re*, as shown in Figure 8, while increasing *Re* leads to decreasing heights. When discretizing domains into multiples mesh elements, the aspect ratio for a single element should be around 1 to ensure mesh quality. When it comes to cells near walls, aspect ratios are connected with Δs and chord-wise spacings Δc. As shown in Figure 9, more nodes along the wall are distributed with smaller Δs to keep aspect ratios compliant with the requirement of solvers. Thus, a decrease in *Re* leads to a decrease in the number of cells. To sum up, we suggest that using a minimum *Re* which is achievable in experiments, ensures full-turbulent conditions, and can be solved with fewer mesh elements.

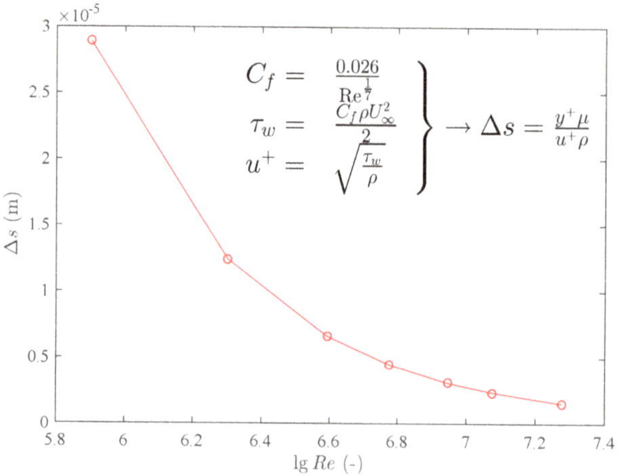

Figure 8. Different meshes' height of first layer with varying *Re* and $y^+ = 1$.

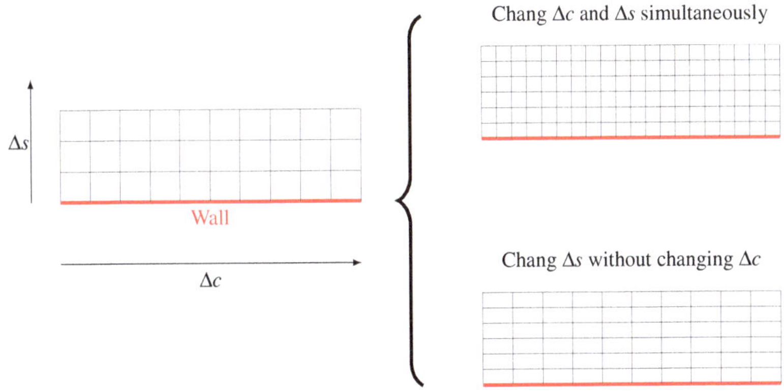

Figure 9. Impacts of height of first mesh Δs on aspects ratios.

To determine the threshold of the Reynolds number, above which the turbulence develops completely and force coefficients do not change significantly, cases with varying *Re* are simulated in this section. Based on the flat-plate boundary layer theory and the calculation method in [33], the height of meshes' first layer Δs above the rudder surface depends on y^+ and *Re*. According to the application scope of the *k*-ω SST model, y^+ is set as 1 for all cases, and corresponding heights are shown in Figure 8.

Common benchmark wind tunnel tests are carried out at *Re* in the range of 1.00×10^5 [34] to 1.00×10^7 [25]. As low-Reynolds-number RANS analysis is still challenging [35,36] and high-Reynolds-number simulations may be expensive in computation time, the present work tests the NACA 0012 profile at *Re* in a range of $2.00 \times 10^5 \sim 1.89 \times 10^7$. Among these aerodynamic experimental results, limited cases with $Re = 3.94 \times 10^6 \sim 1.89 \times 10^7$ and a small Mach number can be taken as validation results of incompressible water simulations, as the compressibility effects of a fluid with a Mach number smaller than 0.20 is small. Table 4 presents the case settings and results of various *Re* under distinct α series, as specific values of α are listed in [25].

Figure 10 presents calculation results of hydrodynamic coefficients versus the same rudder angles for different *Re*, which indicates that the lift curve rises with an increase of the *Re* while the drag curve decreases. Compared to that of the lift coefficients, the drag coefficients are more sensitive to changes in *Re*. The drag coefficient under 15° at $Re = 1.89 \times 10^7$ is about half of the value at $Re = 4.00 \times 10^5$, whereas the lift coefficient is 1.20 times larger. The differences in lift and drag forces between low and high *Re* increase with an increasing angle of attack. Consistent with findings by Ladson [25] and Molland and Turnock [37], $Re = 6.00 \times 10^6$ can be considered as a threshold value for the mesh generation, above which little variation may be found. Moreover, the y^+ along the chord for four cases under small and large rudder angles with varying *Re* are shown in Figure 11. y^+ values for the case with $Re > 5.97 \times 10^6$ are roughly between 0 and 1, indicating that the *k*-ω SST turbulence model is applicable for the case. Since ship rudders in propeller slipstream usually experience full turbulence with large *Re*, low *Re* conditions are not of interest in current studies. In this paper, we set *Re* of the following simulations as 6.00×10^6.

Table 4. Comparison of results of hydrodynamic coefficients under different *Re* with that of experimental benchmark data adopted from [25] in relative differences.

Re	$\%\Delta C_D(\alpha \approx 0.00°)$	$\%\Delta C_L(\alpha \approx 4.18°)$	$\%\Delta C_D(\alpha \approx 4.18°)$	$\%\Delta C_L(\alpha \approx 8.22°)$	$\%\Delta C_D(\alpha \approx 8.22°)$	$\%\Delta C_L(\alpha \approx 13.10°)$	$\%\Delta C_D(\alpha \approx 13.10°)$	$\%\Delta C_L(\alpha \approx 15.20°)$	$\%\Delta C_D(\alpha \approx 15.20°)$
3.94×10^6	32.26	3.33	49.48	1.69	24.03	−0.67	41.61	−1.99	39.59
5.97×10^6	25.67	1.97	17.53	1.29	44.17	−1.84	50.87	−3.28	44.76
6.00×10^6	25.57	2.01	17.36	1.32	43.61	−1.64	49.30	−2.91	42.49
8.86×10^6	22.10	1.73	20.07	0.09	26.34	−2.32	51.85	−3.33	49.86
1.19×10^7	38.37	−0.47	17.54	0.00	25.81	−1.57	41.30	−3.24	32.23
1.89×10^7	47.95	1.13	5.67	−0.88	9.18	−1.51	25.84	−1.99	19.82

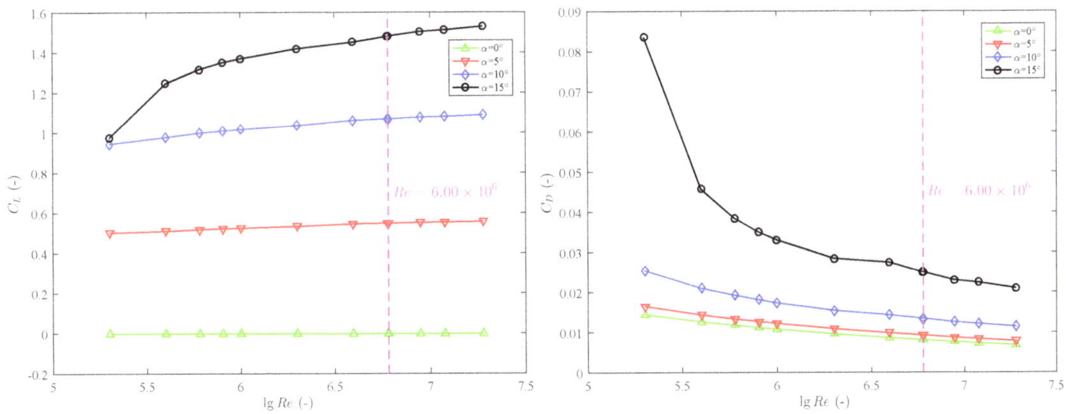

Figure 10. Impacts of Re on lift and drag coefficients.

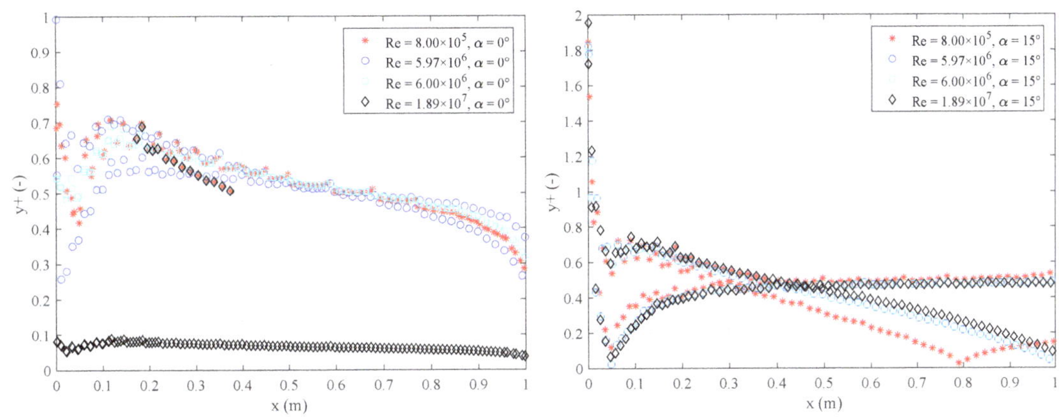

Figure 11. y^+ along rudder chord with varying Re.

5.2. Impacts of Domain Sizes

A C-mesh is applied to investigate the effects of the domain size on RANS solutions by conducting simulations with various sizes against varying cell numbers in this section. Test parameters and related relative differences are given in Table 5. C_{in} and C_{out} are doubled for each case compared with that of the previous one, while the narrowest domain extends 7.5 c upstream and 15 c downstream, which is 16 times smaller than the largest ones, i.e., 120 c and 240 c. Since equivalent mesh nodes correspond to different spacing variations in a variety of domains, different nodes distributions are applied for each domain to determine reasonable meshing strategies.

Pressure distributions in the whole domain and around the rudder are presented in Figure 12. From Figure 12a–c, regions with great gradient changes near the rudder can be observed even for the smallest domain. Comparing Figure 12d–f, neither high-pressure regions near the stagnation points nor regions near the trailing edges show significant distinctions. The impact of domain size on pressure distributions around the rudder is relatively small, which corresponds to minor differences of C_L mainly derived from pressure integration.

Compared with that of C_L, accuracy of C_D is more sensitive to domain sizes in Table 5. With an increase of the domain size, the accuracy of the prediction of C_D is more significantly improved than C_L. Small domain sizes show poor accuracy in C_D but obtained C_L is acceptable for engineering applications. The relative differences of C_D for $\alpha = 15.20°$ drop from 55.19% to 34.79%, indicating that boundary effects can evidently change viscosity components around the rudder, which mainly contribute to drag forces. On the other hand, larger domains, like 60 $c/120$ c and 120 $c/240$ c, tend to show high and stable precision with more cells. However, further expansion of the domain size may dramatically increase the number of cells and computation time.

Table 5. Comparison of results of C-mesh based on various domain sizes with that of experimental benchmark data adopted from [25] in relative differences.

C_{in}/C_{out}	NOC	$\%\Delta C_L(13.10°)$	$\%\Delta C_D(13.10°)$	$\%\Delta C_L(15.20°)$	$\%\Delta C_D(15.20°)$	$\%\Delta C_L(AVE)$	$\%\Delta C_D(AVE)$
7.5 $c/15$ c	160,293	\	\	\	\	\	\
	203,144	−1.75	68.46	−2.57	55.19	2.16	61.82
	251,065	−1.46	65.74	−21.04	\	\	\
	423,225	−2.10	71.32	−3.28	59.88	2.69	65.60
	524,605	−1.65	67.15	−2.38	53.52	2.01	60.33
15 $c/30$ c	160,293	\	\	\	\	\	\
	203,144	\	\	−2.60	43.80	\	\
	251,065	−1.31	50.88	\	\	\	\
	423,225	−1.96	56.69	\	\	\	\
	524,605	−1.50	52.33	−2.40	42.50	1.95	47.41
30 $c/60$ c	160,293	\	\	\	\	\	\
	203,144	−1.45	47.99	−2.49	39.87	1.97	43.93
	251,065	−1.19	44.95	−1.87	34.19	1.53	39.57
	423,225	−1.83	50.77	−3.27	44.66	2.55	47.71
	524,605	−1.37	46.32	−2.33	37.85	1.85	42.08
60 $c/120$ c	160,293	−0.83	41.54	−1.28	30.84	1.06	36.19
	203,144	−1.35	45.53	−2.40	37.90	1.87	41.72
	251,065	\	\	\	\	\	\
	423,225	−1.75	48.12	−3.22	42.64	2.49	45.38
	524,605	−1.29	43.63	−2.28	35.76	1.78	39.70
120 $c/240$ c	160,293	−0.92	42.50	\	\	\	\
	203,144	−1.33	44.90	−2.37	37.43	1.85	41.16
	251,065	−1.06	41.43	−1.85	32.64	1.45	37.03
	423,225	−1.71	46.89	\	\	\	\
	524,605	−1.24	42.38	−2.25	34.79	1.75	38.59

From the perspective of efficiency, a domain size that eliminates boundary effects and meanwhile requires fewer mesh nodes is desired for following investigations. Table 5 shows that a domain size of $C_{in} = 30\,c$ and $C_{out} = 60\,c$ with about 2.50×10^5 cells achieves a good balance in both accuracy and efficiency, and Table 6 shows that the validation process is achieved for C_L obtained from the case (ϕ_2). This domain size is large enough to obtain sufficiently accurate estimates of lift and drag, and the number of cells is still manageable by common desktop computers. Thus, it is applied for the remainder of the simulations in this paper. Table 5 indicates that even more accurate predictions can be obtained with larger domains. Thus, we suggest applying a large domain when computation power and time are available. If the lift is the only purpose, a small domain size of 15 $c/30$ c may be more favorable as it requires less computation power and time than a large domain.

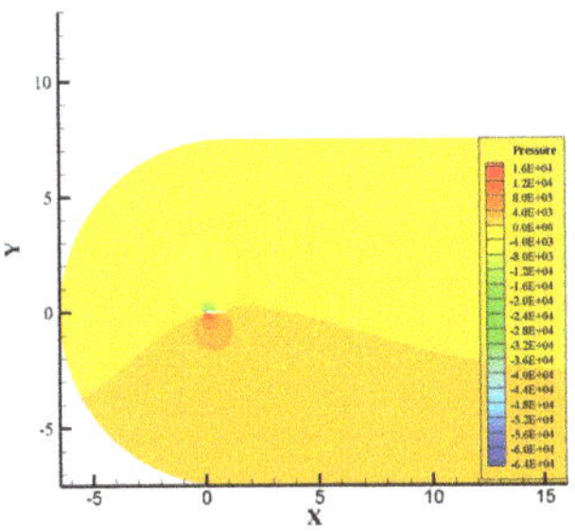

(a) 7.5 $c/15$ c, Overall view.

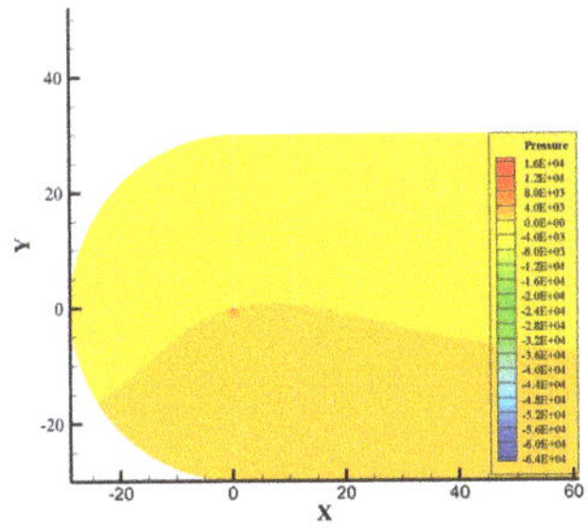

(b) 30 $c/60$ c, Overall view.

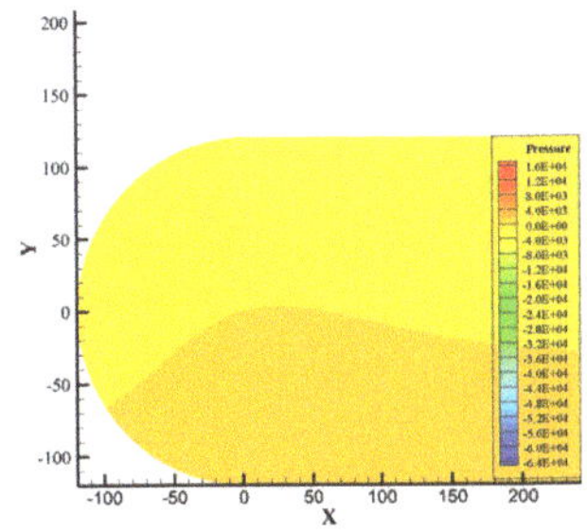

(c) 120 $c/240$ c, Overall view.

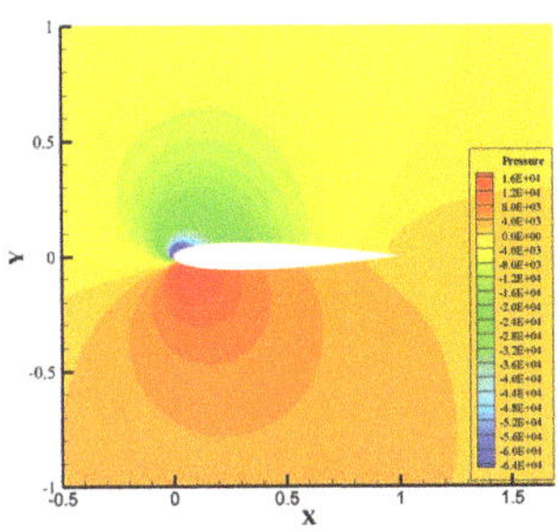

(d) 7.5 $c/15$ c, View around the rudder.

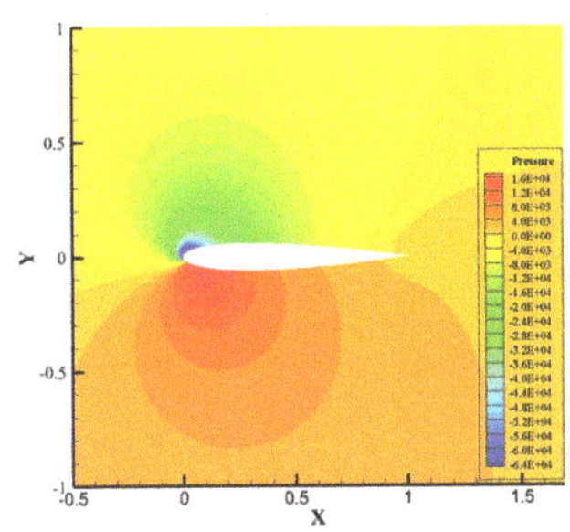

(e) 30 $c/60$ c, View around the rudder.

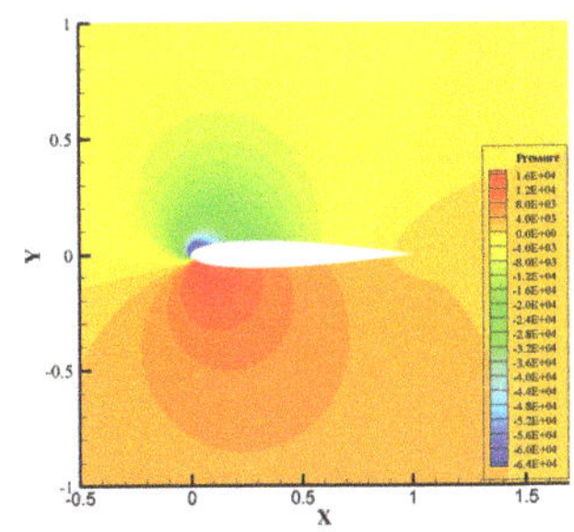

(f) 120 $c/240$ c, View around the rudder.

Figure 12. Pressure contours of different domain sizes with $\alpha = 15.20°$.

Table 6. Uncertainty estimations for 4 cases (30 $c/60$ c, NOC = 203,144, 251,065, 423,225, 524,605, $\alpha = 15.2°$) based on mesh dependency.

Solution	C_L			C_D			C_M		
	p	$\%U_G$	$\%\|E\|$	p	$\%U_G$	$\%\|E\|$	p	$\%U_G$	$\%\|E\|$
ϕ_1	2.00	10.02	2.49	2.00	0.21	39.87	2.00	0.12	2.13
ϕ_2		16.18	1.87		0.54	34.19		0.12	0.81
ϕ_3		17.73	3.27		0.62	44.66		0.21	6.39
ϕ_4		11.57	2.33		0.37	37.85		0.09	1.91

5.3. Impacts of Node Distributions

After determining the test Re and the domain size, raw meshes are ready for the mesh independence study for different mesh types. The common procedure is to carry out simulations on an initial mesh with a residual error in a range of $1.00 \times 10^{-4} \sim 1.00 \times 10^{-5}$. After that, refine the mesh globally to around 2 or $\sqrt{2}$ times the initial mesh. Next, run simulations with the refined mesh and compare solutions obtained from the coarse mesh and the fine mesh. Repeat the refinement until the results do not significantly change with a finer mesh. The mesh independence study requires at least 3 solutions to evaluate the convergence of certain inputs [26]. Considering the computation time, it is always better to use the smallest number of cells. In this paper, mesh resolutions are changed by conducting local rather than global refinements, while the grid refinement factor R_G is still around $\sqrt{2}$ for target parameters.

Since the flow near boundaries of the rudder profile changes more violently than that in the far-field and has larger effects on rudder hydrodynamics, near-field node distributions are studied by fixing far-field node distributions for four mesh types. After that, the impacts of far-field distributions are studied with proper near-field distributions.

In this section, with a domain size of 30 $c/60$ c, impacts of near-field node distributions are investigated for four mesh types, as shown in Figures 13–16. Meshes in different regions are categorized into two parts, which are located in the far-field and near the rudder surface. Near-field meshes are defined as those located along the rudder chord and within the boundary layers regions. The domain for C-mesh in Figure 13 consists of a semicircle and a rectangle, and nodes distributed parallel to the radius are defined as in radial directions, and the rudder in O-mesh is surrounded by radial nodes in Figure 15. The nodes extending from the trailing edge to the outlet of the boundary are defined as in wake directions. Patterned after the definitions above, node distributions for H-mesh are shown in Figure 14, and nodes from the leading edge to the inlet are in inlet directions. For Hybrid-mesh in Figure 16, considering quite different mesh generation methods compared with that of structured meshes, meshes in the refinement area near the rudder are treated as near-field meshes while others are in far-field.

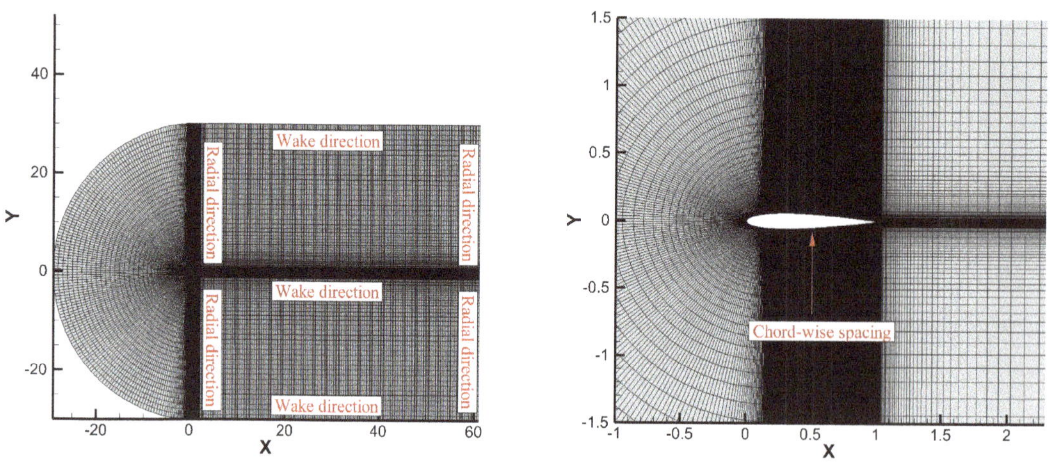

(a) C-mesh, far-field. (b) C-mesh, near-field.

Figure 13. Node distributions for C-mesh.

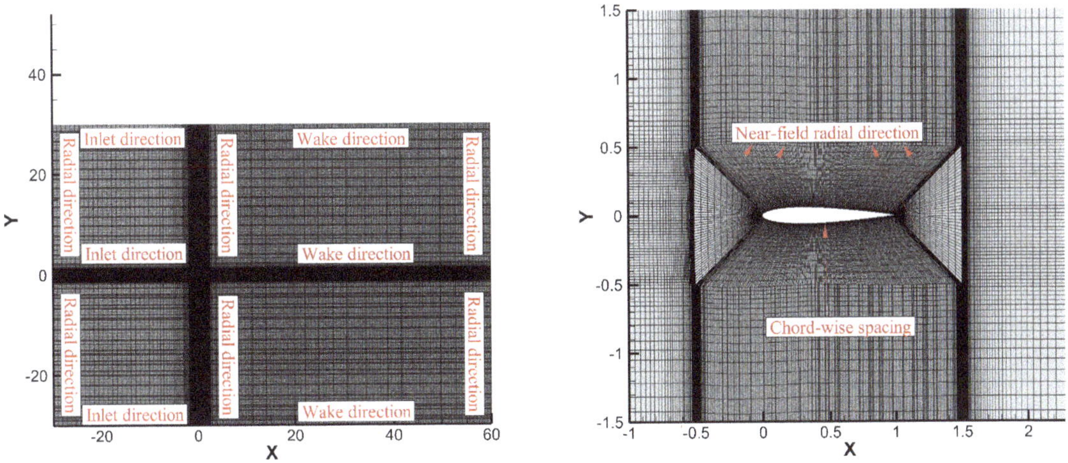

(a) H-mesh, far-field.

(b) H-mesh, near-field.

Figure 14. Node distributions for H-mesh.

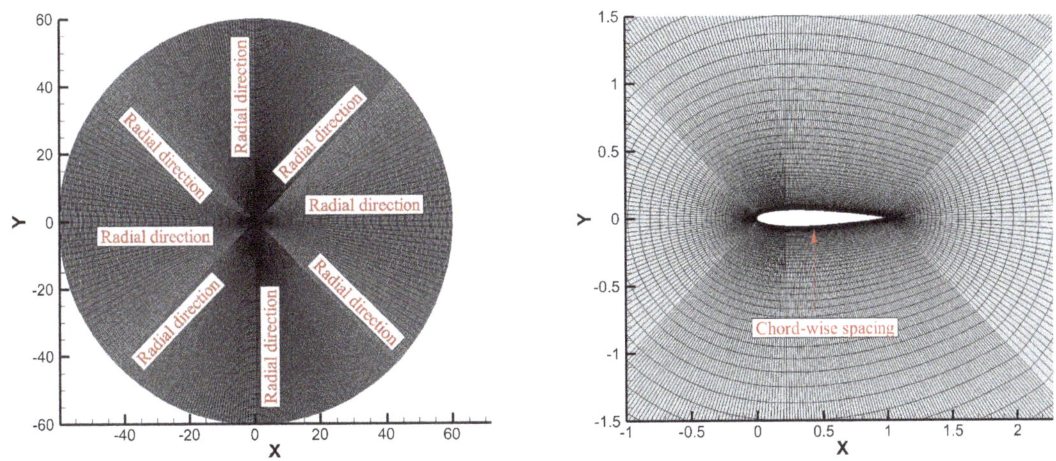

(a) O-mesh, far-field.

(b) O-mesh, near-field.

Figure 15. Node distributions for O-mesh.

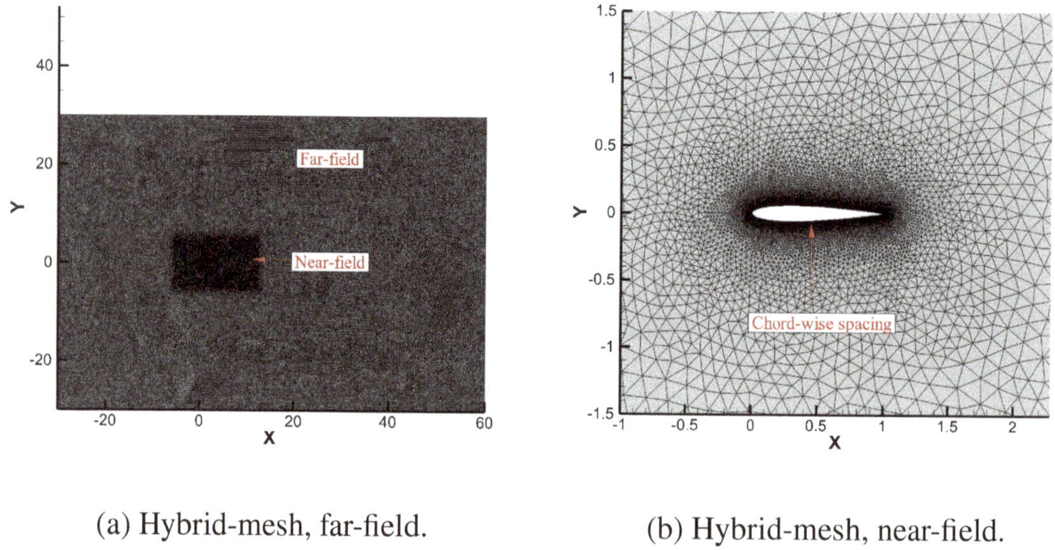

Figure 16. Node distributions for Hybrid-mesh.

5.3.1. Impacts of Near-Field Node Distributions

In this part, impacts of near-field node distributions are studied by changing chord-wise spacings and layer-wise growth rates for C-mesh, H-mesh, O-mesh, and Hybrid-mesh. The near-chord refinement is performed by changing the number of nodes in the chord-wise direction for structured meshes and chord-wise spacings for Hybrid-mesh with R_G around $\sqrt{2}$. For structured meshes, a reference setting is coarsened or refined to perform parametric studies. The quantitative relation between a coarsened or refined case and the reference case can be expressed as:

$$\log_{R_G} \frac{N_{c_i}}{N_{c_0}} = \log_{R_G} N_{c_{i0}} = i, \quad (i = -2, -1, \dots) \tag{27}$$

where i denotes the level of refinement with respect to the reference case, and N_{c_i} is the number of nodes along the chord. Taking C-mesh for example, the rudder is wrapped by denser cells in chord-wise directions, as shown in Figure 13. As for cells in the boundary layer region of the rudder, much attention should be paid to capture violently changing velocity gradients near the wall. The layer-wise growth rate is investigated to provide a detailed view of mesh resolutions in boundary layer regions.

Cells around the rudder with varying chord-wise spacings for C-mesh are shown in Figure 17. As the chord-wise spacing decreases in Table 7, the number of cells with large aspect ratios in the wake increases, which introduces the difficulty of convergence, especially under large angles of attack. The accuracy of C_D is slightly improved with denser chord-wise node distributions, while variations of C_L are even smaller. Larger deviations are obtained for C_M for small α, probably due to measurement errors in experiments induced by small absolute values. For $\alpha = 4.18°$ and $\alpha = 8.22°$, better mesh resolution along the rudder lead to lower precision, which are related to convergence problems due to the fact that small flow features are being solved. Since calculation results tend to stabilize without diffusing after $\log_{R_G} N_{c_{i0}} = 2$ and 5.48×10^5 mesh quadrilaterals show relatively high accuracy, the case with $\log_{R_G} N_{c_{i0}} = 3$ is selected for the following simulations on C-mesh, and corresponded uncertainty estimations are shown in Table 8.

Table 7. Comparison of results of C-mesh based on various chord-wise spacings with that of experimental benchmark data adopted from [25] in relative differences with 670 nodes in chord-wise direction for reference setting.

$\log_{R_G} N_{ci0}$	NOC	$\%\Delta C_L(4.18^\circ)$	$\%\Delta C_D(4.18^\circ)$	$\%\Delta C_M(4.18^\circ)$	$\%\Delta C_L(8.22^\circ)$	$\%\Delta C_D(8.22^\circ)$	$\%\Delta C_M(8.22^\circ)$	$\%\Delta C_L(13.10^\circ)$	$\%\Delta C_D(13.10^\circ)$	$\%\Delta C_M(13.10^\circ)$	$\%\Delta C_L(15.20^\circ)$	$\%\Delta C_D(15.20^\circ)$	$\%\Delta C_M(15.20^\circ)$	$\%\Delta C_L(AVE)$	$\%\Delta C_D(AVE)$	$\%\Delta C_M(AVE)$
−2	169,860	1.85	18.10	−60.52	1.02	47.98	−22.60	\	\	\	\	\	\	\	\	\
−1	203,534	1.98	17.39	−62.91	1.37	43.79	−30.17	−1.28	46.15	6.94	−2.10	36.68	0.52	1.68	36.00	25.13
0	251,065	1.99	17.36	−62.48	1.42	42.89	−30.23	\	\	\	\	\	\	\	\	\
1	318,413	1.98	17.47	−61.99	1.28	44.17	−26.44	−1.81	50.89	13.49	\	\	\	\	\	\
2	413,475	2.01	17.36	−62.58	1.32	43.61	−27.09	−1.64	49.30	11.47	−2.91	42.49	4.26	1.97	38.19	26.35
3	548,022	2.03	17.15	−63.17	1.44	42.09	−29.86	−1.05	43.73	4.52	−1.76	34.33	−1.74	1.57	34.32	24.82
4	738,295	2.03	17.15	−62.90	1.42	42.16	−28.96	−1.12	44.26	5.42	−1.88	35.22	−1.12	1.61	34.70	24.60
5	1,007,091	2.02	17.15	−62.63	1.39	42.23	−28.06	−1.18	44.79	6.31	−1.56	34.66	−0.84	1.54	34.71	24.46

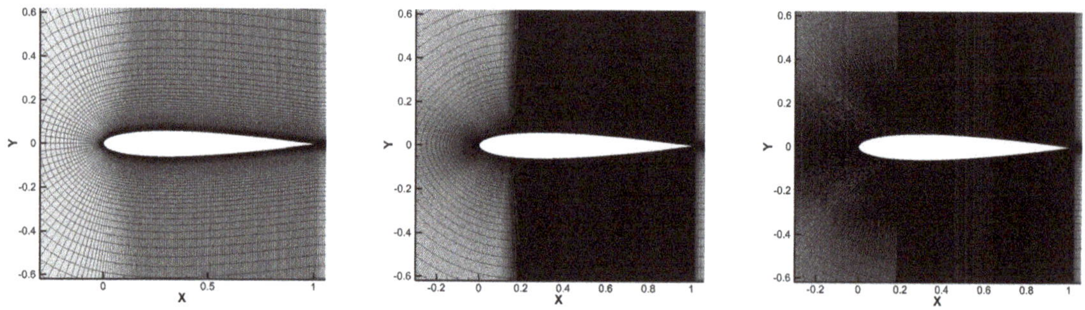

(a) $\ln r = -2$. (b) $\ln r = 1$. (c) $\ln r = 4$.

Figure 17. Cells around the rudder with varying chord-wise spacings for C-mesh.

Table 8. Uncertainty estimations for C-mesh ($\log_{R_G} N_{c_{i0}} = 2, 3, 4, 5$, $\alpha = 15.2°$) with varying chord-wise spacings based on mesh dependency.

Solution	C_L			C_D			C_M		
	p	$\%U_G$	$\%\lvert E\rvert$	p	$\%U_G$	$\%\lvert E\rvert$	p	$\%U_G$	$\%\lvert E\rvert$
ϕ_1	2.00	2.02	2.91	1.04	1.28	42.49	1.23	0.70	4.26
ϕ_2		0.46	1.76		1.11	34.32		0.59	1.74
ϕ_3		0.06	1.88		0.72	35.22		0.37	1.11
ϕ_4		0.04	1.56		0.67	34.65		0.34	0.84

For H-mesh, besides chord-wise nodes, nodes distributed along with radial directions in the refinement region also count in near-field node distributions. Therefore, chord-wise nodes and radial nodes near the rudder are both refined for H-mesh. Due to discrepancies in refinement regions in Table 9, the NOC of H has a more rapid growth rate compared with C-mesh. Generally, variations of the accuracy of hydrodynamic coefficients are consistent with the increase in the number of cells, and relatively greater improvements can be observed for C_D and C_M taking $\log_{R_G} N_{c_{i0}} = 2$ as the dividing line. For the NOC less than 3.00×10^5, C-mesh is a better choice for predictions of $\alpha < 8.22°$ when C-mesh and H-mesh share the same order of magnitude. Considering that near-field node distributions of H-mesh are able to capture more precise flow details around the rudder with denser near-field node distributions, H-mesh with fine mesh resolutions can be applied with enough computational resources. In this part, to balance the accuracy and efficiency, $\log_{R_G} N_{c_{i0}} = 2$ is selected for following simulations on H-mesh, and corresponded uncertainty estimations are shown in Table 10.

Table 9. Comparison of results of H-mesh based on various chord-wise spacings with that of experimental benchmark data adopted from [25] in relative differences with 386 nodes in chord-wise direction for reference setting.

$\log_{R_G} N_{ci0}$	NOC	$\%\Delta C_L(4.18^\circ)$	$\%\Delta C_D(4.18^\circ)$	$\%\Delta C_M(4.18^\circ)$	$\%\Delta C_L(8.22^\circ)$	$\%\Delta C_D(8.22^\circ)$	$\%\Delta C_M(8.22^\circ)$	$\%\Delta C_L(13.10^\circ)$	$\%\Delta C_D(13.10^\circ)$	$\%\Delta C_M(13.10^\circ)$	$\%\Delta C_L(15.20^\circ)$	$\%\Delta C_D(15.20^\circ)$	$\%\Delta C_M(15.20^\circ)$	$\%\Delta C_L(AVE)$	$\%\Delta C_D(AVE)$	$\%\Delta C_M(AVE)$
−2	178,209	2.53	18.68	−83.92	1.08	61.15	−34.30	−3.47	84.99	23.30	−6.28	81.61	13.11	3.34	61.61	38.66
−1	208,245	2.61	17.19	−83.64	1.48	50.20	−39.06	−2.55	64.66	19.18	−4.60	60.33	10.91	2.81	48.09	38.20
0	255,852	2.73	17.41	−85.96	1.82	45.53	−45.29	−1.45	51.54	8.66	\	\	\	\	\	\
1	333,306	2.84	17.90	−88.34	1.97	43.93	−48.35	−1.05	47.09	4.81	−2.09	37.97	2.96	1.99	36.72	36.11
2	462,288	2.84	17.75	−87.83	2.05	42.18	−49.32	−0.67	41.65	1.45	−1.30	30.54	−0.37	1.72	33.03	34.74
3	683,862	2.87	17.24	−87.77	2.08	41.06	−49.42	−0.64	40.21	1.63	−1.28	29.36	−0.09	1.72	31.97	34.73
4	1,079,760	2.89	17.06	−88.12	2.13	40.32	−50.54	−0.42	37.89	−1.11	−0.85	26.08	−2.60	1.57	30.34	35.59

Table 10. Uncertainty estimations for H-mesh ($\log_{R_G} N_{C_{i0}} = 1,2,3,4$, $\alpha = 15.2°$) with varying chord-wise spacings based on mesh dependency.

Solution	C_L			C_D			C_M		
	p	$\%U_G$	$\%\lvert E\rvert$	p	$\%U_G$	$\%\lvert E\rvert$	p	$\%U_G$	$\%\lvert E\rvert$
ϕ_1	2.00	2.20	2.08	1.12	0.42	37.96	0.95	0.21	2.96
ϕ_2		0.80	1.30		0.31	30.54		0.16	0.37
ϕ_3		0.07	1.28		0.21	29.35		0.13	0.09
ϕ_4		0.08	0.85		0.14	26.08		0.10	2.60

For O-mesh in Table 11, the errors of C_L and C_D are larger than C and H-mesh because the mesh type fails to capture the wake after the trailing edge. Further, variations of layer-wise growth rates are incapable of improving the situation. Therefore, from the perspective of efficiency, further investigations are not carried out on O-mesh considering its obvious defects.

Various chord-wise spacings are applied for Hybrid-mesh, and the results are in Table 12. Compared with that of structured meshes, the calculation accuracy of C_L is as good as C and H-mesh, but that of C_D is worse than C and H-mesh while still better than O-mesh. C_L tends to be stable when the NOC is more than 2.5×10^5, and C_D results are gradually improved with smaller chord-wise spacings. In the following part, a chord-wise spacing of 2.23×10^{-4} c is selected, and validation for C_L is achieved as indicated in Table 13.

Table 11. Comparison of results of O-mesh based on various chord-wise spacings to that of experimental benchmark data adopted from [25] in relative differences with 296 nodes in chord-wise direction for reference setting.

$\log_{R_G} N_{c_{i0}}$	NOC	$\%\Delta C_L(4.18^\circ)$	$\%\Delta C_D(4.18^\circ)$	$\%\Delta C_M(4.18^\circ)$	$\%\Delta C_L(8.22^\circ)$	$\%\Delta C_D(8.22^\circ)$	$\%\Delta C_M(8.22^\circ)$	$\%\Delta C_L(13.10^\circ)$	$\%\Delta C_D(13.10^\circ)$	$\%\Delta C_M(13.10^\circ)$	$\%\Delta C_L(15.20^\circ)$	$\%\Delta C_D(15.20^\circ)$	$\%\Delta C_M(15.20^\circ)$	$\%\Delta C_L(AVE)$	$\%\Delta C_D(AVE)$	$\%\Delta C_M(AVE)$
−2	116,012	7.66	−0.45	−87.30	25.07	73.38	16.88	−54.32	1016.56	−430.26	−76.12	597.97	−111.56	40.79	422.09	161.50
−1	163,852	3.48	15.45	−77.29	22.23	100.94	2.36	−2.78	70.68	4.52	−48.31	540.63	−67.68	19.20	181.93	37.96
0	193,752	−0.39	28.12	−82.61	20.05	132.55	3.36	−3.35	84.92	5.53	−3.99	66.75	−3.11	6.95	78.08	23.65
1	331,292	−1.24	34.48	−96.77	19.16	142.69	−8.57	−3.80	106.31	−6.21	−5.00	96.24	−9.14	7.30	94.93	30.17
2	472,800	−4.00	44.66	−87.64	17.65	163.98	−11.24	−3.90	110.78	−11.54	−4.52	90.11	−11.43	7.52	102.38	30.46
3	666,770	\	\	\	17.96	156.92	−6.51	−4.74	124.83	−16.92	−52.91	842.17	−653.75	\	\	\
4	944,840	−7.22	48.92	−111.48	\	\	\	\	\	\	−8.03	91.73	17.80	\	\	\

Table 12. Comparison of results of Hybrid-mesh based on various chord-wise spacings with that of experimental benchmark data adopted from [25] in relative differences.

Δc	NOC	$\%\Delta C_L(4.18^\circ)$	$\%\Delta C_D(4.18^\circ)$	$\%\Delta C_M(4.18^\circ)$	$\%\Delta C_L(8.22^\circ)$	$\%\Delta C_D(8.22^\circ)$	$\%\Delta C_M(8.22^\circ)$	$\%\Delta C_L(13.10^\circ)$	$\%\Delta C_D(13.10^\circ)$	$\%\Delta C_M(13.10^\circ)$	$\%\Delta C_L(15.20^\circ)$	$\%\Delta C_D(15.20^\circ)$	$\%\Delta C_M(15.20^\circ)$	$\%\Delta C_L(AVE)$	$\%\Delta C_D(AVE)$	$\%\Delta C_M(AVE)$
2.23×10^{-3}	165,801	3.89	26.16	−139.77	2.93	58.92	−102.81	−0.45	66.54	−27.97	−1.88	56.56	−17.50	2.29	52.05	72.01
1.12×10^{-3}	220,230	2.81	24.61	−98.77	1.97	56.27	−65.66	−0.28	57.21	−19.72	−0.39	43.43	−20.39	1.36	45.38	51.13
7.88×10^{-4}	265,546	3.15	23.68	−111.24	2.33	53.73	−77.53	−0.26	55.59	−20.22	−1.03	42.59	−12.24	1.69	43.90	55.31
5.88×10^{-4}	314,070	3.11	23.18	−108.70	2.23	52.39	−71.74	−0.60	55.97	−13.90	−0.88	42.25	−13.51	1.71	43.45	51.96
4.46×10^{-4}	374,853	2.93	23.23	−102.73	2.09	53.57	−66.59	−0.64	55.73	−12.36	−1.02	41.83	−12.42	1.67	43.59	48.53
3.15×10^{-4}	476,662	2.89	22.13	−100.24	2.14	52.34	−66.90	−0.60	54.14	−11.16	−1.35	42.18	−8.66	1.74	42.70	46.74
2.23×10^{-4}	620,995	2.73	22.04	−94.71	1.84	51.30	−57.25	−0.98	54.76	−4.10	−1.38	41.61	−6.55	1.73	42.42	40.66

Table 13. Uncertainty estimations for Hybrid-mesh ($\Delta c = 5.88 \times 10^{-4}\ c, 4.46 \times 10^{-4}\ c, 3.15 \times 10^{-4}\ c$, $2.23 \times 10^{-4}\ c$, $\alpha = 15.2°$) with varying chord-wise spacings based on mesh dependency.

Solution	C_L			C_D			C_M		
	p	$\%U_G$	$\%\|E\|$	p	$\%U_G$	$\%\|E\|$	p	$\%U_G$	$\%\|E\|$
ϕ_1	2.00	8.08	0.88	1.21	0.13	42.24	2.00	0.15	13.50
ϕ_2		6.58	1.02		0.12	41.83		0.06	12.42
ϕ_3		5.58	1.35		0.12	42.18		0.03	8.66
ϕ_4		5.96	1.38		0.08	41.60		0.004	6.54

No obvious variations can be observed for both C_L, C_D and C_M in Table 14. The insensitivity to layer-wise growth rates for C-mesh is probably caused by good chord-wise mesh resolutions. A rate of 1.15 encounters convergence problems, while results obtained from $r \leqslant 1.10$ are stable. Therefore, $r = 1.10$ is selected.

For H-mesh in Table 15, the accuracy of C_L and C_D has subtle improvements within 1%, while variability of C_M is slightly larger. The trend shows that a slower growth rate of mesh height leads to better results, but, remarkably, too small rates may reduce calculation efficiencies and contribute to mesh quadrilaterals with too large aspect ratios, which are disadvantageous to the convergence of CFD results. In this case, $r = 1.05$ is well balanced in accuracy and computation time.

For Hybrid-mesh in Table 16, the connection between the structured region and the unstructured region is influenced by layer-wise growth rates. When $r = 1.03$, relatively large aspect ratios in boundary layer regions increases the difficulty of convergence. Therefore, $r = 1.10$ is selected as a value of transition of convergent conditions.

5.3.2. Impacts of Far-Field Node Distributions

With selected near-field node distributions, nodes in far-field regions need to be determined to match variations of cell sizes near wall boundaries to present global meshing strategies. For structured meshes, similar to Section 5.3.1, nodes in different directions are changed according to characters of their topological structures, which can be expressed as:

$$\log_{R_G} \frac{N_{r_i}}{N_{r_0}} = \log_{R_G} N_{r_{i0}} = i, \quad (i = -2, -1, \ldots, 4) \tag{28}$$

$$\log_{R_G} \frac{N_{w_i}}{N_{w_0}} = \log_{R_G} N_{w_{i0}} = i, \quad (i = -2, -1, \ldots, 4) \tag{29}$$

where N_{r_i} and N_{w_i} denote the number of nodes in radial and wake directions. For Hybrid-mesh, cell sizes in far-field regions are varied to change mesh properties.

Table 14. Comparison of results of C-mesh based on various layer-wise growth rates with that of experimental benchmark data adopted from [25] in relative differences.

r	NOC	$\%\Delta C_L(4.18^\circ)$	$\%\Delta C_D(4.18^\circ)$	$\%\Delta C_M(4.18^\circ)$	$\%\Delta C_L(8.22^\circ)$	$\%\Delta C_D(8.22^\circ)$	$\%\Delta C_M(8.22^\circ)$	$\%\Delta C_L(13.10^\circ)$	$\%\Delta C_D(13.10^\circ)$	$\%\Delta C_M(13.10^\circ)$	$\%\Delta C_L(15.20^\circ)$	$\%\Delta C_D(15.20^\circ)$	$\%\Delta C_M(15.20^\circ)$	$\%\Delta C_L(AVE)$	$\%\Delta C_D(AVE)$	$\%\Delta C_M(AVE)$
1.20	548,022	1.96	17.06	−63.80	1.31	43.17	−28.81	−1.17	46.09	4.41	−1.79	35.84	−2.75	1.56	35.54	24.94
1.15	548,022	\	\	\	\	\	\	\	\	\	\	\	\	\	\	\
1.10	548,022	2.03	17.17	−63.06	1.42	42.31	−29.35	−1.17	44.87	5.98	−2.01	35.40	−0.40	1.66	34.94	24.70
1.05	548,022	2.03	17.16	−62.98	1.42	42.31	−29.26	−1.18	44.88	6.15	−2.02	35.42	−0.30	1.66	34.94	24.67
1.03	548,022	2.03	17.16	−63.19	1.43	42.29	−29.67	−1.15	44.80	5.60	−1.95	35.21	−0.81	1.64	34.86	24.82

Table 15. Comparison of results of H-mesh based on various layer-wise growth rates that of experimental benchmark data adopted from [25] in relative differences.

r	NOC	$\%\Delta C_L(4.18^\circ)$	$\%\Delta C_D(4.18^\circ)$	$\%\Delta C_M(4.18^\circ)$	$\%\Delta C_L(8.22^\circ)$	$\%\Delta C_D(8.22^\circ)$	$\%\Delta C_M(8.22^\circ)$	$\%\Delta C_L(13.10^\circ)$	$\%\Delta C_D(13.10^\circ)$	$\%\Delta C_M(13.10^\circ)$	$\%\Delta C_L(15.20^\circ)$	$\%\Delta C_D(15.20^\circ)$	$\%\Delta C_M(15.20^\circ)$	$\%\Delta C_L(AVE)$	$\%\Delta C_D(AVE)$	$\%\Delta C_M(AVE)$
1.20	46,288	2.86	16.54	−87.45	2.05	41.22	−48.90	−0.72	41.52	2.02	−1.38	30.81	−0.10	1.75	32.52	34.62
1.15	46,288	2.87	17.12	−87.99	2.08	41.55	−49.74	−0.61	41.14	0.50	−1.17	29.96	−1.48	1.68	32.44	34.93
1.10	46,288	2.84	17.75	−87.82	2.05	42.20	−49.30	−0.67	41.66	1.46	−1.30	30.55	−0.37	1.72	33.04	34.74
1.05	46,288	2.86	17.52	−87.96	2.07	41.74	−49.57	−0.63	41.04	0.94	−1.23	29.92	−0.78	1.70	32.55	34.81
1.03	46,288	2.87	17.40	−88.05	2.08	41.60	−49.77	−0.60	40.87	0.51	−1.18	29.74	−1.18	1.68	32.40	34.88

Table 16. Comparison of results of Hybrid-mesh based on various layer-wise growth rates with that of experimental benchmark data adopted from [25] in relative differences.

r	NOC	$\%\Delta C_L(4.18^\circ)$	$\%\Delta C_D(4.18^\circ)$	$\%\Delta C_M(4.18^\circ)$	$\%\Delta C_L(8.22^\circ)$	$\%\Delta C_D(8.22^\circ)$	$\%\Delta C_M(8.22^\circ)$	$\%\Delta C_L(13.10^\circ)$	$\%\Delta C_D(13.10^\circ)$	$\%\Delta C_M(13.10^\circ)$	$\%\Delta C_L(15.20^\circ)$	$\%\Delta C_D(15.20^\circ)$	$\%\Delta C_M(15.20^\circ)$	$\%\Delta C_L(AVE)$	$\%\Delta C_D(AVE)$	$\%\Delta C_M(AVE)$
1.20	607,987	2.30	21.80	−81.67	1.23	50.70	−36.72	−1.64	54.72	7.37	−2.14	44.18	−2.79	1.83	42.85	32.14
1.15	611,491	2.24	22.13	−80.52	1.36	51.11	−41.85	−1.16	54.54	−1.17	−1.43	41.82	−6.33	1.55	42.40	32.47
1.10	620,995	2.73	22.03	−94.78	1.84	51.32	−57.42	−0.98	54.76	−4.09	−1.87	43.50	−4.28	1.86	42.90	40.14
1.05	652,233	2.87	17.74	−98.81	1.78	48.03	−54.62	−1.24	54.52	−0.37	−1.69	41.02	−5.83	1.89	40.33	39.91
1.03	660,281	3.37	6.31	−109.96	2.16	38.29	−63.27	−1.00	50.00	−3.25	\	\	\	\	\	\

Number of cells against radial nodes vary over a considerable extent from nearly 3.00×10^5 to more than 1.00×10^6 for C-mesh and H-mesh, as shown in Tables 17 and 18. However, such variation has few impacts on magnitudes of any hydrodynamic coefficients, indicating that sufficient mesh resolution along the rudder chord has successfully solved the flow in regions with great gradients. In other words, relatively few nodes in radial directions are capable of ensuring the calculation accuracy. Therefore, the reference case already achieved mesh independence for both C-mesh and H-mesh along with radial directions, and the case is still utilized for benchmark in the following studies.

Similar to impacts of nodes in radial directions, variations of C_L, C_D and C_M against changing cell numbers induced by cell sizes in wake directions are very small, as shown in Tables 19 and 20. Compared with the trend of accuracy variations in Section 5.3.1, impacts of node distributions in the near-field on CFD results are stronger than that in the far-field, which is reasonable considering the decaying disturbance of the object on the flow field with increased distances from the wall boundary. As long as mesh resolutions in the near-field are handled carefully, fewer mesh elements are needed in the far-field. Mesh elements in unstructured regions for Hybrid-mesh are changed as shown in Table 21, and the magnitude of results variations are the same as C-mesh and H-mesh. Further decreasing element sizes of unstructured meshes are memory-consuming, and divergent cases occur when the size is below 0.0625 c. Therefore, the body element size of 0.125 c is selected.

Tables 22–26 show uncertainty estimations for C-mesh, H-mesh and Hybrid-mesh with varying far-field node distributions. Most uncertainty quantities in Tables 22–25 are relatively small, corresponding to the fact that further increase in radial or wake directions has little impact on the calculation accuracy.

Table 17. Comparison of results of C-mesh based on various radial nodes with that of experimental benchmark data adopted from [25] in relative differences with 150 nodes in radial directions for reference setting.

$\log_{R_G} N_{ri0}$	NOC	$\%\Delta C_L(4.18^\circ)$	$\%\Delta C_D(4.18^\circ)$	$\%\Delta C_M(4.18^\circ)$	$\%\Delta C_L(8.22^\circ)$	$\%\Delta C_D(8.22^\circ)$	$\%\Delta C_M(8.22^\circ)$	$\%\Delta C_L(13.10^\circ)$	$\%\Delta C_D(13.10^\circ)$	$\%\Delta C_M(13.10^\circ)$	$\%\Delta C_L(15.20^\circ)$	$\%\Delta C_D(15.20^\circ)$	$\%\Delta C_M(15.20^\circ)$	$\%\Delta C_L(AVE)$	$\%\Delta C_D(AVE)$	$\%\Delta C_M(AVE)$
−2	272,172	1.89	17.08	−63.18	1.16	46.20	−19.87	−1.27	47.33	5.08	−1.90	37.84	−2.64	1.55	37.11	22.69
−1	386,190	2.02	17.17	−64.08	1.41	42.79	−30.40	−1.09	45.27	3.84	−1.75	35.82	−2.59	1.57	35.26	25.23
0	548,022	2.03	17.17	−63.06	1.42	42.31	−29.35	−1.17	44.86	5.98	−2.01	36.06	−0.40	1.66	35.10	24.70
1	776,058	2.03	17.17	−63.06	1.42	42.31	−29.35	−1.17	44.86	5.98	−2.01	36.06	−0.40	1.66	35.10	24.70
2	1,099,722	\	\	\	\	\	\	\	\	\	\	\	\	\	\	\
3	1,555,794	2.04	17.12	−63.04	1.43	42.15	−29.31	−1.17	44.64	6.04	−2.00	35.88	−0.36	1.66	34.95	24.69
4	2,203,122	2.04	17.10	−63.02	1.44	42.08	−29.29	−1.16	44.54	6.06	−2.00	35.80	−0.34	1.66	34.88	24.68

Table 18. Comparison of results of H-mesh based on various radial nodes to experimental benchmark data adopted from [25] in relative differences with 150 nodes in radial directions for reference setting.

$\log_{R_G} N_{ri0}$	NOC	$\%\Delta C_L(4.18^\circ)$	$\%\Delta C_D(4.18^\circ)$	$\%\Delta C_M(4.18^\circ)$	$\%\Delta C_L(8.22^\circ)$	$\%\Delta C_D(8.22^\circ)$	$\%\Delta C_M(8.22^\circ)$	$\%\Delta C_L(13.10^\circ)$	$\%\Delta C_D(13.10^\circ)$	$\%\Delta C_M(13.10^\circ)$	$\%\Delta C_L(15.20^\circ)$	$\%\Delta C_D(15.20^\circ)$	$\%\Delta C_M(15.20^\circ)$	$\%\Delta C_L(AVE)$	$\%\Delta C_D(AVE)$	$\%\Delta C_M(AVE)$
−2	314,088	2.84	17.74	−87.76	2.05	42.20	−49.31	−0.67	41.66	1.45	−1.30	30.52	−0.36	1.72	33.03	34.72
−1	375,344	2.84	17.75	−87.82	2.05	42.20	−49.30	−0.67	41.65	1.45	−1.30	30.55	−0.37	1.72	33.04	34.74
0	462,288	2.84	17.75	−87.83	2.05	42.20	−49.30	−0.67	41.65	1.45	−1.30	30.55	−0.38	1.72	33.04	34.74
1	584,800	2.84	17.75	−87.83	2.05	42.20	−49.31	−0.67	41.65	1.45	−1.30	30.55	−0.38	1.72	33.04	34.74
2	758,688	\	\	\	2.05	42.19	−49.30	−0.67	41.65	1.45	−1.30	30.54	−0.38	\	\	\
3	1,002,712	\	\	\	2.05	42.19	−49.31	−0.67	41.64	1.45	−1.30	30.53	−0.37	\	\	\
4	1,351,488	\	\	\	\	\	\	−0.67	41.64	1.45	−1.30	30.54	−0.37	\	\	\

Table 19. Comparison of results of C-mesh based on various wake nodes with that of experimental benchmark data adopted from [25] in relative differences with 300 nodes in wake directions for reference setting.

$\log_{R_G} N_{wi0}$	NOC	$\%\Delta C_L(4.18^\circ)$	$\%\Delta C_D(4.18^\circ)$	$\%\Delta C_M(4.18^\circ)$	$\%\Delta C_L(8.22^\circ)$	$\%\Delta C_D(8.22^\circ)$	$\%\Delta C_M(8.22^\circ)$	$\%\Delta C_L(13.10^\circ)$	$\%\Delta C_D(13.10^\circ)$	$\%\Delta C_M(13.10^\circ)$	$\%\Delta C_L(15.20^\circ)$	$\%\Delta C_D(15.20^\circ)$	$\%\Delta C_M(15.20^\circ)$	$\%\Delta C_L(AVE)$	$\%\Delta C_D(AVE)$	$\%\Delta C_M(AVE)$
−2	503,322	3.54	32.94	−41.90	1.42	42.31	−29.37	−1.17	44.86	5.97	−2.01	36.07	−0.41	2.03	39.04	19.41
−1	521,798	2.03	17.17	−63.07	1.42	42.31	−29.37	−1.17	44.86	5.97	−2.01	36.07	−0.41	1.66	35.10	24.70
0	548,022	2.03	17.17	−63.07	1.42	42.31	−29.36	−1.17	44.86	5.97	−2.01	36.06	−0.41	1.66	35.10	24.70
1	584,974	2.03	17.17	−63.07	0.39	41.54	−4.57	−1.17	44.86	5.98	−2.01	36.06	−0.40	1.40	34.91	18.51
2	637,422	2.03	17.16	−63.07	1.42	42.30	−29.35	−1.17	44.84	5.98	\	\	\	\	\	\
3	711,624	2.03	17.16	−63.07	1.42	42.29	−29.35	−1.18	44.83	6.01	−2.01	36.04	−0.38	1.66	35.08	24.70
4	816,222	\	\	\	\	\	\	\	\	\	\	\	\	\	\	\

Table 20. Comparison of the results of H-mesh based on various wake nodes with that of experimental benchmark data adopted from [25] in relative differences with 300 nodes in wake directions for reference setting.

$\log_{R_G} Nw_{i0}$	NOC	$\%\Delta C_L(4.18^{\circ})$	$\%\Delta C_D(4.18^{\circ})$	$\%\Delta C_M(4.18^{\circ})$	$\%\Delta C_L(8.22^{\circ})$	$\%\Delta C_D(8.22^{\circ})$	$\%\Delta C_M(8.22^{\circ})$	$\%\Delta C_L(13.10^{\circ})$	$\%\Delta C_D(13.10^{\circ})$	$\%\Delta C_M(13.10^{\circ})$	$\%\Delta C_L(15.20^{\circ})$	$\%\Delta C_D(15.20^{\circ})$	$\%\Delta C_M(15.20^{\circ})$	$\%\Delta C_L(AVE)$	$\%\Delta C_D(AVE)$	$\%\Delta C_M(AVE)$
−2	405,738	2.84	17.75	−87.83	2.07	41.74	−49.57	−0.67	41.65	1.45	−1.30	30.55	−0.38	1.72	32.92	34.81
−1	429,112	2.84	17.75	−87.78	2.07	41.74	−49.57	−0.67	41.65	1.45	−1.31	30.57	−0.35	1.72	32.93	34.79
0	462,288	2.84	17.75	−87.80	2.07	41.74	−49.57	−0.67	41.65	1.45	−1.31	30.57	−0.35	1.72	32.93	34.79
1	509,036	2.84	17.74	−87.79	2.07	41.74	−49.57	−0.67	41.65	1.45	−1.31	30.56	−0.35	1.72	32.92	34.79
2	575,388	2.84	17.75	−87.80	2.07	41.74	−49.57	−0.67	41.65	1.45	−1.31	30.57	−0.35	1.72	32.93	34.79
3	669,261	\	\	−2296.89	2.07	41.74	−49.57	−0.67	41.63	1.46	−1.31	30.54	−0.33	\	\	\
4	801,588	\	\	−8910.58	2.07	41.74	−49.57	−0.67	41.63	1.47	−1.31	30.54	−0.33	\	\	\

Table 21. Comparison of the results of Hybrid-mesh based on various element sizes with that of experimental benchmark data adopted from [25] in relative differences.

Element Size	NOC	$\%\Delta C_L(4.18^{\circ})$	$\%\Delta C_D(4.18^{\circ})$	$\%\Delta C_M(4.18^{\circ})$	$\%\Delta C_L(8.22^{\circ})$	$\%\Delta C_D(8.22^{\circ})$	$\%\Delta C_M(8.22^{\circ})$	$\%\Delta C_L(13.10^{\circ})$	$\%\Delta C_D(13.10^{\circ})$	$\%\Delta C_M(13.10^{\circ})$	$\%\Delta C_L(15.20^{\circ})$	$\%\Delta C_D(15.20^{\circ})$	$\%\Delta C_M(15.20^{\circ})$	$\%\Delta C_L(AVE)$	$\%\Delta C_D(AVE)$	$\%\Delta C_M(AVE)$
0.17678c	621,117	2.71	22.02	−94.43	1.84	51.32	−57.41	−0.99	54.35	−3.76	−1.41	41.77	−6.38	1.74	42.37	40.49
0.12500c	719,043	2.59	21.88	−90.64	1.80	51.40	−55.08	−0.72	53.35	−6.48	−1.10	40.96	−8.89	1.55	41.90	40.27
0.08839c	919,255	2.77	21.42	−96.73	2.02	51.73	−61.37	−0.76	54.01	−6.93	−1.24	40.11	−7.05	1.70	41.82	43.02
0.06250c	1,314,763	2.70	21.71	−93.17	1.88	51.95	−57.74	−0.86	54.26	−5.91	−1.16	42.14	−9.81	1.65	42.51	41.66
0.04419c	2,106,077	2.70	20.96	−91.30	1.92	50.73	−57.52	−0.58	52.55	−9.67	−1.41	41.24	−7.21	1.65	41.37	41.43

Table 22. Uncertainty estimations for C-mesh with varying radial node distributions ($\log_{R_G} N_{C_{i0}} = -2, -1, 0, 1$) based on mesh dependency.

Solution	C_L			C_D			C_M		
	p	$\%U_G$	$\%\|E\|$	p	$\%U_G$	$\%\|E\|$	p	$\%U_G$	$\%\|E\|$
ϕ_1	0.23	2.87	1.90	2.00	0.06	37.84	0.56	0.33	2.63
ϕ_2		2.69	1.75		0.04	35.81		0.23	2.58
ϕ_3		1.77	2.00		0.02	36.06		0.18	0.40
ϕ_4		1.27	2.00		0.02	36.06		0.16	0.40

Table 23. Uncertainty estimations for H-mesh with varying radial node distributions ($\log_{R_G} N_{C_{i0}} = -2, -1, 0, 1$) based on mesh dependency.

Solution	C_L			C_D			C_M		
	p	$\%U_G$	$\%\|E\|$	p	$\%U_G$	$\%\|E\|$	p	$\%U_G$	$\%\|E\|$
ϕ_1	2.00	0.0002	1.30	2.00	0.0003	30.56	2.00	0.0001	0.38
ϕ_2		0.0006	1.30		0.0002	30.55		0.0001	0.37
ϕ_3		0.0004	1.30		0.0001	30.55		0.0001	0.37
ϕ_4		0.0007	1.30		0.0001	30.55		0.0001	0.37

Table 24. Uncertainty estimations for C-mesh with varying wake node distributions ($\log_{R_G} N_{C_{i0}} = -2, -1, 0, 1$) based on mesh dependency.

Solution	C_L			C_D			C_M		
	p	$\%U_G$	$\%\|E\|$	p	$\%U_G$	$\%\|E\|$	p	$\%U_G$	$\%\|E\|$
ϕ_1	2.00	0.003	2.00	2.00	0.001	36.06	2.00	0.001	0.42
ϕ_2		0.002	2.00		0.001	36.06		0.001	0.42
ϕ_3		0.001	2.00		0.001	36.06		0.001	0.42
ϕ_4		0.001	2.00		0.001	36.06		0.001	0.42

Table 25. Uncertainty estimations for H-mesh with varying wake node distributions ($\log_{R_G} N_{C_{i0}} = -2, -1, 0, 1$) based on mesh dependency.

Solution	C_L			C_D			C_M		
	p	$\%U_G$	$\%\|E\|$	p	$\%U_G$	$\%\|E\|$	p	$\%U_G$	$\%\|E\|$
ϕ_1	2.00	0.0102	1.30	2.00	0.0002	30.55	2.00	0.0006	0.38
ϕ_2		0.0057	1.31		0.0006	30.57		0.0002	0.35
ϕ_3		0.0032	1.31		0.0006	30.57		0.0001	0.35
ϕ_4		0.0017	1.31		0.0007	30.56		0.0001	0.35

Table 26. Uncertainty estimations for Hybrid-mesh (Element size = 0.17678 c, 0.12500 c, 0.08839 c, 0.06250 c, $\alpha = 15.2°$) with varying far-field node distributions based on mesh dependency.

Solution	C_L			C_D			C_M		
	p	$\%U_G$	$\%\|E\|$	p	$\%U_G$	$\%\|E\|$	p	$\%U_G$	$\%\|E\|$
ϕ_1	2.00	0.14	1.41	2.00	0.16	41.77	0.86	0.63	6.38
ϕ_2		0.17	1.10		0.15	40.96		0.61	8.88
ϕ_3		0.14	1.24		0.24	40.11		0.57	7.04
ϕ_4		0.12	1.16		0.20	42.14		0.43	9.81

5.4. Impacts of Mesh Types

Simulations of different mesh types with varying mesh properties are conducted in Sections 5.1–5.3. Generally, the accuracy of C_L is sufficient (within 5% deviations compared with experimental data) for engineering applications except for O-mesh, while discrepancies of C_D between CFD and EFD results are relatively large under high angles of attack. Physically, the lift and drag coefficient are nondimensional forms of lift and drag force, respectively, while the former is mainly generated by the pressure difference between the windward and leeward surfaces, the inviscid component, and the latter primarily contributed by the frictional shear stress, the viscous component. The precise prediction of viscous stress relies on a more exact forecasting model in regions of separated flow, along with an advanced understanding of rudder hydrodynamics, the development of the discretization, and numerical methods [38]. In other words, it is inevitable to observe excessive error of C_D with the RANS method and limited computational resources. However, selecting the appropriate mesh type can improve the accuracy of both C_L and C_D to a certain extent in the current framework. Therefore, based on the simulation results obtained above, the impacts of different mesh types on rudder hydrodynamics are summarized in this section. To evaluate the efficiency and accuracy of those mesh types, the calculation accuracy of C_D versus changing numbers of cells in the near-field and far-field is shown in Figures 18 and 19, while blue dotted lines connect minimum errors under each α and red lines mark selected mesh settings for each type. Since there is a common trend between impacts of radial and wake distributions on calculation accuracy for C-mesh and H-mesh, only radial node distributions are selected to perform the estimation.

In Figure 18, for the NOC below 3.00×10^5, C-mesh shows higher accuracy than other mesh types. However, H-mesh with better mesh resolution around the rudder achieves better performance under large angles of attack. What's more, it can be foreseen that a further increase in nodes may lead to more accurate results if computational power is not an issue. According to the tendency of Hybrid-mesh, the accuracy of C_D remained stable, and smaller chord-wise spacing cannot make up for defects of unstructured mesh in solving rudder flows.

In Figure 19, even though variations of C_D of Hybrid-mesh are slightly larger than C-mesh and H-mesh, results obtained from these mesh types settle in stable levels with fixed chord-wise spacings. Although making further efforts in far-field mesh resolutions can still improve the accuracy, the consequent mesh elements with small size may dramatically increase computation time, as well as bring difficulties with convergence. C-mesh and H-mesh with NOC around 5.00×10^5 are selected, which are about 2.00×10^5 less than Hybrid-mesh in the number of cells and show higher accuracy especially under high rudder angles, proving that structured meshes have higher efficiency in computations than hybrid meshes.

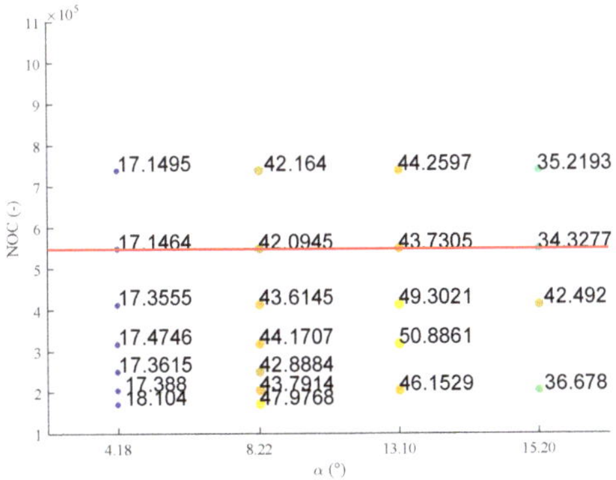

(a) C-mesh.

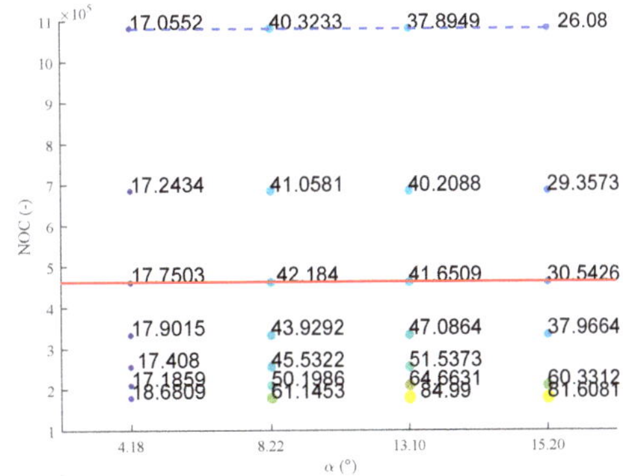

(b) H-mesh.

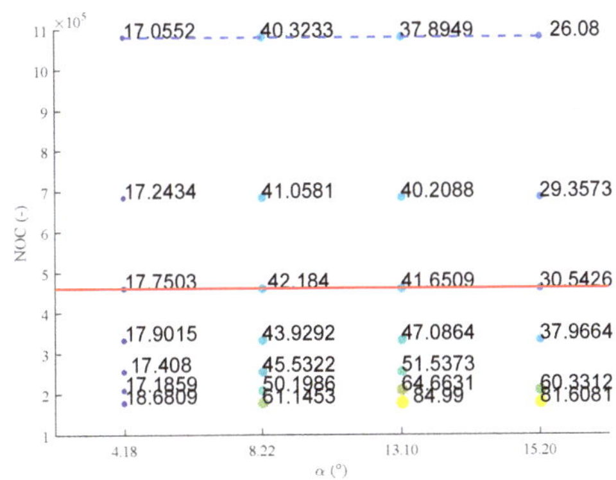

(c) Hybrid-mesh.

Figure 18. Calculation deviations of C_D compared with experimental benchmark data adopted from [25] varying with numbers of cells, while changing chord-wise spacings in the near-field (blue dotted lines connect minimum errors under each α and red lines mark selected mesh settings for each type).

Based on the principle of balancing efficiency and accuracy, recommended mesh properties for four mesh types investigated in Sections 5.1–5.3 are presented in Table 27, and performance of these settings are compared with that of other codes (Langley Research Center [19]) in Table 28, which coincide with H-mesh best. The results of these CFD codes are achieved using the same block-structured C-mesh in a domain of 500 c around the profile and 500 c behind it, which is much larger than the domain in this paper. Hence, the appearance of differences in Table 27 is reasonable, and relatively good results are achieved by the current method efficiently. Structured meshes of C-type and H-type can achieve accurate C_L predictions, while O-mesh fails to capture flow features in wake directions because of the difficulties of actualizing clustering grid points in the wake while maintaining a reasonable spacing at the trailing edge [5]. The concentration degree of mesh elements in regions with a large change of solution gradients for H-mesh is greater than that of C-mesh, contributing to higher accuracy in C_D predictions when the NOC is above 3.00×10^5. Some unexpected divergent cases can be observed for C-mesh and H-mesh, and no general rules can be derived to explain why that happens, though it probably relates to the processing mechanism of the solver. In systematic parameter studies, it is recommended to fine-tune mesh properties to obtain convergent results. Hybrid-mesh can be generated more simply with less manual intervention, and calculation results are acceptable but still inferior to C-mesh and H-mesh. However, the high adaptability of unstructured meshes to complex geometries extends the scope of application of Hybrid-mesh, which means that such a meshing strategy can be easily transformed into sophisticated configurations like high-lift rudders.

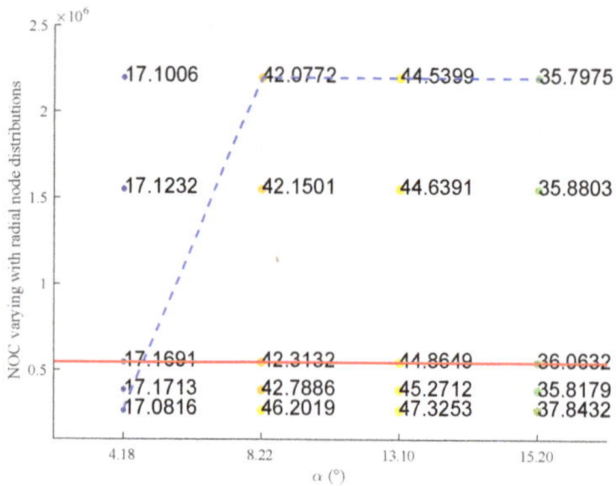

(a) C-mesh.

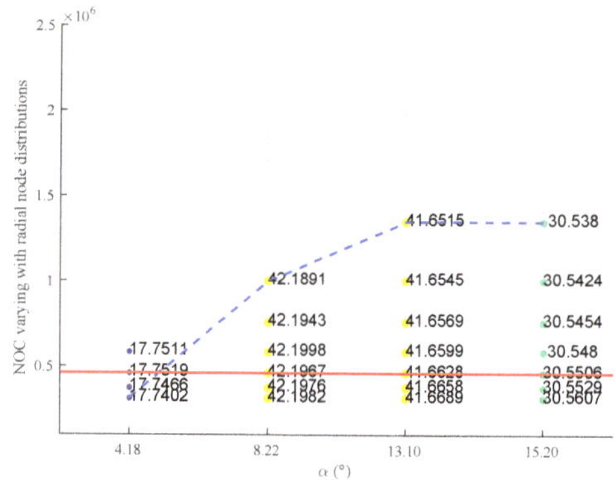

(b) H-mesh.

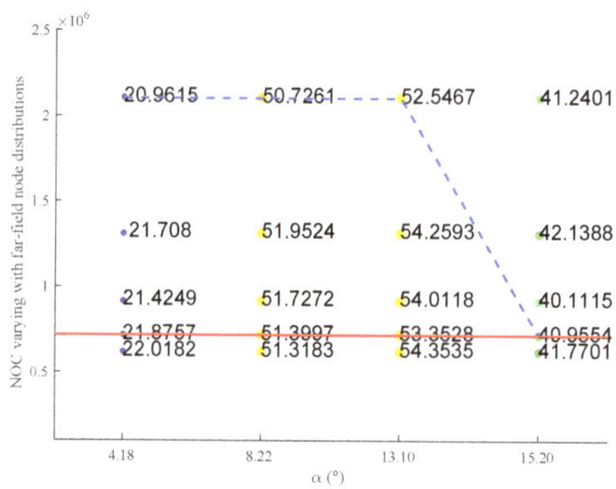

(c) Hybrid-mesh.

Figure 19. Calculation deviations of C_D compared with experimental benchmark data adopted from [25] varying with numbers of cells, while changing radial node distributions in near-field (blue dotted lines connect minimum errors under each α and red lines mark selected mesh settings for each type).

Table 27. Recommended mesh properties for different mesh types.

Mesh Properties \ Mesh Types	C-Mesh	H-Mesh	O-Mesh	Hybrid-Mesh
Re	6.00×10^6	6.00×10^6	6.00×10^6	6.00×10^6
Domain size	$30\ c/60\ c$	$30\ c/60\ c$	$60\ c/60\ c$	$30\ c/60\ c$
Chord-wise spacing	1895 nodes	692 nodes	658 nodes	$1.78 \times 10^{-3}\ c$
Layer-wise growth rate	1.10	1.05	\	1.10
Far-field node distribution	Radial-150 nodes Wake-300 nodes	Radial-150 nodes Wake-300 nodes	Radial-\ Wake-\	Element size-0.125 c
Number of cells	5.48×10^5	4.62×10^5	\	7.19×10^5

Table 28. Comparison of mesh types with recommended settings to that of other CFD codes adopted from [19] in relative differences.

Codes	$\alpha(°)$	C_L			C_D		
		C-Mesh	H-Mesh	Hybrid-Mesh	C-Mesh	H-Mesh	Hybrid-Mesh
CFL3D	0	\	\	\	0.91%	1.44%	3.47%
	10	−0.57%	−0.01%	0.05%	7.88%	7.18%	16.02%
	15	−1.37%	−0.72%	−0.63%	11.05%	6.80%	15.94%
FUN3D	0	\	\	\	1.03%	1.57%	3.60%
	10	−1.13%	−0.58%	−0.53%	6.42%	5.72%	14.45%
	15	−1.64%	−0.99%	−0.90%	8.31%	4.17%	13.09%
NTS	0	\	\	\	0.91%	1.44%	3.47%
	10	−0.45%	0.11%	0.17%	6.59%	5.89%	14.63%
	15	−1.58%	−0.93%	−0.84%	12.67%	8.36%	17.64%

6. Further Validations

To further validate the proposed mesh strategy, more 2D rudder profiles are studied in this section, i.e., NACA 0006, NACA 0009, NACA 0015, and NACA 0018, and the calculated force coefficients are compared with corresponding experimental data. The test conditions are shown in Table 29.

Table 29. More profiles studied.

Profile	Re	Reference
NACA 0006	6.00×10^6	Abbott and Doenhoff [39]
NACA 0009	6.00×10^6	Abbott and Doenhoff [39]
NACA 0015	3.26×10^6	Jacobs and Sherman [40]
NACA 0018	3.26×10^6	Jacobs and Sherman [40]

By applying H-mesh and Hybrid-mesh with recommended domain size and nodes distributions, obtained C_L and C_D are compared with EFD results before stall for tested rudder profiles as shown in Figure 20. For profiles with relatively small thickness ratios, such as NACA 0006 and NACA 0009, Hybrid-mesh failed to obtain credible force coefficients. As for H-mesh, in such cases, the mesh type can achieve acceptable C_L calculations, although the scope of α which leads to convergent solutions is limited. Nevertheless, from the perspective of structural strength, thin profiles are generally not used for rudder designs, so no further efforts are made to modify current mesh strategies to adapt thin profiles in this paper. For NACA 0015 and NACA 0018, when $\alpha < 5°$, some divergence may occur for H-mesh, while Hybrid-mesh shows good convergence. After that, CFD results for H-mesh and Hybrid-mesh agree with EFD ones well, while H-mesh shows better accuracy for α around 15°, demonstrating that the proposed meshing strategy can be used for other rudder profiles.

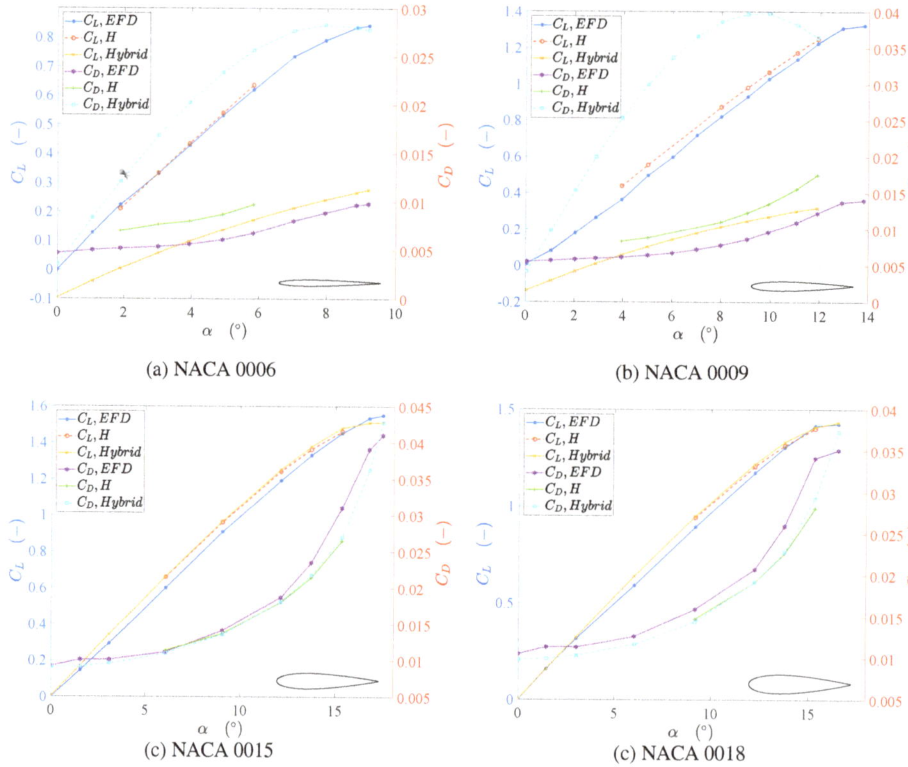

Figure 20. Comparison of results of H and Hybrid mesh with experimental benchmark data adopted from [39,40] for NACA 0006, NACA 0009, NACA 0015 and NACA 0018.

7. Conclusions

This paper describes a comprehensive approach to identifying suitable mesh properties for CFD simulations of airfoil-shaped ship rudders. The goal is to build up an efficient mesh with a sufficient resolution. The suggested mesh properties are summarized in Table 30 for different research objectives. Important conclusions drawn from this work include:

1. The mesh type affects the efficiency of mesh generation and the accuracy of the CFD solutions. The structured mesh gives more accurate drag predictions with fewer cells compared with that of the hybrid mesh. However, an unstructured mesh is more suitable for complex rudder profiles; it can also provide comparable accurate lift predictions with the structured mesh.
2. The domain size has to be reasonably large to fully develop the flow. The drag prediction is more influenced by the domain size than the lift prediction. An increase in domain size leads to an improvement in prediction accuracy.
3. Lift coefficients increase while drag coefficients decrease with an increase of Reynolds numbers in a range of $2 \times 10^5 \sim 1.89 \times 10^7$. The drag coefficient is more sensitive to changes in the Reynolds number than the lift coefficient. A Reynolds number of 6.00×10^6 is suggested as a benchmark, above which few changes in lift and drag coefficients were found.

4. The boundary conditions applied in this study are velocity-inlet, no-slip wall, and pressure-outlet. RANS simulation results with these conditions are in good agreement with the published experimental data [25,39,40].
5. In all cases, C_D prediction accuracy remains poor compared with that of C_L.

Table 30. Recommended mesh properties for different objective.

Objective	Properties	Settings
Accurate prediction for both C_L and C_D	Mesh type	H-mesh
	Re	Actual Reynolds number whenever possible
	Domain size	As large as possible or at least 30 *c* around and 60 *c* after the profile
	Chord-wise spacing	More than 692 nodes for $c = 1$ m
	Layer-wise growth rate	1.05
	Number of radial nodes	150 nodes
	Number of wake nodes	300 nodes
	Number of mesh cells	No less than 4.62×10^5
Accurate C_L only with less mesh number	Mesh type	C-mesh
	Re	6×10^6
	Domain size	30 *c* around and 60 *c* after the profile
	Chord-wise spacing	670 nodes for $c = 1$ m
	Layer-wise growth rate	1.10
	Number of radial nodes	150 nodes
	Number of wake nodes	300 nodes
	Number of mesh cells	About 2.51×10^5
Acceptable C_L and C_D with simple meshing strategy	Mesh type	Hybrid-mesh
	Re	6×10^6
	Domain size	30 *c* around and 60 *c* after the profile
	Chord-wise spacing	2.23×10^{-4} *c*
	Layer-wise growth rate	1.10
	Element size	0.125 *c*
	Number of mesh cells	About 7.19×10^5

We realize that the above conclusions are based on a case study of 2D ship rudders. For unconventional rudder types, such as flap rudders and fishtail rudders, the conclusions should not be applied directly. For instance, the gap of the flap rudder and the concave part of the fishtail rudder may require an additional number of cells. What's more, attention should be paid to mesh elements in span-wise directions for 3D rudders. Considering larger calculation errors with larger angles of attack, extra efforts are also needed in applications of other CFD methods, such as Large-Eddy Simulation (LES) and Direct Numerical Simulation (DNS), which commonly require a much finer mesh than the RANS method. Therefore, further study is required to determine the most suitable mesh for different rudder types and meshing strategies in 3D configurations.

Author Contributions: Conceptualization, S.L. and J.L.; methodology, S.L. and J.L.; software, S.L. and J.L.; validation, S.L.; formal analysis, R.H.; investigation, R.H.; resources, J.L.; data curation, S.L.; writing—original draft preparation, S.L. and J.L.; writing—review and editing, R.H.; visualization, S.L.; supervision, J.L.; project administration, J.L.; funding acquisition, J.L. All authors have read and agreed to the published version of the manuscript.

Funding: This research was funded by National Natural Science Foundation of China (62003250), Research on Intelligent Ship Testing and Verification (2018473), Program of Marine Economy Development Special Fund (Six Marine Industries) under Department of Natural Resources of Guangdong Province (Project No. GDNRC[2021]59).

Institutional Review Board Statement: Not applicable.

Informed Consent Statement: Not applicable.

Data Availability Statement: Data supporting reported results can be found in Ladson 1988, Abbott 1959, Jacobs 1937.

Conflicts of Interest: The authors declare no conflict of interest.

Abbreviations

The following abbreviations are used in this manuscript:

AVE	Average error
NOC	Number of cells
GCI	Grid convergence index
CF	Correction factor
α	Angle of attack of the rudder
Δs	Height of the meshes' first layer
Δc	Chord-wise spacing around a rudder profile
ΔC_L	Relative difference of the calculated lift coefficient and benchmark data
ΔC_D	Relative difference of the calculated drag coefficient and benchmark data
ΔC_M	Relative difference of the calculated moment coefficient and benchmark data
δ	Thickness of a flat plate boundary layer
E	Comparison error
ϕ_i	Solution of ith mesh
c	Chord length of a rudder profile
N_c	Number of nodes in chord-wise
N_r	Number of nodes in radial directions
N_w	Number of nodes in wake directions
p	Order of accuracy
p_{th}	Theoretical order of accuracy
Re	Reynolds number
R_G	Grid refinement factor
r	Layer-wise growth rate in boundary layer regions
U_G	Grid uncertainty
U_V	Validation uncertainty
y^+	Dimensionless wall distance

References

1. Liu, J.; Hekkenberg, R. Sixty years of research on ship rudders: Effects of design choices on rudder performance. *Ships Offshore Struct.* **2017**, *12*, 495–512. [CrossRef]
2. Liu, J.; Quadvlieg, F.; Hekkenberg, R. Impacts of the rudder profile on manoeuvring performance of ships. *Ocean. Eng.* **2016**, *124*, 226–240. [CrossRef]
3. Badoe, C.E.; Phillips, A.B.; Turnock, S.R. Influence of drift angle on the computation of hull-propeller-rudder interaction. *Ocean. Eng.* **2015**, *103*, 64–77. [CrossRef]
4. Van Nguyen, T.; Ikeda, Y. Hydrodynamic characteristic of rudder sections with high lift force, Part 2: The wedge tail shapes. *J. Jpn. Soc. Nav. Archit. Ocean. Eng.* **2014**, 171–174.
5. Lutton, M.J. Comparison of C-and O-grid Generation Methods Using a NACA 0012 Airfoil. Master's Thesis, Air Force Institute of Technology, Kaduna, Nigeria, 1989.
6. Basha, W. Accurate drag prediction for transitional external flow over airfoils. Master's Thesis, Concordia University, Montreal, QC, Canada, 2006.
7. Stuck, A.; Turnock, S.; Bressloff, N. *An Evaluation of the RANS Method for the Prediction of Steady Ship Rudder Performance Compared to Wind Tunnel Measurements*; Technical Report; University of Southampton: Southampton, UK, 2004.
8. Turnock, S.R.; Holroyd, N.J.; Date, J.C. Appendage design for the America's Cup using CFD. In Proceedings of the European Congress on Computational Methods in Applied Sciences and Engineering (ECCOMAS–CFD 2001), Swansea/Wales, UK, 4–7 September 2001.
9. Kim, H.J.; Kim, S.H.; Oh, J.K.; Seo, D.W. A proposal on standard rudder device design procedure by investigation of rudder design process at major Korean shipyards. *J. Mar. Sci. Technol.* **2012**, *20*, 450–458.
10. Fujii, H. Experimental researches on rudder performance (1). *J. Zosen Kiokai* **1960**, *107*, 105–111. (In Japanese) [CrossRef]
11. Fujii, H.; Tsuda, T. Experimental researches on rudder performance (2). *J. Zosen Kiokai* **1961**, *109*, 31–42. (In Japanese) [CrossRef]
12. Fujii, H.; Tsuda, T. Experimental researches on rudder performance (3). *J. Zosen Kiokai* **1962**, *111*, 51–58. (In Japanese) [CrossRef]

13. Nienhuis, U. *Passieve Manoeuvreerhulpmiddelen: Open Water Proeven met Roer*; Technical Report; Maritime Research Institute Netherlands (MARIN): Wageningen, The Netherlands, 1987. (In Dutch)
14. Liu, J.; Hekkenberg, R.; Quadvlieg, F.; Hopman, H.; Bingqian, Z. An integrated empirical manoeuvring model for inland vessels. *Ocean. Eng.* **2017**, *137*, 287–308. [CrossRef]
15. Wasberg, C.E.; Reif, B.A.P. *Hydrodynamical Simulations in Fluent*; Technical Report; Norwegian Defence Research Establishment (FFI): Kjeller, Norway, 2010.
16. Tveiterås, V. Numerical Study of the Interaction of Flow over Two Airfoils in Relative Motion. Master's Thesis, Norwegian University of Science and Technology, Trondheim, Norway, 2011.
17. Narsipur, S.; Pomeroy, B.; Selig, M. CFD analysis of multielement airfoils for wind turbines. In Proceedings of the 30th AIAA Applied Aerodynamics Conference, New Orleans, LA, USA, 25– 28 June 2012; p. 2781.
18. Eleni, D.C.; Athanasios, T.I.; Dionissios, M.P. Evaluation of the turbulence models for the simulation of the flow over a National Advisory Committee for Aeronautics (NACA) 0012 airfoil. *J. Mech. Eng. Res.* **2012**, *4*, 100–111.
19. Langley Research Center. Turbulence Modeling Resource. Available online: http://turbmodels.larc.nasa.gov/naca0012_val.html (accessed on 10 September 2014).
20. Nordanger, K.; Holdahl, R.; Kvarving, A.M.; Rasheed, A.; Kvamsdal, T. Implementation and comparison of three isogeometric Navier-Stokes solvers applied to simulation of flow past a fixed 2D NACA 0012 airfoil at high Reynolds number. *Comput. Methods Appl. Mech. Eng.* **2015**, *284*, 664–688. [CrossRef]
21. Siddiqui, M.S.; Rasheed, A.; Kvamsdal, T. Numerical assessment of RANS turbulence models for the development of data driven Reduced Order Models. *Ocean. Eng.* **2020**, *196*, 106799. [CrossRef]
22. Liseikin, V.D. *Grid Generation Methods*, 2nd ed.; Springer: Berlin/Heidelberg, Germany, 2009.
23. Aftosmis, M.; Gaitonde, D.; Tavares, T.S. Behavior of linear reconstruction techniques on unstructured meshes. *AIAA J.* **1995**, *33*, 1–51. [CrossRef]
24. Stern, F.; Wilson, R.V.; Coleman, H.W.; Paterson, E.G. *Verification and Validation of CFD Simulations*; Technical Report; Iowa Institute of Hydraulic Research: Iowa City, IA, USA, 1999.
25. Ladson, C.L. *Effects of Independent Variation of Mach and Reynolds Numbers on the Low-Speed Aerodynamic Characteristics of the NACA 0012 Airfoil Section*; Technical Report; Langley Research Center: Hampton, VA, USA, 1988.
26. International Towing Tank Conference. ITTC Recommended Procedures and Guidelines 7.5-03-01-01: Uncertainty Analysis in CFD Verification and Validation. 2008. Available online: https://ittc.info/media/4184/75-03-01-01.pdf (accessed on 5 September 2021).
27. Islam, H.; Soares, C.G. Uncertainty analysis in ship resistance prediction using OpenFOAM. *Ocean. Eng.* **2019**, *191*, 105805. [CrossRef]
28. Roache, P.J. *Verification and Validation in Computational Science and Engineering*; Hermosa: Albuquerque, NM, USA, 1998; Volume 895.
29. Celik, I.B.; Ghia, U.; Roache, P.J.; Freitas, C.J.; Coleman, H.; Raad, P.E. Procedure for estimation and reporting of uncertainty due to discretization in CFD applications. *J. Fluids Eng.* **2008**, *130*.
30. Stern, F.; Wilson, R.; Shao, J. Quantitative V&V of CFD simulations and certification of CFD codes. *Int. J. Numer. Methods Fluids* **2006**, *50*, 1335–1355.
31. Eça, L.; Hoekstra, M. A procedure for the estimation of the numerical uncertainty of CFD calculations based on grid refinement studies. *J. Comput. Phys.* **2014**, *262*, 104–130. [CrossRef]
32. De Luca, F.; Mancini, S.; Miranda, S.; Pensa, C. An extended verification and validation study of CFD simulations for planing hulls. *J. Ship Res.* **2016**, *60*, 101–118. [CrossRef]
33. Pointwise. Y+ Calculator. 2020. Available online: http://www.pointwise.com/yplus/ (accessed on 5 September 2021).
34. Thieme, H. *Design of Ship Rudders (Zur Formgebung von Shiffsrudern)*; Technical Report; Shipbuilding Institute, University of Hamburg: Washington, DC, USA, 1965.
35. Catalano, P.; Tognaccini, R. RANS Analysis of the low-Reynolds number flow around the SD7003 airfoil. *Aerosp. Sci. Technol.* **2011**, *15*, 615–626. [CrossRef]
36. Tang, L. Reynolds-Averaged Navier-Stokes simulation of low-Reynolds-number airfoil aerodynamics. *J. Aircr.* **2008**, *45*, 848–856. [CrossRef]
37. Molland, A.F.; Turnock, S.R. *Marine Rudders and Control Surfaces: Principles, Data, Design and Applications*, 1st ed.; Elsevier Butterworth-Heinemann: Oxford, UK, 2007.
38. Roberts, T.W. Prospects for drag prediction using computational fluid dynamics. In Proceedings of the Aerospace Technology Conference and Exposition, Long Beach, CA, USA, 1 October 1990.
39. Abbott, I.H.; Doenhoff, A.E.V. *Theory of Wing Sections. Including a Summary of airfoil Data*; Dover Publications, Inc.: New York, NY, USA, 1959.
40. Jacobs, E.N.; Sherman, A. *Airfoil Section Characteristics as Affected by Variations of the Reynolds Number*; Technical Report Archive and Image Library; UNT Libraries Government Documents Department: Denton, TX, USA, 1937.

Journal of
Marine Science and Engineering

MDPI

Article

Research on PID Parameter Tuning and Optimization Based on SAC-Auto for USV Path Following

Lifei Song [1,2], Chuanyi Xu [1,2], Le Hao [1,2], Jianxi Yao [1,2] and Rong Guo [1,2,*]

1 Key Laboratory of High Performance Ship Technology, Wuhan University of Technology, Ministry of Education, Wuhan 430070, China
2 School of Naval Architecture, Ocean and Energy Power Engineering, Wuhan University of Technology, Wuhan 430070, China
* Correspondence: grong2006@whut.edu.cn

Abstract: Unmanned surface vessels (USVs) are required to follow a path during a task. This is essential for the USV, especially when following a curvilinear path or considering the interference of waves, and this work has been proven to be complicated. In this paper, a PID parameter tuning and optimizing method based on deep reinforcement learning were proposed to control the USV heading. Firstly, the Abkowite dynamics model with three degrees of freedom (DOF) is established. Secondly, the guidance law on the line-of-sight (LOS) method and the USV heading control system of the PID controller are designed. To satisfy the time-varying demand of PID parameters for guiding control, especially when the USV moves in waves, the soft actor–critic auto (SAC-auto) method is presented to adjust the PID parameters automatically. Thirdly, the state, action, and reward functions of the agent are designed for training and learning. Finally, numerical simulations are performed, and the results validated the feasibility and validity of the feasibility and effectiveness of the proposed method.

Keywords: unmanned surface vehicle; deep reinforcement learning; parameter tuning; path-following control

Citation: Song, L.; Xu, C.; Hao, L.; Yao, J.; Guo, R. Research on PID Parameter Tuning and Optimization Based on SAC-Auto for USV Path Following. *J. Mar. Sci. Eng.* **2022**, *10*, 1847. https://doi.org/10.3390/jmse10121847

Academic Editors: Carlos Guedes Soares and Serge Sutulo

Received: 24 October 2022
Accepted: 15 November 2022
Published: 1 December 2022

1. Introduction

A USV is a kind of ship which navigates on the water and is controlled by an automated algorithm or a remote device. In recent years, with the rapid development of marine science and control theory, the USV has been widely used in marine rescue and marine monitoring due to its advantages of speed and economy [1,2]. Path following is the foundation for a USV to perform tasks and is also the reflection of the intelligence of a USV.

In recent years, artificial intelligence methods, especially reinforcement learning (RL) technology, have effectively improved the accuracy of heading control for USVs. Reinforcement learning is a machine learning technique in which the agents gain knowledge about the specified scenario with training and learning from interactions with the environment directly; it can be combined with the concept of hierarchical neural networks (HNNs) in deep learning (DL) to form various types of deep reinforcement learning (DRL) methods, such as deep Q-learning network (DQN) [3], deep deterministic policy gradient (DDPG) [4], asynchronous advantage actor–critic (A3C) [5] and soft actor–critic (SAC) [6], etc. These algorithms have achieved unprecedented success in many challenging areas. In particular, reinforcement learning techniques are also used in the USV field. Gonzalez-Garcia et al. [7] proposed a USV guidance control method based on DDPG that combined sliding mode control. By training the heading command, the path following of the USV is realized. The results showed that the performance is improved, and the control stability is enhanced. Zhao et al. [8] proposed a deep Q-learning (DQL) method based on the adaptive gradient descent function to guide USV navigation. The results showed that the algorithm performs well in reducing the complexity and improving the accuracy of the path following. Wang et al. [9] proposed an optimal trajectory tracking control algorithm based on deep reinforcement learning, with which the tracking error can converge; its effectiveness and superiority are proven through simulation. The SAC algorithm is an RL algorithm proposed by Haarnoja et al., based on AC. The core idea of the algorithm is that entropy information is combined with the original reward to encourage exploration; the behavior strategy to maximize the reward with entropy is trained. The algorithm can preserve the randomness of the behavior strategy to the maximum extent and improve the agent's ability to perceive the environment. Zheng et al. [10] proposed a linear active disturbance rejection control strategy based on the SAC algorithm, which performed the tracking on both a straight and a circular trajectory for a USV under the wave effect.

The path-following control system comprises guidance laws and controllers [11]. Line-of-sight (LOS) guidance law [12,13] has been widely used in the design of a path-following controller for USVs due to its simplicity, efficiency, and ease of implementation. The law is independent of the controller and does not depend on the mathematical model of the system. For example, in reference [14], the virtual control law that uses the tracking error is calculated to design the guidance law; in contrast, in reference [15], the visual angle of the desired path is calculated for the USV heading controller. The path-following controller can be designed with self-adaption [16,17], backstepping [18,19], sliding mode [20,21], and other control methods. However, the robustness of adaptive control methods needs to be improved. The backstepping algorithm is highly dependent on the model and needs accurate model parameters. Although sliding mode control does not require a high-accuracy model, its chattering problem is difficult to eliminate [22]. Moreover, due to the high cost of the experiment, the difficulty of adjusting the control parameters, the nonlinearity of the motion model, and the uncertainty of the environment, it is difficult to guarantee the stability of the system. Therefore, it is difficult for these control algorithms to show their advantages in practical applications.

Therefore, it is significant to propose an adaptive control method with a simple system, good robustness, and low requirement on model accuracy. PID control is widely used in a ship's heading control. Miao et al. [23] designed a self-adapting expert S-PID controller for a mini-USV to perform heading control and verified the effectiveness and reliability of the designed motion control system through experiments. Based on the LOS guidance law, Zhu et al. [24] conducted a simulation analysis on three control algorithms, incremental PID, fuzzy PID, and variable-length-ratio-fuzzy PD, the results showed that the third has more anti-interference advantages than the other two. Fan et al. [25] designed a track control algorithm combining LOS and fuzzy adaptive PID and conducted a real boat test. The results show that the algorithm reduces the influence of time-varying drift angle on track control, however, some steering overshoot and position deviation still emerged at the corner of the path.

At present, the existing algorithms are not practical in automatically optimizing the PID parameters, and little research has been done on the application of the RL method to the heading control of USVs. For this reason, a PID parameter optimization method based on the SAC algorithm for USVs is proposed to achieve adaptive heading control. The remainder of this paper is organized as follows. In Section 2, the USV kinematic model based on Abkowite is introduced and the path-following guidance system is designed. Then the PID parameter optimization algorithm based on SAC is proposed. In Section 3, the simulation training process and the simulation results under three different working conditions are performed to verify the feasibility and effectiveness of the proposed algorithm. Finally, the concluding remarks are provided in Section 4.

2. The Design of the DRL

In this section, the overall design diagram for the proposed system is presented, and, to solve the tedious process of PID parameter tuning, an adaptive SAC-PID control method is introduced to solve the mechanical tuning problem of PID parameters. The overall flow diagram of the proposed method is given in Figure 1.

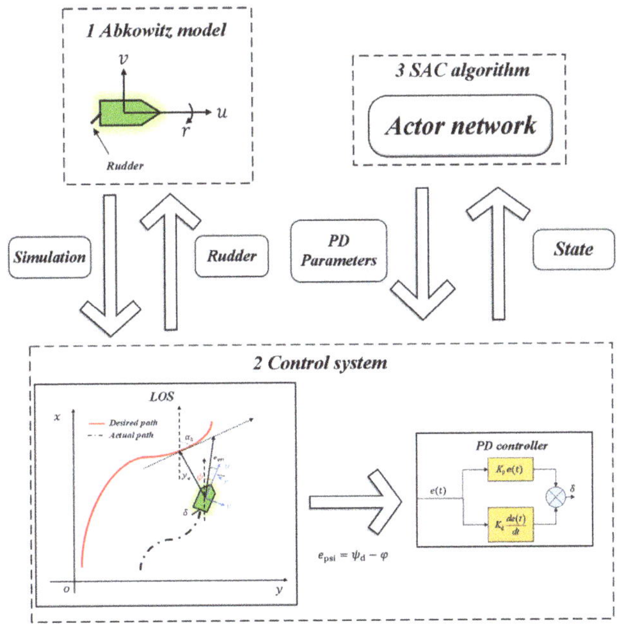

Figure 1. The overall flow diagram for the proposed method.

As shown in Figure 1, the Abkowitz holistic model is established, and considering the integral saturation condition of the PID controller, the control system is based on the LOS guidance and the PD controller is designed to control the USV heading. To obtain reasonable PD parameters, a neural network based on the SAC algorithm is established, and the agent is trained to interact with the abovementioned control system. The final network is used for the experimental simulation in which the required control parameters are inputted for the control parameters by online transmission.

2.1. Dynamic Model of USV

Since the USV is a kind of ship that navigates on water, a hydrodynamic model with three degrees of freedom can be used to describe the relationship between the USV's motion state and the power, external force, fluid force, and corresponding torque on the USV in the local coordinate system of the USV. After ignoring the pitch, roll, and heave of the USV, the two-dimensional coordinate description of ship motion is shown in Figure 2, where $o - xy$ is the global coordinate system (Earth coordinate system), and $o_0 - x_0y_0$ is the local coordinate system of the USV.

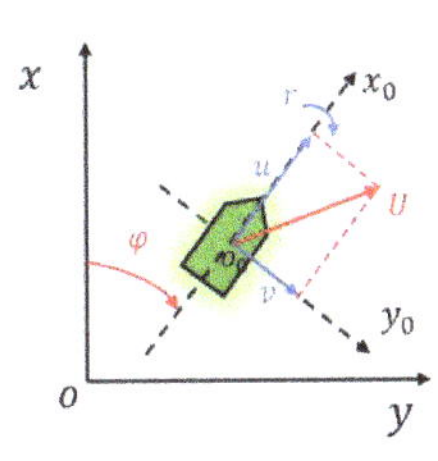

Figure 2. USV manipulation motion coordinate system. u, v, and r are the velocities on the three degrees of freedom (forward, lateral, and head) of the ship motion, respectively. ψ is the heading angle.

The motion equation of the USV with three degrees of freedom can be described by the following equation:

$$\begin{cases} X = m(\dot{u} - rv - x_G r^2) \\ Y = m(\dot{v} + ru + x_G \dot{r}) \\ N = I_z \dot{r} + m x_G (\dot{v} + ru) \end{cases} \tag{1}$$

where m is the total weight of the USV, x_G is the longitudinal coordinate of the center of gravity, and I_z is the moment of inertia of the vertical axis over the center of gravity of the USV. X, Y, and N are the hydrodynamic and torque components of the USV on the 3-DOF of the three motion directions of u, v, and r, respectively.

The integral-type Abkowitz model refers to the motion model proposed by Abkowitz, which takes the ship as an entirety under force to process the mechanical analysis. In theory, the hydrodynamic derivative of the infinite order derived from the Taylor expansion equals the real value. Still, a large amount of research shows that the accuracy is satisfied for engineering applications when the hydrodynamic is derived from the third derivative [26,27]. The third-order Taylor series expansion model is shown in Equation (2). Δu represents the difference between the current speed and the designed speed, which can be described as Equation (3).

$$\begin{cases} m(\dot{u} - rv - x_G r^2) = X_u \Delta u + X_{uu}(\Delta u)^2 + X_{uuu}(\Delta u)^3 + X_{\dot{u}}\dot{u} + X_{vv}v^2 + X_{rr}r^2 + X_{vr}vr \\ +X_{\delta\delta}\delta^2 + X_{v\delta}v\delta + X_{r\delta}r\delta \\ m(\dot{v} + ru + x_G\dot{r}) = Y_v v + Y_{vvv}v^3 + Y_{\dot{v}}\dot{v} + Y_r r + Y_{rrr}r^3 + Y_{\dot{r}}\dot{r} + Y_{vrr}vr^2 + Y_{vvr}v^2 r + Y_\delta \delta \\ +Y_{\delta\delta\delta}\delta^3 + Y_{u\delta}u\delta + Y_{v\delta\delta}v\delta^2 + Y_{vv\delta}v^2\delta + Y_{r\delta\delta}r\delta^2 + Y_{rr\delta}r^2\delta \\ I_z\dot{r} + m x_G(\dot{v} + ru) = N_v v + N_{vvv}v^3 + N_{\dot{v}}\dot{v} + N_r r + N_{rrr}r^3 + N_{\dot{r}}\dot{r} + N_{vrr}vr^2 + N_\delta \delta \\ +N_{\delta\delta\delta}\delta^3 + N_{u\delta}u\delta + N_{v\delta\delta}v\delta^2 + N_{vv\delta}v^2\delta + N_{r\delta\delta}r\delta^2 + N_{rr\delta}r^2\delta \end{cases} \tag{2}$$

$$\Delta u = u - u_0 \tag{3}$$

Move all terms that are proportional to translational accelerations $\dot{u}$, $\dot{v}$, and $\dot{r}$ in Equation (2) to the left of the equation, while the inertial force, the lift force, and the drag force of the fluid on the hull have been moved to the right of the equation. Then the Abkowitz mathematical model of ship motion can be given as shown in Equation (4). The USV model parameters are shown in Table 1. Table 2 contains the hydrodynamic coefficients.

$$\begin{cases} (m - X_{\dot{u}})\dot{u} = F_X(u, v, r, \delta) \\ (m - Y_{\dot{v}})\dot{v} + (m x_G - Y_{\dot{r}}) = F_Y(u, v, r, \delta) \\ (m x_G - N_{\dot{v}})\dot{v} + (I_z - N_{\dot{r}})\dot{r} = F_N(u, v, r, \delta) \end{cases} \tag{4}$$

where $X_{\dot{u}}, Y_{\dot{v}}, Y_{\dot{r}}, N_{\dot{v}}$, and $N_{\dot{r}}$ are the hydrodynamic derivatives.

Table 1. Main parameters of the USV.

Parameters	Values
USV Weight m	3.273 t
Length L	7.0 m
Width b	1.27 m
Draught Depth T	0.455 m
Motor Rated Speed t_r	16.687°/s
Designed speed u_0	1.242 m/s
Moment of inertia I_z	10.0178 Kg·m^2
Center of gravity x_G	0.2485 m

Table 2. Hydrodynamic Coefficients.

Parameters	Values	Parameters	Values
$X_{\dot{u}}$	−0.00135	$X_{v\delta}$	0.001609
$Y_{\dot{v}}$	−0.014508	$X_{r\delta}$	−0.001034
$Y_{\dot{r}}$	−0.001209	Y_v	−0.019

Table 2. *Cont.*

Parameters	Values	Parameters	Values
$N_{\dot{v}}$	−0.000588	Y_{vvv}	−0.129
$N_{\dot{r}}$	−0.000564	Y_r	0.005719
X_u	−0.0022	Y_{rrr}	−0.000048
X_{uu}	0.0015	Y_{vrr}	−0.02429
X_{uuu}	0.0	Y_{vvr}	0.0211
X_{vv}	0.00159	Y_δ	0.00408
X_{rr}	0.000338	$Y_{\delta\delta\delta}$	−0.003059
X_{vr}	0.01391	$Y_{u\delta}$	−0.00456
$X_{\delta\delta}$	−0.00272	$Y_{v\delta\delta}$	0.00326
N_v	−0.007886	$Y_{vv\delta}$	0.003018
N_{vvv}	0.000175	$Y_{r\delta\delta}$	−0.002597
N_r	−0.003701	$Y_{rr\delta}$	0.000895
N_{rrr}	−0.000707	N_δ	−0.001834
N_{vrr}	0.00372	$N_{\delta\delta\delta}$	0.001426
$N_{u\delta}$	0.00232	$N_{v\delta\delta}$	−0.001504
$N_{vv\delta}$	−0.001406	$N_{r\delta\delta}$	0.001191
$N_{rr\delta}$	−0.000398		

2.2. Path Following

The USV path-following control problem is defined as controlling the USV to the desired path S. The line-of-sight (LOS) guidance law is a classical and effective navigation algorithm that does not depend on any dynamic control model and is insensitive to high-frequency white noise. The guidance law can efficiently calculate the desired course and pass it to the controller to achieve the control goal in real time owing to the reason that the desired course is only related to the desired route and the real-time position of the USV. LOS guidance algorithms can be divided into two types, which are based on the lookahead distance and the enveloping circle. The former type is adopted in this paper. The figure of LOS guidance law is shown in Figure 3.

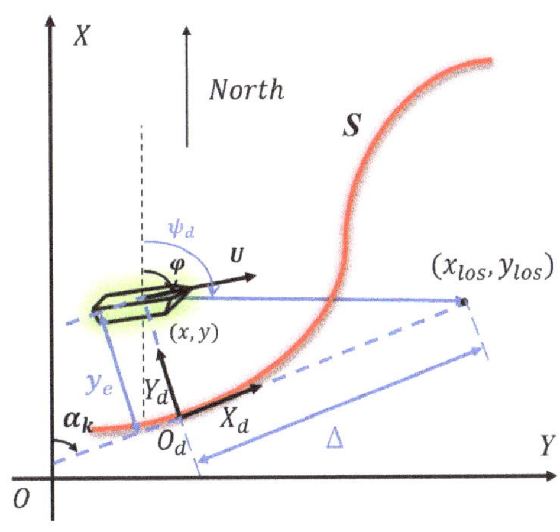

Figure 3. LOS guidance law. $o-xy$ is the global coordinate system. $o_d-x_dy_d$ is the carrier coordinate system. ψ_d is the desired heading angle. φ is the angle between the bow of the USV and the vertical axis of the global coordinate system. y_e is the lateral error about path tracking. α_k is the tangential angle at the o_d point on the desired path. U is the actual velocity of the USV. Δ is the lookahead distance, usually set as an integer multiple of the ship's length. In this paper, we take the multiple as 2.

The PID controller is a linear regulator that compares the desired heading angle $\psi_d(t)$ with the actual heading angle $\varphi(t)$ to form the heading angle deviation $e(t)$:

$$e(t) = \psi_{\mathrm{d}}(t) - \varphi(t) \tag{5}$$

The desired rudder angle can be expressed as Equation (6):

$$\delta(k) = K_{\mathrm{p}}e(k) + K_{\mathrm{i}}\sum_{i=0}^{k} e(i) - K_{\mathrm{d}}(e(k) - e(k-1)) \tag{6}$$

Considering the integral saturation condition of the PID controller, the PD parameters are adjusted to ensure the USV quickly tends to the desired track and keeps the USV navigating within the error range. Therefore, Equation (6) can also be expressed as Equation (7),

$$\delta = K_{\mathrm{p}}e + K_{\mathrm{d}}(e - e') \tag{7}$$

where e' is the error at the last time. The neural network is performed to produce the appropriate PD parameters.

2.3. SAC Algorithm

Figure 4 clearly describes the interaction process between the reinforcement learning agent and the environment, which is also called the Markov Decision Process (MDP) [25]. $(\boldsymbol{S}, \boldsymbol{A}, \rho, r)$ is an important tuple in MDP, in which $\boldsymbol{S}$ is all the states in the environment, $\boldsymbol{A}$ is the set of all the actions, ρ represents the probability density of the next state, and $s_{t+1} \in \boldsymbol{S}$ is given the current state $s_t \in \boldsymbol{S}$, and the action $a_t \in \boldsymbol{A}$. r is a bounded immediate payoff at each time when one state transfers to another. $\rho_\pi(s_t, a_t)$ represents the state-action distribution generated by policy π. At time t, the agent obtains the state s_t from the environment and inputs it into the policy π to obtain the action a_t. The action a_t is executed and the reward r_t of the current step is obtained, while the agent enters the next state s_{t+1}. γ represents the discount factor, so the total reward at time t can be described as Equation (8). The state value function, shown in Equation (9), can be used to evaluate the quality of the current state, while the state-action-value function, shown in Equation (10), can be used to represent whether the action made in the current state is of high quality. The transformation between these two can be described as Equations (11) and (12).

$$R_t = \sum_{k=0}^{\infty} \gamma^k r_{t+k} \tag{8}$$

$$V_\pi(s) = E_\pi\left[\sum_{k=0}^{\infty} \gamma^k r_{t+k} \middle| s_t = s\right] = E[R_t | s_t = s] \tag{9}$$

$$Q_\pi(s,a) = E[R_t | s_t = s, a_t = a] \tag{10}$$

$$V_\pi(s) = E[Q_\pi(s,a) | s_t = s] \tag{11}$$

$$Q_\pi(s,a) = R_{t+1} + \gamma \sum_{s_{t+1} \in S} P_{ss'}^a V_\pi(s_{t+1}) \tag{12}$$

where $P_{ss'}^a = P[s_{t+1} = s' | s_t = s, a_t = a]$, and P is a state transition probability matrix. The optimization objective in reinforcement learning is to maximize the long-term reward R. According to the MDP solution process, the optimal strategy π is the policy that maximizes the reward R, which can be described as Equation (13).

$$\pi = \arg\max \sum_{t=0}^{T} E_{(s_t,a_t)\sim\rho_\pi}\left[\gamma^t r(s_t, a_t)\right] \tag{13}$$

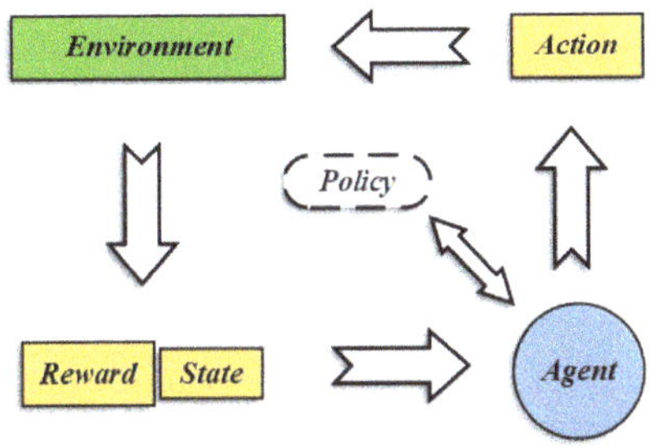

Figure 4. Schematic diagram for reinforcement learning.

Compared with other strategies, the core idea of SAC is that entropy information which encourages the agent to explore for maximizing the entropy reward is combined with the original reward. Thus, Equation (13) can be updated as Equation (14),

$$\pi = \arg\max \sum_{t=0}^{T} E_{(s_t,a_t)\sim\rho_\pi}\left[\gamma^t r(s_t,a_t) + \alpha H(\pi(s_t))\right] \tag{14}$$

where α is the entropy coefficient, which controls the weight between the entropy term and the revenue term and also influences the randomness of the strategy. H is entropy, which represents the randomization of the current policy, expressed as Equation (15).

$$H(P) = E_{x\sim \mathrm{P}}[-\log P(x)] \tag{15}$$

In this paper, the SAC algorithm is mainly composed of five networks, including two value networks (one V network, one target-V network), two action-value networks (Q network), and one actor network (π network). The V network is used to calculate the value of the value function. The Q network is used to calculate the value of the action-value function. The network π outputs the policy value that guides the action of the agent. The overview of the SAC system is shown in Figure 5.

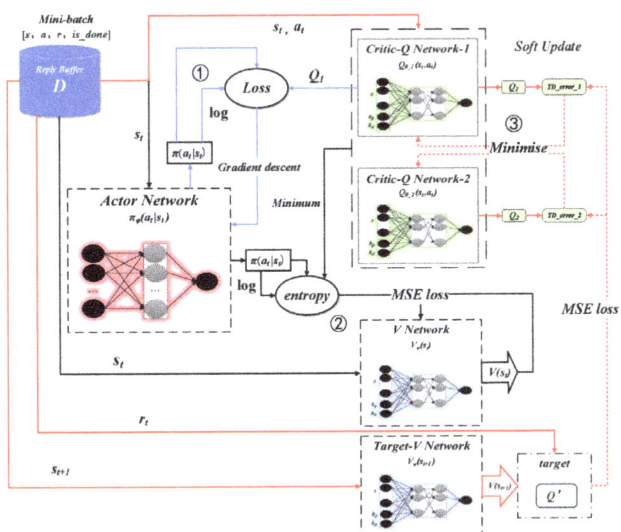

Figure 5. SAC architecture.

2.3.1. Training and Updating of the Actor Network

The strategy π is a Gaussian distribution with mean μ and covariance σ calculated by the neural network. The sampling of each policy $\pi_\phi(\cdot|s_t)$ is a function calculated from the state s, policy parameter ϕ, and independent noise $\xi \in N(0,1)$, which can be described as Equation (16). The loss function of the actor network training is given as Equation (17).

$$\widetilde{a}_\phi(s,\xi) = \tanh(\mu_\phi(s) + \sigma_\phi(s) \odot \xi) \tag{16}$$

$$Loss = E_{\xi \in \mathrm{N}}\left[\alpha \log\ \pi_\phi(\widetilde{a}_\phi(s,\xi)|s) - Q(s,\widetilde{a}_\phi(s,\xi))\right] \tag{17}$$

Compared with other RL methods, obtaining action $\widetilde{a}$ to calculate the *Loss* instead of selecting the action from the sampled *mini-batch* data, the actor network is reused to predict all of the possible actions. The optimization objective of the actor network can be expressed as Equation (18). The gradient calculation formula of the actor network is expressed as Equation (19).

$$J_\pi(\phi) = \underset{\xi \in N}{E}\left[\alpha \log\ \pi_\phi(\widetilde{a}_\phi(s,\xi)|s) - \min_{i=1,2} Q_{\theta\_i}(s,\widetilde{a}_\phi(s,\xi))\right] \tag{18}$$

$$\hat{\nabla}_\phi J_\pi(\phi) = \nabla_\phi \alpha \log(\pi_\phi(a_t|s_t)) + \left(\nabla_{a_t} \alpha \log(\pi_\phi(a_t|s_t)) - \nabla_{a_t} \min_{i=1,2} Q_{\theta\_i}(s_t,a_t)\right) \nabla_\phi \widetilde{a}_\phi(s,\xi) \tag{19}$$

2.3.2. Training and Updating of V Networks

As shown in Figure 5, the V network is updated with the *mini-batch*, which is the data sampled from the experience pool. The combination of the probability $\pi(a_t|s_t)$ of performing action a_t in the current state s_t, the probability $\log(\pi(a_t|s_t))$ after taking the logarithm, and the minimum value of the state-action value Q_1 and Q_2 is taken as the true value of the V network. The MSE method is adopted for loss function calculation and V network training. The objective function can be expressed as Equation (20):

$$J_V(\psi) = E_{s_t \sim D}\left[\frac{1}{2}\left(V_\psi(s_t) - E_{a_t \sim \pi_\phi}\left[\min_{i=1,2} Q_{\theta\_i}(s_t,a_t) - \alpha \log \pi_\phi\left(a_t\middle|s_t\right)\right]\right)^2\right] \tag{20}$$

where ψ is the parameter in the V network. D is the experience pool. $a_t \sim \pi_\phi$ means that instead of sampling from the experience pool, the actions are sampled according to the current policy. The gradient calculation formula of the V network is expressed as Equation (21).

$$\hat{\nabla}_\psi J_V(\psi) = \nabla_\psi V_\psi(s_t)(V_\psi(s_t) - Q_\theta(s_t,a_t) + \log \pi_\phi(a_t|s_t)) \tag{21}$$

2.3.3. Training and Updating of Critic-Q Network

As shown in Figure 5, the Q network is updated with the *mini-batch*, which is the data sampled from the experience pool. $Q' = r_t + \gamma V(s_{t+1})$ is used to calculate the true value of the state s_t, and Q_1 and Q_2 at the same action a_t are used to estimate the predictive value of the state s_t. The objective function can be expressed as Equation (22):

$$J_Q(\theta) = E_{(s_t,a_t) \sim D}\left[\frac{1}{2}(Q_\theta(s_t,a_t) - Q'(s_t,a_t))^2\right] \tag{22}$$

where θ is the parameter in the Q network. $Q'(s_t,a_t)$ is presented as Equation (23):

$$Q'(s_t,a_t) = r(s_t,a_t) + \gamma E_{s_{t+1} \sim P}\left[V_{\overline{\Psi}}(s_{t+1})\right] \tag{23}$$

where $\overline{\Psi}$ is the parameter of the target-V network in the state s_{t+1}. The gradient calculation formula for the Q network is expressed as Equation (24).

$$\hat{\nabla}_\theta J_Q(\theta) = \nabla_\theta Q_\theta(s_t,a_t)(Q_\theta(s_t,a_t) - r(s_t,a_t) - \gamma V_{\overline{\Psi}}(s_{t+1})) \tag{24}$$

Leaving the entropy coefficient α unchanged would be unreasonable because constant change in reward would negatively affect the whole training process. Therefore, it is necessary to automatically adjust α. To improve the learning speed and improve the stability of the agent, this article designed a neural network to adaptively adjust the size of the entropy coefficient α based on the theory of reference [26]. Specifically, when the agent enters a new solution area where the agent's exploration ability should be enhanced to find the best action, α should increase so that the agent will not be trapped in the local optimum. When the agent has almost finished the exploration in a solution area where the learning ability of the agent should be improved, to accumulate experience from the best action, α should be decreased. The optimization function maximizes the expected return under the constraint of the minimum expected entropy, which can be expressed as Equation (25):

$$\begin{gathered}\max_{\pi_{0:T}} E_{\rho_\pi}\left[\sum_{t=0}^{T} \gamma^t r(s_t, a_t)\right] \\ s.t.\ E_{(s_t,a_t)\sim\rho_\pi}[-\log(\pi_t(a_t|s_t))] \geq H_0, \forall t\end{gathered} \tag{25}$$

where H_0 is a constant, representing the preset minimum entropy value. To solve Equation (25), the Lagrange multiplier method is performed to transform the optimization problem into the primal problem and its dual problem. Then the final optimization result can be obtained as Equation (26).

$$\alpha_t^* = \underset{\alpha_t}{\operatorname{argmin}} \sum_{t=0}^{T} E_{\rho_t^*}[-\alpha_t \log \pi_t^*(a_t|s_t) - \alpha_t H_0] \tag{26}$$

where ρ_t^* indicates the state-action pair of the optimal policy. Then the network for α is setup and the stochastic gradient descent is performed as in Equation (26), which can be dubbed Equation (27).

$$\begin{gathered}\hat{\nabla}_\alpha J(\alpha) = \nabla_\alpha E_{a_t\sim\pi_t}[-\alpha_t \log \pi_t(a_t|s_t) - \alpha_t H_0] \\ \alpha \leftarrow \alpha - \eta \nabla_\alpha J(\alpha)\end{gathered} \tag{27}$$

where a_t is derived from the current policy $\pi_t(s_t)$, but s_t is selected from the *mini-batch*. The Adam algorithm is used for optimization and the learning rate lr_α is set to 0.0001.

2.3.4. The Design of State, Action Space, and Reward

During path following, the USV will be disturbed by wind, waves, currents, and other environmental factors. To make the output parameters of the agent more accurate, the environmental information should be considered in the state space design as much as possible. Based on the USV model constructed in Section 2, the state space is defined as Equation (28).

$$s = \left[u, v, r, \varphi, y_e, \alpha_k, \delta, \dot{\delta}, e, \dot{e}\right] \tag{28}$$

Similarly, the action space is defined as $a = [K_p, K_d]$, and K_p ranging from $[-0.5, 0.5]$, and K_d ranges from $[-50, 50]$. The reward function r has two parts, one is r_{psi}, the other is r_{y_e}; r_{psi} and r_{y_e} are designed as shown in Equations (30) and (31).

$$r = r_{psi} + r_{y_e} \tag{29}$$

$$r_{psi} = \begin{cases} 0, & e \leq 0.1 \\ -e - 0.1e', & e \geq 0.1 \end{cases} \tag{30}$$

$$r_{y_e} = \begin{cases} 0, & y_e \leq 1 \\ -0.1, & y_e \geq 1 \end{cases} \tag{31}$$

The design of the actor network and the critic network are shown in Figures 6 and 7. It can be seen that they share the same structure. The dimension of the input layer in the actor network is set as 10. The hidden layer consists of two layers with 400 and 300 neuron nodes, respectively. The dimension of the output layer is set to 2. The dimension of the input layer in the critic network is set as 12. The hidden layer includes two layers with 400 and 300 neuron nodes, respectively. The dimension of the output layer is set to 1. In order to prevent gradient saturation and gradient disappearance, ReLU is used as the activation function of hidden layers in both actor and critic, and tanh is adopted as the activation function of the output layer.

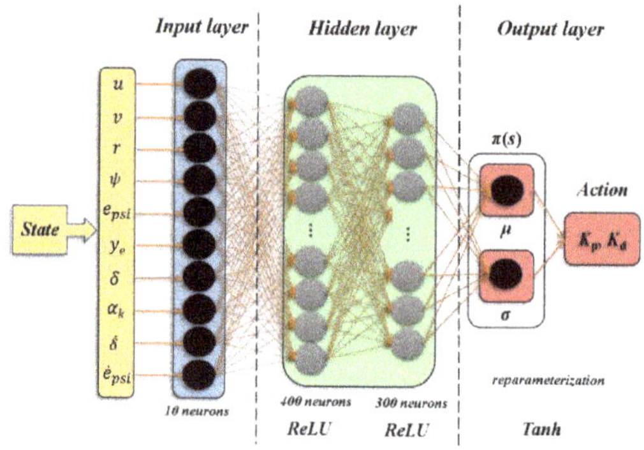

Figure 6. Actor network architecture.

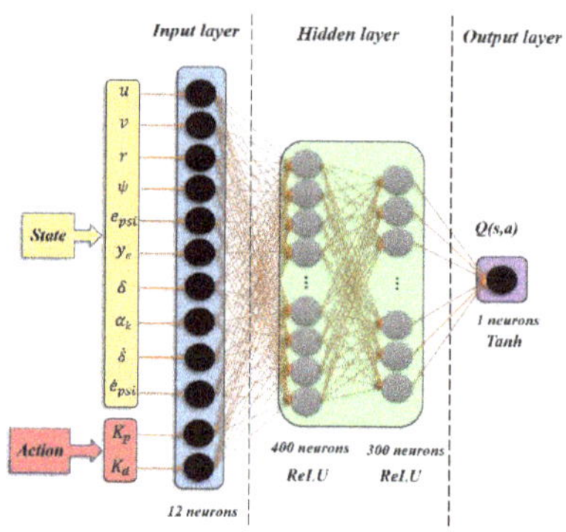

Figure 7. Critic network architecture.

3. Training and Simulation Results

3.1. Network Training

The algorithm code was written based on Pycharm (Jetbrains, Czech Republic) and TensorFlow 2.0.1 (Google Brain, USV) and was used as the framework, and the code runs on a computer with 8GB RAM. The maximum time step of each training was 2500 and the number of training runs was set to 1500. On average, it takes 154 min to complete the training for each path following. The Adam optimizer based on gradient descent was used to learn the parameters of the deep neural network during training. To test the superiority of the SAC algorithm, training and learning based on the DDPG algorithm were performed for comparison. The hyperparameters of the agents are shown in Table 3.

At the initial times of the network training, the strategy was almost random, and the agent could not learn much useful experience, so the effect of following the desired path was not good. Figure 8 shows the training process diagram. To better highlight the average performance and fluctuation range of the algorithm, it was designed as a mean–variance curve. The vertical coordinate was designed as the average return per 10 training sessions. The return was calculated after each training, and the parameters of each network were updated for N_T times according to the reward. Note that the reward represents an immediate return on the action taken. The return is the sum of the immediate returns after the training.

Table 3. Hyperparameters of the algorithms.

Agent	DDPG	SAC	SAC-Auto
Discount factor γ	0.99	0.99	0.99
Hidden layer 1	400	400	400
Hidden layer 2	300	300	300
Activation function	ReLU	ReLU	ReLU
Batch size	100	100	100
Experience pool capacity	106	106	106
τ	0.0001	0.0001	0.0001
lr_a	0.001	0.001	0.001
lr_c	0.001	0.001	0.001
α	none	0.2	auto

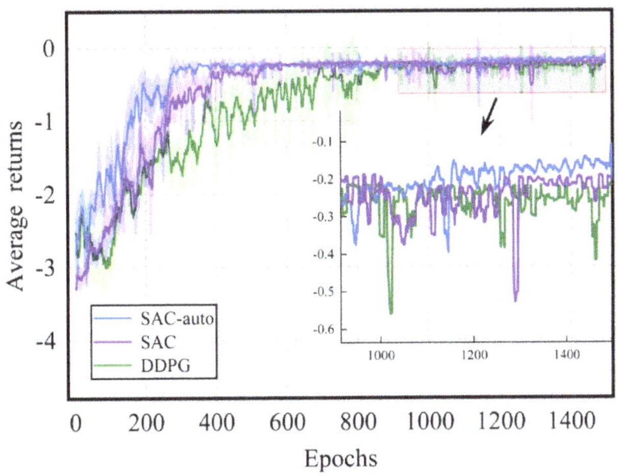

Figure 8. Training process.

It can be seen from Figure 8 that all three algorithms can converge. SAC-auto converges faster than the DDPG. The average return and mean square deviation (MSD) after training are shown in Table 4. The higher average return and smaller MSD indicate that the agent based on the SAC algorithm can better complete the path-following task and has a better stability.

Table 4. Comparison of results after training.

Agent	DDPG	SAC	SAC-Auto
Average return	−0.263	−0.248	−0.190
MSD	0.116	0.078	0.05

3.2. Simulation Results

In this section, the effectiveness of the proposed method is verified by the linear and circular path-following simulation in the simulated wind and wave environment, where $u_0 = 1.242$ m/s, $v_0 = 0$ m/s, and $r_0 = 0$ m/s at the beginning. Under the same guidance law, the abovementioned control method with three trained RL parameters is compared with the adaptive PID parameter controller [28].

In this paper, in order to verify the anti-interference and navigation stability of the system, the interference of $\left[-0.2 \times 10^3,\ 0.2 \times 10^3\right]$N was added to the transverse force, and the interference of $\left[-0.2 \times 10^3,\ 0.2 \times 10^3\right]$N·m was added to the turning moment. The transverse disturbing force is shown in Figure 9. The turning disturbing moment is shown in Figure 10.

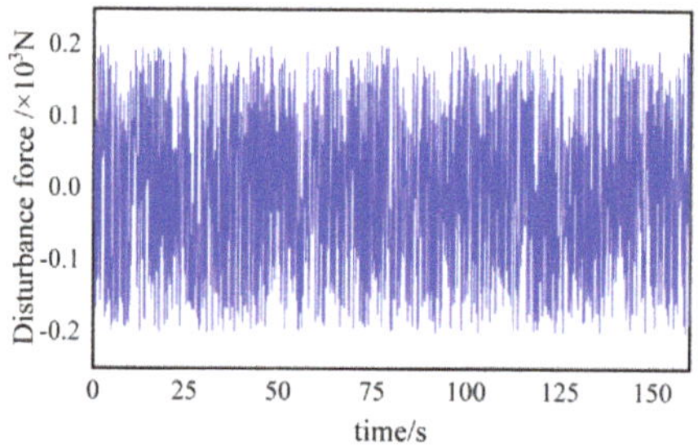

Figure 9. Transverse disturbing force.

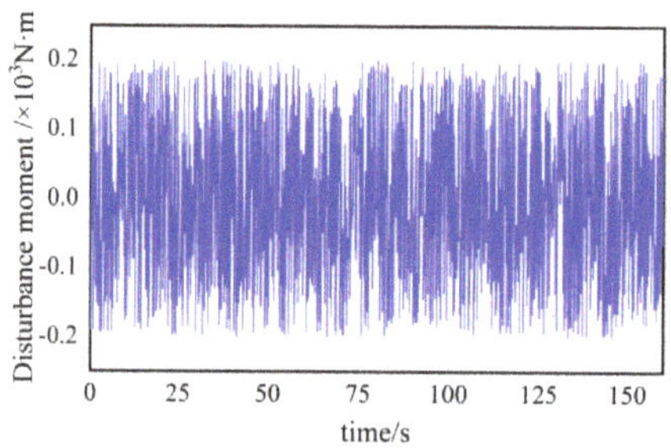

Figure 10. Turning disturbing moment.

3.2.1. Linear Path Following

The linear reference path was designed as the line segment between the points (20, 20) and (160, 20). The initial position of the USV is (0, 0), and its initial heading is parallel to the path. Figure 11 is the path following the comparison figures between the controllers mentioned above. Figures 12 and 13 are the comparisons of the heading angle and rudder angle during the path-following process. The path-tracking errors of the four controllers are shown in Table 5. The controller based on the reinforcement learning algorithm takes about 2 to 3 s to complete a path-following task, which is not much different from the calculation time in the literature [28], but the actual time still needs to prevail.

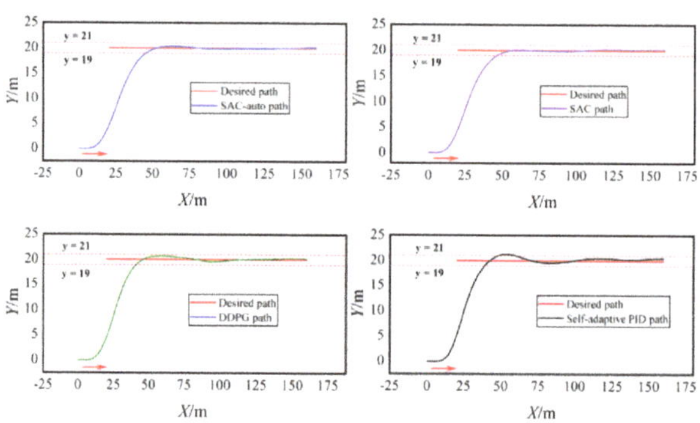

Figure 11. Path following.

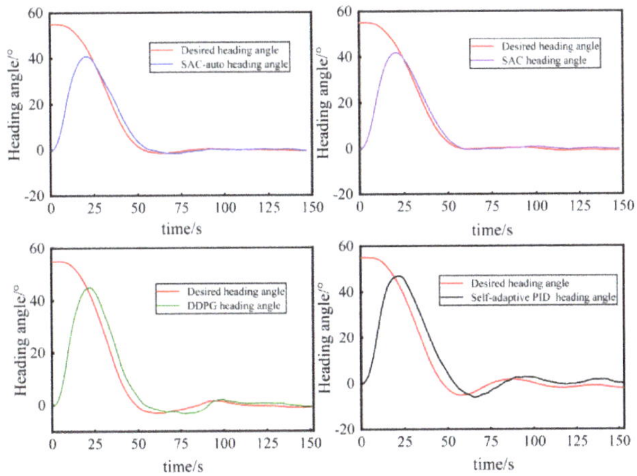

Figure 12. Heading angle.

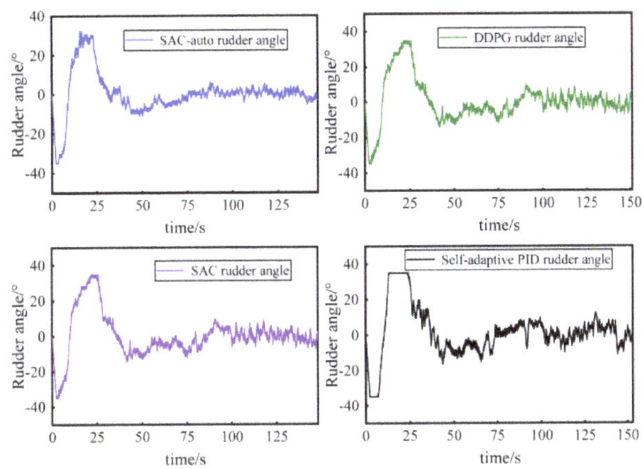

Figure 13. Rudder Angle.

Table 5. Data comparison table of four algorithms.

Controller	Transverse Error Mean/(m)	Heading Deviation Mean/(°)	Average Operation Time per Step/(ms)
SAC-auto	0.128	0.221	1.63
SAC	0.136	0.376	1.65
DDPG	0.292	0.888	1.61
Self-adaptive PID	0.449	2.085	1.89

It can be seen from Figure 11 that in the presence of a disturbing force, the tracking trajectories obtained with four controllers finally approached the desired paths within the specified error range. The rudder angle is rapidly operated to overcome the disturbance of waves. Compared to the adaptive controller, the RL-based controller has less overshoot during steering and produces smoother trajectories. Compared to the DDPG algorithm controller, the SAC-auto controller has better performance in both heading control and rudder maneuvering. According to Table 5, compared to the SAC controller, the steady-state performance of the SAC-auto controller is improved, with which, when the desired direction changes, the improved parameter can provide the USV with a fast adjustment to the desired direction and a stable path-following effect on the desired path, and the average deviation of the direction angle when stable is limited to 0.5°. It can also be concluded from Figure 13 that the fluctuation of the rudder angle with SAC-auto is the smoothest, and the maximum fluctuation of the rudder angle is less than 5°, indicating that the rudder is slightly frayed and the control gear can be well-protected.

The parameter curves of K_p and K_d output by SAC-auto are shown in Figure 14. Figure 15 shows the transverse force and turning moment of the USV. It can be seen that under the control of the SAC-auto algorithm, the transverse force and turning moment of the USV fluctuate less; the reason is that the maneuvering fluctuation of the rudder angle is smaller.

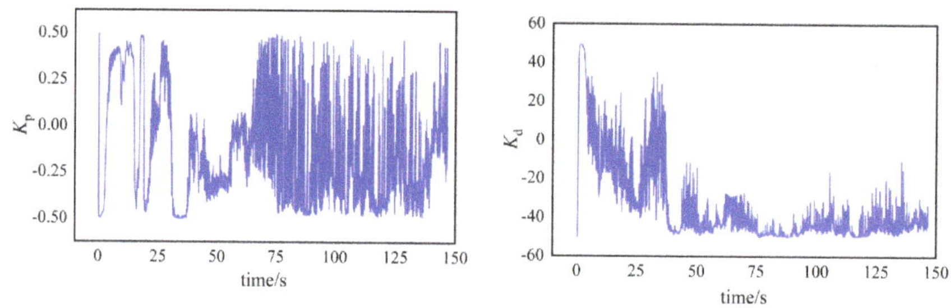

Figure 14. K_p, K_d curves.

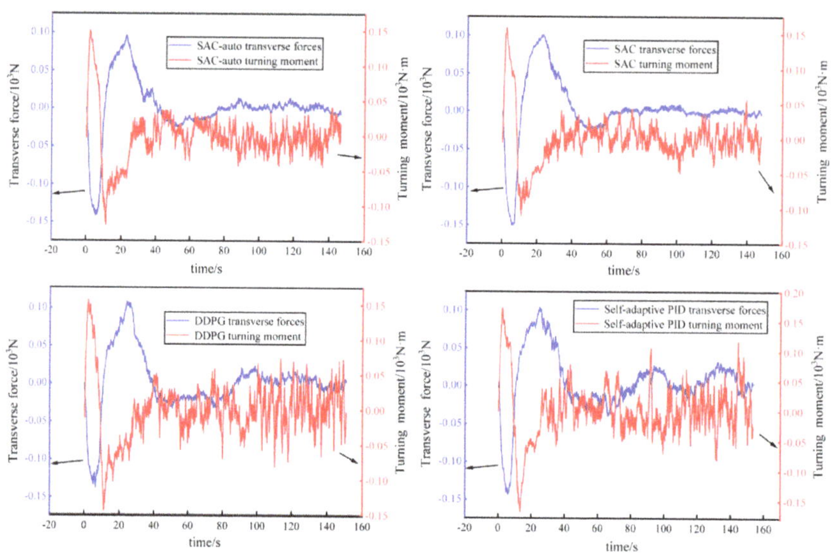

Figure 15. Transverse forces and turning moments.

3.2.2. Curve and Polyline Path Following

The above simulation results verify the feasibility and anti-interference ability of the SAC-auto algorithm when following a straight path. In order to verify the performance of the algorithm when following other desired paths which are more complex, and to inspect whether the SAC-auto algorithm can adaptively produce appropriate PD parameters, the path-following simulations of zigzag and turning are performed.

Scenario 1. The desired path of the zigzag is designed as the polyline between points (20, 20), (100, 100), (180, 20), and (260, 100). The initial position of the USV is located at (0, 0), and the initial heading is parallel to the Y-axis.

Scenario 2. The turning path is a circle with points (0, 0) as the center and 40m as the radius. The initial position of the USV is (0, 0), and the initial heading is parallel to the Y-axis.

Figure 16 shows the graphs for path following, rudder angle, heading angle, and curves of K_p and K_d for different algorithms in the tracking process.

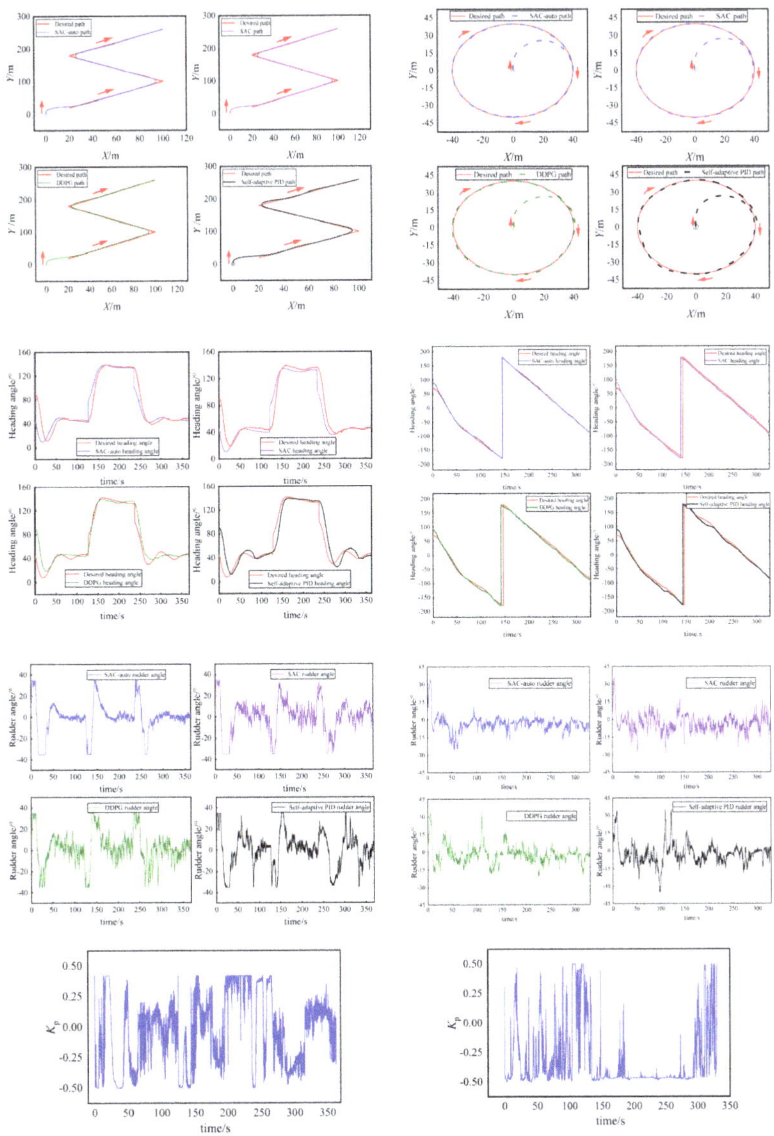

Figure 16. *Cont.*

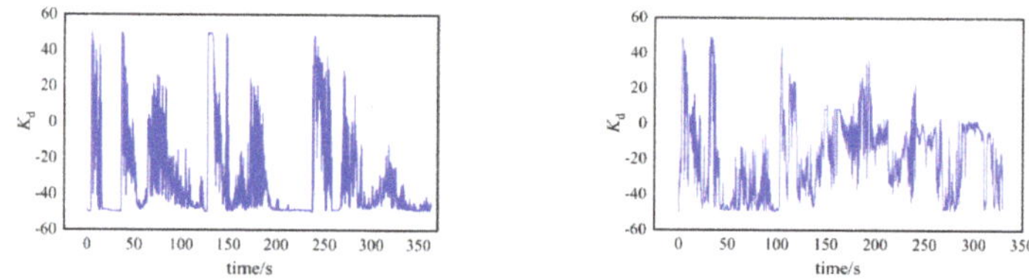

Scenario 1 Scenario 2

Figure 16. Zigzag and circle path following, heading angle, rudder angle, and K_p, K_d curves.

It can be seen from the simulations that the SAC-auto controller has a better performance in both heading control and rudder maneuvering under the condition of path following of zigzag and turning when considering the disturbance of the wave. The fluctuation of the rudder angle with SAC-auto is the smoothest, indicating that the SAC-auto control method performs well when following a complex desired path.

4. Conclusions

The classic adaptive PID control method used for path following does not perform well under complex conditions such as following a curvilinear path or considering wave interference. Concerning this issue, this paper presents a path-following control method based on SAC for PID parameter setting. First, a 3-DOF USV dynamics model based on Abkowite was established. Second, the guidance system using the line-of-sight method and the USV heading control system in the PID controller was designed. Third, the SAC algorithm was then used, and the state space, action space, and reward function were designed for the training of the RL on the path-following scenarios; the SAC is promoted to adaptively and rapidly adjust the PID parameter during the simulation. Finally, the algorithm was proven in the simulation experiments under path following in a straight line, zigzag, and turning with disturbance of wave scenarios, which verify the feasibility and robustness of the proposed method. In further research, the experiment should be conducted.

Author Contributions: Data curation, C.X.; funding acquisition, R.G.; methodology, L.S.; software, C.X.; visualization, C.X.; oringin draft, L.H.; writing—original draft, C.X., and L.S.; writing—review and editing, J.Y. All authors have read and agreed to the published version of the manuscript.

Funding: This research received no external funding.

Institutional Review Board Statement: Not applicable.

Informed Consent Statement: Not applicable.

Data Availability Statement: Not applicable.

Conflicts of Interest: The authors declare no conflict of interest.

References

1. Xu, F.C.; Xie, Y.L.; Liu, X.C.; Chen, X.; Han, W. Research Status and Key Technologies of Intelligent Technology for Unmanned Surface Vehicle System. In Proceedings of the International Conference on Sensing, Diagnostics, Prognostics, and Control (SDPC), Bejing, China, 5–7 August 2020; IEEE: New York, NY, USA, 2020; pp. 229–233.
2. Song, L.F.; Shi, X.Q.; Sun, H.; Xu, K.K.; Huang, L. Collision avoidance algorithm for USV based on rolling obstacle classification and fuzzy rules. *J. Mar. Sci. Eng.* **2021**, *9*, 1321. [CrossRef]
3. Mnih, V.; Kavukcuoglu, K.; Silver, D.; Rusu, A.A.; Veness, J.; Bellemare, M.G.; Graves, A.; Riedmiller, M.; Fidjeland, A.K.; Ostrovski, G.; et al. Human-level control through deep reinforcement learning. *Nature* **2015**, *518*, 529–533. [CrossRef] [PubMed]
4. Silver, D.; Lever, G.; Heess, N.; Degris, T.; Wierstra, D.; Riedmiller, M. Deterministic policy gradient algorithms. In Proceedings of the International Conference on Machine Learning, PMLR 2014, Bejing, China, 22–24 June 2014; pp. 387–395.
5. Mnih, V.; Badia, A.P.; Mirza, M.; Graves, A.; Harley, T.; Lillicrap, T.P.; Silver, D.; Kavukcuoglu, K. Asynchronous methods for deep reinforcement learning. In Proceedings of the International Conference on Machine Learning, PMLR 2016, New York, NY, USA, 20–22 June 2016; pp. 1928–1937.

6. Haarnoja, T.; Zhou, A.; Hartikainen, K.; Tucker, G.; Ha, S.; Tan, J.; Kumar, V.; Zhu, H.; Gupta, A.; Abbeel, P.; et al. Soft actor-critic algorithms and applications. *arXiv* **2018**, arXiv:1812.05905.
7. Gonzalez-Garcia, A.; Castañeda, H.; Garrido, L. USV Path-Following Control Based on Deep Reinforcement Learning and Adaptive Control. In Proceedings of the Global Oceans 2020: Singapore–US Gulf Coast, Online, 5–30 October 2020; IEEE: New York, NY, USA, 2020; pp. 1–7.
8. Zhao, Y.J.; Qi, X.; Ma, Y.; Li, Z.X.; Malekian, R.; Sotelo, M.A. Path following optimization for an underactuated USV using smoothly-convergent deep reinforcement learning. *IEEE Trans. Intell. Transp. Syst.* **2020**, *22*, 6208–6220. [CrossRef]
9. Wang, N.; Gao, Y.; Liu, Y.J.; Li, K. Self-learning-based optimal tracking control of an unmanned surface vehicle with pose and velocity constraints. *Int. J. Robust Nonlinear Control* **2022**, *32*, 2950–2968. [CrossRef]
10. Zheng, Y.M.; Tao, J.; Sun, Q.L.; Sun, H.; Chen, Z.Q.; Sun, M.W.; Xie, G.M. Soft Actor–Critic based active disturbance rejection path following control for unmanned surface vessel under wind and wave disturbances. *Ocean. Eng.* **2022**, *247*, 110631. [CrossRef]
11. Feng, Z.; Pan, Z.S.; Chen, W.; Liu, Y.; Leng, J.X. USV Application Scenario Expansion Based on Motion Control, Path Following and Velocity Planning. *Machines* **2022**, *10*, 310. [CrossRef]
12. Moreira, L.; Fossen, T.I.; Soares, C.G. Path following control system for a tanker ship model. *Ocean Eng.* **2007**, *34*, 2074–2085. [CrossRef]
13. Lekkas, A.M.; Fossen, T.I. A time-varying lookahead distance guidance law for path following. *IFAC Proc. Vol.* **2012**, *45*, 398–403. [CrossRef]
14. Liu, T.; Dong, Z.P.; Du, H.W.; Song, L.F.; Mao, Y.S. Path following control of the underactuated USV based on the improved line-of-sight guidance algorithm. *Pol. Marit. Res.* **2017**, *24*, 3–11. [CrossRef]
15. Yang, Z.K.; Zhong, W.B.; Feng, Y.B.; Sun, B. Unmanned surface vehicle track control based on improved LOS and ADRC. *Chin. J. Ship Res.* **2021**, *16*, 121–127.
16. Liu, W.; Liu, Y.; Bucnall, R. A robust localization method for unmanned surface vehicle (USV) navigation using fuzzy adaptive Kalman filtering. *IEEE Access* **2019**, *7*, 46071–46083. [CrossRef]
17. Do, K.D. Global robust adaptive path-tracking control of underactuated ships under stochastic disturbances. *Ocean Eng.* **2016**, *111*, 267–278. [CrossRef]
18. Dong, Z.; Wan, L.; Li, Y.; Liu, T.; Zhang, G. Trajectory tracking control of underactuated USV based on modified backstepping approach. *Int. J. Nav. Archit. Ocean Eng.* **2015**, *7*, 817–832. [CrossRef]
19. Zhou, W.; Wang, Y.; Ahn, C.K.; Cheng, J.; Chen, C. Adaptive fuzzy backstepping-based formation control of unmanned surface vehicles with unknown model nonlinearity and actuator saturation. *IEEE Trans. Veh. Technol.* **2020**, *69*, 14749–14764. [CrossRef]
20. Ashrafiuon, H.; Muske, K.R.; McNinch, L.C.; Soltan, R.A. Sliding-mode tracking control of surface vessels. *IEEE Trans. Ind. Electron.* **2008**, *55*, 4004–4012. [CrossRef]
21. Sun, Z.J.; Zhang, G.Q.; Qiao, L.; Zhang, W.D. Robust adaptive trajectory tracking control of underactuated surface vessel in fields of marine practice. *J. Mar. Sci. Technol.* **2018**, *23*, 950–957. [CrossRef]
22. Guo, B.Z.; Zhao, Z.L. *Active Disturbance Rejection Control for Nonlinear Systems: An Introduction*; John Wiley & Sons: Hoboken, NJ, USA, 2016.
23. Miao, R.; Dong, Z.; Wan, L.; Zeng, J. Heading control system design for a micro-USV based on an adaptive expert S-PID algorithm. *Pol. Marit. Res.* **2018**, *25*, 6–13. [CrossRef]
24. Wang, J.H.; Zhao, M.K. Simulation of path following optimization control of unmanned surface vehicle. *Comput. Simul.* **2016**, *33*, 362–367.
25. Fan, Y.S.; Guo, C.; Zhao, Y.S.; Wang, G.F.; Shi, W.W. Design and verification of straight-line path following controller for USV with time-varying drift angle. *Chin. J. Sci. Instrum.* **2016**, *37*, 2514–2520.
26. Xu, P.F.; Cheng, C.; Cheng, H.X.; Shen, Y.L.; Ding, Y.X. Identification-based 3 DOF model of unmanned surface vehicle using support vector machines enhanced by cuckoo search algorithm. *Ocean Eng.* **2020**, *197*, 106898. [CrossRef]
27. Budak, G.; Beji, S. Controlled course-keeping simulations of a ship under external disturbances. *Ocean Eng.* **2020**, *218*, 108126. [CrossRef]
28. Wang, L.; Xiang, J.L.; Wang, H.D. Local Path Planning Algorithm for Unmanned Surface Vehicle Based on Improved Bi-RRT. *Shipbuild. China* **2020**, *61*, 21–30.

MDPI
St. Alban-Anlage 66
4052 Basel
Switzerland
Tel. +41 61 683 77 34
Fax +41 61 302 89 18
www.mdpi.com

Journal of Marine Science and Engineering Editorial Office
E-mail: jmse@mdpi.com
www.mdpi.com/journal/jmse

www.ingramcontent.com/pod-product-compliance
Lightning Source LLC
Chambersburg PA
CBHW061548220926
51850CB00005B/283
9783036581064